Green
Health

Green
Health

An A-to-Z Guide

The SAGE Reference Series on
Green Society
Toward a Sustainable Future

OLADELE OGUNSEITAN, GENERAL EDITOR
University of California, Irvine

PAUL ROBBINS, SERIES EDITOR
University of Arizona

Los Angeles | London | New Delhi
Singapore | Washington DC

Los Angeles | London | New Delhi
Singapore | Washington DC

FOR INFORMATION:

SAGE Publications, Inc.
2455 Teller Road
Thousand Oaks, California 91320
E-mail: order@sagepub.com

SAGE Publications Ltd.
1 Oliver's Yard
55 City Road
London EC1Y 1SP
United Kingdom

SAGE Publications India Pvt. Ltd.
B 1/I 1 Mohan Cooperative Industrial Area
Mathura Road, New Delhi 110 044
India

SAGE Publications Asia-Pacific Pte. Ltd.
33 Pekin Street #02-01
Far East Square
Singapore 048763

Publisher: Rolf A. Janke
Assistant to the Publisher: Michele Thompson
Senior Editor: Jim Brace-Thompson
Production Editors: Kate Schroeder, Tracy Buyan
Reference Systems Manager: Leticia Gutierrez
Reference Systems Coordinator: Laura Notton
Typesetter: C&M Digitals (P) Ltd.
Proofreader: Kris Bergstad
Indexer: Judy Hunt
Cover Designer: Gail Buschman
Marketing Manager: Kristi Ward

Golson Media
President and Editor: J. Geoffrey Golson
Author Manager: Ellen Ingber
Editors: Mary Jo Scibetta, Kenneth Heller
Copy Editors: Barbara Paris, Tricia Lawrence, Mary Beth Curran

Printed in the United States of America

Library of Congress Cataloging-in-Publication Data

Green health : an A-to-Z guide / Oladele Ogunseitan, editor.

p. cm. — (The Sage reference series on green society: toward a sustainable future)
Includes bibliographical references and index.

ISBN 978-1-4129-9688-4 (hardback) — ISBN 978-1-4129-7459-2 (ebk)

1. Health attitudes. 2. Health behavior. 3. Green movement. I. Ogunseitan, Oladele.

RA776.9.G738 2011 613—dc22 2011006528

11 12 13 14 15 10 9 8 7 6 5 4 3 2 1

Contents

About the Editors

Green Series Editor: Paul Robbins

Paul Robbins is a professor and the director of the University of Arizona School of Geography and Development. He earned his Ph.D. in Geography in 1996 from Clark University. He is general editor of the *Encyclopedia of Environment and Society* (2007) and author of several books, including *Environment and Society: A Critical Introduction* (2010), *Lawn People: How Grasses, Weeds, and Chemicals Make Us Who We Are* (2007), and *Political Ecology: A Critical Introduction* (2004).

Robbins's research centers on the relationships between individuals (homeowners, hunters, professional foresters), environmental actors (lawns, elk, mesquite trees), and the institutions that connect them. He and his students seek to explain human environmental practices and knowledge, the influence nonhumans have on human behavior and organization, and the implications these interactions hold for ecosystem health, local community, and social justice. Past projects have examined chemical use in the suburban United States, elk management in Montana, forest product collection in New England, and wolf conservation in India.

Green Health General Editor: Oladele Ogunseitan

Oladele Ogunseitan is a professor and the chair of the Department of Population Health and Disease Prevention, Program in Public Health at the University of California (UC), Irvine, where he is also a professor of Social Ecology. Since 2009, he has served as the co-director of the UC Irvine Global Health Framework Program which is funded by the Fogarty International Center of the National Institutes of Health (NIH). In 2010, he began serving as the director of Research Education, Training and Career Development for the NIH-funded Institute for Clinical and Translational Science. He directs the Research and Education in Green Materials Program, originally funded as a lead campus–component of the UC-systemwide Toxic Substances Research and Teaching Program.

His research interests are in the environmental burden of disease, health, and development and microbial ecotoxicology. He earned his doctorate at the Center for Environmental Biotechnology at the University of Tennessee and his Masters of Public Health at the University of California, Berkeley, where he also earned a Certificate in International Health. He is an alumni faculty fellow at the Belfer Center for Science and International Affairs at the Kennedy School of Government, Harvard University. He served as a Macy Foundation fellow at the Marine Biological Laboratory in Woods Hole, Massachusetts, and Institute Faculty for Vulnerability Assessments at the International Institute for Applied Systems Analysis in Laxenburg, Austria.

Introduction

"Health is a state of complete physical, mental and social well-being and not merely the absence of disease or infirmity." This definition of health, adopted by 61 countries at the inception of the World Health Organization (WHO), has not been modified since 1948. It is difficult to argue against an exercise toward all-encompassing conceptualization of human health. The challenge remains how to translate the robust concept into collective action against human vulnerability and into strategies that support sustainable institutions and policies designed to maintain the complete state of health.

This volume aims to contribute to the translation of conceptual health into action by expatiating "new" understandings of the interconnectedness of human health, social systems, and nature. These understandings are captured by appending the term *green,* which, in contemporary scholarship, has become the quintessential prefix for products and processes that evoke cautious implementation of human desires and activities toward the reduction of adverse impacts on environments and ecosystems.

The roots of green health, or the ecological model of public health, reach farther into history than the definition offered by the WHO. In 1920, Charles-Edward Winslow (1877–1957) defined public health as "the science and art of preventing disease, prolonging life and promoting health through the organized efforts and informed choices of society, organizations, public and private, communities and individuals." *Green Health* is about making informed choices in healthcare regarding alternative materials to reduce toxic exposures; choices about energy resources that minimize health-damaging pollution without compromising the integrity of biological diversity; choices about safer and effective medications; choices about preventative strategies and health-protective policies that cut across various sectors of government and sociocultural diversity; and choices about investments in green infrastructures—green transportation, green buildings, green information technologies, and green agriculture and nutrition. It is not possible to fully discuss green health without exploring these topics (see Figure 1, p. x).

The rationale for new interests in characterizing the linkages between green development and the collective vision of public health is perhaps best revealed through the assessment of global and regional burden of disease. From an ecological point of view, the human species has been remarkably successful, with an extremely high population growth rate and increasing life expectancy in most regions of the world over the past few centuries. However, the success seems to have come at a cost that is not sustainable in terms of environmental pollution, resource depletion, and the emergence of new or resurgence of old diseases in the human population. The costs are well documented in this volume, for example through articles "Children's Health" (by Rebecca A. Malouin), "Cardiovascular Diseases" (by Richard Wills), and "Severe Acute Respiratory Syndrome" (S. Harris Ali).

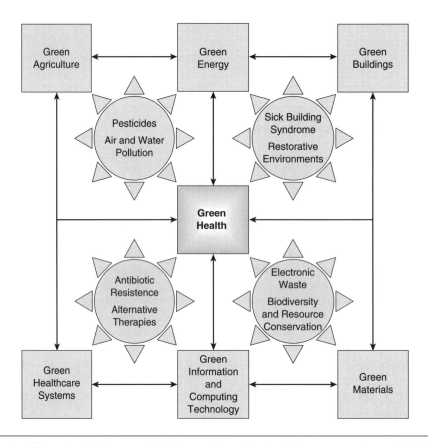

Figure I   Green health is essentially the product of societal investments in sustainable practices across various sectors, including agriculture, energy, building infrastructures, materials development, information technology, and healthcare delivery systems. The large-scale problems associated with current unsustainable practices are reflected in a growing disease burden attributable to preventable risk factors such as toxic exposures (for examples, see articles on lead [Pb], manganese, mercury, and arsenic by Paul Richard Saunders, Abhijay P. Karandikar, Hueiwang Anna Cook Jeng, and Erin E. Stamper, respectively), inadequate experience of natural landscapes (see article on topophilia by J. Aaron Hipp), development of multiple antibiotic resistance in pathogen populations (see related articles by Gautam J. Desai and Alfonso J. Rodriquez-Morales), the depletion of biodiversity (see article on genetically engineered crops by Carrie Nicole Wells), and energy resources (see article on nuclear power by Jen Schneider).

The WHO's global burden of disease assessments showed, for example, that neuropsychiatric disorders account for approximately 28 percent of the burden attributable to noncommunicable diseases, the highest level in that category—and approximately 14 percent of the total burden of disease globally. In the articles "Mental Health" (Gareth Davey) and "Mental Exercises" (Sylvie Boiteau), we learn that public health strategies focused on prevention of mental diseases at the community level have lagged behind the pace of increases in the prevalence of disease and disability. We can certainly do more to promote the "green lifestyle" as a strategy for preventing and managing mental disabilities. Recent research supports the view that access to "green open space" and "green buildings" in urban areas may be helpful, but as pointed out in the articles "Urban Green" (Stephen T. Schroth) and "Topophilia" (J. Aaron Hipp), the implementation of placed-based disease prevention strategies will require concerted efforts on the part of healthcare providers, urban designers, and policy makers to reinvent the way we address deficits in mental health status.

The emergence of green health as a conceptual and actionable framework in public health is also supported by recent data from the environmental burden of disease assessments. Estimates of the portion of the global burden of disease attributable to environmental risk factors indicate that 25 to 33 percent of human diseases can be controlled by improving the quality of environments in which human beings live and work. These kinds of assessments probably underestimate the influences of the environment on human health, as discussed by Merrill Singer in the article "Ecosyndemics," whereby environmental conditions promote the clustering and adverse interaction of two or more diseases in a population. In this context, indoor and outdoor air pollution, infected and toxic water resources, and climate change continue to act independently and interactively to exert unacceptable adverse impacts on human health worldwide. In many cases, adoption of "green development" strategies will considerably reduce if not eliminate the sources of risk factors associated with the environmental burden of disease.

For example, green energy (wind, solar, etc.) replacements for the current heavy dependence on petrochemical and coal fuels will contribute to the alleviation of air pollution and climate change. A global move toward green energy will also reduce contamination of water resources with mercury, lead, and other toxic pollutants. Effective dissemination of green technology can make water supplies safer and more reliable in developing countries plagued by communicable waterborne diseases. There is also some optimism that green technological innovation, appropriate government regulation, and public education—as highlighted in the articles "Green Chemistry" (Schroth), "California's Green Chemistry Initiative" (Natalia Milovantseva), and "Education and Green Health" (Joel H. Kreisberg)—will ameliorate some of the problems that have emerged in industrialized countries.

The articles in this volume were selected to capture new thoughts and ideas on how the greening of human societies influence public health and how choices made in the pursuit of health can further support the quest for greenness. The scope spans the broad range of interactions among industrial sectors and some of the consequences of current "non-green" products and processes on human health, as depicted in Figure 1. "How green is green?" is one of the enduring questions in sustainability. The solution is to derive an internally consistent and broadly applicable metric for greenness. The article "Metrics of Green Health" addresses the challenges inherent in developing such measures. The more philosophical questions in the genre of "How green *should* green be?" are addressed by articles focused on the adverse consequences of unsustainable practices. For example, in the article titled "Government Role in Green Health," Kreisberg argues that because there are currently no generic government standards for integrating green products processes, the presumed additional costs might lead businesses to prematurely reject greening initiatives.

The perspectives offered by the authors in this volume will by no means be universally endorsed as the "final word" on green approaches to health, but the perspectives are invariably provocative, spurring the reader to learn more or to take specific action. In many cases, the articles present the "first word" on the topics presented. The field of green health is potentially vast and the rate of advances in product and process development is rapid. These "first word" articles will serve as templates for further developments, refinements, creative inventions, and interventions that should advance the global course of green health.

Oladele Ogunseitan
General Editor

Reader's Guide

Agriculture and Nutrition

Animal Products
Biological Control of Pests
Chemical Pesticides
Fast Food
Fertilizers
Food Allergies
Genetically Engineered Crops
Home-Grown Food
Obesity
Organic Produce
Supplements
Wine and Other Alcohols

Air Resources

Airborne Diseases
Air Filters/Scrubbers
Asthma
Climate Change
Indoor Air Quality
Ozone
Particulate Matter
Regional Dust
Smog
Smoking
Ultraviolet Radiation

Built Environment

Cities
Fungi and Sick Building Syndrome
Highways
Occupational Hazards
Radon and Basements
Recreational Space

Rural Areas
Solid Waste Management
Suburbs
Topophilia
Urban Green

Consumer Products

Automobiles (Emissions)
Cell Phones
Computers and Printers (Ink)
Dry Cleaning
Ergonomics
Fabrics
Hobby Products
Lighting
Microwave Ovens
Paper Products
Pest Control
Plastics in Daily Use
Radiation Sources

Energy Resources

Alternative Energy Resources (Solar)
Biodiesel
Clean Coal
Electricity
Firewood and Charcoal
Hydroelectricity
Lead Sources and Health
Light Bulbs
Manganese Sources and Health
Mercury Sources and Health
Methane/Biogas
Nuclear Power
Petrochemicals

Green Health Economic Issues

Cost-Benefit Analysis for Alternative
    Products
Emergency Rooms
Healthcare Delivery
Health Disparities
Health Insurance Reform
Nursing, Lack of
Pharmaceutical Industry Reform

Green Strategies for Chronic Disease Prevention and Control

Cancers
Cardiovascular Diseases
Degenerative Diseases
Immune System Diseases
Injuries
Kidney Diseases
Liver Diseases
Lung Diseases
Mental Exercises
Metabolic Syndrome Diseases
Musculoskeletal Diseases
Neurobehavioral Diseases
Oral Diseases
Physical Activity and Health
Reproductive System Diseases
Skin Disorders

Green Strategies for Infectious Disease Prevention and Control

Acquired Immune Deficiency Syndrome
Biological Weapons
Bird Flu
Gastroenteritis
International Travel
Malaria
Methicillin-Resistant *Staphylococcus
    Aureus*
Seasonal Flu
Severe Acute Respiratory Syndrome
Sexually Transmitted Diseases
Stomach Ulcers and *Helicobacter Pylori*
Streptococcus Infections
Tuberculosis
Vaccination/Herd Immunity

Healthcare Industry

Antiseptics
Children's Health
Dental Mercury Amalgams
Health Insurance Industry
Hospitals (Carbon Footprints)
Infectious Waste
Low-Level Radioactive Waste
Men's Health
Mental Health
Nosocomial Infections
Women's Health

Legal and Policy Issues and Metrics

Biomedicine
California's Green Chemistry Initiative
Calorie Labeling for Restaurants
Centers for Disease Control and
    Prevention (U.S.)
Education and Green Health
Environmental Protection Agency (U.S.)
Fast Food Warnings
Government Role in Green Health
Green Chemistry
Industrial Ecology
International Policies
Metrics of Green Health
Personal Consumer Role in Green Health
Phaseout of Toxic Chemicals
Private Industry Role in Green Health
Taxation of Unhealthy Products
United Nations Environment Programme
World Health Organization's
    Environmental Burden of Disease

Pharmaceutical and Personal Care Products

Advertising and Marketing
Antibiotic Resistance
Antibiotics
Anti-Cholesterol Drugs
Anti-Depressant Drugs
Caffeine
Hormone Therapy
Pain Medication

List of Articles

List of Contributors

Ackom, Emmanuel Kofi
University of British Columbia

Ali, S. Harris
York University

Bagher-Ebadian, Hassan
Independent Scholar

Bellestri, Tani
Independent Scholar

Bezbatchenko, Annie W.
New York University

Birdsall, Kate
Michigan State University

Boiteau, Sylvie R.
University of Massachusetts Medical School

Boslaugh, Sarah
Washington University in St. Louis

Campbell, Keely E.
Knox College

Carlin, Sarah E.
Knox College

Carr, David L.
Independent Scholar

Collishaw, Brianna R.
Knox College

Currier, Connie
Michigan State University

Davey, Gareth
University of Chester

Davidson, Michele
George Mason University

Derrington, Stephen M.
Kansas City University of Medicine and Biosciences, College of Osteopathic Medicine

Desai, Gautam J.
Kansas City University of Medicine and Biosciences, College of Osteopathic Medicine

Dooley, Bridget G.
Knox College

Filiberto, David M.
Independent Scholar

Gonshorek, Daniel O.
Knox College

Graham, Daniel J.
University of Minnesota

Guha, Mohua
International Institute for Population Sciences

Guimaraes, Alessandra
University of Sint Eustatius School of Medicine

Harper, Gavin D. J.
Cardiff University

Helfer, Jason A.
Knox College

Hibbert, Kathleen
University of California, Irvine

Hipp, J. Aaron
Washington University in St. Louis

Hughes, Kathryn J.
Yale University

Jackson, Jacqueline F.
Old Dominion University

Jarvie, Michelle Edith
Independent Scholar

Jeng, Hueiwang Anna Cook
Old Dominion University

Kaminski, Michael J.
Knox College

Kang, Daniel Hsing Po
University of California, Irvine

Karandikar, Abhijay P.
Pines Health Services

Karner, Luke L.
Knox College

Kilfoy, Briseis A.
*Yale University School of
   Medicine*

Kopnina, Helen
University of Amsterdam

Kreisberg, Joel H.
Teleosis Institute

Kte'pi, Bill
Independent Scholar

Lowerson, Victoria
University of California, Irvine

Malouin, Rebecca A.
Michigan State University

Martens, Samantha
Michigan State University

McClain, Rance
Independent Scholar

McDavid, Maurice J.
Knox College

Mendez, Roberto Carlos
Old Dominion University

Merten, Sarah
Michigan State University

Meyers, Alexandra
Independent Scholar

Milovantseva, Natalie
University of California, Irvine

Mize, Adam J.
Knox College

Mostert, Monique
Independent Scholar

Nagy, Lisa Lavine
*The Preventive and Environmental
   Health Alliance, Inc.*

Nanas, Elizabeth
*Wayne State University, Hong Kong
   University of Science & Technology*

Neu, Denese M.
HHS Planning & Consulting, Inc.

Ogunseitan, Oladele
University of California, Irvine

Olson, Elizabeth A.
University of California, Merced

Padula, Alessandra
University of L'Aquila

Pandit, Rahul
Utrecht University

Petricca, Kadia
University of Toronto

Ramar, Cassandra
*Kansas City University of Medicine and
   Biosciences, College of Osteopathic
   Medicine*

Rodriguez-Morales, Alfonso J.
*Universidad de Los Andes, Universidad
   Central de Venezuela*

Saunders, Paul Richard
*Canadian College of Naturopathic
   Medicine*

Schneider, Jen
Colorado School of Mines

Schroth, Stephen T.
Knox College

Singer, Merrill
University of Connecticut

Stamper, Erin E.
Independent Scholar

Stephenson, Patricia
Independent Scholar

Sun, Gang
University of California, Davis

Taylor, Tracey A. H.
Kansas City University of Medicine and Biosciences

Turner, Claire C.
Knox College

Vadrevu, Krishna Prasad
Ohio State University

Vera, Yadira
Universidad Simón Bolivar

Wells, Carrie Nicole
Clemson University

Wills, Richard
Independent Scholar

Wipper, Michael A.
Knox College

Wysocki, Amanda
University of Washington

Yu, Liang
Old Dominion University

Green Health Chronology

1796: English scientist Edward Jenner discovers the concept of vaccination.

1819: German chemist Friedrich Ferdinand Runge discovers the chemical compound of caffeine.

1894: The pulmonary plague, once known as the Black Death, reaches its third pandemic when the disease spreads to Hong Kong and Canton.

1905: Harold Antoine Des Voeux coins the term *smog*, a portmanteau of *smoke* and *fog*.

1918–1919: An influenza pandemic results in the deaths of approximately 100 million people worldwide.

1921: A tuberculosis vaccine, named the Bacille Calmette-Guerin vaccine, is developed.

1925: The Geneva Protocol for the Prohibition of the Use in War of Asphyxiating, Poisonous or Other Gases, and of Bacteriological Methods of Warfare is passed, prohibiting the use of biological weapons in war and peacetime.

1928: Sir Alexander Fleming discovers penicillin, an antibiotic that treats bacterial infections and sexually transmitted diseases such as chlamydia.

1950–2000: The percentage of the world's population living in cities increases from 30 percent to 47 percent.

1961: The U.S. state of California enacts a law requiring all new cars to have "positive crankcase ventilation."

1963: U.S. Congress passes the Clean Air Act, enabling the federal government to enforce regulations designed to lessen harmful exposure to airborne contaminants.

1963: Maurice Hilleman develops a vaccine to prevent the onset of measles.

1969–2009: The number of new cases of asthma increases by 200 percent.

1970: The World Health Organization (WHO) publishes the pamphlet *Health Aspects of Biological and Chemical Weapons.*

1970: The U.S. Environmental Protection Agency (EPA) is established.

1972: The United Nations Environment Programme is established.

1976: U.S. Congress passes the Toxic Substances Control Act.

1977: The solvent benzene is banned for use in pesticides after it is discovered that it causes leukemia.

1977: U.S. Congress passes the Clean Water Act.

1983: Two research groups led by Robert Gallo and Luc Montagnier verify that the acquired immune deficiency syndrome (AIDS) virus is caused by infection with the human immunodeficiency virus (HIV).

1984: The WHO releases a report showing that 30 percent of new and remodeled buildings have such poor air quality that workers are prone to developing symptoms related to sick building syndrome.

1986: California passes Proposition 65, requiring any material containing substances known to cause cancer to have a warning label accurately explaining the risks.

1987: The Montreal Protocol, an international agreement designed to prevent the depletion of the ozone layer, is signed.

1990: The Intergovernmental Panel on Climate Change (IPCC) releases its first report.

1990: U.S. Congress passes the Nutrition Labeling and Education Act, requiring fast food companies to print labels providing information regarding the nutritional content of their food.

1991: The world's first genetically engineered food crop, the Flavr Savr tomato, is developed by the California company Calgene and approved by the U.S. Food and Drug Administration (FDA).

1994: U.S. Congress passes the Dietary Supplement Health and Education Act (DSHEA), requiring manufacturers of dietary supplements to clearly and accurately label their products.

1995: Reports indicate that the percentage of drug-resistant cases of tuberculosis stands at 52 percent.

1995–2002: The total number of people living in the United States who are diagnosed with AIDS drops by 70 percent.

November 2002–July 2003: Severe acute respiratory syndrome (SARS) infects 8,096 people, with 774 deaths reported.

November 2003–July 2008: Approximately 400 new cases of bird flu infection are reported to the WHO.

2004: Congress passes the American Jobs Creation Act, which provides funding and tax credits to firms that produce biodiesel and biodiesel blends.

2005: Reports are released concluding the nearly one-third of Americans are obese.

2005: In his book *Lost Child in the Woods,* Richard Louv coins the term *nature deficit disorder* to describe how children are developing behavioral problems as a result of spending less time outdoors than previous generations.

2005: The Nobel Prize is awarded to a group of researchers who developed a way for chemical syntheses to have a less harmful impact on the environment.

2006: New York City enacts legislation requiring restaurants to label accurately the amount of calories in their dishes.

2006: The FDA formally requests that food manufacturers state the amount of trans fats in their products, causing many fast food companies to alter their recipes in order to avoid being labeled as unhealthy.

2007: California enacts laws that spur the creation of its Green Chemistry Initiative.

2007: In the Supreme Court case of *Massachusetts v. Environmental Protection Agency,* the Court rules 5–4 that the EPA has the legal authority to regulate the emission of heat-trapping gases in automobile emissions.

2008: The Environmental Working Group conducts an experiment showing that four of the most popular brands of bottled water in the United States are contaminated with bacteria.

2008: Reports are released concluding that Americans generate approximately 250 million tons of garbage per year.

July 27, 2008: As the FBI begins to narrow its suspect list for the 2001 anthrax attacks that resulted in the deaths of five people, Bruce Edward Ivins, a renowned microbiologist, commits suicide after being told of the Justice Department's impending investigation against him, thus implicating himself.

August 6, 2008: Federal prosecutors working on the 2001 anthrax attacks case announce that Bruce Edward Ivins was the sole perpetrator of the crimes.

2009: The G-8 group of industrialized nations enters into a pact to reduce greenhouse gas emissions by 80 percent by the year 2050.

2009: The U.S. National Oceanic and Atmospheric Administration (NOAA) reports that projected carbon dioxide levels will most likely cause a thousand-year drought beginning in 2050.

2010: The largest offshore oil spill in U.S. history occurs as an explosion rocks the Deepwater Horizon oil rig, spilling approximately 35,000 to 60,000 barrels of oil per day into the Gulf of Mexico and severely damaging the region's environment.

2010: The World Meteorological Organization reports that the decade lasting from 2000 to 2009 was the warmest on record since temperatures were first officially recorded in the 1850s.

March 2011: The nuclear debate is reignited around the world as a result of the Japanese nuclear emergency at the Fukushima Daiichi power plant, which is located about 150 miles north of heavily populated Tokyo. The leak of radiation, brought on as a result of Japan's 9.0 earthquake and subsequent tsunami, prompted safety reviews of nuclear power plants in China and the European Union. In the United States, President Barack Obama asked the Nuclear Regulatory Commission to do a comprehensive review of domestic plants in light of the events in Japan.

ACQUIRED IMMUNE DEFICIENCY SYNDROME

Acquired immune deficiency syndrome (AIDS) is a disease of the human immune system caused by the human immunodeficiency virus (HIV). HIV progressively decreases the effectiveness and resiliency of the immune system, and as a result, infected individuals become susceptible to opportunistic infections such as pneumonia, tuberculosis, Kaposi's sarcoma, and other infections normally fought off by the immune system.

AIDS is now a global pandemic. By 2001, AIDS was the world's fourth leading cause of death, after heart disease, stroke, and acute respiratory infection. With the release of the World Health Organization's (WHO) and UNAIDS World AIDS report 2008, it was estimated that 33.2 million people are living with the disease worldwide, with an annual AIDS death toll hovering around 2 million people.

While AIDS is a worldwide problem, it continues to thrive in areas of pervasive poverty, with roughly 75 percent of the aforementioned deaths occurring in sub-Saharan Africa (SSA) and many other low-income countries. The unfortunate consequence of its high prevalence within several SSA countries has been the noted decline in life expectancy. In addition, AIDS has resulted in huge social and economic impacts to many low-income countries. Economic productivity at a household and national level, as well as a burgeoning orphan epidemic in Africa have been some of the major social and economic consequences of the pandemic.

As issues of climate change and environmental awareness become more of a concern, correlation to the impacts on AIDS will be needed as current research in the area remains scarce. Researchers do suspect, however, that the transmission of HIV, the precursor to AIDS, may be influenced by climate change and yield new HIV infections as environmental alterations foster socioeconomic consequences such as increased urbanization and overcrowding.

Origin, Transmission, and Disease Stages

Although researchers suggest that HIV existed back in the 1930s, the first AIDS diagnosis was not made until 1981. HIV and AIDS first appeared in the United States as a disease in the gay male population and was initially labeled as GRID (gay-related immune deficiency). Individuals who had received blood transfusions also exhibited symptoms of early

1

infection. On July 27, 1982, the term *AIDS*—referring to *Acquired Immune Deficiency Syndrome*—was used for the first time. Since then, all countries to date have reported cases of infected individuals. However, HIV and AIDS have particularly devastated SSA and many low-income countries. For instance, prevalence data for sub-Saharan countries has hovered between 15 and 30 percent or greater for HIV infected adults aged 15 to 49 years.

Transmission of HIV occurs through contact of a mucous membrane or the bloodstream with a bodily fluid containing HIV such as semen, preseminal or vaginal fluids, blood, or breast milk. Thus, common pathways for transmission have included anal, vaginal, or oral sex; blood transfusions; contaminated hypodermic needles; childbirth; and exchange between mother and baby during childbirth, breastfeeding, or with any of the fluids previously mentioned.

In 1990, the WHO outlined four stages of HIV: (1) incubation period, (2) acute infection, (3) latency stage, and (4) AIDS. The initial incubation period upon infection is asymptomatic and usually lasts between two and four weeks. The second stage, acute infection, lasts an average of 28 days and can include symptoms such as fever, swollen lymph nodes, sore throat, rash, muscle pain, malaise, and mouth sores. The third, latency, stage shows few or no symptoms and can last anywhere from two weeks to 20 years and beyond. The final stage, AIDS, displays symptoms of various opportunistic infections (typically associated with the decline of CD4 [or white blood cell] count below a threshold of 200 cells per μL).

Economic and Social Impacts of AIDS

As previously mentioned, the social and economic impacts of AIDS, particularly in many low-income countries, have been overwhelming. The AIDS epidemic has emerged not only as a health crisis but also as a major threat to development and human society. Among households, the direct costs of HIV and AIDS can be measured in the lost income of those who die or who lose their jobs because of their illness. And in the long run, household savings fall, income consumption of healthcare and funeral costs increases, and expenditure patterns are distorted as families struggle to cope with the demands of the sick and dying.

On a national level, this has translated into overwhelming economic impacts that have hindered country productivity. This has largely been the case in Africa and Asia, where the major effects are a reduction in the labor supply due to the loss of young adults in their most productive years, as well as the increased costs to the already understaffed and underfunded health system, which is now struggling to provide prevention and treatment services.

At a social level, AIDS has been primarily responsible for leaving vast numbers of children across Africa orphaned (without one or both parents). Currently, it is estimated that 15 million children have been orphaned or made vulnerable as a result of AIDS. This has resulted in numerous social crises as many of these children are forced to undertake a parental role for younger siblings or enter care under extended family, community, or institutional networks. To address these issues, many nongovernmental organizations (NGOs) have played a large role in providing support and capacity to care for orphans and other vulnerable children.

Prevention and Control

Since early pathways of transmission were understood, promoting education campaigns took center stage. To date, health organizations and governments have promoted education, safe-sex interventions, and needle exchange programs in an attempt to slow the spread of the virus. While numerous studies have supported the benefits of health education to promote cautious sexual behavior, behavioral studies have revealed that many social and cultural factors also play a key role in the spread of the virus.

There is currently no publicly available vaccine or cure for HIV or AIDS. The only known method of prevention is avoiding exposure to the virus or, if that fails, an antiretroviral treatment administered after a highly significant exposure, called post-exposure prophylaxis (PEP). PEP has a very demanding four-week schedule of dosage and very unpleasant side effects, including diarrhea, malaise, nausea, and fatigue. Yet, while the antiretroviral treatment reduces both mortality and morbidity associated with HIV infection, the high cost and routine treatment course required have made accessibility difficult for marginalized populations and infected individuals from low-income settings. More than half the global population with AIDS remains untreated by antiretrovials, and HIV prophylaxis is used in only 9 percent of pregnancies among HIV-positive women.

Climate Change and AIDS

As climate change has resulted in extreme weather conditions and environmental degradation, researchers have begun to speculate on its long-term impacts on health. However, current research correlating climate change to AIDS incidence has been scarce. This is most likely attributed to the complexity of understanding how weather and climate impact infectious disease, such as HIV, that is largely propagated by human behavior. In 2004, the United Nations Secretary-General's High-Level Panel on Threats, Challenges and Change outlined six clusters of threats to a more secure world, and of those, environmental degradation and infectious disease were two.

In 2007, UNAIDS released a joint working paper on AIDS and climate change. In it, the HIV and Climate Change Complex (HACC) conceptual framework was proposed, which outlines key links between HIV, AIDS, and climate change. Researchers suspect that with environmental changes leading to increased mobility of populations and harsher weather conditions, the likelihood for increased vulnerability and susceptibility to HIV infection and AIDS also will be increased.

For instance, the influence of weather on social habits—such as overcrowding in slums and lack of economic productivity due to failed crops, which perhaps may foster increased risky sexual behavior—may pose a greater vulnerability to infection. Additionally, for people already infected with HIV, harsh weather conditions such as drought, extreme cold, or natural disasters may result in unfavorable consequences on food production, further impairing diet and nutrition. This may reduce the body's general resistance to infections, leaving patients susceptible to advanced effects of AIDS and a comprised immune system.

As climate change continues to cause environmental degradation, additional research and analysis will be required at social, behavioral, and epidemiological levels to guide our understanding on the impacts of AIDS.

See Also: Climate Change; Immune System Diseases; Sexually Transmitted Diseases.

Further Readings

Kallings, L. O. "The First Postmodern Pandemic: 25 Years of HIV/AIDS." *Journal of Internal Medicine*, 263/3 (2008).

McMichael, A. J., M. McKee, V. Shkolnikov, and T. Valkonen. "Global Trends in Life Expectancy: Convergence, Divergence—or Local Setbacks?" *The Lancet*, 363 (2004).

UNAIDS: The Joint United Nations Programme on HIV/AIDS, and the World Health Organization. (December 2007). "2007 AIDS Epidemic Update." http://data.unaids.org/pub/EPISlides/2007/2007_epiupdate_en.pdf (Accessed August 2010).

United Nations Development Programme (UNDP), UN Food and Agriculture Organization (FAO), and the National Center for Atmospheric Research (NCAR). "Climate and HIV/AIDS: A Hotspots Analysis for Early Warning Rapid Response Systems." (2004). http://www.fao.org/nr/climpag/pub/Climate%20and%20HIV-AIDS%202004%20Gommes.PDF (Accessed August 2010).

United Nations Environment Programme (UNEP) and UNAIDS: The Joint United Nations Programme on HIV/AIDS. "Climate Change and AIDS: A Joint Working Paper." (2007). http://data.unaids.org/pub/BaseDocument/2008/20081223_unep_unaids_joint_working_paper_on_cca_en.pdf (Accessed August 2010).

Kadia Petricca
University of Toronto

ADVERTISING AND MARKETING

The advertising and marketing of all health products and products that make health claims are strictly regulated in the United States by the Food and Drug Administration (FDA) or the Federal Trade Commission (FTC), depending on the nature of the product.

Dietary supplements and foods that are marketed with health claims are regulated by the FDA as foods rather than as drugs. Dietary supplements are defined by law as products that are not intended as conventional food; are ingested in pill, capsule, tablet, powder, or liquid form; and contain a vitamin, mineral, nontobacco plant product, amino acid, or a concentrate, metabolite, constituent, or extract of any of the above. Three substances that do not fit this description are additionally regulated as dietary supplements under U.S. federal law: the steroids pregnenolone and dehydroepiandrosterone (DHEA) and the hormone melatonin. Under the Dietary Supplement Health and Education Act of 1994 (DSHEA), such supplements are not subject to tests of their safety and effectiveness, nor must they pass an approval process such as the one drugs must. Manufacturers of such supplements are allowed to make "structure or function claims" on labels and in advertising—claims of specific health benefits, though these may not include claims of treating, curing, preventing, or diagnosing any disease, and must include a disclaimer to that effect.

A side effect of the Dietary Supplement Health and Safety Act has been widespread consumer misconceptions about the FDA's responsibilities in the matter of dietary supplements. Polls almost 10 years after the act's passage showed that most Americans believed that potential side effects of dietary supplements had to be listed on the label and that supplements were subject to government approval before sale and marketing—in essence, these respondents erroneously believed that dietary supplements were subject to regulations similar to those of over-the-counter medications. Because of this belief in a regulatory environment, advertising for such supplements may carry a stronger sense of authority and trustworthiness—because of the audience's sense that the product and the statements about the product have been subject to a greater degree of scrutiny than in fact they have.

However, such ads are regulated to some degree, just not as strictly as the public believes. A dietary supplement cannot be marketed as a treatment or cure for any specific disease or condition; such claims would qualify them as drugs, subject to testing and regulation as such. The lines between dietary supplements that merely "promote good health" or perform similarly vague goods, dietary supplements with specific function claims such as "promoting healthy sleep patterns" (function claims being permissible if they can be substantiated with reliable research), and substances with druglike claims targeting specific diseases are actually somewhat blurry. In particular, many dietary supplements are sold because of a widespread belief in their efficacy in either reducing the risk of cancer or treating the disease, and they may even be marketed as such, even though this clearly exceeds the bounds of a mere function claim. The DSHEA was passed in part because of significant lobbying by the health freedom movement, an unofficial collection of groups that are critical of the pharmaceutical industry and perceive regulation of nonpharmaceutical health remedies as restricting their freedom of choice. Such groups have also made it difficult to target unsafe or ineffective health remedies and either ban them or charge them with fraud, as this too is seen as simply playing into "Big Pharmaceutical's" hand.

Drugs and Cosmetics

The advertising and promotion of prescription drugs is regulated, and regularly reviewed, by the FDA, which requires that such advertisements disclose the risks of the drug—which in print results in the "fine print" disclaimers usually at the bottom of an advertisement page, and in radio or television ads takes the form of a narrator listing the litany of potential side effects (as discovered during the clinical trials that are part of the approval process). In nearly every case, a prescription drug may only be advertised for the specific medical use for which it was approved. Though at first glance this may sound trivial, a growing number of drugs are actually prescribed for "off-label" purposes, from the use of Viagra by men with no erectile difficulty to the use of blood pressure medication in preventing migraines to the widespread prescription of antipsychotics and anticonvulsants in treating nonpsychotic psychiatric conditions such as bipolar disorder. Indeed, in psychiatry, there are disorders for which there are only off-label remedies or for which off-label remedies constitute the bulk of prescriptions.

The FDA also regulates cosmetic products, in which area it is concerned principally with safety and advertising: products that have not undergone thorough safety testing must be so disclaimed on their labels. So-called cosmeceuticals are cosmetic products that make structure or function claims and are subject to premarket testing of safety and efficacy, just as drugs are. The term is an industry category with no special legal meaning: under federal law, a product sold for its medical benefits is a drug, regardless of whether it happens to also be a cosmetic. Consumers generally do not think of such cosmetics as drugs because they are applied topically rather than ingested. Typical examples include antiaging moisturizing creams and other skin treatments. Because of the cost to the producer incurred by extensive efficacy testing and clinical trials and the delay that the process causes in bringing a product to market, there are fewer such "cosmeceuticals" on the market than first glance suggests; a great many products that seem to promise medical-like benefits actually bear their FTC-required disclaimers indicating that the product has no medical or drug properties, only a cosmetic effect. Certain tropes of advertising and turns of phrases have become common in order to suggest that products "create a youthful glow" without actually going so far as to make function claims of antiaging, wrinkle reduction, and the like.

Misleading marketing of cosmetic products that emphasizes the presence of a certain active ingredient is sometimes called "angel dusting." The practice is found in advertising for cosmetics, food, dietary supplements—essentially any product category that suggests medical benefits without claiming them and thus needing to face FDA regulation. The basic tactic of angel dusting is to draw attention to the presence in the product of a biologically active ingredient that the customer knows is beneficial in other contexts, but that in the case of the angel-dusted product does not occur in sufficient quantity to have a therapeutic effect—and therefore need not be regulated as a drug. In many cases, labeling laws do not require disclosing the amount of an ingredient in a product; certain substances must be disclosed by amount, such as caffeine and sugar in food products, but in many cases it is sufficient to simply list the ingredients in decreasing order of concentration, which makes it difficult to tell when the active ingredient is present only in minuscule quantities. The alternative medical practice of homeopathy operates, from an advertising perspective, on the same principle as angel dusting: because it is based on an obsolete understanding of science, homeopathy prescribes remedies in which the active ingredient is diluted well beyond the point of therapeutic benefit. Though homeopathic remedies are safe, this is because they have no proven efficacy that cannot be explained by the placebo effect. Despite this, homeopathic remedies are classified as drugs in the United States and are regulated accordingly, usually as over-the-counter products.

In addition to such homeopathic remedies, there are approximately 100,000 individual over-the-counter drug products, created from the 800 FDA-approved over-the-counter drug ingredients; though the FDA regulates these ingredients, the FTC regulates their advertising and marketing.

Nutraceuticals

A growing category of health product is the "nutraceutical": food products marketed with claims of health benefits. Though this includes dietary supplements, much of the growth has been in the area of "functional foods." Functional foods are foods—not just dietary supplements or ingredients to add to foods—that are promoted as beneficial to health in ways apart from their supply of nutrients. Though the term is new, many packaged cereals such as those from the Kellogg Company have been marketed as functional foods since their introduction at the end of the 19th century: cereals fortified with additional vitamins and minerals and intended to promote good health. Older functional foods include fermented foods with live cultures, such as yogurt, sauerkraut, kimchee, and miso, which have probiotic benefits, promoting the growth of beneficial bacteria in the human body. Though all such products possess these benefits, only in recent years have they been marketed for such benefits, with some brands actively associating themselves with probiotics or even inoculating their products with more bacteria or with patented cultures of such bacteria.

When such foods have not undergone testing or have not been evaluated by the FDA, they bear the typical disclaimer: "These statements have not been evaluated by the Food and Drug Administration. This product is not intended to diagnose, treat, cure, or prevent any disease." But the content of the marketing itself is still clear, and certain foods and food products become synonymous with the health claims associated with them in the public ear—superfruits and green tea especially—to the extent that pomegranate and acai, first catapulted into fame by the health claims associated with them, have become available as jelly bean flavors and in other contexts in which they could provide no nutritive or medical benefit whatsoever, regardless of the strength and validity of their health claims. Green tea, of course, has entered the American consciousness in a similar manner, with green tea drinks often touted for their antioxidant effects and marketed with more health-conscious statements and imagery than black tea drinks have been.

Functional foods and nutraceuticals are distinct from medical foods, defined in the 1988 Orphan Drug Act Amendments as specially formulated foods intended for the dietary management—not the prevention, treatment, or cure—of a disease that causes specific nutritional needs. Such foods are used under medical supervision and are regulated by the FDA. Because the customer base is so small, advertising for medical food products tends to be limited in scope (taking the form of pamphlets distributed at the doctor's office, for instance, or ads in magazines on managing the disease in question). The growing prevalence of type 2 diabetes, peanut allergy, and gluten intolerance in the U.S. population suggests that ads for foods tailored for such conditions may become more common.

Any food product may be marketed with a health angle, even if it is not a functional food as such, and under the Nutrition Labeling and Education Act of 1990, when these foods are marketed to the consumer with such claims, they must meet FDA definitions. For instance, terms like *high fiber* and *low fat* have specific definitions that must apply to the product. Furthermore, a food product may not be marketed as healthy if its healthiness is not demonstrable, though this is a gray area that the FDA only occasionally cracks down on; noted examples are sugary junk food breakfast products marketed to children as "part of a nutritious breakfast" in which the nutritious aspects of the breakfast shown in the advertising materials are all derived from the other items being served, not the product being advertised. Drinks may not be called juice or juice drinks if they do not contain it and must declare their percentage of juice content in order to attempt to cut down on fruit-flavored sugar-water drinks being mistaken by naive consumers as healthy and nutritious. None of these requirements apply to restaurants, though in the years since the act's passage, there have been numerous attempts to hold restaurants to a new standard of disclosure, whether on menus or in advertising materials or both.

Tobacco and Alcohol

The government also regulates the advertising of many nonfood, noncosmetic, nonmedicinal products for health reasons, notably tobacco and alcohol. In the last few decades, regulations have continually tightened around tobacco marketing and advertising in an attempt to prevent tobacco companies from marketing to children and teenagers, with each further restriction passing when it has become clear that the previous restriction did not sufficiently curtail teen smoking. Product mascots like Joe Camel and loyalty programs like Camel Cash were banned because of the cartoon character's appeal to children, as were tobacco ads on television. More recently, the FDA has banned tobacco companies from sponsoring sports or entertainment events; has prohibited the sale of cigarettes in amounts smaller than a pack of 20, the sales of singles and so-called kiddie packs (half- or quarter-sized packs) believed to have made smoking more affordable and appealing to children and young teens; has outlawed not only free samples of cigarettes but any giveaway (purchase incentive) of nontobacco items with the purchase of tobacco; banned vending machines (already banned tout court in many states) from all-ages facilities; and banned various flavored tobacco products in response to a multiyear trend in the cigarette industry toward sweet, candy-flavored cigarettes.

The Federal Trade Commission has similarly explored the possibility of restricting alcohol advertising because of the possibility of targeting underage drinkers, but unlike the tobacco industry, the American alcohol industry (which has a long history of having its activities restricted by law) is largely self-regulating. Because of the Joe Camel issue in the tobacco industry, the use of cartoon characters is discouraged by industry self-regulatory bodies, though not condemned outright; drinking must be portrayed in a responsible manner, and drunkenness itself cannot be used as a selling point (in contrast with foreign markets, it is never even portrayed in an obvious manner in American advertising); and advertisements are restricted to media where 70 percent of the audience is over 21, based on available demographic analyses (such as the Nielsen reports for television shows). It is largely for this reason that alcohol ads are permitted on television while tobacco ads are not.

See Also: Bottled Water; Calorie Labeling for Restaurants; Children's Health; Fast Food Warnings; Men's Health; Obesity; Pharmaceutical Industry; Women's Health.

Further Readings

Grossman, Elizabeth. *Chasing Molecules: Poisonous Products, Human Health, and the Promise of Green Chemistry*. Washington, DC: Shearwater Press, 2009.

Kessler, David. *A Question of Intent*. Jackson, TN: PublicAffairs, 2002.

Layzer, Judith A. *The Environmental Case: Translating Values Into Policy*. Washington, DC: CQ Press, 2005.

Malerba, Larry. *Green Medicine*. Berkeley, CA: North Atlantic Press, 2010.

Wargo, John. *Green Intelligence*. New Haven, CT: Yale University Press, 2009.

Bill Kte'pi
Independent Scholar

Airborne Diseases

Airborne diseases refers to the grouping term for those human diseases that are caused by organisms that can be transmitted by the air (airborne transmission). This is an important environmental aspect for the global public health that should be considered in the new perspective of creating a more ecological green world, where such environmental threats should be significantly reduced or even eliminated. There are many diseases that can additionally be transmitted by other ways, such as blood-borne, vertical, or congenital methods. Airborne transmission occurs when droplet nuclei (evaporated droplets) less than 5 µm in size are disseminated in the air. These droplet nuclei can remain suspended in the air for some time. Droplet nuclei are the residuals of droplets, and when suspended in the air, they dry and produce particles ranging in size from 1 to 5 µm. Diseases spread by this mode include open/active pulmonary tuberculosis (TB) (due to *Mycobacterium tuberculosis*), measles, chicken pox (varicella-zoster virus, VZV), pulmonary plague (*Yersinia pestis*), and hemorrhagic fevers such as Lassa fever.

Droplet transmission occurs when there is adequate contact between the mucous membranes of the nose and mouth or conjunctivae of a susceptible person and large particle droplets (>5 μm). Droplets are usually generated by the infected person during coughing, sneezing, or talking, or when healthcare workers perform procedures such as tracheal suctioning. Diseases transmitted by this route include pneumonias (bacteria), pertussis (*Bordetella pertussis*), diphtheria (*Corynebacterium diphteriae*), severe acute respiratory syndrome (SARS) (coronavirus, SARS-CoV), influenza, mumps, and meningitis (bacteria).

In the case of tuberculosis, infection with the tubercle bacillus is usually by direct airborne transmission from person to person. For measles, which is a highly contagious infection, transmission is primarily by large respiratory droplets, increasing during the late winter and early spring in temperate climates, and after the rainy season in tropical climates. In varicella, or chicken pox, transmission is via droplets, aerosol, or direct contact, and patients are usually contagious from a few days before rash onset until the rash has crusted over. In the case of pulmonary plague, direct person-to-person transmission does occur when respiratory droplets may transfer the infection from the patient to others in close contact, but this is considered an exception. Plague is a zoonotic disease affecting rodents and is usually transmitted by fleas from rodents to other animals and to humans. Lassa fever virus is carried by rodents and transmitted by excreta, either as aerosols or by direct contact.

On the other hand, diseases such as pertussis (whooping cough), a highly contagious acute bacterial disease involving the respiratory tract and caused by *Bordetella pertussis*, is transmitted by direct contact with airborne discharges from the respiratory mucous membranes of infected persons. In the case of diphtheria, transmission is from person to person through droplets and close physical contact. For SARS, transmission of SARS-CoV is primarily from person to person. Its transmission to noninfected hosts apparently occurs mainly during the second week of illness, which corresponds to the peak of virus excretion in respiratory secretions and stool, and when cases with severe disease start to deteriorate clinically. Most cases of human-to-human transmission have occurred in the healthcare setting, in the absence of adequate infection-control precautions. In the case of influenza, respiratory transmission occurs mainly by droplets disseminated by unprotected coughs and sneezes. Short-distance airborne transmission of influenza viruses may occur, particularly in crowded enclosed spaces. Hand contamination and direct inoculation of virus is another possible source of transmission. For mumps, transmission occurs as in measles, as a consequence of respiratory droplets.

Measles

Remarkable progress has been made in reducing measles incidence and mortality as a consequence of implementing the measles mortality reduction strategy of the World Health Organization (WHO) and United Nations Children's Fund (UNICEF). The revised global measles mortality reduction goal set forth in the WHO-UNICEF Global Immunization Vision and Strategy for 2006–2015 is to reduce measles deaths by 90 percent by 2010 compared to the estimated 757,000 deaths in 2000. The possibility of measles eradication has been discussed for almost 40 years (as of 2010), and measles meets many of the criteria for eradication. Global measles eradication will face a number of challenges to achieve and sustain high levels of vaccine coverage and population immunity, including population growth and demographic changes, conflict and political instability, and public perceptions of vaccine safety. To achieve the measles mortality reduction goal, continued progress needs to be made in delivering measles vaccines to the world's children.

Measles is a highly contagious disease that was responsible for high infant mortality before the advent of an effective vaccine in 1963. In immunocompetent individuals, measles virus infection triggers an effective immune response that starts with innate responses and then leads to successful adaptive immunity, including cell-mediated immunity and humoral immunity. The virus is cleared, and lifelong protection is acquired. However, changing epidemiology of measles due to vaccination as well as severe immunodeficiency has created new pockets of individuals vulnerable to measles.

Chicken Pox

Infections with varicella zoster virus (VZV) are common viral infections associated with significant morbidity. Diagnosis and management is complex, particularly in immunocompromised patients and during pregnancy.

VZV has a global distribution. Varicella (chicken pox), the manifestation of the primary infection with VZV, is highly contagious—96 percent of susceptible subjects exposed to it develop the disease. About 90 percent of primary infections occur in children under the age of 10 years. Less than 5 percent of people develop the disease after the age of 15 years. Notably, the prevalence of primary VZV infection is lower in tropical and subtropical countries than in Europe and North America. Therefore, individuals from tropical and subtropical countries immigrating to Europe or North America are at increased risk of primary VZV infection in adulthood. VZV is shed in respiratory secretions and cutaneous lesions. Transmission is airborne or by direct contact of skin and mucosa with the contents of the blisters. The portal of entry is the upper respiratory mucosa and the conjunctiva.

Herpes zoster is a very common disease that can be severe. The impact of herpes zoster on older patients is not fully appreciated by some physicians. Herpes zoster can sometimes have devastating consequences on patient quality of life, particularly when postherpetic neuralgia develops. Herpes zoster incidence has been remarkably consistent over time and throughout the world. Given the severity of the disease and the considerable lifetime risk, more effective management of herpes zoster and postherpetic neuralgia is essential. Over the past 20 years, there has been progress in compressing morbidity to the very end of life through improved diet, exercise, and healthcare. The next challenge will be to postpone disabilities due to aging by using strategies such as vaccination to prevent diseases common in elderly patients. Prevention of herpes zoster and postherpetic neuralgia should be one important goal for promoting healthy aging.

Pulmonary Plague

The plague is an infectious bacterial disease; *Yersinia pestis* is the etiological agent. This infection has a high fatality rate when not treated. Plague has caused a number of pandemics throughout human history. Its major morbidity has occurred in three huge pandemics since the sixth century, with millions of deaths, and numerous smaller epidemics and sporadic cases. Referring to specific clinical symptoms of pulmonary plague, the disease became known as the Black Death. This pandemic probably originated in central Asia and began spreading westward along major trade routes. Upon its arrival in the eastern Mediterranean, the disease quickly spread, especially by sea traffic, to Italy, Greece, and France and later throughout Europe by land. Until the 18th century, many European cities were frequently affected by other great plague epidemics. The worldwide spread of the third pandemic began when the plague reached Hong Kong and Canton in 1894. In the following years, the role of rats and fleas and their detailed role in the transmission of plague was discovered and experimentally verified. Today the plague is still endemic in many countries of the world.

Plague is a zoonosis that primarily affects rodents; humans and other animals (e.g., domestic cats) are accidental hosts. The natural ecosystem of *Y. pestis* depends largely on the flea and rodent interaction, with seasonal variability noted based on environmental conditions. Infected fleas bite their rodent hosts, inoculating the rodent. Mortality in these animals remains lower than in other nonrodent mammals, and disease is passed from infected rodent to flea and the life cycle continues. Transmission to humans occurs via bites from fleas from infectedrodents, infected animal scratches or bites, exposure to infected humans, and bioterrorism. Transmission via infected fleabite is the most common mode, with squirrels, rabbits, domestic cats, and prairie dogs being the most common animals of transmission. Large rodent or other animal die-off, particularly in more susceptible species, may herald a large epidemic in nature. Plague is found worldwide, and in the United States, endemic disease is found largely in the western states.

Hemorrhagic Fevers

The hemorrhagic fever viruses include a number of geographically distributed viruses found worldwide, including the Ebola and Marburg viruses, Rift Valley fever, Crimean Congo hemorrhagic fever, Lassa fever, yellow fever, and dengue fever. The Ebola and Marburg viruses are in the family Filoviridae. Although any of the many viral hemorrhagic fevers can cause a febrile respiratory illnesses, Marburg and Ebola viruses serve as a classic template for viral hemorrhagic fevers. Marburg virus has a single species, whereas Ebola has four different species, which vary in virulence in humans. Transmission appears to occur through contact with nonhuman primates and infected individuals. Settings for transmission have included vaccine workers handling primate products, nonhuman primate food consumption, nosocomial transmission, and laboratory worker exposure. The use of viral hemorrhagic fever in bioterrorism has also been postulated, largely based on its high contagiousness in aerosolized primate models. The exact reservoir for the virus was initially thought to be with wild primates, but recently bats have been labeled as the reservoir, passing the infection on to nonhuman primates in the wild.

Severe Acute Respiratory Syndrome (SARS)

SARS is caused by a novel coronavirus that was first detected in 2003. This novel virus was finally understood to have evolved from a horseshoe bat coronavirus and spread to humans after investigations showed remarkable genetic similarity between the bat and human strains. Thousands of cases occurred worldwide in the initial epidemic in 2003, but the epidemic abated and new cases have not been reported since. The clinical presentation is initially characterized by fever, chills, rigors, malaise, nausea, and shortness of breath, with respiratory failure developing later. The symptoms occurred on average seven days after contact. Pneumonia appeared to develop approximately eight days after onset of fever, and 45 percent of patients developed hypoxemia. About 20 percent of patients went on to develop severe lung injury and acute respiratory distress syndrome (ARDS) that required mechanical ventilation. Development of ARDS from onset of fever is bimodal, with peaks at 11 and 20 days. The global fatality rate was 11 percent, and most of the deaths were in patients over 65. No deaths were reported in children.

See Also: Acquired Immune Deficiency Syndrome; Antibiotics; Bird Flu; Centers for Disease Control and Prevention (U.S.); Tuberculosis; Vaccination/Herd Immunity.

Further Readings

Franco-Paredes, Carlos, Alfonso J. Rodriguez-Morales, and José Ignacio Santos-Preciado. "Clinical and Epidemiological Aspects of Influenza." *Latin American Medical Student Science and Research*, 11 (2006).

Johnson, Robert, Janet McElhaney, Biagio Pedalino, and Myron Levin. "Prevention of Herpes Zoster and Its Painful and Debilitating Complications." *International Journal of Infectious Diseases*, 11 (2007).

Kempfa, W., P. Meylanb, S. Gerberc, C. Aebid, R. Agostie, S. Büchnerf, B. Coradig, J. Garwegh, H. H. Hirschi, C. Kindj, U. Lauperk, S. Lautenschlagerl, P. Reusserm, C. Ruefn, W. Wunderlio, D. Nadalp, and Swiss Herpes Management Forum. "Swiss Recommendations for the Management of Varicella Zoster Virus Infections." *Swiss Medical Weekly*, 137 (2007).

Moss, W. J. "Measles Control and the Prospect of Eradication." *Current Topics in Microbiology and Immunology*, 330 (2009).

Naniche, D. "Human Immunology of Measles Virus Infection." *Current Topics in Microbiology and Immunology*, 330 (2009).

Rodriguez-Morales, Alfonso J. "Ecoepidemiology and Satellite Epidemiology: New Tools in the Management of Problems in Public Health." *Peruvian Journal of Experimental Medicine and Public Health*, 22 (2005).

Sandrock, Christian E. "Severe Febrile Respiratory Illnesses as a Cause of Mass Critical Care." *Respiratory Care*, 53 (2008).

World Health Organization. *International Travel and Health*. Geneva: World Health Organization, 2009.

Zietz, B. P. and H. Dunkelberg. "The History of the Plague and the Research on the Causative Agent *Yersinia pestis*." *International Journal of Hygiene and Environmental Health*, 207 (2004).

Alfonso J. Rodriguez-Morales
Universidad de Los Andes, Universidad Central de Venezuela

Air Filters/Scrubbers

Air filters and scrubbers play an important role in health and safety control in new industries. If dust or other hazardous particulate matter generated by industries is released and diffused in the air, it can easily travel to different parts of the building and colonize on walls, ducts, clothing, or furniture and contaminate clean areas. If released externally (e.g., via a smokestack) they can travel to distant geographic regions, spreading air pollution far from the original source. Air filters and scrubbers can help remove the hazardous particles from the air and bring the level of these particles in the air to a safe level, continuously cleaning the air and providing a clean, healthy, and productive environment. Air filters also are used in the home to remove irritants and allergens such as tobacco smoke, pet dander, and mold spores from the air.

The performance of air filter machines (also known as air cleaners or air purifiers) are measured by the CFM unit, which denotes the cubic feet of air purified or filtered per minute. Therefore, a machine with performance of 1,000 CFM can filter 1,000 cubic feet of air per minute.

Air filters are customarily classified as either industrial or nonindustrial. The distinction is in the load they handle: industrial air filters are capable of removing a high concentration of dust and pollutants from the air, while nonindustrial air filters are limited to filtering a low concentration of pollutants from the air. Another important consideration is the size of the pollutant to be removed: for instance, dust particles may be 10 micrometers in diameter (still small enough to be inhaled and thus harm human health), while the particles found in smoke and haze (from such sources as gas emitted from an industrial plant) may be less than 2.5 micrometers in diameter.

There are several different techniques used by air filters to purify the air, with the major categories being the following:

- HEPA filters
- Chemical filters
- Pressure filters
- Electrical filters

A High Efficiency Particulate Air (HEPA) filter can remove at least 99.97 percent of airborne particles 0.3 micrometers in diameter and remove larger particles with even higher efficiency. Particles of this size are the most difficult to filter, and also represent the most penetrating particle size (i.e., they pose the greatest danger for inhalation). HEPA filters are composed of a mat including a set of randomly arranged fibers, typically made from fiberglass with diameters between 0.5 and 2.0 microns. The air space between HEPA filter fibers is greater than 0.3 micrometers. HEPA filters are the most effective air filtration filters for removing contaminants from the air in new and high technology industries.

A HEPA filter has a fan that forces air to flow through the filter where the particulate matter can be trapped. The greater the airflow is through the filter, the cleaner the environment. The HEPA filter first was used in the Manhattan Project in 1940 to prevent the spread of airborne radioactive contaminants and is still used in the nuclear industry today, often in conjunction with activated carbon filters. This system allows the removal of radioactive gases as well as particles before exhausted air is released into the atmosphere. In biomedical applications, HEPA filters are used to prevent the spread of airborne bacterial and viral organisms and, therefore, infection. Typically, medical-use HEPA filtration systems also incorporate high-energy ultraviolet light units to kill off live bacteria and viruses trapped by the filter media. Modern airliners also use HEPA filters to reduce the spread of airborne pathogens in recirculated air.

In chemical filters contaminated air is passed through a compartment filled with a chemical solution, which is designed based on the type of pollutant in the air. The chemical solution reacts with the pollutant and chemically absorbs the pollutant from the air.

Oil-based chemical solutions are frequently used in chemical filters because they can easily absorb dust and most of the pollutants in the air.

In pressure filters, a negative pressure helps the air scrubbers capture the contaminated air and produce clean noncontaminated air. Air always flows from areas of higher pressure to areas of lower pressure so creating and maintaining negative pressure means that the air will flow into a room or building rather than outward, preventing airborne contaminants such as mold, dust, and odors from escaping and contaminating other parts of the work area through openings or leaks in the walls. Negative air pressure rooms are used in hospitals to isolate patients with highly infectious, airborne diseases such as tuberculosis. This is an example of negative air pressure preventing contaminated air from leaving the room even when the door is opened.

In electrical filters, the contaminated air is passed through a very strong electrical field, which polarizes the pollutants in the air and allows the newly polarized particles to be absorbed by a set of negatively charged plates known as the collection section. A prefiltering compartment helps remove bigger particles and increases the accuracy of the electrical filters. The absorbed particles are removed in the cleaning compartment. Electrical filters are efficient at filtering dust and pollutants of various sizes that are produced in a high concentration and play an important role in cement industries where high levels of the pollutants must be removed from the air in a fixed amount of time. However, they have a smaller CFM compared to the other filtering machines, meaning electric filters can filter a lower volume of air per minute.

See Also: Airborne Diseases; Carbon Filters; Indoor Air Quality.

Further Readings

Brown, R. C. *Air Filtration: An Integrated Approach to the Theory and Applications of Fibrous Filters.* Oxford, UK: Pergamon, 1993.

Raymer, Paul H. *Residential Ventilation Handbook: Ventilation to Improve Indoor Air Quality.* New York: McGraw-Hill Professional, 2009.

Sandrock, Christian E. "Severe Febrile Respiratory Illnesses as a Cause of Mass Critical Care." *Respiratory Care,* 53 (January 2008).

Hassan Bagher-Ebadian
Independent Scholar

Alternative Energy Resources (Solar)

The sun's rays may have been the very first energy source utilized by man. There is evidence that primitive societies captured the heat energy from the sun to heat their homes and for cooking, two uses that remain common today in some parts of the world. Legend has it that in the third century B.C.E., Archimedes used a giant lens to concentrate the sun's rays and set fire to an invading Roman fleet, utilizing a principle that today is used to generate solar energy. The use of the sun's rays as a light source is also ancient, and the principle of using sunlight to light the interiors of buildings is enshrined in law in some countries.

For instance, Great Britain recognizes a "right to light" principle. This law limits construction that would cut off the source of natural daylight to other buildings that have traditionally enjoyed access to that light. More recently, Antoine Lavoisier developed a solar furnace in France in the late 18th century; solar water heaters were developed in the United States in the 1890s; and solar powered water pumps were used for irrigation in Egypt from the early 20th century. Harnessing solar radiation to produce electricity became possible with the development of photovoltaic cells in the 1950s, although the photoelectric effect had been described in 1839. Photovoltaic cells were used to provide power to spacecraft in the 1960s. By the 1970s, solar power was beginning to be seen as a reasonable supplement or alternative to other methods of generating power.

The amount of electricity that photovoltaic solar cells generate depends on the amount of sunlight and the angle at which that sunlight hits the panels. Here, President Obama views a photovoltaic array of solar panels with the base commander during a tour of the Nellis Air Force Base in May 2009.

Source: White House/Pete Souza

Types of Solar Power

There are three methods in common use today to capture the energy of the sun's rays for practical use. The first is through passive or active thermal systems, which use the sun's energy to heat buildings, water, and so on. Passive solar heating requires only that a building or installation be designed to capture the sun's rays through orientation, placement of windows, and so on, and requires no additional energy to operate the system. Active solar heating, which is generally more efficient but also more complex and requires the use of electrical or mechanical equipment, uses collectors to capture the energy of the sun. The captured energy is then transmitted to the building through means such as a pump or fan. Both methods are used to heat water. A simple example of passive solar water heating is a painted black barrel placed on a roof—similar to systems used by some campers to heat water in a black bag for use in bathing and dishwashing—while sophisticated systems involve collectors and pumps. These latter systems are commonly used in Europe and Asia.

Photovoltaic cells, also called solar cells, are used to convert solar energy to electricity. The first working photovoltaic cells were developed at Bell Laboratories in the United States in 1954. Although originally considered a curiosity, solar cells proved useful as a power source for spacecraft in the 1960s and 1970s since they were a nonpolluting alternative to power generated from coal or petroleum. The heart of a solar cell is a wafer of semiconductor material (generally silicon) that has been treated so it is positive on one side and negative on the other. Electrical conductors are attached to both sides to form an electrical circuit to capture the energy created when the sun's rays strike the cell and knock electrons loose from the atoms in the semiconductor material. This creates electricity, which can be used as power such as a light. More electricity may be created by connecting multiple solar cells in a module; multiple modules form an array. Modules and arrays produce direct current and are designed to output electricity at a particular voltage (e.g., 12 volts). In general, the greater the area exposed to the sun, the greater the electricity generated.

Concentrated solar power or central solar power (CSP) systems concentrate the sun's rays—often with the use of a solar tracker to follow the sun's movement during the day—with lenses or mirrors, and then they focus it on photovoltaic cells or use it to generate heat. In the former case, CSP is a more efficient and less expensive way to produce energy from photovoltaic cells. In the latter case, the sun's energy can be used to operate a turbine in a manner similar to a conventional power plant. An example of this is heating water to produce steam to power a turbo generator to produce electricity. CSP technology is relatively recent and is developing rapidly, and thus it currently exists in several forms. A parabolic trough system uses a parabolic reflector to concentrate light onto a receiver and heat a molten fluid that is then used as a heat source to generate power. An example of this type of power is the Solar Energy Generating Systems (SEGS) complex in the Mojave Desert in California. A dish engine, or Dish Sterling system, uses a parabolic reflector to heat a fluid and power a Stirling Engine (heat engine). This is currently the most efficient method of generating power from sunlight. An example of heat engine solar technology is the "Big Dish" in Whyalla, South Australia. A solar chimney is among the simplest of CSP designs and uses a greenhouse-like structure to heat the interior air that is then used to run an air turbine.

Current Use of Solar Power

Solar generation of electricity is a growing but still minor contributor to total electricity generation in most of the world. In the last 15 years, the demand for solar power has grown by about 25 percent annually. The pace has accelerated in recent years. In 2000, the global photovoltaic capacity was estimated at 1.2 GWp (gigawatt peak: giga = billion; Watt Peak: a unit for measuring the output of a photovoltaic cell or module under standard conditions), while in 2008 it was estimated at 6.5 GWp, a growth rate of about 35 percent annually. In the European Union (EU) capacity more than doubled from 2007 to 2008, from 1825.6 MWp (M = mega = million). Within the EU, Spain is the leader in solar power with a total of 2,670.9 MWp, followed by Germany with 1,505.0 MWp. The other countries are far behind: Italy is third with 197.3 MWp, followed by the Czech Republic with 50.3 MWp. Spain also is the leader in per-capita solar energy with 75.2 Wp/person, followed by Germany (65.2 Wp/person) and Luxembourg (50.5 Wp/person). Almost all solar electrical capacity in Europe is connected to the electrical grid, with a small amount used for stand-alone systems to power isolated sites, roadside safety devices, and the like.

Japan has been an early adopter of photovoltaic technology thanks in part to several government programs, including the Sunshine Project of 1974 and the New Sunshine Project of 1993. This latter effort announced a target of installing 70,000 solar roofs in Japan. In 1997, the New Energy Law offered subsidies and incentives to both homeowners and businesses to adopt solar technologies. As of 2008, Japan had 2,149 MWp of solar power installed. China is a world leader in manufacturing photovoltaic cells but has been slower to adopt their use. Implementation is expected to increase, and government incentives such as those provided in the 2009 Golden Sun program, which subsidizes construction of solar power projects in remote locations, will probably speed the adoption of solar power. India is rapidly embracing solar power for a variety of purposes, including cooking, heating water, and running pumps, and was the first country to establish a Ministry of Renewable Energy. In 2009, the Indian government announced a plan to produce 20 GWp of solar energy by 2020.

Solar power currently supplies a small amount of Australia's total energy use (115 MWp installed as of 2009), but the country has inaugurated several projects to increase its use. The Mildura Solar Concentrator Power Station, expected to be operational in 2013, will be the largest photovoltaic power station in the world. The United States, despite being involved in much of the early research involved in making solar power possible, creates less than 0.1 percent of its electricity from solar power. However, the use of solar power is rapidly increasing. Major installations in the United States include the SEGS facility in California, the largest solar plant in the world, and the Nevada Solar One solar thermal plant near Boulder City, Nevada. Further growth is expected, due in part to many government programs encouraging development and use of solar power. In California, electric utilities will be required to produce 20 percent of their power using renewable sources by 2017.

Advantages and Disadvantages

Solar power has a number of advantages, which has resulted in its adoption in countries around the world and in nations at all levels of technological development. First of all, solar radiation is a renewable and virtually inexhaustible resource. Some experts have calculated that the world's electricity needs could be fulfilled by the installation of CSP systems on less than 1 percent of the world's desert terrain. Solar power also is a very clean source of energy, producing few greenhouse gases. It has minimal environmental impact other than what is required to manufacture and transport the necessary equipment. Generation of power from the sun also requires relatively little land and can often be combined with buildings used for other purposes, such as installing solar panels on a home or office building. Solar panels require practically no upkeep once installed. Finally, solar power installations can be freestanding, so a remote location can supply its own electricity needs without having to be hooked up to a grid.

There are also disadvantages to solar power. Although technology is constantly improving, photovoltaic cells are still relatively inefficient, and the initial cost of installing a solar power system is high. Photovoltaic cells only work in the sunlight, so some form of storage or alternative energy source is required at night and during cloudy days. Finally, power from photovoltaic cells is direct current and must be converted to alternating current if the solar source is connected to an electrical grid. Some of these disadvantages can be mitigated through public policy. Many electrical grid systems allow homeowners to sell excess solar electricity generated during the day, while the homeowner can purchase power from the grid to use at night. Various locations have instituted tax credits or other subsidies to mitigate the high initial cost of solar installation or conversions.

See Also: Electricity; Government Role in Green Health; Personal Consumer Role in Green Health.

Further Readings

Environmental Protection Agency. "Auxiliary and Supplemental Power Fact Sheet: Solar Power." (October 2007). http://www.epa.gov/owmitnet/mtb/solar_final.pdf (Accessed August 2010).

Knier, Gil. "How Do Photovoltaics Work?" (2002). http://science.nasa.gov/science-news/science-at-nasa/2002/solarcells (Accessed August 2010).

Photovoltaic Barometer: "Eurobserver." (March 2009). http://www.eurobserver.org/pdf/baro190.pdf (Accessed August 2010).

Sioshansi, Fereidoon P., ed. *Generating Electricity in a Carbon-Constrained World.* Burlington, MA: Elsevier Academic Press, 2010.

Smith, Zachary A. and Katrina D. Taylor. *Renewable and Alternative Energy Resources: A Reference Handbook.* Santa Barbara, CA: ABC-CLIO, 2008.

Sarah Boslaugh
Washington University in St. Louis

ANIMAL PRODUCTS

Animal products are any products produced by or taken from an animal, especially animals raised or slaughtered for that purpose. (Fossil fuels and fertilizer with animal remains are generally not considered animal products.) Though the term is most frequently used to refer to foods, because of the number of people in the world who avoid animal products or avoid them in certain circumstances—from vegetarians and vegans to those keeping kosher or halal diets—the term is also applicable to the many other products derived from animals.

Animal fibers, for instance, include silk, fur, and feathers. Animal fibers vary greatly, as demonstrated by the difference between the two most common animal fibers used commercially: wool from domestic sheep and silk from the mulberry silkworm. Even wool can vary dramatically according to the species of sheep (or other animal: the hair of some species of goats, alpacas, and rabbits is also called wool, as sometimes are the undercoats of the American bison and the musk ox). In addition to textiles, animal fiber has other uses. Horsehair is commonly used for brushes and other applications. Spider silk is used to treat wounds and to assist in mammalian neuronal regeneration by acting as a biocompatible artificial nerve conduit, and has often been used as a thread in optical instruments.

Feathers are best known for their use in consumer products as filling for sleeping bags, bedding and mattresses, and winter coats because of their excellent capacity for trapping heat. But they are actually extremely important in industry, where the feather waste produced by the poultry industry—which is slow to decompose—is used to culture microbes and produce enzymes.

Sinew, the tough fibrous tissue that connects muscle to bone, has a natural ability to withstand tension and so was an important animal product in the preindustrial era, when strong cordage was needed. Sinew provided the most important fiber for tool building and other construction purposes in harsh environments like those of the circumpolar peoples, where the local ecosystem provided no other natural fibers suitable for human use.

Catgut, similarly, has long been used not only to string musical instruments and tennis racquets but as surgical sutures. Despite the name, catgut is a tough, flexible fibrous cord manufactured from the intestines of sheep, goats, hogs, horses, or mules; the word is probably a contraction of "cattlegut."

Animal hide and skin provide leather (especially from cattle hide), which is processed through means that raise environmental concerns even among people who have no ethical objection to the use of animal products or the raising of animals for food. The chemicals used in the tanning process and the hydrogen sulfide and ammonia used in the dehairing and deliming processes can result in considerable soil, air, and water pollution, with the wastewater of a tannery containing extremely high levels of chromium, sulfides, and biological waste, as well as the pesticides used to prevent insect infestations while transporting the untreated hides before tanning. The chromium levels in particular are considered a danger because of the accumulation resulting from the volume of production done at most leather treatment sites.

Lanolin or "wool fat" is a greasy yellow substance secreted by the sebaceous glands of wool-bearing mammals like sheep and is responsible for their ability to shed water from their coats. Quantity of lanolin production varies by species, but like wool, lanolin can be harvested by humans without doing any harm to the sheep. A fatty wax that forms highly stable emulsions with water, lanolin has been used in waterproofing since prehistory and has been used for rust-proofing, ship-finishing, and lubrication since the Industrial Revolution. Medical-grade lanolin, free of impurities, is hypoallergenic and bacteriostatic and is given to breastfeeding mothers to treat sore, cracked nipples. It is also commonly used to treat skin irritations including chapped lips, diaper rash, itches, dry skin or calluses, burns and cuts, and is an ingredient in many shaving creams and gels. Because it is so easily absorbed by the skin, lanolin is used as the base for ointments both in folk remedies and in the cosmetics and medical industries; as it is absorbed, the skin also absorbs the medicinal chemicals that have been mixed with the ointment. Shaving cream with lanolin is traditionally used by baseball players to "break in" their gloves.

A waxy, bilious secretion of sperm whale intestines, possibly created in order to ease the digestive passage of hard objects like beaks, which are too difficult to digest, ambergris is often erroneously believed to be a product of whale vomit; it is actually usually passed through the anus. Masses of it can be found floating on the surface of the sea or washing up on the coast; most of the commercial ambergris harvest comes from the Bahamas and other Caribbean islands (for reasons yet unknown, a disproportionate amount of ambergris is found in the Southern Hemisphere, even though sperm whales swim in both hemispheres). As it ages, ambergris acquires a sweet scent like tobacco, and it was historically used in perfumery; it has also been used for flavoring food and cigarettes, and to treat headaches, colds, and epilepsy. Dissolved in wine, it was used as an aphrodisiac, and studies have shown that it has both an aphrodisiac and pain-killing effect in rats.

Despite coming from very different animals, castoretum—produced by the castor sacs of beavers—was used in many of the same applications as ambergris in medieval times, notably perfumery. It was also a medicinal ingredient for a long period of time, used to treat fever and headaches. Today, its principal use outside of perfumery is as an additive to enhance the flavor and scent of cigarettes.

Mink oil, similar to human sebum and produced by minks in their thick fatty subdermal layer, is a significant source of palmitoleic acid, with a number of cosmetic and medical applications. There are some indications that palmitoleic acid may also have some application in fighting obesity.

The gallstones of animals come to market as a by-product of meat processing and are extremely valuable, selling for $30 or more per gram. The gallstone market is primarily driven by their usage in traditional folk medicines, especially for the reduction of fever.

Shellac is a resin excreted by females of the lac bug species in India, Thailand, and Mexico. Harvested and sold as dry flakes, it is dissolved in alcohol to make liquid shellac, which has a wide variety of uses including the creation of record albums before the 1950s;

use as a sealant, varnish, or primer; and use as an odor blocker. Before synthetic plastics, shellac was used to manufacture many of the items now made of plastic, like inkwells, picture frames, and dentures. Because shellac is alkaline, it can be used to coat medical pills to produce a timed release; it is also commonly used to coat apples in order to preserve them, and when used as a food additive, it is commonly called "beetlejuice" in the industry.

Beeswax is produced by honeybees and has been used by humankind since prehistory as a modeling and molding material as well as in shipbuilding, the preparation of mummies, and the stabilizing of explosives in the black powder era. In the modern age, in addition to its use as a candle wax, about 60 percent of commercial beeswax is used by the cosmetics and pharmaceuticals industries. It is sometimes found in skin-care products, for instance, and is the key ingredient in bone wax. Bone wax is softened beeswax that is used to stop bone bleeding during surgeries, blocking holes and causing bone hemostasis. Though there are adverse reactions that can result, including an increased infection rate, which presents a danger during cardiac surgery, bone wax is cheap and easy to use and is considered the best option at this time.

Until recently, tallow—rendered beef or mutton fat—was used in most high-end shaving soaps. Today, most have moved on to glycerin formulas, other than the specialty soaps from the boutique company The Art of Shaving. Tallow was also used as a cooking fat—even McDonald's cooked its french fries in a nearly all-tallow fat blend until 1990—but has fallen out of favor as vegetable oils and vegetable-based shortenings have become more popular. It is still used in industry as a lubricant and soldering flux and in animal feed.

Bone char, produced by heating animal bones to between 750 and 900 degrees Fahrenheit, has a high adsorptive capacity for heavy metals and is thus useful for purifying and defluoridating water and other filtration tasks.

Ground bones become bone meal, usually used as a slow-release plant fertilizer because of its phosphorous content. When thus used, bone meal is an organic fertilizer that can help prevent blossom end rot in tomato plants. Because of concerns over bovine spongiform encephalopathy (BSE), bone meal was eliminated from livestock feed in the 1990s in most of the Western world. Blood meal—dried powdered blood manufactured as a slaughterhouse by-product—can also be used as a fertilizer and is one of the highest organic sources of nitrogen. Farmers have also long used it in their gardens to ward off rabbits and other produce-stealing pests. Meat and bone meal, which is about 50 percent protein and 10 percent fat, has a variety of uses. Though no longer used in animal feed for ruminant animals because of BSE concerns, it is still used in feed for monogastric animals like horses, rabbits, pigs, cats, and dogs. But the most interesting use of meat and bone meal has been as a burning fuel: though it has only two-thirds the energy value of coal, it is available as a by-product of animal slaughter and is therefore sustainable.

Casein is the predominant protein in cow milk and has a number of commercial uses. In addition to its use as a food additive in baby food, packaged foods, dairy products, pet food, and to clarify wine, casein has a long history of usage in the manufacture of glue and paints and in the treatment of leather and rubber products. When used in food, casein's presence may be to add nutritive value or may be for functional purposes such as improving emulsification or texture, enhancing whipping or foaming, thickening, or increasing binding with water (as in powdered drink mixes, for example).

A great number of chemical compounds are harvested from coral colonies by the pharmaceuticals industry. Such compounds are used in medicines to treat cancer and AIDS and to manage severe chronic pain. Coral products are also used to assist in bone grafting, and there are indications that the bamboo coral species may be usable as living bone implants.

Although man-made sponges (typically made of cellulose fibers or foamed plastic polymers) have largely replaced the sponges of the Porifera phylum as household sponges, natural sponges remain a critical animal-product resource. In particular, they and the microbial symbionts that dwell in them contain chemicals with antiviral, antibacterial, anticarcinogenic, and antifungal properties.

A number of animal hormones have commercial uses in a broad variety of applications. For instance, animal-derived thyroid hormones, administered orally because of the ease with which they are absorbed by the gut, are used to treat hypothyroidism in humans. While synthetic hormones are available, they are not always as effective as those derived from pigs, which have been in medical use since the 19th century. Melatonin, a popular hormone sold as a dietary supplement in the United States, is used in treating sleep disorders, mood disorders, Alzheimer's disease, and as an over-the-counter treatment to prevent migraines or aid with sleep. Although synthetic melatonin is more common, some melatonin on the market is derived from animal pineal glands.

In addition to human medical uses, animal hormones are regularly used in the raising of animals. Many concerns have been raised worldwide over the use of hormones in animal feed, for instance. Although the U.S. Department of Agriculture has declared that the use of such hormones poses no danger to the quality and safety of the meat or milk of the animal, many groups and individuals disagree, and the European Union has banned the import of beef containing any of six hormones that are legal and widely used in the United States, resulting in an ongoing controversy over the question of whether this is a violation of international trade agreements. Growth hormones, a subclass of peptide hormones, are especially an object of concern in food, though most such hormones are synthetically created, such as bovine somatotropin, first synthesized in 1994 using recombinant DNA technology.

See Also: Antibiotics; Fertilizers; Methane/Biogas.

Further Readings

Grossman, Elizabeth. *Chasing Molecules: Poisonous Products, Human Health, and the Promise of Green Chemistry*. Washington, DC: Shearwater Press, 2009.
Kirby, David. *Animal Factory: The Looming Threat of Industrial Pig, Dairy, and Poultry Farms to Humans and the Environment*. New York: St. Martin's, 2010.
Shore, Laurence S. and Amy Pruden. *Hormones and Pharmaceuticals Generated by Concentrated Animal Feeding Operations: Transport in Water and Soil (Emerging Topics in Ecotoxicology)*. New York: Springer, 2009.

Bill Kte'pi
Independent Scholar

ANTIBIOTIC RESISTANCE

Since the discovery of penicillin by Sir Alexander Fleming in 1928, antibiotics have been used to treat bacterial infections. However, the liberal use of antibiotics has led to many drugs becoming ineffective. The ability of a microorganism to survive, despite antibiotic treatment, is known as antibiotic resistance. Bacteria use many forms of resistance to ensure survival, including both genetic and nongenetic changes.

The rapid increase in the amount of resistance is a problem in healthcare today. In the past, methicillin-resistant *Staphylococcus aureus* (MRSA) infections were only seen in hospitals or other institutional settings; however, patients now acquire infections in the community. This is cause for concern, because most MRSA infections cannot be treated with standard antibiotics and require treatment with a broad-spectrum antibiotic. There has also been a steady climb in the number of nosocomial (acquired in the hospital) multidrug-resistant infections. Some of these are more difficult to treat than MRSA infections and can have serious adverse effects on already ill patients. These changes have begun to limit the treatments available in both nosocomial and community-acquired infections. Other factors leading to antibiotic resistance include the widespread use of consumer products such as antibacterial cutting boards, tissues, and other products containing antibiotics.

Mechanism of Resistance

There are multiple genetic, idiopathic (unknown), and iatrogenic (caused by a physician) factors contributing to the crisis of antibiotic resistance.

There are four main mechanisms by which bacteria acquire resistance:

- Enzymatic degradation of the drug (the bacteria break down the drug)
- Modification of the drug's target
- Decreased permeability (bacteria change borders to block antibiotics from entering)
- Active export of the drug (bacteria pump antibiotics out of cells)

Bacterial and human cells differ in a few key ways. Bacteria have a cell wall, a structure not present in human cells, and it is this wall that is the target of antibiotics. However, there are multiple ways for bacteria to protect themselves. To synthesize cell walls, bacteria use an enzyme called a penicillin binding protein (PBP). Some beta-lactam antibiotics, such as penicillin, bind to the PBPs to prevent synthesis of the wall, eventually leading to cell destruction. Bacteria can fight this attack in two different ways. Some bacteria can produce beta-lactamase, an enzyme that destroys the active portion of the antibiotic. Other bacteria produce modified PBPs, which allow beta-lactam antibiotics to bind without being able to inhibit cell wall synthesis.

Bacterial cell membranes are permeable to different components in their environment. By altering what can/cannot pass through the cell membrane, destruction with an antibiotic can be fought off because the drug will not be able to enter inside the cell in a concentration strong enough to be effective. Multidrug-resistant pumps actively export drugs that enter the bacteria. Bacteria take in surrounding materials and pump out the antibiotic from within the cell.

Genetically Mediated Antibiotic Resistance

Most mechanisms of bacterial resistance to antibiotics are due to genetic modification, including chromosome-mediated, plasmid-mediated, or transposon-mediated resistance. Clinically, plasmid-mediated resistance is of greatest concern because of the opportunity for plasmids to code for resistance to multiple antibiotics. Chromosome-mediated resistance occurs through spontaneous mutations in the DNA of the bacterium. A mutation in a gene that codes for the binding site of the antibiotic, or the transport proteins that allow the drug to enter the cell, can make the bacteria resistant.

Plasmid-mediated resistance may allow for resistance to more than one antibiotic. Resistance plasmids (R factors) are DNA segments that can be transferred between bacteria. They carry genes that code for different enzymes that can be used to alter the bacteria in ways discussed above. Plasmids can replicate independently of bacterial chromosomes and can be transferred to bacteria of both the same and different species. Plasmid transfer occurs through conjugation, which allows for the transfer of genetic material from direct cell-to-cell contact.

Transposons are genes that are transferred within larger pieces of DNA and move easily from one site to the next. They can cause mutations in the genes into which they insert or alter the expression of nearby genes. Unlike plasmids, transposons cannot replicate independently. A standard transposon has three main components: a gene that codes for an enzyme, a transposase, that allows the transposon to be removed and reinserted into the bacterial DNA; a gene that codes for a repressor to regulate the synthesis of the transposase; and a third gene that codes for drug resistance.

Nongenetic Modes of Resistance

Nongenetic-based modes of resistance do not rely on alterations in bacterial DNA or require a plasmid or transposon; they are dependent on the environment surrounding the bacteria. If the source of an infection is unreachable, for example, due to the formation of an abscess, an antibiotic will not be effective. The presence of a foreign body, such as a prosthetic valve or catheter, increases the likelihood of contracting a bacterial infection resistant to multiple antibiotics.

Iatrogenic (physician-caused) factors also add to the problem. Multiple antibiotics are given when one would suffice. Unnecessarily long courses of antibiotics are used, and at times, physicians prescribe antibiotics for viral infections. In some countries, patients can obtain antibiotics over the counter, which leads to problems with incorrect selection of antibiotics as well as to improper or incomplete dosing.

Part of the problem is also caused by indiscriminate use of antibiotics in animals. The use of antibiotics may allow development of antibiotic resistance.

Antibacterial products are marketed as a way to stop the spread of infections; however, their widespread use is worsening the challenges of antibiotic resistance. This is evident in the large number of consumer products that are labeled "antibacterial" and promoted heavily in the mainstream media.

Prevention of Resistance

The excessive use of broad-spectrum antibiotics is forcing experts to consider tighter control of antibiotic use. Currently, there are minimal data proving that limiting antibiotic use will decrease the amount of bacterial resistance, but the idea has promise. Cycling antibiotics (removing certain antibiotics temporarily from use) for a certain period of time may help to decrease the amount of resistance.

Technology has also allowed more precision in the diagnosis and treatment of bacterial infections. Tests can often identify the exact cause of an infection as well as which antibiotics will work in treating it. This permits the preservation of broad-spectrum drugs for infections that actually require them.

Conclusion

Antibiotic resistance is an ongoing concern. Through genetic and nongenetic means, bacteria continue to change, rendering drugs that worked in the past useless. There is no clear-cut answer to the prevention of antibiotic resistance, but with the current prevention strategies, there is hope that the amount of resistance will decrease.

See Also: Antibiotics; Nosocomial Infections; Vaccination/Herd Immunity.

Further Readings

Archer, G., et al. "Treatment and Prophylaxis of Bacterial Infections." In *Harrison's Principles of Internal Medicine*, 17th ed. New York: McGraw-Hill Professional, 2008.
King, M. D., et al. "Emergence of Community-Acquired Methicillin-Resistant *Staphylococcus aureus* USA 300 Clone as the Predominant Cause of Skin and Soft-Tissue Infections." *Annals of Internal Medicine*, 144/309 (2006).
Levinson, W. "Antimicrobial Drugs: Resistance." In *Review of Medical Microbiology and Immunology*, 10th ed., by W. Levinson. New York: McGraw-Hill Medical, 2008. http://www.accessmedicine.com/content.aspx?aID=3326466 (Accessed July 2009).
Smith, W., et al. "Antibiotic Cycling." http://www.uptodate.com (Accessed July 2009).

Gautam J. Desai
Cassandra N. Ramar
Kansas City University of Medicine and Biosciences,
College of Osteopathic Medicine

ANTIBIOTICS

The impact of infectious diseases remains a significant burden of morbidity and mortality in developed and developing countries worldwide. Early in the history of human societies, the extreme burden of infectious diseases was related to ineffective treatments to eliminate or curtail pathogens in affected people. In the mid-1900s, Alexander Fleming discovered—by accident—strains of *Penicillium* fungus that inhibited the growth of bacteria, changing medical practice forever. He brought the principles to humanity for the development of chemicals (antibiotics) that were able to control (bacteriostatic) or to eliminate (bactericide) microbes.

Antibiotics are antimicrobial drugs that affect the growth of bacteria and other organisms by different cellular mechanisms at different structural components at a given moment and place or at different moments and places. Antibiotics can also be defined as "a substance produced by or a semisynthetic substance derived from a microorganism and able in dilute solution to inhibit or kill another microorganism."

Doctors depend on antibiotics to treat illnesses caused by bacteria. If there is any doubt whether the infection is bacterial or viral in origin, clinicians may be tempted to prescribe antibiotics just to be on the safe side, to eliminate the risk of a life-threatening bacterial infection. Treating viral illnesses or noninfective causes of inflammation with antibiotics is ineffective, however, and contributes to the development of antibiotic resistance, toxicity, and allergic reactions, leading to increasing medical costs. A major factor behind unnecessary use of antibiotics is, of course, incorrect diagnosis. For this reason, timely and accurate information on whether the infection is bacterial in origin is highly beneficial in the fight against antibiotic resistance.

Unnecessary prescription of antibiotics could also represent an environmental issue, particularly when used in animals, where significant quantities of antibiotics could enter the environment. This is an environmentally important aspect for global public health that should be considered in the new perspectives of a more ecologically green world in which such environmental threats should be significantly reduced or even eliminated.

Use of antibiotics, however, is currently very important because their availability for proper use is associated with a significant reduction of infections, particularly those caused by bacteria. This is also related to countries' development. The majority of developed countries have undergone a prototypical epidemiologic and demographic transition. Host–infectious agent relations have evolved over centuries, but these transitions are largely attributed to a decrease in the burden and mortality from infectious diseases. As a result, life expectancy has radically changed over the past century, with a significant increase in longevity. Although we welcome these improvements in health status, we never imagined that prolongation of life due to fewer infectious diseases would have disadvantages in future generations. It has become evident that inflammatory disorders, including both autoimmune and allergic diseases, have increased in prevalence to epidemic proportions, particularly in more affluent, industrialized countries, over the past 40 years. At the same time, the rate of allergic diseases in the developing world has not changed over the same period of time. There is current scientific evidence that this tendency can be partially explained by the significant decreased incidence of infectious diseases.

We now live in a single epidemiological world system without defined boundaries. With travel and commerce among different regions in the world, diseases that exist in one region can easily spread to others. Without widespread and intelligent use of currently available antibiotics, antimicrobial resistance may threaten the national security of many countries. In order to prevent emergence of antimicrobial resistance, certain global, regional, or local measures have been taken.

The resistance of the infectious agents of certain diseases to drugs specifically used to combat them is a phenomenon confirmed by scholars who have observed the appearance of bacterial strains with susceptibility profiles highly resistant to drugs previously used. From the perspective of global public health, this resistance constitutes an increasing problem that has worsened through the misuse or abuse of antibiotic agents. This is expressed in the Declaration on Resistance to Antibiotics presented by the World Medical Association in its 48th General Assembly held in South Africa in 1996.

In Resolution WHA51.17 adopted by the World Health Organization (WHO) in 1998, member states were urged to do the following:

- Promote the appropriate and effective use of antibiotics with regard to cost
- Prohibit issuing antibiotics without a prescription from a skilled health professional
- Limit the excessive use of antibiotics in animal husbandry with regard to products destined for human consumption
- Promulgate or strengthen legislation to impede the manufacture, sale, and distribution of counterfeit antibiotics, as well as the sale of antibiotics in parallel markets
- Strengthen health services and their surveillance capacity in order to obtain compliance with current legislation

The WHO Global Strategy for Containment of Antimicrobial Resistance, published in 2001, strengthens these concepts and urges governments to ensure the implementation of the recommended measures. According to WHO criteria, and to experts on the subject, the factors that have contributed importantly to the development of the problem are as follows:

- Prescriptions for antibiotics being issued by unqualified people
- Prescriptions for antibiotics being issued either indiscriminately or excessively by qualified professionals
- The exaggerated or incorrect use of antibiotics in hospitals
- Self-medication and misguided perceptions on the part of poorly informed patients
- Patient noncompliance to regimens or prescribed doses
- Inadequate or misleading advertising and promotion
- The sale of antibiotics in unauthorized parallel markets
- A lack of legislation regulating the use of antibiotics and enforcing existing standards

As a result, certain factors—not technoscientific, but rather sociocultural ones—can be observed that directly affect this problem and have been considered in studies conducted in Argentina, Brazil, Chile, Cuba, Ecuador, Mexico, and Uruguay. Certain studies provide evidence to prove that this problem is, indeed, present in the region of the Americas. They underline, among other points, that in Latin America the lack of regulatory legislation makes it possible to distribute and sell antibiotics freely without a prescription, which subsequently makes the task of controlling this practice especially difficult in the region, there being no regulatory standards on antibiotics. This means that consumers can obtain them anywhere, making self-medication a serious problem.

Latin America presents a very broad and varied spectrum with respect to drafting legislation aimed at ensuring the health of the population. Drug control, licensing for the medical and pharmaceutical professions, and scientific, technical, and research activities—as well as those pertaining to trade—are all normative components of considerable importance within what is commonly called "sanitary law" or "health legislation."

This legislation has been developed on a par with the growth both of commerce and of the pharmaceutical industry. Also worth emphasizing is the influence of international organizations specializing in health and, nowadays, also of international economic nuclei within the region such as the Southern Cone Common Market (MERCOSUR), the Andean Community (CAN), and the Central American Integration System. These groups have contributed important sets of standards for various subjects concerning drugs, as well as resolutions that have been ratified by the states involved and incorporated into current legislation now in effect in each country.

See Also: Acquired Immune Deficiency Syndrome; Airborne Diseases; Antibiotic Resistance; Centers for Disease Control and Prevention (U.S.); Seasonal Flu; Tuberculosis; Vaccination/Herd Immunity.

Further Readings

Cook, Gordon and Alimuddin Zumla. *Manson's Tropical Diseases*. London: Saunders, 2003.

Da Silva, Paulo Sergio Lucas, et al. "The Product of Platelet and Neutrophil Counts (PN Product) at Presentation as a Predictor of Outcome in Children With Meningococcal Disease." *Annals of Tropical Paediatrics*, 27/1 (2007).

Fine, Paul E. "Herd Immunity: History, Theory, Practice." *Epidemiologic Reviews*, 15 (2003).

Folch, Erick, et al. "Infectious Diseases, Non-Zero-Sum Thinking, and the Developing World." *American Journal of Medical Sciences*, 326/2 (2003).

Franco-Paredes, Carlos, et al. "Cardiac Manifestations of Parasitic Infections Part 3: Pericardial and Miscellaneous Cardiopulmonary Manifestations." *Clinical Cardiology*, 30/6 (2007).

Franco-Paredes, Carlos, Ildefonso Tellez, and Carlos del Rio. "Inverse Relationship Between Decreased Infectious Diseases and Increased Inflammatory Disorder Occurrence: The Price to Pay." *Archives of Medical Research*, 35 (2004).

John, T. Jacob and Reuben Samuel. "Herd Immunity and Herd Effect: New Insights and Definitions." *European Journal of Epidemiology*, 16 (2000).

Nuutila, Jari and Esa-Matti Lilius. "Distinction Between Bacterial and Viral Infections." *Current Opinion in Infectious Diseases*, 20 (2007).

Sosa, Anibal. "Resistencia a Antibióticos en América Latina." Boston: Association for the Prudent Use of Antibiotics (APUA), 1998.

World Health Organization, 51st World Health Assembly. "Emerging and Other Communicable Diseases, Antimicrobial Resistance." Resolution WHA51.17 (1998).

Alfonso J. Rodriguez-Morales
Universidad de Los Andes, Universidad Central de Venezuela

ANTI-CHOLESTEROL DRUGS

Cholesterol is a waxy, fatlike substance made in the liver and other cells of the human body. It is also found in foods derived from animals, such as dairy products, eggs, and meat. Cholesterol assists cell walls in the production of hormones, vitamin D, and the bile acids that aid fat digestion. While the human body requires a certain amount of cholesterol to function properly, too much cholesterol can lead to health problems, such as cardiovascular disease. The U.S. Centers for Disease Control and Prevention (CDC) estimates that about 17 percent of the U.S. population has high cholesterol.

Cholesterol travels through the blood attached to a protein in units called lipoproteins, which are classified as high density, low density, or very low density, depending on how much protein there is in the unit in relation to fat:

- Low-density lipoproteins (LDL): also called "bad" cholesterol; can cause buildup of plaque on artery walls. The more LDL there is in the blood, the greater the risk of heart disease.
- High-density lipoproteins (HDL): also called "good" cholesterol; helps rid the body of bad cholesterol in the blood. The higher the level of HDL cholesterol, the better. If your levels of HDL are low, your risk of heart disease increases.
- Very-low-density lipoproteins (VLDL): similar to LDL cholesterol in that it contains mostly fat and not much protein.

A variety of factors can affect one's cholesterol levels, including the following:

- Diet
- Smoking

- Weight
- Exercise
- Age and sex
- Diabetes
- Heredity
- Certain medications and other conditions

When treating high cholesterol, the main goal is to lower one's LDL while raising HDL. There are only two ways to lower LDL: eating a healthy diet and taking cholesterol-lowering drugs; these medications are most effective in people who eat a low-cholesterol diet, and the most common are statins, niacin, and selective cholesterol absorption inhibitors.

Statins block the action of a chemical in the liver necessary for making cholesterol. They work by inhibiting an enzyme called HMG-CoA reductase, which controls cholesterol production in the liver. Several types of statins exist and are sold under various brand names, including Lipitor and Crestor. Though millions of Americans rely on statins to reduce their cholesterol, a growing body of scientific research indicates that the side effects of these drugs, and increasing questions regarding their efficacy, warrant careful consideration when prescribing or taking these drugs. Some of the side effects include insomnia, constipation or diarrhea, headaches, loss of appetite, loss of sensation in the nerve endings of the hands and feet, muscle pain (myopathy), irritability, memory loss, depression, liver dysfunction, kidney failure, and cataracts.

While most researchers agree that these side effects are, indeed, related to statins, many argue that the benefits of statins outweigh the potential consequences. Related to that dispute is the fact that many researchers disagree about who actually benefits from statins, with skeptics arguing that the results of clinical trials are often interpreted erroneously in order to support the use of statins. For instance, a statin trial that was published in the *New England Journal of Medicine* in 2005 reported that nearly half of the 18,000 people in the trial were excluded because they had some illness or did not respond well to the drug. Further, skeptics argue that statins have not been shown to prevent premature death among men over 65, nor have any controlled studies shown that statins benefit women who do not have preexisting heart disease. Complicating matters further is the fact that the side effects of statins are often associated with aging, making them difficult to recognize.

Drugs other than statins have also come under fire for not performing as well as their manufacturers suggested, such as Zetia. Unlike statins, which work by inhibiting cholesterol production in the liver, Zetia is a selective cholesterol absorption inhibitor and is designed to reduce cholesterol by blocking its absorption in the small intestine. Zetia and similar drugs, which are often used in conjunction with statins, not only have been found to not reduce plaque in the arteries but also may increase cancer risks. Further, clinical studies suggest that this class of drugs does not have any added benefits for patients already taking statins. Studies have also found that a prescription version of niacin performed significantly better at slowing the buildup of plaque and has none of the side effects associated with Zetia and other similar drugs.

Ongoing clinical trials are necessary for determining which anti-cholesterol drugs perform best, and with the least side effects, and more information is needed to determine who benefits from which drugs and why. Further, the clinical trials that are conducted need to be carefully studied, with their findings scrupulously parsed by nonbiased scientists and researchers. For now, it seems that eating a healthy diet, exercising, and dropping habits such as cigarette smoking are the safest, nondrug approaches for controlling cholesterol.

When these lifestyle behaviors are not enough to keep cholesterol in check, consumers would be wise to diligently seek out information regarding the anti-cholesterol drugs that they are prescribed.

See Also: Cardiovascular Diseases; Pharmaceutical Industry; Physical Activity and Health.

Further Readings

"Are Cholesterol Drugs a Good Idea for Healthy People?" *Time* (March 31, 2010). http://wellness.blogs.time.com/2010/03/31/are-cholesterol-drugs-a-good-idea-for-healthy-people (Accessed May 2010).
"Statins: Side Effects of Anti-Cholesterol Drugs Questioned." *Daily Mail* (June 27, 2006). http://www.dailymail.co.uk/health/article-392689/Statins-Side-effects-anti-cholesterol-drugs-questioned.html (Accessed May 2010).
WebMD. "Niacin Tops Zetia in Cutting Artery Plaque." http://www.webmd.com/cholesterol-management/news/20091116/niacin-tops-zetia-in-cutting-artery-plaque (Accessed May 2010).

Tani Bellestri
Independent Scholar

ANTI-DEPRESSANT DRUGS

Each year in the United States, 26.2 million people suffer from a diagnosed mental disorder. According to the National Institute of Mental Health, approximately 6.7 percent of all Americans, or 14.8 million individuals, suffer from major depressive disorder, and another 1.5 percent suffer from dysthymic disorder, a form of chronic mild depression that exists for a period of more than two years. Women are twice as likely to suffer from depression as men. It is estimated that 10 percent of all Americans take prescription anti-depressants. While 80 percent of depression suffers receive relief of symptoms from medication, psychotherapy, and/or support groups, 50 percent of those with unsuccessful medication treatment will discontinue their medications due to unacceptable side effects and/or financial strain. It is estimated that 29 to 46 percent of individuals that use anti-depressants alone to combat depression will show only partial or no improvement in symptoms even when adequate anti-depressants are prescribed. Many individuals seek out herbal or dietary remedies due to reluctance to take prescription medications or dissatisfaction with prescription medications.

Prescription anti-depressant medications account for more than 27 million prescriptions annually. Selective serotonin reuptake inhibitors or serotonin-specific reuptake inhibitors (SSRIs) are the most commonly prescribed anti-depressant pharmacy agent. SSRIs are used to treat depression, anxiety, eating disorders, chronic pain, and post-traumatic stress disorder. SSRIs inhibit the reuptake of serotonin in the brain. Drugs in this category include citalopram, dapoxetine, escitalopram, fluoxetine, indalpine, paroxetine, and sertraline. Serotonin-norepinephrine reuptake inhibitors, selective serotonin reuptake enhancers, and serotonin-norepinephrine-dopamine reuptake inhibitors are also serotonergic anti-depressants that are used in the treatment of depression.

Use of complementary and alternative therapies has risen dramatically in the last three decades. According to the National Center for Complementary and Alternative Medicine, the use of complementary medicine has risen to 38.3 percent of adults and 11.8 percent of children in the United States. American Indian and Alaskan Native adults use complementary medicine the most, followed by white women with higher education and income levels. According to researchers, it is estimated that Americans spend $5.1 billion annually out-of-pocket for complementary therapies.

Complementary therapies have been used to treat mild to moderate depression. According to some, severe depression should be treated by a qualified practitioner using more traditional therapies that typically require traditional anti-depressant drugs. B complex vitamins, including B-1, B-2, B-3, B-6, and B-12, are known to promote serotonin, a brain chemical known for its anti-depressant qualities. Serotonin is a brain chemical that transmits signals between nerve cells and causes blood vessels to narrow. Alterations in serotonin levels can affect and alter mood. Low serotonin levels have been directly linked to mood disorders, including depression. B complex vitamins can be obtained by taking a B-complex vitamin or a multivitamin that contains 20 to 50 milligrams (mg) of B-1, B-2, B-3, and B-6, and 50 mg of B-12. B complex vitamins can be taken in addition to a traditional anti-depressant without undue side effects.

The most widely used nontraditional herbal medication for depression management is St. John's wort from the *Hypericum perforatum* plant. Recent studies from the Cochrane Collaboration have shown that the effectiveness of St. John's wort is equivalent to traditional anti-depressants commonly prescribed without the side effects. The standardized dosage recommendations are 900 to 1,500 mg daily. St. John's wort contains cytochrome P450s and the drug transporter P-glycoprotein and may interfere with conventional drugs such as anti-cancer drugs, anti-HIV agents, anti-inflammatory drugs, antimicrobial agents, cardiovascular drugs, central nervous system agents, hypoglycemic agents, immune-modulating agents, oral contraceptives, proton pump inhibitors, respiratory medications, and statins. St. John's wort should be used with caution in individuals who have hypertension and should not be taken in combination with other amino-acid supplements.

Another holistic treatment modality is the supplement 5-hydroxytryptophan (5-HTP). 5-HTP is a form of the amino acid L-tryptophan, a precursor to the neurotransmitter serotonin that is extracted from the *Griffonia simplicifolia* plant. 5-HTP stimulates serotonin production, which increases levels in the blood and brain, combating depression. The recommended dosage is 150 to 300 mg per day in divided doses. 5-HTP is often combined with St. John's wort for their combined anti-depressant properties; however, it should not be taken in conjunction with prescribed anti-depressants. It is also contraindicated for pregnant and lactating mothers.

S-adenosylmethionine (SAMe) is produced naturally in the body from the essential amino acid methionine and adenosine triphosphate (ATP) and is associated with increased availability of serotonin, dopamine, and other neurotransmitters that affect mood. Some individuals may make inadequate amounts of SAMe, making supplementation an effective treatment for depressive symptoms. Supplementation of SAMe is an effective adjunct to traditional therapy such as anti-depressants or other herbal or dietary therapies. Individuals in one study that added SAMe supplementation to their anti-depressant medications found a 50 percent improvement in symptoms, with 43 percent going into remission. Possible side effects include constipation, nausea, heartburn, and headaches. Other side effects include a reduction in blood sugar, dry mouth, thirst, and increased urination. The typical dosage is 400 to 800 mg per day in two dosages on an empty stomach. Because it can cause insomnia,

SAMe should not be taken in the evening. Due to lack of scientific studies in pregnant women, SAMe is not recommended for pregnant or nursing women and should be used with caution in individuals with bipolar disorder since it can cause mania or hypomania.

Supplementation with omega-3 fish oils that contain eicosapentaenoic acid (EPA) and docosabexaenoic acid (DHA) has been closely linked to mood stabilization. Omega-3 is a polyunsaturated fatty acid that enables serotonin to more effectively permeate the cell membrane. Most individuals find it difficult to get a significant quantity of omega-3 unless they eat a diet high in flaxseed, fish, or supplement with fish oil capsules. Countries that have higher rates of fish consumption have significantly less depression than those where fish consumption is low. New Zealand, where fish intake equals 40 pounds annually per individual, compared to Japan that averages 150 pounds of seafood per year per individual, has 60 times greater the incidence of depression. Concern over diet has led to a reduction of omega-3 in recent years. Many Americans obtain omega-3 from red meat and eggs, which are often seen as unhealthy. Eating large quantities of fish also has health concerns related to increased mercury levels. A healthy diet should contain 5 grams of essential fatty acids divided between omega-3 and omega-6. When omega-3 is unavailable in the body, omega-6 provides similar properties, although it is less effective with neurotransmitters than omega-3s. Omega-3s have been effective in helping to stabilize mood disorders in individuals with depression and bipolar disorders. The standardized dosage for omega-3 is 1,500 to 3,000 milligrams per day. The main source of omega-3 is fish oil capsules. Cod liver oil in high concentrations is not recommended because it can cause an excess of vitamin A. Flaxseed oil is another source of omega-3s that is less effective but may be more desirable to vegans. It should be noted that both fish oil and flaxseed have a blood-thinning effect and should not be combined with aspirin, heparin, Coumadin, or other medications with anticoagulant properties. Omega-3s can be used as an adjunct to both traditional anti-depressants and herbal depression remedies.

Anti-depressant remedies include herbal and dietary sources. The use of complementary and alternative medicine may offer effective treatment for depressive disorders without unwanted side effects. Individuals on anti-depressant drugs should consult their physician before beginning a complementary therapy.

See Also: Centers for Disease Control and Prevention (U.S.); Mental Health; Pharmaceutical Industry.

Further Readings

Barnes, Patricia, Barbara Bloom, and Richard Nahin. "Complementary and Alternative Medicine Use Among Adults and Children: United States, 2007." *National Health Statistics Reports,* 12 (December 10, 2008).

DeSmet, Peter. "Herbal Remedies." *New England Journal of Medicine,* 347/25 (2002).

Mischoulon, D. "Update and Critique of Natural Remedies as Antidepressant Treatments." *Psychiatric Clinics of North America,* 30/1 (2007).

National Center for Complementary and Alternative Medicine. http://nccam.nih.gov (Accessed April 2010).

National Institute of Mental Health. "The Numbers Count: Mental Disorders in America." http://www.nimh.nih.gov/health/publications/the-numbers-count-mental-disorders-in-america/index.shtml (Accessed April 2010).

Stengler, Mark. "Depression: When the Sadness Won't Go Away." *Bottomline Natural Healing*, 5/1 (January 2009).

Michele Davidson
George Mason University

Antiseptics

Antiseptics are antimicrobial substances that are applied to living tissues and that tend to reduce the possibility of infection or microbial colonizing. Antiseptics are generally distinguished from antibiotics by the latter's ability to be transported through the lymphatic system to destroy bacteria within the body, and from disinfectants, which destroy microorganisms found on nonliving objects. Some antiseptics are true germicides, capable of destroying microbes (bactericidal), whereas others are bacteriostatic and only prevent or inhibit their growth. Antibacterials are antiseptics that have the proven ability to act against bacteria. Microbicides that kill virus particles are called viricides or antivirals.

Antiseptics are particularly important during medical procedures and are of utmost importance in medical hand washing. Appropriate hand washing can minimize microorganisms acquired on the hands by contact with body fluids and contaminated surfaces. Hand washing breaks the chain of infection transmission and reduces person-to-person transmission. All healthcare personnel and family caregivers of patients must practice effective hand washing. Patients and primary caregivers need to be instructed in proper techniques and situations for hand washing. Compliance with hand washing is, however, frequently suboptimal.

Reasons for this include lack of appropriate equipment; low staff-to-patient ratios; allergies to hand-washing products; insufficient knowledge among staff about risks and procedures; the time required; and casual attitudes among staff toward bio-safety.

The role of hands in the transmission of hospital infections has been well demonstrated and can be minimized with appropriate hand hygiene. Hand washing is the single most important means of preventing the spread of infection. Hands should be washed between patient contacts and after contact with blood, body fluids, secretions, excretions, and equipment or articles contaminated by these.

For hand washing, the following facilities are required:

- Running water: large washbasins with hands-free controls that require little maintenance and with anti-splash devices
- Products: dry soap or liquid antiseptic, depending on the procedure
- Suitable material for drying of hands: disposable towels, reusable sterile single-use towels, or roller towels that are suitably maintained

Specific antiseptics recommended for hand disinfection usually include 2 to 4 percent chlorhexidine, 5 to 7.5 percent povidone iodine, 1 percent triclosan, or 70 percent alcoholic hand rubs. Alcoholic handrubs are not a substitute for hand washing, except for rapid hand decontamination between patient contacts.

For surgical scrub (surgical care), training is needed in the current procedure for preparation of the hands prior to surgical procedures. Scrubbing of the hands for 3 to 5 minutes is sufficient. The recommended antiseptics are 4 percent chlorhexidine or 7.5 percent povidone iodine.

Equipment and products are not equally accessible in all countries or healthcare facilities. Flexibility in products and procedures and sensitivity to local needs will improve compliance. In all cases, the best procedure possible should be instituted.

See Also: Acquired Immune Deficiency Syndrome; Airborne Diseases; Centers for Disease Control and Prevention (U.S.); Seasonal Flu; Tuberculosis.

Further Readings

Cook, Gordon and Alimuddin Zumla. *Manson's Tropical Diseases*. London: Saunders, 2003.

Da Silva, Paulo Sergio Lucas, et al. "The Product of Platelet and Neutrophil Counts (PN Product) at Presentation as a Predictor of Outcome in Children With Meningococcal Disease." *Annals of Tropical Paediatrics*, 27/1 (2007).

Fine, Paul E. "Herd Immunity: History, Theory, Practice." *Epidemiologic Reviews*, 15 (2003).

Folch, Erick, et al. "Infectious Diseases, Non-Zero-Sum Thinking, and the Developing World." *American Journal of Medical Sciences*, 326/2 (2003).

Franco-Paredes, Carlos, et al. "Cardiac Manifestations of Parasitic Infections Part 3: Pericardial and Miscellaneous Cardiopulmonary Manifestations." *Clinical Cardiology*, 30/6 (2007).

Franco-Paredes, Carlos, Ildefonso Tellez, and Carlos del Rio. "Inverse Relationship Between Decreased Infectious Diseases and Increased Inflammatory Disorder Occurrence: The Price to Pay." *Archives of Medical Research*, 35 (2004).

John, T. Jacob and Reuben Samuel. "Herd Immunity and Herd Effect: New Insights and Definitions." *European Journal of Epidemiology*, 16 (2000).

Nuutila, Jari and Esa-Matti Lilius. "Distinction Between Bacterial and Viral Infections." *Current Opinion in Infectious Diseases*, 20 (2007).

Sosa, Anibal. "Resistencia a Antibióticos en América Latina." Boston: Association for the Prudent Use of Antibiotics (APUA), 1998.

World Health Organization. *Guidelines on Prevention and Control of Hospital Associated Infections*. Geneva, Switzerland: WHO, 2009.

World Health Organization. *Practical Guidelines for Infection Control in Health Care Facilities*. Geneva, Switzerland: WHO, 2000.

World Health Organization, 51st World Health Assembly. "Emerging and Other Communicable Diseases, Antimicrobial Resistance." Resolution WHA51.17 (1998).

Alfonso J. Rodriguez-Morales
Universidad de Los Andes, Universidad Central de Venezuela

ARSENIC POLLUTION

Arsenic pollutes waters in more than 70 counties around the world, and tens of millions of people are potentially exposed to and experience adverse health effects that result from acute or chronic contact with the element. Although direct ingestion or dermal contact with contaminated water is the main source of exposure, arsenic can also be held by soil or taken up by plants when contaminated water is used to irrigate fields. While some plants are more resistant to the uptake of arsenic, others extract it more readily. Differences could potentially be due to geologic considerations, soil chemistry characteristics, differing root systems depths, climatic differences, or extent or source of contamination.

Although arsenic water contamination is a global-scale problem, areas in southeast Asia, especially Bangladesh and West Bengal, are among those that have experienced the most dramatic effects of such pollution. While arsenic is a natural constituent of the Earth's crust and therefore present in bedrock, human development and corresponding disruption of bedrock for water extraction can release the otherwise trapped arsenic into groundwater. Such was the case when the Bangladesh Department of Public Health Engineers partnered with the United Nations Children's Fund (UNICEF) and the World Bank to address microbial-contaminated surface water by sinking millions of tube wells across Bangladesh.

Arsenic Sources

Arsenic is present as a natural constituent of the Earth's crust and can be released through volcanic eruptions, weathering, or some industrial processes. As such, it can be found in bedrock and soil and, to some extent, in water and air. Different mechanisms of action have been proposed to explain how arsenic can migrate from bedrock and become bioavailable through drinking water. These explanations rely on solid–solution interactions in which arsenic bonded to minerals in the soil, sediments, and rocks interacts with the flow of groundwater in the presence of oxygen or other reactive agents and optimal pH conditions.

In Bangladesh, arsenic-contaminated bedrock was derived from the natural, historical shifting and sifting of sediment from the Himalaya Mountains into the Delta region. Sediment traveled down the channels of the major regional rivers and was unable to be swept away efficiently by relatively weak ocean currents in the Bay of Bengal and thus accumulated. However, prior to drilling, this arsenic had remained separated from aquifers by a layer of clay or silt that acted as a cap, preventing water and oxygen from interacting with the bedrock. The drilling of tube wells disrupted this cap and, in combination with other human-industrial activities, prompted the widespread contamination witnessed today.

One theory regarding the mechanism of pollution involves pyrite oxidation, in which drilling or irrigation extraction forces air to interact with previously undisrupted land. The oxygen within air causes a decomposition of the bonds that keep arsenic immobile, and thus it is released into the water. Another theory involves oxyhydroxide reduction. This latter theory supports the idea that arsenic has a high affinity to other particles that are present in the soils and sediments of the floodplains of Bangladesh, such as iron or manganese oxyhydroxides.

While attached to these agents, arsenic is contained and unable to contaminate groundwater, but when conditions catalyze the dissociation of arsenic from this interaction—through drilling, mining, or excessive flooding—then arsenic is released from bondage and comes to contaminate the groundwater. In its organic form, arsenic is essentially nontoxic because it is attached to carbon and hydrogen and is thus contained; however, when it reacts with oxygen, chlorine, or sulfur, it enters into an inorganic form and can have dire health consequences. In the past, the inorganic form of arsenic has been used to treat wood and, additionally, to combat parasitic agents in veterinary settings.

Permissible Limits and Health Effects

While the World Health Organization sets advisable limits of arsenic in drinking water at 10 micrograms/liter, some individual countries have national standards that differ. For example, Bangladesh has set a more liberal national standard of 50 micrograms/liter. Although this leads to discrepancies in reported statistics on locations exceeding permissible limits, some areas grossly exceed either standard—concentrations in Bangladesh have been sampled to range from less than 0.5 to 3,200 micrograms/liter.

While a high dose of inorganic arsenic, identified as 600 micrograms/kilogram of body weight per day, has proven fatal to humans, lower levels will cause acute and chronic ailments throughout all body systems. Some effects are witnessed shortly following exposure, whereas other ailments can take up to 20 years to become observable. Acute effects include nausea, vomiting, headaches, weakness, delirium, hypotension, shock, anemia, and leucopenia. Chronic effects—which are more pertinent to the situation in Bangladesh—include gastrointestinal effects, anemia, peripheral neuropathy, skin lesions in the forms of flaking, both hypo- and hyper-pigmentation, keratosis (hardening of skin and development of wart-like nodules), gangrene of the extremities, vascular lesions, and liver or kidney damage.

Arsenic is also implicated in cancers of the skin, bladder, kidney, prostate, liver, and lung, and because arsenic can pass through the placental barrier, it also contributes to higher rates of spontaneous abortion and lower than normal birth weights. In addition to these physical presentations of disease, arsenic has been implicated in contributing to cognitive deficits.

Certain subpopulations are more affected by arsenic than others. Irrespective of the geographic distribution of the contamination, worldwide children are at higher risk than their adult counterparts. Although the former ingest less water in absolute terms, the proportional difference of water intake between children and adults is less than the proportional difference between their body weights. Since dose is calculated as a fraction of one's body weight, the daily dose of children frequently exceeds the daily dose of adults. With this said, because many of the adverse effects of chronic arsenic exposure are latent, symptoms of pollution poisoning can be more readily observable in adults, who have been exposed for a longer duration.

Mechanism of Effects

Once ingested, enzymes begin methylating inorganic arsenic into various methylated derivatives. These methylated metabolites are thought to be responsible for the deleterious effects of arsenic poisoning. To support this, one study found that chimpanzees—a species that lacks the genetics to produce the enzyme responsible for methylation of arsenic—did not exhibit the phenotype specific to arsenosis, while humans, rats, and mice, all of which do produce said enzyme, exhibited said phenotype. Because arsenic cannot be destroyed but merely changes form depending on environmental conditions and chemical reactions, the implications of tapping into a new source of arsenic, as was the case in Bangladesh, will have lasting health consequences.

Mitigation and Prevention

Eliminating or reducing exposure to an environmental hazard can occur at various levels along the exposure pathway. According to the hierarchy of controls model, the most effective means of controlling a hazard is to address the source of that hazard; however, in the case of arsenic, because this can prove expensive and often lacks feasibility and political will, controlling at steps along the exposure pathway can also meet with success. Last in the hierarchy is controlling the hazard at the level of the individual, which shifts the onus to the individual. Although this last method is usually the easiest to conceptualize and implement, its effectiveness is limited by difficulties in disseminating information to the greater population in such a way that people not only gain knowledge but also are motivated to change their attitudes and behaviors in response.

With this said, the relevant strategies that involve mitigating arsenic pollution at the source involve utilizing a different source of drinking water. This includes sinking wells that will access untapped aquifers at a depth greater than 200 meters into water tables or digging shallow wells of only 20 to 30 meters. Rainwater collection has also been suggested as a means to obtain drinking, bathing, and cooking water. However, extreme weather patterns in areas of south and southeast Asia lead to challenges in proper storage and filtration.

In areas with contaminated wells, immediate mitigation work has been done to label pipes containing high levels of arsenic for the purpose of warning people in affected communities. However, one of the main limitations to this approach is the fact that arsenic pollution is not a spatially stagnant issue. Although groundwater will ultimately follow gravity and make its way to the sea or ocean, the routes taken by such waters are modifiable depending on ongoing human development, water extraction, and the natural shifting of sediment and land. The complexity of these input variables is a major hindrance to the assurance of accurate groundwater modeling. As such, a pipe labeled as safe may not be so, and that labeled as contaminated may be safe. Since 74 percent of the people in Bangladesh live in rural regions, typical Western media methods do not attain the same degree of effectiveness as they would in a more developed urban region. Therefore, reliance on individual behavior change poses feasibility challenges.

Because arsenic has an affinity for adsorption onto other minerals, including aluminum, iron, and magnesium, these salts can be used as water treatment to leach the arsenic out of water. However, this results in a highly contaminated sludge whose disposal warrants differential considerations.

Stigma and Justice

Mitigation efforts have met with varying levels of success, due in part to the tasteless, odorless, and colorless nature of arsenic, and in part to prevalent misconceptions regarding the epidemiology of contamination. Although, in Bangladesh, arsenic-related illnesses were documented as early as the 1980s, the government did not officially recognize the existence of arsenic contamination until 1993. The lack of urgency and timeliness of mitigation allowed communities to develop their own theories of causation and prompted the stigmatization of victims.

The international media has stigmatized arsenic and the pollution of drinking water in Bangladesh to the international community. The international perspective seems to revolve around the unfair contamination of the water and incorporates empathy for the victim. In contrast, within the country, the historical lack of knowledge and consciousness regarding the source of pollution and daily exposures have led to many misconceptions and to the projection of blame onto the individual—which has subsequently led to social divisions. For example, due to its presentation of skin lesions and gangrenous extremities, arsenic poisoning has been misperceived among some as a form of leprosy, a highly contagious disease. As a result, the victims have frequently been forced to hide away, becoming outcasts of society and fueling the cycle of misinformation, ignorance, and stigmatization of arsenic pollution poisoning.

Conclusion

Arsenic-contaminated water continues to be a persistent problem. Exacerbated by its odorless and colorless nature and the latent period of related illnesses, arsenic and effective mitigation techniques remain misunderstood by many.

No simple or singular solution will remedy the situation in Bangladesh or other areas adversely affected by arsenic pollution. Decades of history have revealed that for an effort to be effective, the strategy has to be comprehensive enough to address the hazard at different levels throughout the exposure pathway and to engage the community in a culturally sensitive and appropriate way. In addition, cooperation between the various stakeholders and a unified strategic plan are essential to effective public health messaging to those affected, especially those in rural communities.

See Also: Groundwater; Phaseout of Toxic Chemicals; Recycled Water; Waterborne
   Diseases.

Further Readings

British Geological Survey (BGS). "Arsenic in Groundwaters Across the World." In *DPHE/ BGS Phase 2: Bangladesh Report* (2001). http://www.bgs.ac.uk/arsenic/bphase2/Reports/ ChapWorldReview.pdf (Accessed August 2009).

Li, J., S. B. Waters, Z. Drobna, V. Devesa, M. Styblo, and D. J. Homas. "Arsenic (+3 Oxidation State) Methyltransferase and the Inorganic Methylation Phenotype." *Toxicology and Applied Pharmacology*, 204/2 (2005).

Moinuddin, Mustafa. "Drinking Death in Groundwater: Arsenic Contamination as a Threat to Water Security." *ACDIS Occasional Paper: Program in Arms Control, Disarmament, and International Security.* Champaign: University of Illinois at Urbana-Champaign, May 2004. http://www.acdis.uiuc.edu/research/OPs/Moinuddin/cover.html (Accessed August 2009).

Mount, Jeffrey. *California Rivers and Streams.* Berkeley: University of California Press, 1995.

Ravenscroft, P., H. Brammer, and K. Richards. *Arsenic Pollution: A Global Synthesis.* Oxford, UK: Wiley-Blackwell, 2009.

UNICEF—United Nation Children's Fund. "Arsenic Primer: Guidance for UNICEF Country Offices on Investigation and Mitigation of Arsenic Contamination." New York: UNICEF, 2008. http://www.unicef.org/wes/files/Arsenic_primer_28_07_2008_final.pdf (Accessed August 2009).

Erin E. Stamper
Independent Scholar

ASTHMA

Children under 18 years of age are at special risk of developing asthma; of the over 17 million people in the United States who suffer from asthma, about one-third are children.

Source: iStockphoto.com

Asthma is one of the most prevalent chronic diseases in the world. Its name is derived from the Greek word *aazein*, which means panting or breathless. It has been recognized as a health condition for centuries. Early references to it are found in ancient Greek medical texts like the Corpus Hippocraticum and the extensive medical writings of Galen dating to before 200 C.E. Asthma was first clinically described in modern bio-medicine during the period just prior to the turn of the 20th century. Since that time its medical definition has been revised extensively. At first, bio-medicine viewed the disease as a neurotic affliction, but today it is broadly understood as a physical disease characterized by episodic inflammation and narrowing of small airway passages in the lungs that result in difficulty in breathing. Recognized symptoms of asthma include the following:

- Coughing, especially at night or early morning, that interferes with sleep
- Wheezing upon exhalation
- Chest tightness and a sense of pressing on the chest
- Shortness of breath

Because of the varied expression of this disease and the range of severity across patients, a person having these symptoms does not necessarily have asthma. Similarly, being free of these symptoms does not mean a person is free of asthma. Diagnosis is usually based on review of a set of clinical factors, including reported symptoms, patient history, physical examination, and the results of a lung function test. This test employs an instrument called a spirometer to measure the volume of air a patient is able to process while breathing in and out. It also measures how rapidly the patient can take in and exhale air. Although asthma is a treatable (i.e., controllable) disease, it is not currently curable. While asthma is mild in some patients, in others it can be fatal. Because asthma is affected by air quality and living conditions, in a world of rapidly changing environmental conditions and social arrangements, there is good reason to believe that asthma rates will grow over the coming decades.

Asthma and Society

The distribution of asthma reflects prevailing health and living conditions within and across nations. As of this writing, at least 300 million around the globe suffer with asthma and rates of this environment-sensitive respiratory disease have been skyrocketing in recent years. Over the last 40 years, the global rate of asthma has increased by 50 percent every decade. Rates vary by country, with the highest prevalence found in New Zealand, where almost 30 percent of the population has been diagnosed with the disease. The prevalence in highly developed countries generally ranges between 10 percent and 20 percent. Although its frequency has been rising, especially in urban areas, asthma is less common in developing countries, as would be expected given the role of industrial air pollutants in triggering this condition. Nonetheless, the World Health Organization estimates that about a quarter of a million people, mostly in lower- and lower-middle-income countries, annually die of asthma. In other words, because of limited access to medical treatment for asthma, rates of affliction may be lower but rates of mortality higher in less developed nations. Thus, asthma patients who are admitted to hospitals with acute symptoms are disproportionately from economically disadvantaged groups.

Epidemiology of Asthma

It is estimated that asthma is the cause of about one in every 250 deaths worldwide, with an enormous social cost in terms of what are termed *disability adjusted life years* (DALYs), a term that refers to the total number of years of productive life that are lost due to disability. It is projected that the proportion of the world's population that is urban dwelling will increase from about 50 percent currently to almost 60 percent by 2025 and that environmental triggers are likely to increase during this period as well. Thus, it is expected that the number of people in the world suffering with asthma will continue to climb over the next several decades, reaching 400 million by 2025. Importantly, while the total health and social burden of asthma in many countries should make this disease a top public health priority, often this is not the case. Moreover, from a global health standpoint, trends in asthma-related health policies and practices appear to be headed in the wrong direction.

In the United States, the estimated cost to society from asthma is approximately $6 billion a year, with hospitalization accounting for more than half of asthma-related expenditures. While asthma affects about 8 percent of the U.S. population, as elsewhere in the world, its prevalence has been steadily climbing, especially among ethnic minorities living in impoverished urban neighborhoods. As is commonly the case with health conditions, including diseases that are directly affected by environmental conditions like air pollution, the poor and marginalized suffer the greatest frequencies and gravest consequences. In other words, it is quite literally the case that the poor do not breathe the same air as the wealthy.

Commonly, three distinct factors—subordinate ethnicity, poverty, and residence in the inner city—are closely related in the U.S. population, but disaggregation, although difficult, is necessary to assess fully the role each of these factors plays in the development of asthma. Regarding ethnicity, national research has found that the rate of childhood asthma for individuals aged 6 months to 11 years is 3 percent among whites, compared to over 7 percent among African Americans. Although the rate of asthma varies by age, gender, and location of residence, even when these factors are controlled statistically, ethnic differences remain significant. Among Puerto Ricans in the United States, the rate of asthma is 125 percent higher than it is among non-Hispanic whites, while among American Indians, Alaska Natives, and African Americans, asthma prevalence is 25 percent higher than among whites. Moreover, mortality from asthma is as much as three times higher among ethnic minorities compared to whites.

Although poverty and low socioeconomic status have been found to be strong predictors of asthma, the precise ways in which lower income increases asthma severity as well as the specific biological pathways involved remain to be fully clarified. Heightened inflammatory response, a component of asthmatic reactions, is known to be associated with poverty and lower socioeconomic status and is believed to be a consequence of enduring high levels of everyday stress. A likely pathway involves specific body changes, such as decreased capacity for the regulation of inflammation, caused by physical stressor experiences associated with impoverishment (e.g., fear of living in an unsafe environment).

Children and Asthma

Children are at special risk of developing asthma. Of the over 17 million people in the United States who suffer from asthma, about a third are children under 18 years of age, who comprise only about one-fourth of the total population. Asthma, in fact, is one of the most common chronic illnesses of childhood in the nation. In the United States, Puerto Rican children have an especially high rate of this breath-robbing condition and, because of asthma, they are more likely to miss school than children of other ethnic groups from the same neighborhoods. Research shows that low-income Puerto Rican families with young children with asthma often lack the necessary information, training in asthma management, and medical resources for good asthma control. The problem, however, is not simply one of the healthcare and public health systems' failing to adequately prepare Puerto Rican parents in asthma control and management. Treatment studies show that Puerto Rican and African American children with diagnosed asthma are less likely to receive beta-2-agonists, the standard medicine in asthma management, than are white children with asthma. Further, Puerto Rican children receive fewer inhaled steroids, the usual treatment for frequent asthma episodes, than white children.

Asthma and the Environment

There are multiple triggers of asthma episodes and more than one trigger is often at work in an asthma outbreak. In the urban environment, both outdoors and in their homes, low-income children are exposed to heavier concentrations of various air pollutants that may play a critical role in linking socioeconomic and ethnic inequality and rates of childhood asthma. Moreover, there is a higher number of environmental hazards and greater level of pollution in areas inhabited by low-income ethnic minorities than in wealthier areas. Thus, it is said of asthma that genetics loads the gun but environmental conditions pull the trigger, especially environmental conditions that have been reshaped by human activity. The human impact on the environment and the dark shadow it casts over the air we breathe have expanded exponentially, especially so in the years since World War II.

One of the anthropocentric environmental pollutants that play a role in asthma is nitrogen dioxide (NO_2), a molecule that is produced in abundance by motor vehicles and thermal power plants. Research shows that a high level of exposure to NO_2 during the week prior to the onset of a respiratory viral infection is associated with increased asthma severity. The level of air pollution is strongly influenced by automobile traffic patterns, with areas that are within 500 feet of heavily used roads having higher levels of air pollution and higher rates of asthma among local residents. In these risk zones, a variety of pollutants are found that are linked to respiratory and other health problems, including NO_2, particulate matter (soot), and volatile organic compounds.

It is expected that increased global warming will exacerbate rates of asthma. Rising temperatures caused by greenhouse gases and increasing levels of carbon dioxide (CO_2) in the atmosphere caused by the use of fossil fuels are stimulating expanded pollen production by ragweed, mugwort, and other pollenous plants. Spring, the season when tree pollen reaches peak levels, has been starting 10 to 15 days earlier over the past three decades, a trend expected to continue as a result of global warming. Pollen is a significant outdoor trigger of asthma episodes, as is mold, which spreads more rapidly at higher temperatures. Higher smog levels are also promoted by rising temperatures. Smog is formed when sunlight heats pollutants in the air. Global warming appears to impact not only the kinds of complex pollutants that are being formed in the atmosphere but also how and where they are dispersed. Moreover, warmer temperatures contribute to the growing frequency of significant forest and other wildfires, events that increase the level of asthma-stimulating particulate matter in the air.

Among indoor pollutants, environmental tobacco smoke (i.e., secondhand smoke exposure) is closely linked with the rise in asthma prevalence. In population surveys in the United States, reported rates of exposure to secondhand smoke are similar among whites, African Americans, and Latinos. This research has shown that cigarette consumption of over a pack a day was reported by about 30 percent of white mothers and 17 percent of African American mothers. Research among emergency room patients has found that over 50 percent of the children admitted with asthma symptoms had at least one smoker living in their homes and almost 40 percent had elevated levels of cotinine in their urine, a by-product that suggests heavy exposure to tobacco smoke. Similarly, a study of patients at urban medical centers found that almost 60 percent of the families with an asthmatic member reported that at least one individual living in the household was a smoker, just under 40 percent of the primary caretakers in these households were smokers, and urinary cotinine was elevated in almost half of the children. Additionally, these studies strongly suggest that exposure to tobacco smoke in the home is more common in inner-city homes, reflecting general patterns of increased smoking cessation among middle- and upper-class families but not among the poor.

Other environmental conditions associated with poverty, such as high levels of indoor dampness and dust mite and cockroach exposure, have attracted increasing scientific attention as possible contributors to the severity of reactions of asthma-sensitive individuals. Other indoor triggers include hair and skin cells shed by mammalian pets (especially cats); fungi or molds living on walls and other moist household surfaces and objects; fuel burned by appliances including gases and particulate matter given off by stoves, furnaces, water heaters, space heaters, and fireplaces; compounds found in household sprays (e.g., hair spray) and in home furnishings, floor coverings, and decorations; medications like aspirin or other nonsteroidal anti-inflammatory drugs; and substances in foods, such as sulfites. Exercise can also trigger an asthma episode.

Infections are an additional asthma trigger and there is growing evidence of synergistic interaction between allergens (e.g., tree pollens, pollution) and pathogens in the development and health impact of asthma. Several studies of children admitted to the hospital with acute asthma flare-ups have tested them for various nasal pathogens and found that these children are at a significantly higher risk both of exposure and sensitivity to allergens and of having a higher respiratory pathogenic load than other child patients. Further, multivariate analysis has shown that comorbidity with both allergies and nasal pathogens increases the risk of hospital admission by almost 20 times. Rhinoviruses (a cause of the "common cold") have been identified in a number of studies as the most frequent respiratory pathogen, accounting for over 80 percent of viruses detected in children with asthma.

Evidence is mounting, in short, that asthma outbreaks often are a product of the syndemic interaction of environmental allergens and pollutants, on the one hand, and natural viral infection, on the other. Moreover, once an individual develops asthma, respiratory pathogens have been found to precipitate as much as 85 percent of asthma attacks in children and 44 percent in adults. Consequently, respiratory infections are now seen as a primary factor in acute asthma events across the life course. While viral respiratory infections appear to be the most common infectious trigger of acute asthma, some studies have found other types of pathogens such as *Mycoplasma pneumoniae*, a bacteria, may also contribute to asthma attacks. Bacteria (e.g., *Chlamydophila pneumoniae*) have also been found to be associated with adult onset asthma and with both new-onset wheezing and severity of asthma. Several experimental studies have suggested that a primary pathway of asthma episodes is the enhancement of air pathway inflammation in sensitized individuals simultaneously exposed to allergens and infection by one or more respiratory pathogens. In patients with asthma, antibiotic treatment for *C. pneumoniae* has been found to be effective in reducing asthma symptoms.

Beyond issues like environmental quality and infection, several other aspects of the physical and social environment, including psychosocial stressors, have been found to set off asthma attacks. Urban research has shown, for example, that street violence may be associated with the occurrence and increased severity of asthma among inner-city children. Children exposed to violence in their neighborhoods, such as hearing gunshots or witnessing physical violence, are twice as likely as children who are not exposed to violence to experience wheezing, a common asthma symptom, and to use bronchodilator asthma medication for wheezing and almost three times as likely to be diagnosed with asthma.

Overcoming Barriers to Asthma Intervention

Efforts to lower asthma rates must confront a number of issues. One of these issues is cultural understanding. Asthma beliefs and practices have been found to vary among ethnic groups. One important variation related to ethnicity is the different descriptors used to portray an asthma attack. A multicity cross-cultural comparison of four different Latino groups in the United States, Mexico, and Guatemala, for example, found that while there was broad agreement about the major respiratory signs of asthma, including wheezing, cough, chest noise, and rapid or difficult breathing, only Puerto Ricans included symptoms like chest pain, decreased activity, increased blood pressure, chest congestion, fast heartbeat, red, tired, or dark eyes, and difficulty breathing or talking to describe asthma attacks. Failure to understand this kind of information can lead to public health efforts that are overly generalized and thus fail to be effective with specific populations, such as Spanish-language prevention materials that assume that the term *asthma* means the same thing to all Latinos.

Indeed, there are multiple barriers to effective global asthma intervention and prevention, including the following:

- Low public health prioritization by international health organizations and funders in light of other pressing health problems, including other respiratory diseases
- Social inequality in living conditions and access to treatment
- Inadequate health and social infrastructure in developing nations, including medications
- Efforts by the tobacco industry to protect tobacco smoking from government regulation and control

- Unsafe working conditions and occupational exposures to toxic substances
- Limitations on available asthma data and research on range of environmental factors, including a clear understanding of the role of various indoor and outdoor pollutants and programs designed to monitor the prevalence of asthma
- Lack of a standardized definition of asthma among biomedical providers
- Inadequacies in public understanding and knowledge about the disease and methods to limit its impact
- Cross-cultural differences in understandings about the disease
- Worsening environmental conditions, including global warming
- Absence of usable international asthma treatment and prevention guidelines and cost-effective disease management models in light of limits on existing infrastructures

These barriers must be addressed to reverse current global trends in the frequency and severity of asthma, especially among those with the least resources and greatest health burden already. Undertaking such work is a critical component of reducing health and social disparities associated with asthma.

See Also: Children's Health; Cities; Ecosyndemics; Environmental Illness and Chemical Sensitivity; Health Disparities; Indoor Air Quality.

Further Readings

Beam Dowd, Jennifer, Anna Zajacova, and Allison Aiello. "Early Origins of Health Disparities: Burden of Infection, Health, and Socioeconomic Status in U.S. Children." *Social Science & Medicine*, 68 (2009).

Caninoa, Glorisa, Daphne Koinis-Mitchell, Alexander N. Ortegac, Elizabeth L. McQuaid, Gregory K. Fritz, and Margarita Alegría. "Asthma Disparities in the Prevalence, Morbidity, and Treatment of Latino Children." *Social Science & Medicine*, 63 (2006).

Chauhan, A. J., Hazel M. Inskip, Catherine H. Linaker, Sandra Smith, Jacqueline Schreiber, Sebastian L. Johnston, and Stephen T. Holgate. "Personal Exposure to Nitrogen Dioxide (NO_2) and the Severity of Virus-Induced Asthma in Children." *The Lancet*, 361 (2003).

Global Initiative for Asthma. "Global Strategy for Asthma Management and Prevention" (2008). http://www.ginasthma.org (Accessed December 2009).

McConnell, Rob, Kiros Berhane, Frank Gilliland, Stephanie J. London, Talat Islam, W. James Gauderman, Edward Avol, Helene G. Margolis, and John M. Peters. "Asthma in Exercising Children Exposed to Ozone: A Cohort Study." *The Lancet*, 359 (2002).

Merrill Singer
University of Connecticut

Automobiles (Emissions)

The automobile is one of the most significant inventions to have influenced the growth and development of societies and the trajectory of human civilization. The automobile provide humans with mobility, independence, and convenience. Also, it has liberated the

Traffic jams in Los Angeles are a common occurrence—and the reason why California made the first effort at controlling pollution from automobiles in 1961. The standards set in the 1966 Clean Air Act were similar to the standards set by California.

Source: iStockphoto.com

average person from the limitations of time and place, opened up new opportunities, and offered new experiences. Moreover, it is one of the pivotal elements of our economy. The automobile also affects the environment in many ways. Cars and their associated infrastructure use resources, consume energy, and emit pollutants on a substantial scale. Mobile sources are among the largest contributor of certain air pollutants. This entry examines the history of engines, the impact of automobiles on human health and the environment, and the resultant societal policies and regulations.

History of Engines

In 1876, one of the most important landmarks in engine design came from Nicolaus August Otto, who invented an effective gas motor engine with the first practical four-stroke internal combustion engine. An internal combustion engine uses the explosive combustion of fuel to push a piston within a cylinder. The movement of the piston turns a crankshaft that then turns the car wheels via a chain or a drive shaft. Various types of fuel can be used for internal combustion engines, including gasoline (or petrol), diesel, and kerosene. In 1885, a German mechanical engineer, Karl Benz, designed and built the world's first practical automobile to be powered by an internal combustion engine and received the first patent for a gas-fueled car on January 29, 1886. The first commercial car manufacturers of gasoline-powered vehicles were bicycle producers Charles and Frank Duryea, who built their first motor vehicle in 1893, in Springfield, Massachusetts. However, Ransom E. Olds and his Olds Motor Vehicle Company dominated this era of automobile production and had a large-scale production line running by 1902. Within a year of this, Cadillac (formed from the Henry Ford Company) was producing cars in the thousands. By 1905 an assortment of power technologies was being used by hundreds of producers all over the Western world.

For decades, steam-, electricity-, and gasoline-powered autos competed for dominance, with the gasoline internal combustion engines achieving dominance around 1910. Around this same time, many modern advances were attempted, including gas/electric hybrids, multivalve engines, overhead camshafts, and four-wheel drive; some were used in that early era, while others were discarded and reappeared later. Between 1907 and 1912 in the United States, the high-wheel motor buggy gained popularity, with over 75 makers. In 1912, the Ford Model T dominated sales of the high-wheel motor buggy and was the most widely produced and available car until 1927. By the 1930s, most of the mechanical technology used in today's automobiles had been invented, but unfortunately, around 1930, with the beginning of the Great Depression, the number of auto manufacturers declined sharply.

Automobile design reemerged after World War II in 1949 with the introduction of high-compression V8 engines and modern bodies from General Motors' Oldsmobile to Cadillac brands. In the 1950s, engine power and vehicle speeds rose. Car designs became more integrated and artful, and usage of cars spread around the world. The American market began to change in 1960 as Detroit, the car capital of the United States, began to worry about foreign competition when the European makers began to adopt ever-higher technology and the Japanese appeared as a serious car-producing nation. As the foreign market began to perfect small performance cars, big-engine performance cars became the focus of marketing in America, exemplified by pony cars and muscle cars. Everything changed, though, in the 1970s, as the 1973 oil crisis, automobile emission-control regulations, Japanese and European imports, and stagnant innovation greatly impacted the U.S. automobile industry. During the 1970s, the biggest developments in technology were the widespread use of independent suspensions, wider application of fuel injection, and an increasing focus on the safety of automobiles. Since the 1970s, the most notable advances have been the widespread application of front-wheel drive and all-wheel drive, the adoption of the V6 engine configuration, and fuel injection. Body styles have also changed with the hatchback, minivan, pickup truck, and sports utility vehicle dominating today's market. Recent times have also seen changes in rising fuel efficiency and engine output with the development of computerized engine-management systems.

Impact of Automobiles on the Environment and Human Health

The impact that automobiles have on the environment is often hidden but always very real, and as more people understand this impact, it is expected that the sale of "greener" cars (cars that are cleaner and more fuel efficient) will increase, thus minimizing harm to the environment while meeting transportation needs. Automobiles not only cause unhealthy air pollution but also contribute to oil spills and fouling of water supplies, damage habitats, and play a sizable role in global climate disruption. The impact a car has on the environment begins with production and ends with being scrapped in a junkyard. Over the life of an average motor vehicle, however, much of the environmental damage occurs during driving and is greatly associated with fuel consumption, with nearly 90 percent of greenhouse gas production for a typical automobile being due to fuel consumption. Some degree of pollution is associated with most components of a car, much of it due to the energy consumption, air pollution, and releases of toxic substances that occur when automobiles are manufactured, distributed, and disposed of.

As stated above, most of the environmental impact associated with motor vehicles occurs with their use, due to exhaust emissions and pollution associated with combustion of the fossil fuel. In the United States, nearly all of today's automobiles use gasoline or diesel fuel; and when gasoline, diesel, or other fuels are burned, a mix of hazardous, toxic pollutants comes out the tailpipe. Environmental damage ensues with the exploration for, production, and distribution of motor fuel, beginning at the wellhead. Additionally, gasoline and diesel fuel are poisonous to humans, plants, and animals, and their vapors are toxic.

Air pollution is not the only problem associated with petroleum-based fuels. Oil extraction lays waste to many fragile ecosystems, harming tropical forests in South America and southeast Asia, deserts and wetlands in the Middle East, coastal areas in the continental United States, and the tundra and arctic coastal plains of Alaska. Every year millions of gallons of oil are spilled. Sometimes the disasters become well known, such as the 1989 *Exxon Valdez* spill in Prince William Sound, although, most are rarely reported, and the potential for tragic consequences exists nonetheless.

There are a number of ways automobiles can affect water quality. Runoff of oil, dirt, brake dust, deposited vehicle exhaust, road particles, automotive fluids, and deicing chemicals from roadways and parking lots is a leading source of impairment. Groundwater may become tainted by leaks from underground fuel storage tanks and miscellaneous spills that occur during shipping and handling of the 120 billion gallons used each year. Also, improperly disposed waste fluids such as motor oil can contaminate millions of gallons of freshwater because they do not dissolve. Improperly disposed waste fluids as well as trace metals and degreasing agents used on automobiles contaminate drinking water. Contaminated drinking water is known to cause major illness in humans. Toxins and metals may also be absorbed by certain sea life and cause medical problems to people who eat them.

Major Pollutants Associated With Automobiles

Particulate Matter (PM)

Transportation sources account for about 20 percent of directly emitted PM2.5 (2.5 microns in size), with diesel engines being the major source of direct PM emissions from motor vehicles. Soot and smoke coming from exhaust pipes are obvious sources of PM and have been linked to lung problems, from shortness of breath to worsening of respiratory and cardiovascular disease, damage to lung tissues, and various types of cancer. Certain people are particularly vulnerable to breathing air polluted by fine particles, including asthmatics, individuals with the flu and with chronic heart or lung diseases, children, and the elderly. PM also soils and damages buildings and other contacted materials and is known to form a haze that can obscure visibility.

Previous PM regulation was based on counting all particles up to 10 microns in size (PM10), but PM10 standards failed to adequately control the most dangerous, very fine particles. Therefore, in 1996, the U.S. Environmental Protection Agency (EPA) started to regulate PM2.5, which better focuses on the most damaging category. New fuel-injected gasoline vehicles now directly emit very little PM2.5 but indirectly cause significant PM pollution as a result of their nitrogen oxides, sulfur dioxide, and hydrocarbons emissions (see below), not only from tailpipes but also from vehicle manufacturing and fuel refining.

Nitrogen Oxides (NO$_x$)

NO$_x$ refers mainly to two chemicals, nitrogen oxide (NO) and nitrogen dioxide (NO$_2$), which are formed when nitrogen gas reacts with oxygen during the high temperatures that occur during fuel combustion. Motor vehicles account for about one-third of nationwide NO$_x$ emissions, with many of these emissions coming from heavy-duty diesel trucks, cars, and light trucks. NO$_x$ is directly hazardous as an irritant to the lungs that can aggravate respiratory problems. It also reacts with organic compounds in the air to cause ozone, creating high levels of smog. NO$_x$ is a precursor of fine particles, which may cause respiratory problems. Also, it is a precursor of acid rain, which harms lakes, waterways, forests, and other ecosystems and damages buildings and crops. EPA air quality regulations have helped keep emissions from growing as fast as they might have, but transportation-related NO_x emissions continue to increase.

Sulfur Dioxide (SO_2)

Gasoline and diesel fuels contain varying amounts of sulfur, which burns in the engine to produce sulfur dioxide (SO_2) emissions. This gaseous chemical is itself a lung irritant as well as a cause of acid rain. Some of the cleaner, reformulated versions of gasoline have very low sulfur levels, although most gasoline sold nationwide still has too much sulfur; but levels are being reduced under recently established EPA regulations. Although cars and light trucks are not the largest source of SO_2 emissions, due to the high number of these types of vehicles and the fact that gasoline has a high average sulfur content, cars and light trucks cause twice as much fine PM pollution as heavy freight trucks. Producing all gasoline to meet the cleaner, low-sulfur fuel standard would greatly reduce a significant source of PM pollution from all cars and trucks on the road.

Hydrocarbons (HC)

Hydrocarbons are a broad class of chemicals containing carbon and hydrogen. Hydrocarbons that cause various forms of air pollution are also known as volatile organic compounds, due to their ability to readily evaporate into the air. Many forms of HC are directly hazardous, causing irritation to the lungs and other tissues, and they can also cause cancer, contribute to birth defects, and cause other illnesses. Oftentimes hydrocarbons react with NO_x to form ozone smog, especially during daylight hours. Controlling ozone is one of the major environmental challenges in the United States and, although progress has been made over the past several decades, many cities and regions still have smog alerts when ozone levels get too high. Gasoline vapor contains a mix of hydrocarbons and thus pollution is produced whenever gas tanks are filled. Some regions have special nozzles on fuel pumps to help trap such vapors, but vapors seep out due to the imperfect sealing of the fuel tank, pipes and hoses, and other components leading to the engine. Due to the inability of catalytic converters to completely clean up exhaust gases, hydrocarbons are often released through the tailpipe. Overall, transportation is responsible for about 36 percent of man-made HC emissions in the United States.

Ozone

Ozone (O_3) is a highly reactive form of oxygen that occurs naturally in various parts of the atmosphere but is artificially produced in dangerously high concentrations due to emissions from cars, trucks, and other combustion sources. In the stratosphere, ozone helps protect us from ultraviolet radiation, and loss of this protective ozone layer at high altitudes can lead to increased skin cancer. In the lower atmosphere, ozone is a health hazard—it is the main ingredient of the smog that causes pollution alerts in many cities. Inhaling air polluted by ozone damages the lungs, reduces breathing ability, and makes us more susceptible to many other respiratory problems. Ozone can be deadly to individuals with asthma and other lung conditions, as well as to people with heart conditions, but it is also harmful to both adults and children who are otherwise healthy. The risks of shortness of breath, chest pain, lung congestion, and other symptoms caused by ozone are the reasons why public health officials warn the public to stay indoors and avoid strenuous exercise on severe air pollution days. Although cars and trucks do not directly emit ozone, they are a major cause of ozone smog by adding to the amount of HC in the air, and tailpipe NO_x reacts with HC to form ozone. Cities without major industries and power plants still have serious smog problems, most of them attributed to pollution from cars, trucks, and vans.

Carbon Monoxide (CO)

Carbon monoxide is an odorless, colorless, but potentially deadly gas that is created by the incomplete combustion of any carbon-containing fuel. If CO is inhaled, it combines with the hemoglobin in blood and this can result in the impairment of delivery of oxygen to the brain and other parts of the body. The toxicity of CO is well known; it produces symptoms above 100 ppm. CO exposure may have long-term toxicity, and chronic intoxication is manifested by nonspecific symptoms such as headaches, dizziness, and nausea. It is estimated that motor vehicles are responsible for about 60 percent of CO emissions nationwide.

Metals

Metals are widely distributed in the environment. They are not biodegradable, but can be transformed into different chemical forms, often with different valance states. Activities related to industrialization, such as combustion of fossil fuel, can result in some of the transformation processes with motor vehicles. In automobiles, metals are a part of the structure of the automobile: in batteries, in brake pads, as fuel additives to improve engine performance, as fuel additives in conjunction with particle filters to reduce particulate emissions, and as catalysts in catalytic converters that reduce emissions of hydrocarbons, carbon monoxide, and nitrogen oxides. Metals were first added to fuel in order to reduce emissions of concern but, unfortunately, they have the potential of causing deleterious effects themselves or of causing other changes in emissions that may increase toxicity (such as changing particle size distribution). Metals may also be part of particulate matter and contribute to their toxic effects.

Automobile Regulations

Throughout the 1950s and 1960s, various federal, state, and local governments in the United States conducted studies on the sources of air pollution and a significant portion was found to be attributable to the automobile. Until 1959, minimal emission control existed in the United States at the local level. California made the first effort at controlling pollution from automobiles, requiring all new 1961-model cars to have positive crankcase ventilation. The positive crankcase ventilation system draws crankcase fumes into the engine's intake tract so they are burned rather than released unburned from the crankcase into the atmosphere. Crankcase fumes are heavy in unburned hydrocarbons, which are a precursor to photochemical smog. In 1962, New York also required positive crankcase ventilation and by 1964, most new cars sold in the United States were so equipped. Also in the state of California, in 1966, the first legislated exhaust (tailpipe) emission standards were promulgated, followed by the United States as a whole in 1968. It was not until 1966 that the first federal standards under the Clean Air Act were enacted. The standards set in the 1966 Clean Air Act were similar to the standards set in the 1959 California regulations, but successive amendments converted the standards from pollutant concentration values to pollutant mass values.

In 1974, it was concluded that detuning techniques that were used to meet emission standards were seriously reducing engine efficiency and thus increasing fuel usage. New emission standards were created in 1975, forcing the invention of the catalytic converter for after-treatment of the exhaust gas. During this same time, leaded gasoline was the most widely used type of fuel and, unfortunately, after-treatment of the leaded gas was not possible, due to the lead residue contaminating the platinum catalyst. The automobile industry became aware of the need to eliminate leaded fuels and in 1972, General Motors proposed this to the American Petroleum Institute, which proposed the elimination of leaded fuels for 1975 and later model-year cars. Although the elimination of leaded fuels and production and distribution proved to be a major challenge, it was completed successfully in time for the 1975 model-year cars. All cars are now equipped with catalytic converters and the use of unleaded fuel is widespread.

Future Direction

A combination of environmental concerns—air quality, global warming—has led to much emphasis on developing cleaner, energy-efficient automobiles. Now, with technological advances such as computer-regulated engines and battery breakthroughs, the potential for significant progress in addressing environmental issues is more feasible than ever. At the corporate level, large auto manufacturers are finally putting their energies into cars that economize fuel consumption, eliminate tailpipe emissions, and rely on renewable power sources. At the entrepreneurial level, inventors and engineers are finding new markets for products such as long-lasting batteries and laser-guided driver aids. Government has also taken an active role in the transportation revolution, offering tax incentives to consumers and allowing more public-sector investment. Despite advances, the source of power to drive the engine has remained the major problem in developing cleaner, energy-efficient automobiles. A variety of alternative-fuel vehicles have been proposed or sold, including electric cars, compressed-air cars, hydrogen cars, and hybrid cars.

Electric cars are propelled purely by means of an electric motor powered by an onboard battery that can be recharged using a standard electrical outlet or a charging station. Air cars use engines with either a dedicated compressed air engine (which releases compressed air to activate the pistons) or a dual-fuel engine (uses both compressed air and conventional gasoline). Hydrogen cars use hydrogen as the onboard fuel for motive power by converting the chemical energy of hydrogen to mechanical energy (torque) in one of two methods: combustion or electrochemical conversion in a fuel cell. In hydrogen internal combustion engine vehicles, the hydrogen is combusted in engines by fundamentally the same method used in traditional internal combustion engine vehicles. In fuel cell conversion, the hydrogen reacts with oxygen to produce water and electricity, the latter being used to power an electric traction motor. Hybrid cars feature a small, fuel-efficient gas engine combined with an electric motor that assists the engine when accelerating. The electric motor is powered by batteries that recharge automatically while driving. There are currently two types of gasoline-electric hybrid cars: the parallel hybrid and the series hybrid. In a parallel hybrid car, a gasoline engine and an electric motor work together to move the car forward, while in a series hybrid the gasoline engine either directly powers an electric motor that powers the vehicle or charges batteries that power the motor.

Although there is a current push for greener vehicles, there is no one solution to the issue of reducing the effect of automobiles on the environment and human health. Long-term solutions require that vehicles use less-polluting energy sources such as biofuels, propane, natural gas, and/or hydrogen. One problem is that producing alternate fuels and hybrid cars often requires CO_2 emissions that may offset the benefits of improved vehicular design. When ethanol is made from corn, more than 75 percent of its energy value is spent on its production and it still produces carbon dioxide. Although electric cars have offered much promise to this solution, the main limitation is battery technology, as batteries have been found to be heavy, wear out quickly with repeated recharging, and require expensive materials such as lithium. Although electric cars produce little to no air pollution, generating the electricity to power the car is a major source of air pollution. Ultimately, there may be no benefit to the atmosphere if an electric car is recharged with electricity produced by a fossil fuel–burning generator.

Reduction of air pollution must begin with source materials and continue through the use cycle of the vehicle. While it is feasible to use fossil fuels in generation plants with all the latest techniques of emission control and CO_2 recycling, these plants are as yet uncommon and unproven for extensive use. A new infrastructure of nonpolluting, affordable electricity production must be built in order for electric cars to be of any benefit. As of 2010, the best solution was thought to be reduced use of cars and widespread mass transit options, thus ultimately leading to a greener tomorrow.

See Also: Airborne Diseases; Biodiesel; Climate Change; Particulate Matter.

Further Readings

Bellis, M. "The History of Cars and Engines" (1997). http://inventors.about.com/od/cstart inventions/a/Car_History.htm (Accessed May 2009).

Degobert, P. *Automobiles and Pollution*. Warrendale, PA: Society of Automotive Engineers, 1995.

Environmental Protection Agency (EPA). "Devices and Additives to Improve Fuel Economy and Reduce Pollution—Do They Really Work?" (2009). http://www.epa.gov/OMS/consumer/420f09013.htm (Accessed May 2009).

Environmental Protection Agency (EPA). "Pollutants" (2009). http://www.epa.gov/OMS/invntory/overview/pollutants/index.htm (Accessed May 2009).

Fuhs, A. E. *Hybrid Vehicles and the Future of Personal Transportation*. Boca Raton, FL: CRC Press, 2009.

Rae, J. *The American Automobile Industry*. Boston: Twayne Publishers, 1984.

Watson, A. Y., R. R. Bates, and D. Kennedy. *Air Pollution, the Automobile, and Public Health*. Washington, DC: National Academy Press, 1988.

Hueiwang Anna Cook Jeng
Independent Scholar

Jacqueline F. Jackson
Old Dominion University

B

BIODIESEL

Biodiesel is fatty acid alkyl monoesters derived from vegetable oils, animal fats, or recycled cooking oils. Chemical modification of those oils is required to produce a fuel that can be used in an unmodified diesel engine. Through base-catalyzed transesterification, a major chemical modification method, vegetable oils will react with alcohol in the presence of a catalyst to produce an alcohol ester and glycerol. This reaction results in three moles of fatty acid alkyl monoester (biodiesel) and a mole of glycerol that is the by-product of this reaction.

Biodiesel production and use in the United States began in the 1970s. At first, cooperatives and small-batch producers provided a limited supply of fuel to a small number of cars and farm equipment. After the U.S. Department of Agriculture (USDA) established its Bioenergy Program in 2000 and provided direct payments to biofuel producers, biodiesel production increased to 2 million gallons by the end of that year. The enactment of the American Jobs Creation Act of 2004 marked an important turning point for the future production of biodiesel in the United States. The bill allowed distributors of biodiesel and biodiesel blends (mixed with petroleum diesel) to receive a federal excise tax credit. By the end of 2005, biodiesel production climbed to 75 million gallons, a 300 percent increase in one year. It is now clear that the U.S. biodiesel industry is emerging as a significant provider of renewable fuel.

Application and Economics

Biodiesel is most often blended with petroleum diesel in ratios of 2 percent (B2), 5 percent (B5), or 20 percent (B20). It can also be used as pure biodiesel (B100). Biodiesel fuels can be used in regular diesel vehicles without making any changes to the engines, although older vehicles may require replacement of fuel lines and other rubber components. Biodiesel has similar material compatibility to ultralow sulfur diesel (ULSD), so vehicles built to run on that should be compatible with pure biodiesel.

Due to diminishing petroleum reserves and government support, biodiesel attracted attention in the early 2000s as a renewable fuel. However, the rapid increase in biodiesel capacity has had a major effect on vegetable oil and animal fat prices. For example, soybean oil prices on average increased from 23 cents per pound in 2005 to 25 cents per

Though U.S. biodiesel production and use began in the 1970s, it was not until the U.S. Department of Agriculture established its Bioenergy Program in 2000 that biodiesel production reached two million gallons a year. Here, a biodiesel manufacturing plant in Iowa towers over a soybean field.

Source: iStockphoto.com

pound in 2007. Continued growth in biodiesel production has begun to draw down soybean oil stocks and raise prices even more, to 33 cents per pound in mid-2009. The increase in vegetable oil prices and the cost of feedstock may force biodiesel producers to raise prices, resulting in lower demand. If these conditions continue to prevail, the impressive growth that the biodiesel industry has experienced will likely level off.

Environmental and Health Effects

Biodiesel is a biodegradable, nontoxic, almost sulfurless and nonaromatic alternative diesel fuel and is considered a more environmentally friendly fuel than diesel. In general, biodiesel emits mainly carbon monoxide (CO), carbon dioxide (CO_2), oxides of nitrogen, and smoke. When a diesel engine is operated with biodiesel, exhaust emissions could decrease in the following approximate amounts compared to diesel: CO by 20 percent, hydrocarbons (HC) by 30 percent, particulate matter (PM) by 40 percent, and soot emissions by 50 percent. In the United States, most biodiesel is made from soybeans. According to a model developed by the Argonne National Laboratory (ANL), neat (100 percent) biodiesel from soybeans can cut conventional petroleum-based diesel's global warming pollution contribution by more than half.

Due to high biodiesel costs and engine compatibility issues, however, biodiesel is often blended with conventional diesel fuel. That blend likely affects the quantity and the composition of emissions and potential biological effects of the exhaust. The addition of biodiesel to petroleum can shift the composition toward more unregulated pollutants, such as acetaldehyde, acrolein, benzene, 1,3-butadiene, ethylbenzene, *n*-hexane, naphthalene, styrene, toluene, and xylene.

In spite of the aforementioned advantages of emission reduction, the production and use of biodiesel may have serious environmental impact, such as the use of large amounts of water, increase in soil degradation, destruction of forests, and reduction in food production. When soybeans are used as feedstock for fuel, it increases prices and stimulates demand for turning more land into cultivation. When crop-based biofuels contribute to deforestation or other damaging land conversion practices, the pollution benefits can be compromised or even eliminated. The biofuel life cycle analysis shows that the environmental effects on soil degradation and net emissions load from production and use of biodiesel are mainly dependent on the type of vegetable crop and the method of growth and harvest. Some biofuels can be produced without harmful changes in land use, and these have great potential to reduce global warming pollution. Examples include fuels made from biomass waste products or native perennials grown on land not currently used for or well suited to food crops.

Small enough to be inhaled deep into the lungs, diesel exhaust particles can cause or exacerbate a wide variety of health problems: asthma, pneumonia, chronic bronchitis, other respiratory ailments, and heart disease, along with links to lung cancer and premature death. However, there exists limited cytotoxicity and mutagenicity information on the effects of biodiesel exhaust in biologic systems and the mutagenic potential of biodiesel emissions specific to fuel category (e.g., starting oil feedstock), engine type, and operating conditions. In vitro bacterial assays have indicated that most of the mutagenic activity in biodiesel exhaust is contributed by a minority of the soluble organic fraction mass. Overall, there were neither significant differences from an air-exposed control group nor consistent exposure-related differences among most of the health parameters evaluated. To date, only one epidemiologic study has been conducted in which the acute health effects from exposure to biodiesel exhaust fumes were assessed by a questionnaire given to workers who were typically exposed to diesel fumes.

This investigation demonstrated dose-dependent respiratory symptoms after exposure to rapeseed methyl ester and diesel fumes, but there were no significance differences between the two fuels. Biodiesel should require greater due diligence than it has received to date in the United States. In widespread employment of biodiesel as an alternative fuel, there would be several additional issues pertinent to human health: (1) additives to biodiesel fuel are numerous and may affect human health, (2) implementation of new petroleum diesel engine combustion and after-treatment technologies designed to decrease specific exhaust components will require reevaluation of the emissions from biodiesel exhaust, and (3) the use of pesticides on plants with subsequent contamination of feedstocks could possibly affect specific end points of human health and requires further investigation.

Conclusion

Biodiesel is a renewable and biodegradable fuel source that can be a productive part of a broader strategy to address climate change. To achieve its potential benefit, we should promote biofuels that both use land and energy efficiently and minimize human and environmental impacts of switching to energy derived from plants. There should a healthy debate about the consequence of turning food crops or animal feed into biofuels. Overutilizing these limited resources to make fuel may risk transforming a potential solution to reducing dependence on fossil fuels into a major shortage of food supplies and damage to the ecosystem. We need to ensure that renewable resource policies account for this risk and strike the right balance. Many technical challenges remain for establishing the viability of the biofuels industry, including technological improvements, economic feasibility and sustainability, and concerns about human environmental impacts. Further research should focus on development of better and cheaper catalysts, improvements in current technology for producing high-quality biodiesel, conversion of the by-products such as glycerol to useful products such as methanol and ethanol, and development of low-cost photobioreactors.

See Also: Airborne Diseases; Asthma; Climate Change; Fertilizers; Particulate Matter.

Further Readings

Duffield, James A. "Biodiesel: Production and Economic Issues." *Inhalation Toxicology*, 19 (2007).
Sanli, Canakci H. "Biodiesel Production From Various Feedstocks and Their Effect on the Fuel Properties." *Journal of Industrial Microbiology & Biotechnology*, 35 (2008).
Swanson, Kimberly J., Michael C. Madden, and Andrew J. Ghio. "Biodiesel Exhaust: The Need for Health Effects Research." *Environmental Health Perspectives*, 115 (2007).
Vasudevan, Palligarnai T. and Michael Briggs. "Biodiesel Production—Current State of the Art and Challenges." *Journal of Industrial Microbiology & Biotechnology*, 35 (2008).

Hueiwang Anna Cook Jeng
Independent Scholar

Liang Yu
Old Dominion University

BIOLOGICAL CONTROL OF PESTS

By the 1950s, the use of synthetic insecticides to defend food and fiber crops against insects, mites, and disease appeared to offer an easy solution to crop protection. More than 50 years later, while conventional agricultural methods still rely mainly on the use of chemical pesticides, we recognize that there are many problems associated with the use of insecticides, such as the following:

- Some insect pests become resistant
- Nontarget organisms can be affected
- Pest resurgence can occur
- Adverse affects on environment and human health can arise

In response to the consequences of petrochemical pesticide use, researchers and agriculturalists have developed the Integrated Pest Management (IPM) system, an eco-based pest-control strategy that strives to minimize the use of chemical pesticides by embracing such practices as the use of pest-resistant plants, various cultural control methods, and biological control—the use of pests' natural enemies to suppress their populations and minimize their damage to crops. All insect pests have natural enemies, and they are classified as follows:

- *Parasitoids*: species whose immature stage develops on or within a single insect host, ultimately killing the host; many species of wasps and some flies are parasitoids
- *Predators*: free-living species, such as lady bugs, with prodigious appetites for prey
- *Pathogens*: disease-causing organisms including bacteria, fungi, and viruses; they kill or debilitate their host and are relatively specific to certain insect groups

When choosing a natural enemy to combat a pest, there are several characteristics to look for in the natural enemy, such as a high reproductive rate, good searching ability, host specificity, adaptability to different environmental conditions, and the capacity to be synchronized with its host. Because it requires a deep understanding of the pests and their natural enemies as well as of the crops and environment—and of the interrelationships that exist between all of these components—biological control of pests usually involves an active human role. The goal of the biological control of pests is not to completely eliminate all pests, but to reduce their numbers, and thus, the damage they cause.

Biological control of pests involves three broad, overlapping methods:

- Conservation
- Classical biological control
- Augmentation

The conservation of natural enemies is probably the most important and readily available biological control method available to growers. Natural enemies occur in all agricultural systems, and they are adapted to the local environment and to the target pest, and their conservation is generally simple and cost effective. Classical biological control is the practice of importing, and releasing for establishment, natural enemies to control pests. While it typically involves traveling to the area from which a newly introduced, or "exotic," pest originated, and returning with some of its natural enemies, it is sometimes used on native insect pests. New pests are constantly arriving accidentally or intentionally, and they often survive; when they do, it is usually because their enemies were left behind. Introducing some of their natural enemies can be an

Researchers and agriculturalists have developed the Integrated Pest Management system, an eco-based pest-control strategy that strives to minimize the use of chemical pesticides. Here, two entomologists in the San Joaquin Valley bag almonds to exclude navel orangeworm and hang a pheromone dispenser to disrupt the pests' mating.

Source: U.S. Department of Agriculture Agricultural Research Service/Peggy Greb

important way to reduce the amount of harm exotic pests can do. After procuring the natural enemies, they are subjected to a meticulous quarantine process to ensure that no unwanted organisms, such as hyperparasitoids, are present; the natural enemy is then reared, ideally in large numbers, and released. Follow-up studies are conducted to determine if the natural enemy is successfully combating the pest and to assess the long-term benefit of its presence.

Today, many classical biological control programs for insect pests and weeds are under way across the United States and Canada, though one of the earliest successes of classical biological control reaches back to the late 1800s, to the cottony cushion scale, a pest that was destroying the California citrus industry. A predatory insect, the vedalia beetle, and a parasitoid fly were introduced from Australia, and within a few years, the cottony cushion scale was completely controlled by these introduced natural enemies.

Classical biological control is durable and often more economical than the use of only pesticides. Other than the initial costs of collection, importation, and rearing, little expense is required. When a natural enemy is successfully established, it rarely needs additional input, and it continues to kill the pest with no direct help from humans and at no cost. Unfortunately, classical biological control does not always work. It is usually most effective against exotic pests and less so against native insect pests. The reasons for failure are often not known but may include the release of too few natural enemies, poor adaptation of the natural enemy to environmental conditions at the release location, and lack of synchrony between the life cycle of the natural enemy and host pest.

Augmentation is a method of increasing the population of a pest's natural enemy. This can be done by mass producing a pest in a laboratory and releasing it into the field at the proper time or by breeding a better natural enemy that can attack or find its prey more effectively. Another form of augmentation involves modifying the cropping system to supplement the natural enemies, a practice typically referred to as "habitat manipulation." Many adult parasitoids and predators benefit from sources of nectar and the protection provided by refuges such as hedgerows, cover crops, and weedy borders. Recently, in California, it was found that planting prune trees in grape vineyards provides an improved overwintering habitat for a major grape pest parasitoid. Although the tactic appears to hold much promise, only a few examples have been adequately researched and developed.

The National Biological Control Laboratory (NBCL), which is the in-house research arm of the U.S. Department of Agriculture (USDA), is active in developing functional methods of mass propagation, storage, and delivery of advantageous organisms as well as targeted release strategies for IPM. The NBCL breeds and studies only species and organisms that have been approved for release by federal and state officials. Developing the industry that provides beneficial predators, parasites, and microbes for augmenting pests' natural enemies will buttress the battle against agricultural pests.

Biological control takes intensive management and planning. Further, successful use of biological control requires a greater understanding of the biology of both the pest and its enemies than does simply applying pesticides to crops. In some cases, biological control may be more costly than pesticides, and sometimes the results of using biological controls are not as dramatic or quick as the results of using pesticides. While broad-spectrum insecticides may kill a wide range of insects, biological control methods often target one specific pest, which can sometimes be a disadvantage. Before biological control can further advance, there needs to be more intense investigation of indigenous natural enemies and their impact on the pests they attack. With that information, it may be possible to enhance the efficacy of natural enemies through manipulation of the crop habitat, changes in cultural practices, changes in pesticide application practices, and the introduction of new natural enemies. The development of successful biological control programs presents a challenge that also holds great potential for transforming the methods with which we manage the pests that damage food and fiber crops.

See Also: Chemical Pesticides; Government Role in Green Health; Pest Control; Petrochemicals.

Further Readings

Cornell University. "Biological Control: A Guide to Natural Enemies in North America." http://www.nysaes.cornell.edu/ent/biocontrol (Accessed June 2010).

Hajek, Ann E. *Natural Enemies: An Introduction to Biological Control.* New York: Cambridge University Press, 2004.

U.S. Department of Agriculture, Agricultural Research Service. "National Biological Control Laboratory." http://www.ars.usda.gov/Main/docs.htm?docid=7542 (Accessed June 2010).

Tani Bellestri
Independent Scholar

BIOLOGICAL WEAPONS

Biological weapons are infectious agents, such as bacteria, fungi, viruses, parasites, biochemical toxins, and biomolecules (e.g., potentially prions) that can cause extreme morbidity and mortality in plants, animals, and humans when they are intentionally disseminated to induce terror (bioterrorism) and/or related mass casualties under conditions of civil unrest or military campaigns. Biological agents can also be a threat for the environment. This is an important issue for global public health that should be considered in the new perspectives of a more ecological green world in which such environmental threats should be significantly reduced or eliminated.

Historical Development

The development of biological weapons and bioterrorism is not new. In the last 50 years, they have increased in importance as threats to global public health, particularly after the anthrax bioterrorism events following the terrorist attacks on September 11, 2001, in the United States.

It is difficult to identify the origin of biological warfare, but at least one deliberate contamination of water supplies occurred in approximately 600 B.C.E. using the fungi *Calviceps purpurea*. During the siege of Kaffa (now Feodosia, Ukraine) in the 14th century, the attacking Tatar force experienced an epidemic of plague, an infectious disease produced by the bacterium *Yersinia (pseudotuberculosis) pestis* that probably was originally recognized as disease in the sixth century when the Justinian plague occurred in Egypt, Turkey, and parts of Europe. At that time, the biological threat was used as a weapon by catapulting contaminated cadavers of deceased Tatars affected by the plague into the city to cause an outbreak, which led to a retreat of the defending forces and the conquest of Kaffa. In the 14th century, the plague pandemic, also called the Black Death, probably affected 100 million people, with an estimated case fatality rate of 25 to 30 percent. After that, other cities in Europe probably experienced outbreaks, with contaminated humans, rats, and other animals infected with *Y. pestis* intentionally shipped and sailed to Constantinople (Istanbul today), Genoa, Venice, and other Mediterranean ports. Even with this history, it would be an oversimplification to implicate the biological attack as the sole cause of the plague epidemic in Kaffa and other cities in that century. Multiple factors probably played a significant role in those outbreaks, including environmental variables. In history, there are many other examples of bioterrorism and biological weapons used up until the 20th century, when significant advances in microbiology and war industry led to the posing of this threat at a global level. The possibility of this hypothesis places those biological threats in other conventional weapons (e.g., long-range missiles) that can reach other countries in a few minutes or hours, as was suspected with Iraq during the conflicts in the Persian Gulf in the early 1990s.

Contemporary Use in Warfare

The military setting has been one the first arenas involved in planning, developing, and using biological weapons, currently implementing that knowledge in what is called biodefense—complex elements used to prevent and respond to bioterrorist threats and acts. However, in the past, military groups were linked to outbreaks of anthrax and meliodosis that have been recognized as products of biological warfare during World War I (1914–1918). In 1925, the Geneva Protocol for the Prohibition of the Use in War of Asphyxiating, Poisonous or Other Gases, and of Bacteriological Methods of Warfare, prohibited the use of biological weapons—although not their research, production, study, or possession. Before and during World War II (1939–1945), biological weapon activities were intensified, and as a consequence of this, other pathogens, including *Y. pestis, Bacillus anthracis,* and *Burkholderia (Pseudomonas) mallei,* were used as the bacteriological weapon agents *Neisseria meningitides* (an etiology of meningitis), *Salmonella spp., Shigella spp.,* and *Vibrio cholerae* (etiological agents of diarrhea and dysenteric syndromes). Throughout history, and especially during the mid-1970s, the list of biological weapons increased in number and included *Clostridium botulinum* (etiological agent of botulism, a form of food intoxication), *Francisella tularensis* (etiological agent of tularemia), *Brucella suis, Coxiella burnetti,* staphylococcal enterotoxin B, and Venezuelan equine encephalitis (VEE) virus. Since the 1990s, many countries have developed the capacities to create many of the biological weapon agents included in all categories of microorganisms that represent a global public health threat. The intentional delivery of letters with *B. anthracis* (anthrax) following the 2001 terrorist attacks in the United States highlighted these capacities and vulnerabilities and the risk of nations to these biological threats.

National and International Restrictions

In 1970, the World Health Organization (WHO) made technical guidance available to member states (or countries) in the report "Health Aspects of Biological and Chemical Weapons." This report was instrumental in facilitating international consensus on the Biological and Toxin Weapons Convention (BWC) and the Chemical Weapons Convention. In April 1972, the Convention on the Prohibition of the Development, Production and Stockpiling of Bacteriological (Biological) and Toxin Weapons and on Their Destruction was opened for signature in London, Moscow, and Washington, D.C. In March 1975, this convention entered into force. Its depositary governments were the Russian Federation, the United Kingdom, and the United States. Over the intervening years, increasing numbers of states joined the convention, reaching 155 state parties and 16 signatory states by 2010. The convention effectively prohibits the development, production, acquisition, transfer, stockpiling, and use of biological and toxin weapons, and is a key element in the international community's efforts to address the proliferation of weapons of mass destruction.

These conventions, along with the 1925 Geneva Protocol, are the international legal mechanisms banning these classes of weapons. The 1970s report has recently been revised, and its second edition, "Public Health Response to Biological and Chemical Weapons: WHO Guidance," is now available. In addition, some countries have tried to raise the awareness and developed legal and regulatory options in regard to the threat of biological weapons.

Etiological Agents of Bioterrorism and Their Classification

Potential biological agents used for bioterrorism are classified according to the U.S. Centers for Disease Control and Prevention (CDC) in three categories: A, B, and C. Category A includes pathogens that are rarely seen in the United States. High-priority agents include organisms that pose a risk to national security because they can be easily disseminated or transmitted from person to person; result in high mortality rates and have the potential for major public health impact; cause public panic and social disruption; and require special action for public health preparedness. In Category B are those agents that are considered as the second-highest priority, including those agents that are moderately easy to disseminate, resulting in moderate morbidity rates and low mortality rates, and that require specific enhancements of the CDC's diagnostic capacity and enhanced disease surveillance. Finally, Category C comprises those emerging pathogens that could be engineered for mass dissemination in the future because of availability; ease of production and dissemination; and potential for high morbidity and mortality rates and major health impact, representing the third-highest priority group.

In addition to agents originally included in this classification, additional organisms are considered potential biological threats and have recently been included, such as other tick-borne viral hemorrhagic fevers, yellow fever, and multidrug-resistant (MDR) and extremely resistant (XDR) *M. tuberculosis*.

Surveillance and Preparedness

In this setting, fortunately, epidemiological surveillance systems have improved in many ways, including early detection and response against these biological threats. Additionally, many countries, such as the United States, have developed plans of Emergency Preparedness and Response. Surveillance mathematical models to predict different consequences during a bioterrorist attack are readily being developed. Additionally, information about the threats of bioterrorism is being disseminated more widely among infectious diseases and public health physicians and also to primary care clinicians. Additionally, as there is a significant threat of bioterrorism targeted at animals, as well the potential for zoonotic organisms that can be used as biological weapons (e.g., *Brucella, B. anthrax, Y. pestis, Chlamydophila psittaci*, among others), veterinarians are being included in emergency preparedness trainings.

It is also important to mention that there have been significant sums of federal funding in the United States received for bioterrorism preparedness that impact on the preparedness activities undertaken by local health departments. Overall, budget, leadership, and crisis experience are found to be the most important determinants of local preparedness activity, but CDC preparedness funding plays a mediating role by building capacity through hiring a key leadership position—the emergency preparedness coordinator.

Research Activity on Bioterrorism

Epidemiological, clinical, biological, and military research related to bioterrorism and biodefense have significantly increased since 2001, reaching more than 1,000 new scientific publications in 2002 and later gradually decreasing to less than 400 in 2008. However, even new journals exclusively dedicated to the study of bioterrorism and biological weapons have been developed since 2001, such as the journal *Biosecurity and Bioterrorism:*

Biodefense Strategy, Practice, and Science, launched in 2003. From a clinical point of view, it is also important to highlight that diagnostic tools and therapeutic options for these organisms have been improved in quality and availability, including effective prophylactic measures, vaccines, and drugs. The microbiological diagnosis that allows identification of the causal agent is of utmost importance. This constitutes a key point for taking suitable control measures. Environmentally, there are significant differences in the way to deliberate these biological agents and the infective dose of the aerosols, as detailed in Table 1.

Table 1   Infective Dose (Aerosol) of Selected Potential Biological Weapons

Agent	*Dose*
Anthrax	8,000 to 50,000 spores
Botulinum toxin (type A)	0.001 micrograms (µg)/kilogram (kg)
Brucellosis	10–100 organisms
Plague	100–500 organisms
Q fever	1–10 organisms
Smallpox	10–100 organisms
Staphylococcal enterotoxin B	.30 µg/person (incapacitating); 1.7 µg/person (lethal)
Tularemia	10–50 organisms
Viral encephalitides	10–100 organisms
Viral hemorrhagic fevers	1–10 organisms

Prospects for the Future

In most of the world, the public health infrastructure is already stretched to its limits from coping with natural health hazards, especially in developing countries. Against such a background, the additional threat to health services posed by the possible deliberate use of biological agents could be considered as little more than a slight addition to the existing burden of infectious or communicable diseases, such as those that have been mentioned. For deliberate releases or threats of release of such agents, a spectrum of outcomes can be envisaged that ranges between two extremes: relative insignificance at one end to mass destruction of life or mass casualties at the other, which in the case of these biological agents, depends on the virulence and case fatality rate in the different epidemiological scenarios where the biological weapons would be targeted. Whatever the magnitude of such releases might be, however, widespread panic and fear would be certain; the public health system would be overwhelmed; and economic impact would be considerable, even more so in the context of the global economic crisis that is affecting developed and developing countries in a significant way, particularly the health systems.

It is possible that a deliberately caused epidemic might initially be perceived as a natural event, especially if the agent were spread covertly. In particular, the covert release of a biological agent would not have an immediate effect because of the incubation period for the agent used (this period could be considerably variable). In addition, because of the medical profession's lack of experience with the diseases generally associated with such biological agents, the detection of the epidemic could be delayed. During the incubation period, a deliberately caused epidemic could spread to a few or even many cities and countries. The first casualties would be identified by primary healthcare personnel at the periphery of the healthcare system. The implementation of effective disease surveillance and early warning systems in all cities and countries would enable public health officials to track and identify any unusual events as soon as possible.

Fortunately, emerging online systems are helping with these purposes. In the case of a highly contagious agent, only a small window of opportunity would be available between the time the first case is identified and the massive spread of the disease in the susceptible population. If a deliberately induced disease did spread to a few or many countries before it was identified, the subsequent containment of such an outbreak would require a highly coordinated response at local, national, and international levels to be effective. Such outbreaks may in practice prove very difficult to distinguish from natural outbreaks, especially for common communicable diseases or uncommon diseases sharing a significant group of signs and symptoms with common ones, and—because of their security implications—would also require health agencies to collaborate closely with new and unfamiliar partners such as military, security, and law enforcement agencies. To be effective, collaborative arrangements involving all partners would have to be established and tested well before an incident or emergency occurred.

To fulfill the WHO vision of the attainment of the highest possible level of public health by all nations, each WHO member state must, inter alia, have the capacity to respond appropriately and effectively to threats posed by naturally occurring and accidentally or deliberately caused epidemics, with a view to minimizing their negative impact on the health, security, and economic stability of individual member states and the international community as a whole.

Public health emergencies arising from deliberate epidemics—whether they have actually occurred or may occur—fundamentally transform the context in which public health services must be delivered in order to ensure human safety and security. The function of preparedness for deliberate epidemics (PDE) is to facilitate preparations for such contingencies that are attuned to the different risk and threat assessments and levels of preparedness of individual WHO member states.

PDE's program of work contributes to WHO's response to the requests made by member states in resolution WHA55.16 by (1) strengthening global health preparedness and response, (2) providing tools and support to member states for health emergency preparedness and response, (3) developing guidelines, information, and/or training materials on specific risks to public health, and (4) examining the possible development of new tools and mechanisms, research, and technologies with the potential to enhance public health preparedness and response.

See Also: Airborne Diseases; Antibiotics; Bird Flu; Centers for Disease Control and Prevention (U.S.); Tuberculosis; Vaccination/Herd Immunity.

Further Readings

Ablah, Elizabeth, Lindsay Benson, Kurt Konda, Annie M. Tinius, Leslie Horn, and Kristine Gebbie. "Emergency Preparedness Training for Veterinarians: Prevention of Zoonotic Transmission." *Biosecurity and Bioterrorism: Biodefense Strategy, Practice, and Science*, 6 (2008).

Anaya-Velázquez, Fernando. "Bioethics, Bioweapons and the Microbiologist." *Latin American Journal of Microbiology*, 44 (2002).

Avery, George H. and Jennifer Zabriskie-Timmerman. "The Impact of Federal Bioterrorism Funding Programs on Local Health Department Preparedness Activities." *Evaluation & the Health Professions*, 32 (2009).

Brownstein, John S., Clark C. Freifeld, and Lawrence C. Madoff. "Digital Disease Detection—Harnessing the Web for Public Health Surveillance." *New England Journal of Medicine*, 360 (2009).

Buitrago Serna, Mª José, Inmaculada Casas Flecha, José Mª Eiros-Bouza, Raquel Escudero Nieto, Cesare Giovanni Fedele, Isabel Jado García, Francisco Pozo Sánchez, José Miguel Rubio Muñoz, Mª Paz Sánchez-Seco Fariñas, Sylvia Valdezate Ramos, and José Verdejo Ortes. "Biodefense: A New Challenge for Microbiology and Public Health." *Enfermedades Infecciosas y Microbiología Clínica*, 25 (2007).

Christopher, George W., Theodore J. Cieslak, Julie A. Pavlin, and Edward M. Eitzen. "Biological Warfare—A Historical Perspective." *Journal of the American Medical Association*, 278 (1997).

Franco-Paredes, Carlos, Alfonso J. Rodriguez-Morales, and José Ignacio Santos-Preciado. "Bioterrorism Agents: Getting Ready for the Unthinkable." *Revista de Investigación Clínica*, 57 (2005).

Franz, David R., Peter B. Jahrling, Arthur M. Friedlander, David J. McClain, David L. Hoover, W. Russel Bryne, Julie A. Pavlin, George W. Christopher, and Edward M. Eitzen, Jr. "Clinical Recognition and Management of Patients Exposed to Biological Warfare Agents." *Journal of the American Medical Association*, 278 (1997).

Guarner, Jeannette and Sherif R. Zaki. "Histopathology and Immunohistochemistry in the Diagnosis of Bioterrorism Agents." *Journal of Histochemistry and Cytochemistry*, 54 (2006).

Maher, C. and B. Lushniak. "Availability of Medical Countermeasures for Bioterrorism Events: US Legal and Regulatory Options." *Clinical Pharmacology & Therapeutics*, 85 (2009).

Radosavljevic, Vladan, Desanka Radunovic, and Goran Belojevic. "Epidemics of Panic During a Bioterrorist Attack—A Mathematical Model." *Medical Hypotheses* (May 2009).

Simon, Jeffrey D. "Biological Terrorism—Preparing to Meet the Threat." *Journal of the American Medical Association*, 278 (1997).

Temte, J. L. and M. E. Grasmick. "Recruiting Primary Care Clinicians for Public Health and Bioterrorism Surveillance." *Wisconsin Medical Journal*, 108 (2009).

U.S. Centers for Disease Control and Prevention. "Emergency Preparedness and Response." http://www.bt.cdc.gov/ (Accessed May 2009).

Wallin Arūnė, Živilė Lukšienė, Kęstutis Žagminas, and Genė Šurkienė. "Public Health and Bioterrorism: Renewed Threat of Anthrax and Smallpox." *Medicina (Kaunas)*, 43 (2007).

World Health Organization (WHO). *Preparedness for Deliberate Epidemics*. Geneva, Switzerland: WHO, 2004.

World Health Organization (WHO). *Public Health Response to Biological and Chemical Weapons—WHO Guidance*. Geneva, Switzerland: WHO, 2004.

Alfonso J. Rodriguez-Morales
Universidad de Los Andes, Universidad Central de Venezuela

BIOMEDICINE

Human health is both directly and indirectly related to the environments in which we live. The central purpose in biomedicine is to promote human health, and thus the biomedical field has a strong interest in environmental stewardship. There are at least four aspects of biomedicine that relate to environmental health and stewardship: biomedical research; pharmaceuticals; clinical practice of biomedicine by physicians, nurses, and other trained care providers; and public health and epidemiology. Each of these sectors of biomedicine plays a significant role in environmental impacts and outcomes.

Biomedical research is a broad arena that is carried out in academic health centers and institutes, hospitals and healthcare facilities, community-based research facilities, federal government laboratories, and laboratories of for-profit companies or other types of foundations. Hospitals require large amounts of resources and have begun to take steps to reduce their carbon footprints. The National Association of Physicians for the Environment (NAPE) has taken a leadership role in assessing the roles of biomedical research in environmental stewardship. Biomedical research has impacts on air and water quality, climate, agriculture, and the quality of life. A primary objective in supporting positive environmental stewardship is to extend the developments and successes of green buildings to the entire research process.

Green research buildings need to focus on energy efficiency, the use of renewable resources, the impacts on local environments and communities (such as urban sprawl), conservation of resources, controlling indoor environments and toxins, and the ability to access necessary information regarding environmental impacts. Special concerns in biomedical research that constrain the types of resource conservation and recycling that can be implemented are animal research, hazardous wastes, structure of research committees, and regulatory agencies. Research is a heavily regulated industry already, and thus it is not likely feasible to increase regulation on research to include strict environmental impact requirements. The complexities of the regulatory agencies governing research make the addition of regulatory standards prohibitory. With better analyses of the outcomes of environmental impact reduction projects that have been implemented, other research facilities can follow suit. Reward mechanisms and acknowledgment of successful research designs that have reduced environmental impacts can continue to inspire improved biomedical research mechanisms. The economic costs of some innovative research practices may be prohibitory, but these are less substantial when weighed against the benefits in the long term. Biomedical researchers continue to seek funding to help offset immediate costs of innovative research that reduces environmental impacts. Research campaigns targeted at lead researchers and facility managers aim to reduce pollution, promote environmental safety, conserve energy, and manage waste products more efficiently.

In greening the pharmaceutical industry, as in general, the same basic concepts of reduce, reuse, and recycle apply. The reduction of toxic chemicals in the design and production of pharmaceuticals, so-called green chemistry, has made a substantial impact on mitigating this sector's environmental impacts. The potential negative impacts of the pharmaceutical industry are best addressed from both the manufacturing side and the consumer side. The manufacturing industry can have the most direct and substantial impact by implementing many of the research strategies discussed above in the phases of research and development. Additionally, the process of production can have less-damaging environmental impacts by focusing on drugs that will biodegrade, improving the efficacy of drugs in lower doses (so that less is needed), and using postconsumer and recyclable materials for product packaging. In Europe, pharmaceuticals are evaluated for environmental impact by looking at the drug's persistence, bioaccumulation, and toxicity.

Consumers need to know how to properly dispose of unused or unwanted pharmaceuticals, which includes adequate education about proper disposal and a means to do so. Unused medicines should not be flushed down the toilet, which can lead to contamination of the water supply and other environmental damage. Physicians, dentists, veterinarians, pharmacists, and healthcare providers can educate consumers about the proper way to dispose of unused and unwanted medicines and should be willing to receive these medicines for proper disposal. To avoid wastefulness, consumers are encouraged to buy medicines in smaller quantities and to recycle or donate (when possible) unused and unopened pharmaceuticals. Many consumers do not consider the important impacts that the production of pharmaceuticals and improper disposal can have; it behooves the pharmaceutical industry to continue to educate healthcare providers and consumers about these impacts so that individuals are informed about proper environmental stewardship steps that can be taken.

Environmental medicine encompasses two primary branches: (1) physicians who practice clinical medicine, treating individuals who have illnesses related to environmental exposure, and (2) public health practitioners and epidemiologists who look at the population, as a whole, and aim to influence human health through broader-scope activities, such as improving public policies. The developed world has seen a steady decrease in the amount of illnesses that have resulted from exposure to hazardous materials, but that is not the case for areas in the world that are still developing. Thus, in the United States and other developed nations, the focus has shifted toward population and community health issues, which are best addressed from a public health standpoint. The environmental concerns that are central can range from the natural environment (particularly the increases in natural calamities and cosmic ionizing), human-caused environmental degradation issues (like urban sprawl, use of toxic materials, and means of food production), and even include cultural environments that influence health behaviors. Of course, in the field of environmental medicine, it is not only physicians who are primary players; nurses are frequently at the front lines of toxic exposure assessment, community education about hazards, and preventative care.

One final emerging area of biomedicine that is directly concerned with environmental stewardship is "sustainable medicine" or "sustainable healthcare." There is no standard definition for this area yet, but the focus is on a holistic model that views human health as part of social and natural environments. Practitioners of sustainable medicine may incorporate basic principles of homeopathy and approach health as a product of the interactions between all aspects of life and the environments where we live. Another interpretation of sustainable medicine focuses on the natural life cycle in which death is a part of the process of life. Citing the finiteness of the resources on Earth, the pressures placed on those resources by overpopulation and exacerbated by inequitable access to those resources, this branch of medicine emphasizes using biomedical technology to promote quality of life for all human beings. This means using biomedical resources and technology to combat diseases with devastating effects on the natural life cycle throughout the world and not squandering biomedical resources in artificially extending the lives of those who can afford access to them.

See Also: Hospitals (Carbon Footprints); Pharmaceutical Industry; Suburbs.

Further Readings

Kreisberg, Joel. "What Is 'Sustainable Health Care'?" *Symbiosis: Journal of Ecologically Sustainable Medicine*, 2/3 (2004).

Schwartz, B. S., G. Rischitelli, and H. Hu. "The Future of Environmental Medicine in Environmental Health Perspectives: Where Should We Be Headed?" *Environmental Health Perspectives*, 113/9 (2005).

Wilson, S. H., S. Merkle, D. Brown, J. Moskowitz, D. Hurley, D. Brown, B. J. Bailey, et al. "Biomedical Research Leaders: Report on Needs, Opportunities, Difficulties, Education and Training, and Evaluation." *Environmental Health Perspectives*, 108 (2000).

Elizabeth A. Olson
University of California, Merced

BIRD FLU

Influenza, or flu, can be caused by different viruses of influenza: A, B, or C. Among those included in type A are viruses called bird, or avian, influenza viruses; A (H5N1) virus; or other nonhuman influenza subtypes (e.g., H7, H9).

Human infections with highly pathogenic avian influenza A (H5N1) virus occur through bird-to-human transmission; possibly environment-to-human transmission; and, very rarely, limited, nonsustained human-to-human transmission. However, in the first decade of the 2000s, outbreaks of human avian influenza in southeast Asia have been sustainably reported. Direct contact with infected poultry or with surfaces and objects contaminated by their droppings is the main route of transmission to humans. Exposure risk is considered highest during slaughter, defeathering, butchering, and preparation of poultry for cooking. There is no evidence that properly cooked poultry or poultry products can be a source of infection.

Extensive outbreaks in poultry have occurred in parts of Asia, the Middle East, Europe, and Africa

Extensive outbreaks of avian flu in poultry have occurred in parts of Asia, the Middle East, Europe, and Africa since 2003, but only sporadic human infections have appeared to date. Veterinary medical officers evaluate tissue sections (top monitor) from chickens infected with Hong Kong H5N1 influenza.

Source: U.S. Department of Agriculture, Agricultural Research Service/Rob Flynn

since 2003, but only sporadic human infections have occurred to date. Since 1997, when human infections with a highly pathogenic avian influenza A virus subtype H5N1—previously infecting only birds—were identified in a Hong Kong outbreak, global attention has focused on the potential for this virus to cause the next pandemic. Beginning in December 2003, an unprecedented H5N1 epizootic in poultry and migrating wild birds has spread across Asia and into Europe, the Middle East, and Africa. Humans in close contact with sick poultry, and on rare occasion with other infected humans, have contracted the illness. Continued exposure of humans to avian H5N1 viruses increases the likelihood that the virus will acquire the necessary characteristics for efficient and sustained human-to-human transmission through either gradual genetic mutation or reassortment with a human influenza A virus. Between November 2003 and July 2008, nearly 400 human cases of laboratory-confirmed H5N1 infection were reported to the World Health Organization (WHO) from 15 countries in southeast and central Asia, Europe, Africa, and the Middle East. The WHO has declared the world to be in phase three of a pandemic alert period.

As for influenza in general, neuraminidase inhibitors (e.g., oseltamivir, zanamivir) are inhibitory for the virus and demonstrate proven efficacy in vitro and in animal studies for prophylaxis and treatment of H5N1 infection. Studies in hospitalized H5N1 patients, although limited, suggest that early treatment with oseltamivir improves survival and, given the prolonged virus replication, late intervention with oseltamivir is also justified. Neuraminidase inhibitors are recommended for post-exposure prophylaxis in certain exposed persons. At the time of this writing, the WHO does not recommend pre-exposure prophylaxis for travelers, but advice may change depending on new findings. Inactivated H5N1 vaccines for human use have been developed and licensed in several countries but are not yet generally available, although this situation is expected to change. Some countries are stockpiling these vaccines as part of pandemic preparedness. Although immunogenic, the effectiveness of these vaccines in preventing the H5N1 infection or reducing disease severity is unknown. Currently, the WHO does not have a policy on their use. Additionally, it is recommended to avoid contact with high-risk environments in affected countries such as live animal markets and poultry farms, any free-ranging or caged poultry, or surfaces that might be contaminated by poultry droppings. In affected countries, contact with dead migratory birds or wild birds showing signs of disease should be avoided. Consumption of undercooked eggs, poultry, or poultry products should also be avoided. Hand hygiene with frequent washing or use of alcohol rubs is recommended. If exposure to persons with suspected H5N1 illness or with severe, unexplained respiratory illness occurs, individuals should urgently consult health professionals and contact their local health providers or national health authorities for supplementary information.

Since 1918, influenza virus has been one of the major causes of morbidity and mortality, especially among young children. Though the commonly circulating strain of the virus is not virulent enough to cause mortality, the ability of the virus genome to mutate at a very high rate may lead to the emergence of a highly virulent strain that may become the cause of the next pandemic. Apart from the influenza virus strains circulating in humans (H1N1 and H3N2), the avian influenza H5N1, H7, and H9 virus strains have also been reported to have caused human infections; H5N1, H7, and H9 have shown their ability to cross the species barrier from birds to humans and further replicate in humans. The biological and epidemiological aspects of influenza virus and efforts to control the virus globally should be deeply studied, even more because, additionally to these viruses, swine flu emerged as the cause of a declared pandemic in 2009. This new pandemic is affecting more than 180 countries, with more than 180,000 confirmed cases and at least 1,400 deaths (as of July 2009).

See Also: Acquired Immune Deficiency Syndrome; Airborne Diseases; Centers for Disease Control and Prevention (U.S.); Seasonal Flu; Tuberculosis.

Further Readings

Franco-Paredes, Carlos, Alfonso J. Rodriguez-Morales, and José Ignacio Santos-Preciado. "Clinical and Epidemiological Aspects of Influenza." *Latin American Medical Student Science and Research*, 11 (2006).
Khanna, M., P. Kumar, K. Choudhary, B. Kumar, and V. K. Vijayan. "Emerging Influenza Virus: A Global Threat." *Journal of Biosciences*, 33 (2008).
Kieny, Marie Paule, Alejandro Costa, Joachim Hombach, Peter Carrasco, Yuri Pervikov, David Salisbury, Michel Greco, Ian Gust, Marc LaForce, Carlos Franco-Paredes, José Ignacio Santos, Eric D'Hondt, Guus Rimmelzwaan, Ruth Karron, and Keiji Fukuda. "A Global Pandemic Influenza Vaccine Action Plan." *Vaccine*, 24 (2006).
Pappaioanou, Marguerite. "Highly Pathogenic H5N1 Avian Influenza Virus: Cause of the Next Pandemic?" *Comparative Immunology, Microbiology and Infectious Diseases*, 32 (2009).
Rodriguez-Morales, Alfonso J., Wendy Lorizio, Jair Vargas, Livia Fernández, Balbina Durán, Gabriel Husband, Alexis Rondón, Karelys Vargas, Rosa A. Barbella, and Sonia M. Dickson. "Malaria, Tuberculosis, HIV/AIDS and Avian Influenza: Killers of Mankind?" *Journal of the Surgical and Medical Society of the Emergency Hospital Perez de Leon*, 39 (2008).
U.S. Centers for Disease Control and Prevention. "Immunization Schedules." http://www.cdc.gov/vaccines/recs/schedules/default.htm (Accessed April 2009).
World Health Organization. *International Travel and Health*. Geneva, Switzerland: WHO, 2009.

Alfonso J. Rodriguez-Morales
Universidad de Los Andes, Universidad Central de Venezuela

BOTTLED WATER

Bottled water in the United States is regulated by different agencies than those that regulate drinking water. Because of this, there are differences in the reporting requirements for water testing that make it more difficult to determine the safety and content of bottled water when compared to tap water.

In the United States, drinking water is regulated by the U.S. Environmental Protection Agency (EPA). The EPA sets two standards for drinking water in public water supplies: primary and secondary. Primary standards are legally enforceable and established to protect human health. They place concentration limits on specific pollutants within drinking water, known as maximum contaminant levels (MCLs). Secondary standards are not enforceable and are established to guide the aesthetic quality of water.

Water from a public water supply must be tested to ensure that it meets the primary drinking water standards. Water suppliers must notify the public if the water is not safe to drink and will provide directions, such as boiling water, to make it safe. Water suppliers must also produce an annual report on the results of water testing and make the report available to the public. To help ensure public safety, within the United States, water suppliers also treat drinking water with chlorine prior to its leaving the water treatment facility. This residual chlorine is meant to kill any pathogens that may enter the water stream on its way to an individual tap.

Some consumers dislike the taste or smell of residual chlorine in water. Chlorine can also react with other constituents in water to cause disinfection by-products. For these reasons, some consumers prefer to drink bottled water, which is not required to have residual chlorine.

Bottled water is regulated by the Food and Drug Administration (FDA) but is required to meet primary drinking water standards. Bottled water suppliers are not required to make their water-quality testing results available to the public.

There are numerous types of bottled water, such as artesian, well, spring, purified, and mineral. The FDA's Standard of Identity establishes the requirements each type of water must meet; these standards are based on the water source and the methods used to treat the water. It may be difficult for average consumers to understand the differences between types of bottled water without educating themselves. The types are as follows:

- *Artesian water*: Aquifers are underground bodies of water. Confined aquifers have a layer material with extremely low permeability over the top of them, which may cause the water to be under pressure. When a well is drilled through the confining layer into this aquifer, the water level in the well may be higher than the elevation of the top of the aquifer. When this occurs, the water can be labeled as artesian.
- *Fluoridated*: To be labeled as fluoridated, water must have fluoride added, as limited by the FDA. This type includes water labeled "for infants" or "nursery."
- *Groundwater*: This type of water comes from an aquifer where the water source is either equal to or greater than atmospheric pressure. Artesian water is a type of groundwater.
- *Mineral water*: This water is also a type of groundwater that is isolated and contains at least 250 parts per million of total dissolved solids. All minerals must be naturally occurring in this type of water.
- *Purified water*: Also called "demineralized water," this type of water receives treatment prior to bottling, such as distillation or reverse osmosis. Purified water must meet a definition provided in the United States Pharmacopeia.
- *Sparking water*: Sparkling water contains carbon dioxide, but only in the amount present when taken from its source.
- *Spring water*: Spring water comes from the ground and flows naturally to the surface without any pumping.
- *Sterile water*: To be classified as sterile, water must pass tests required by the United States Pharmacopeia.
- *Well water*: This water comes from a hole, or well, in the ground.

Although bottled water is often marketed as pure and safe, this is not necessarily true. In 2008, the Environmental Working Group (EWG) tested 10 popular brands of water in the United States, discovering an average of eight contaminants in each brand, and four brands contaminated with bacteria. Because bottled water producers are not required to make testing results available to the public, consumers have no way of knowing if their brand of bottled water is safe or not. Additionally, according to the EWG, bottled water costs 1,900 times the cost of tap water.

Despite these concerns, bottled water remains an essential life-sustaining supply in areas of water scarcity or for use after natural disasters, when public water supplies may be destroyed or contaminated. In addition, severely ill individuals, such as AIDS patients, may be particularly sensitive to chemicals in tap water that healthy individuals can process, and choose to consume water purified through reverse osmosis.

See Also: Chlorination By-Products; Environmental Protection Agency (U.S.); Groundwater; Reverse Osmosis; Supplying Water; Water Scarcity; Waterborne Diseases.

Further Readings

Clarke, Tony. *Inside the Bottle: An Exposé of the Bottled Water Industry*. Ottawa, Canada: Canadian Centre for Policy Alternatives, 2007.

Great Lakes–Upper Mississippi Board of State and Provincial Public Health and Environmental Managers. "Recommended Standards for Waterworks." Albany, NY: Health Research, Inc., 2007.

Naidenko, Olga, Nneka Leiba, Renee Sharp, and Jane Houlihan. "Bottled Water Quality Investigation: 10 Major Brands, 38 Pollutants." Environmental Working Group, October 2008. http://www.ewg.org/node/27010/related/clip (Accessed February 2010).

Michelle Edith Jarvie
Independent Scholar

C

CAFFEINE

Caffeine, most commonly found in the form of coffee bean or tea leaf infusions, has been used by humans since the Paleolithic period. Caffeine is a part of many pharmacological repertoires around the globe because it is easily harnessed from specific plant products and has known effects on the central nervous system. The industries producing two of the most common sources of caffeine, coffee and tea, are both parts of sustainable development programs in many areas of the developing world.

Chinese legend attributes the discovery of tea by accidentally boiling some tea leaves to Emperor Shennong in 3000 B.C.E. An Ethiopian goatherd is said to have discovered the effects of coffee beans in the ninth century when his goats reportedly seemed to be stimulated after chewing on coffee beans. The kola nut, another common source of caffeine, has been chewed by West African people for increased energy and appetite suppression. Cocoa, also a good source

A tea plantation in Tzaneen, South Africa. The tea industry is making an effort to address the harmful effects of tea production, which include soil erosion, water pollution, and deforestation resulting from the harvesting of firewood to dry tea leaves.

Source: Wikipedia/Rob Hooft

of caffeine, has been commonly used since 600 B.C.E. by the Mayan culture in the Americas, where it was the main ingredient in the traditional beverage *Xocoatle*. Indigenous groups in the Americas used yaupon holly leaves and stems to make a type of infused beverage that also would have had the characteristic effects of the caffeine found in yaupon holly. By the 17th century, "Arabian wine," a type of coffee, was common in the Ottoman Empire. German chemist Friedrich Ferdinand Runge discovered the chemical compound of caffeine in 1819.

Some religious sects regulate the consumption of caffeine, including the Church of Jesus Christ of Latter-day Saints, Seventh Day Adventists, Church of God (Restoration), and Christian Scientists. Hindus may also abstain from caffeine consumption because it can have effects on consciousness, and most Muslims view caffeine as an acceptable substance when used in moderation.

Caffeine is a psychoactive, stimulant drug that is a bitter, white, crystalline alkaloid that acts on the central nervous system and is a diuretic to which the body can develop a tolerance. Caffeine is part of other alkaloid complexes that have similar properties and attributes, including guaranine (from guarana), mateine (from mate), and theine (found in nonherbal teas). It was previously thought that theine was distinct from caffeine, but scientists now concur that theine is identical to the caffeine found in coffee plants. Caffeine is found in many plants and their fruit, including kola nut (used to make cola soda beverages), coffee berries, tea leaves, yerba maté, guarana berries (common in many energy drinks), yaupon holly, and chocolate (from the cocoa bean). Caffeine is extracted to produce a decaffeinated product; the caffeine that is removed can be used to produce caffeine tablets (used to decrease tiredness and to treat apnea in newborn babies) or can be added to hair and skin care products (since some caffeine may be absorbed through the skin). Caffeine is a natural pesticide that is found in the leaves, beans, fruits, and stems of some plants. It also acts as a natural herbicide because it is found in the soil around some plants, thereby decreasing the need for chemical herbicides.

Coffee and tea are the most common sources of caffeine. Coffee is grown in plantations and forests. The caffeine content varies by the variety, processing, and preparation. The process of roasting dry coffee beans decreases the caffeine content in the final beverage, so darker roast coffee beans have lower caffeine levels, generally, than light roast coffee. Coffee is a traditional plant that can be successfully used in sustainable agricultural models, particularly the shade-growing techniques of coffee farming. Coffee is a common product in many agroforestry programs, as well. Agroforestry blends traditional land-use strategies centered on trees, shrubs, and wildlife to farm two or more outputs that can be harmoniously derived from the system. Shade-grown coffee is part of traditional and polyculture farming systems. In some semitropical climates, shade-grown coffee is associated with higher levels of biodiversity and conservation, making coffee an ideal crop for sustainable development programs.

Tea comes from the plant *Camellia sinensis* and is grown on estates or in gardens. The same plant, *Camellia sinensis,* is used to produce black, green, and other types of nonherbal teas. The plant can be grown in tropical and subtropical climates and is grown in many regions of the globe, including China, India, Sri Lanka, Kenya, Indonesia, and Turkey, among other locations. Tea leaves have higher caffeine content than coffee beans, but because of the way that tea is usually prepared and served, the common serving of tea has less caffeine than a typical cup of coffee. The growing conditions and method of processing (how the leaves are dried) have an impact on the caffeine content of tea. In addition to caffeine, tea contains theobromine and theophylline, which have similar effects as caffeine.

Some common issues with the production of tea include soil erosion, overuse of limited water supply, pollution of water supplies, and deforestation caused by harvesting of firewood to dry the tea leaves. Strong efforts in the tea production industry have been made to rectify many of the human and environmental problems often affiliated with the farming and production of tea. The Rainforest Alliance is a prominent actor in the certification of tea brands that are meeting a set of standards regarding environmental impacts, working conditions, and workers' rights. Through the use of renewable resources, organic or low-impact pest management, and conservation of local area habitats, many of the negative environmental impacts can be mitigated. Social justice is prioritized through goals to provide tea estate workers and their families with safety from violence and other basic needs, including minimum wages, healthcare, education, housing, and proper nutrition.

Caffeine is known to increase frontal lobe brain activity and increase memory task performance. Caffeine tends to decrease broad-range thinking while increasing the ability to focus on single tasks. Children experience the same effects from caffeine as adults. By stimulating the central nervous system, caffeine has the ability to make an individual feel less tired, less hungry, more focused and alert, with better physical coordination and ability to have clearer thought processes. Caffeine also increases sport endurance. Individuals may develop a tolerance to caffeine, which means a decrease in the degree of the normal effects of caffeine. Some people also experience withdrawal symptoms, including headache, irritability, inability to focus, sleepiness or inability to sleep, and body and stomach aches. Caffeine is commonly used to treat tension-type headache and has been used as a traditional medicine by many indigenous groups for that purpose. It also increases the effectiveness of pain relievers by reducing headache and increasing the body's rate of absorption of pain medicine. The American Psychiatric Association's *Diagnostic and Statistical Manual of Mental Disorders* lists four categories of caffeine-related illnesses: caffeine intoxication (which can lead to death in extreme cases); caffeine-induced anxiety disorder; caffeine-induced sleep disorder; and caffeine-related disorder, not otherwise specified. There are no links between moderate caffeine consumption and other illness or diseases, such as cardiovascular disease, fibrocystic breast disease, reproductive function, birth defects, cancer, or behavioral concerns in children.

See Also: Children's Health; Dehydration; Men's Health; Mental Health; Women's Health.

Further Readings

American Psychiatric Association. *Diagnostic and Statistical Manual of Mental Disorders: DSM-IV*. Washington, DC: American Psychiatric Association, 1994.
Bolton, Sanford and Gary Null. "Caffeine: Psychological Effects, Use and Abuse." *Orthomolecular Psychiatry*, 10/3 (1981).
Escohotado, Antonio. *A Brief History of Drugs: From the Stone Age to the Stoned Age*. Rochester, VT: Park Street Press, 1999.

Elizabeth A. Olson
University of California, Merced

CALIFORNIA'S GREEN CHEMISTRY INITIATIVE

In response to the societal pressures presented by the growing healthcare burden of toxic chemicals and pollution-related diseases, the state of California launched California's Green Chemistry Initiative (CGCI) in April 2007. The initiative was designed as a set of legislative tools aimed to prevent health problems and environmental hazards associated with toxic chemicals used in consumer products. The initiative's goal is to develop a comprehensive program for managing chemicals used in consumer products. The innovative approach of the initiative lies in developing policies shifting the spotlight from the point of discovering a problem presented to human and environmental health by hazardous substances used in consumer goods to the initial development stage of designing and manufacturing the chemicals destined to be used in everyday products. This approach strengthens the pollution-prevention branch of environmental regulation.

Over 40 billion pounds of different chemicals for use in a multitude of processes and products are produced in or imported to the United States daily. The global production of chemicals is estimated to triple by 2050. During the life cycle of these products, from manufacture to end-of-life disposal, they come into contact with people in the workplace and in homes through direct as well as indirect access by means of air, water, food, and waste streams. Sooner or later, they enter the Earth's ecosystem in different forms and move on to contaminate potable water supplies, accumulate in the animal food chain, and subsequently affect human health. Harmful chemical exposure is especially detrimental during the most sensitive periods of human development, such as fetal and early childhood, when chemicals may enter the developing organ systems of fetuses and infants through a mother's bloodstream and through breast milk. For tens of thousands of chemicals used in commercial manufacturing, the potential health effects are poorly understood. However, there are examples of scientifically proven connections between toxic substances found in consumer products and human health risks. Lead, which can find its way to children through paint on toys and lunch boxes, is known to cause mild retardation in child development, birth defects, hypertension, arteriosclerosis, cardiac arrhythmia, and adverse effects on reproductive and endocrine systems. Arsenic, used in wood products, may contribute to blood vessel damage, skin changes, and reduced nerve function. Mercury, found in some light bulbs, causes impaired neurological development in fetuses and young children.

By launching the Green Chemistry Initiative, the state of California is shifting toward taking a proactive stand in identification of the hazardous potential of chemicals at the initiation of design and production and evaluating environmental impact throughout the life cycle until final disposal—an approach known as cradle-to-grave. Under scrutiny will be all chemicals used in consumer products produced in or imported to California. To implement the Green Chemistry Initiative, the Department of Toxic Substances Control (DTSC) conducted a broad public process to gather ideas and recommendations for the initiative's implementation mechanisms. To obtain scientific and technical advice, the department formed the Green Ribbon Science Panel composed of members from the fields of science, engineering, manufacturing, law, public health, and risk analysis. The DTSC had also been mandated by the initiative to compile a publicly accessible toxics information clearinghouse for the collection, maintenance, and distribution of data on chemical hazard characteristics and their environmental and toxicological end points.

These legislative bills are stipulated in California law (California Health and Safety Code Section 25251-25257.1). An important component of these pieces of legislation is the setting of guidelines for the process of evaluating and regulating chemicals by the DTSC. These guidelines are (1) to evaluate chemicals with a significant adverse impact on public health, (2) to consider a chemical's entire life cycle, (3) to mandate a proposed alternative chemical, (4) to evaluate the priority of the volume of substances used in the state of California and their potential for consumer exposure and effects on infants and children, (5) to include worker safety assessment in public health impact, (6) to consider emissions of air pollutants, ozone forming compounds, particulate matter, toxic air contaminants, and greenhouse gasses, and the contamination of surface water, groundwater, and soil in environmental impact evaluation, and finally, (7) to perform a life cycle assessment including product function, performance and useful life, energy efficiency, resource consumption, water conservation, production energy inputs, and economic impacts. After conducting this comprehensive evaluation for each chemical of interest, DTSC will have to issue recommendations that would result in one of the following: (1) no action, (2) require additional information, (3) require labeling, (4) restrict use, (5) prohibit use, (6) add manufacturer responsibility at the end of life, including recycling or disposal, or if no safer alternative to the chemical exists, and (8) add a requirement that manufacturers fund a green chemistry challenge grant.

Limited only to consumer products, these new pieces of legislation fail to distinguish between substance, preparation, or article. It is not clear what exactly "chemicals of concern" would be; also, through the regulatory process, a heavier burden is placed on the DTSC than on producers. An enormous challenge lies in identifying alternatives and determining how safe they are. Nonetheless, as a proactive development designed to eliminate toxic substances from everyday use and to highlight green materials, the hope is that California's Green Chemistry Initiative will become a dynamic regulatory vehicle that is improved and refined through the democratic process of rule making and emulated on the federal level for creating a unified enforceable policy aimed at phasing out hazardous chemicals and preventing environmental and public health problems nationwide.

See Also: Cost-Benefit Analysis for Alternative Products; Green Chemistry; Industrial Ecology; Phaseout of Toxic Chemicals.

Further Readings

Assembly Bill 1879. http://www.leginfo.ca.gov/pub/07-08/bill/asm/ab_1851-1900/ab_1879_bill_20080911_enrolled.html (Accessed May 2009).
Centers for Disease Control and Prevention (CDC). "Third National Report on Human Exposure to Environmental Chemicals." Atlanta, GA: CDC, 2005. http://www.cdc.gov/exposurereport/report.htm (Accessed May 2009).
Senate Bill 509. http://info.sen.ca.gov/pub/07-08/bill/sen/sb_05010550/sb_509_cfa_20080122_101428_sen_floor.html (Accessed May 2009).
Wilson M., D. Chia, and B. Ehlers. *Green Chemistry in California: A Framework for Leadership in Chemicals Policy and Innovation.* Special Report to the California Senate Environmental Quality Committee and the Assembly Committee on Environmental Safety and Toxic Materials, March 2006. http://coeh.berkeley.edu/greenchemistry (Accessed May 2009).

Natalie Milovantseva
University of California, Irvine

CALORIE LABELING FOR RESTAURANTS

Calorie labeling for restaurants has been gaining attention in the public and legal spheres since New York City became the first municipality to adopt regulations requiring menu labeling in 2006. Since then, many other local and state municipalities have adopted or are contemplating similar laws. Furthermore, purposed federal menu labeling legislation is under review at the time of this writing. Regulations for menu labeling vary by location but generally require restaurants to post calorie information for standard items on the menu board, printed menu, individual food tag, drive-through, or some combination of these. Current legislation applies only to chain restaurants, for which two working definitions have been established: restaurant chains with 15 or more locations nationally or restaurant chains with 20 or more locations nationally.

Calorie labeling was introduced as a tool for helping Americans make healthier eating choices and preventing and reducing obesity. As of 2005, nearly one-third of Americans (72 million) were considered obese. Obesity has been linked to higher risk of chronic diseases such as diabetes, hypertension, and cardiovascular disease, leading to a diminished quality of life, ongoing health concerns, and higher rates of mortality.

Americans eat more food away home today than ever before. Studies estimate that Americans consume approximately one-third of their calories away from home. Since this is a significant portion of many Americans' diets, positive changes could result in significant progress toward a healthier population. Providing caloric and other nutritional information about restaurant food enables consumers to make informed and educated decisions that ideally would lead to healthier eating habits. Healthier eating would, in turn, reduce obesity and chronic disease and help Americans live longer and healthier lives.

New York City was the first municipality to enact menu-labeling laws in December 2006. The New York Board of Health adopted new regulations that amended health code §81.50. The regulations were initially challenged by the New York State Restaurant Association, deemed as preempted based on an existing federal law, revised and again adopted on March 21, 2008. The regulations adopted in 2008 require food service establishments in New York City with 15 or more locations nationally to list calories of standard menu items on the menu board, printed menu, or food tag. Enforcement of these regulations began in May 2008.

The next jurisdiction to adopt menu label regulations was King County (which encompasses Seattle), Washington, in 2008. This was followed by Multnomah County, Oregon, and Westchester County, New York. Next, California became the first state to enact menu-labeling legislation. Oregon, Maine, and Massachusetts followed suit and adopted similar laws soon after. Many other states and municipalities have introduced menu-labeling legislation.

Two proposed menu-labeling acts were introduced in Congress in 2009: the Menu Education And Labeling act (MEAL) and the Labeling Education And Nutrition Act (LEAN). MEAL was originally introduced in Congress in 2003 and would require chain restaurants with 20 or more outlets to provide nutritional information including number of calories and amount of saturated fat, trans fat, and sodium for each standard menu item. Restaurants would be required to list the number of calories on menu boards as well. This act was not passed in the 2003–2004 congressional session. It was reintroduced in the following two sessions and again failed to pass. In 2009, MEAL was introduced again. At the time of this writing, the MEAL Act is still under consideration by Congress. The LEAN Act would similarly apply to chain restaurants with 20 or more outlets nationally. However, restaurants would not be required to list calories on the menu board; instead, restaurants would be required to have caloric information available in the restaurant. Possible locations for caloric information include the menu, menu board, a sign meeting certain requirements, or as a supplement to the menu. Furthermore, LEAN would preempt any current or future local or state laws. LEAN is currently under consideration by Congress as well.

There are a number of criticisms regarding required menu labeling at restaurants. First, there is an argument that consumer behavior will not change despite additional information. Another argument is that proposed legislation is not comprehensive enough to make any difference in the health of the U.S. population. Menu-labeling laws apply only to chain restaurants with 20 or more locations, which comprise only 10 percent of U.S. restaurants. Hence, some believe that requiring menu labeling is not comprehensive enough to make a difference in the eating habits of Americans.

Proponents of menu labeling claim that it will encourage healthier eating habits in Americans. They also posit that restaurants will offer healthier options or make current options healthier, since the caloric information is readily visible and could discourage consumers from ordering certain items.

The outcome of calorie labeling on consumer choice is still unclear. Because laws requiring calorie labeling are relatively new, there has not yet been adequate time to conduct studies on their effect.

See Also: Fast Food; Fast Food Warnings; Government Role in Green Health; Obesity; Personal Consumer Role in Green Health; Physical Activity and Health.

Further Readings

Center for Science in the Public Interest. "Nutrition Policy." http://www.cspinet.org/menu labeling/index.html (Accessed July 2009).

Ludwig, David S. and Kelly D. Brownell. "Public Health Action Amid Scientific Uncertainty: The Case of Restaurant Calorie Labeling Regulations." *Journal of the American Medical Association*, 302/4 (2009).

U.S. Department of Agriculture. "America's Eating Habits: Changes and Consequences." http://www.ers.usda.gov/Publications/AIB750 (Accessed July 2009).

Amanda Wysocki
University of Washington

CANCERS

What is the connection between cancer incidence and the environment? What is the relative role of nature versus nurture, or genetic inheritance or environmental milieu, in cancer incidence? While both factors play an important, often perversely synergistic role in the development of cancer, approximately two-thirds of all cancer occurrences are related to environmental factors. The percentage of cancer cases attributed to environmental factors would be even higher were modifiable lifestyle choices included. What are the two major environmental culprits? Tobacco smoking and food consumption. More than half of all cancers are linked to a history of smoking tobacco or obesity (from a combination of food consumption and a sedentary lifestyle). Other environmental exposures (e.g., to metals and industrial chemicals in the environment) can cause certain types of cancer.

Humans suffer from more than 100 types of cancer. Cancers first emerge inside a human cell. When the body is functioning properly, it occasionally needs to produce more cells. It does so by allowing old cells to die and younger cells to divide and form new cells. When cancer emerges, the cell regulatory mechanism goes awry and cells continue to divide incessantly, ultimately leading to a tumescent mass. A difficulty in preventing and diagnosing cancers is that it may take many years for a tumor to grow sufficiently for detection. Lung cancer patients, for example, do not develop detectable cancer until 20, 30, or more years after starting to smoke. While there remains much to learn about environmental causes of cancer, clear patterns have emerged.

Cancer Patterns

The annual rate of newly reported cancer cases increased during the 1970s and 1980s yet has been slowly decreasing since the mid-1990s. However, this general pattern conceals diverse trends among diverse types of cancers: cancers of the digestive system, from the mouth and throat to the stomach and colon, have decreased, while lung (particularly for women), bladder, prostate, kidney, liver, brain, lymphomas, and melanomas have increased.

Environmental Causes

Because of the complex interaction with endogenous factors, it remains impossible to predict whether exposure to a certain environmental toxin will cause a given person to develop cancer. We can identify genetic and environmental characteristics that increase cancer risk. Over 100,000 chemicals are employed commonly in American households in the form of cleaners, food additives, solvents, pesticides, and a host of other products. Each year, approximately 1,000 more are added to the list. Complicating the matter further, as single substances they may have myriad effects on cancer pathways; when multiplied ensemble, the potential concatenations of cancer-causing etiologies is nearly limitless. Additionally, numerous natural products can promote cancer. While in vitro studies describe potential carcinogens in humans, epidemiological designs following subjects over time to observe potential cause and effect patterns are the most reliable way to research cancer processes.

Further complicating cancer epidemiology, cancers emerge over many months, and often years, and arise from a complex interaction of endogenous (or genetic) and exogenous (or environmental) factors. There may be little we can do to control endogenous sources themselves, but given that the majority of cancers are related to environmental factors, limiting exposure to these sources can greatly decrease morbidity and mortality from a large range of cancers. In addition to smoking and poor diets, exposure to cancer-inducing viruses and bacteria, drugs, hormones, radiation, and food, air, or waterborne chemicals all increase the risk of developing cancer.

How do we know what portion of cancer cases, and under what circumstances, we can attribute to environmental factors? Although research remains limited, there is a lot we do know. A notable effect of smoking having decreased in the United States in recent decades is the concomitant decline in rates of cancers of the lung, bladder, mouth, colon, kidney, throat, voice box, esophagus, lip, stomach, cervix, liver, and pancreas. Another way scientists can identify carcinogenic sources is by observing how groups who switch from one diet to another have markedly different cancer rates. People who have migrated from Asia to the United States, for example, experienced low rates of several cancers, most notably prostate and breast cancer in their native countries. Following migration to the United States, prostate and breast cancer soared among these migrant populations to ultimately match levels observed in the general U.S. population. Conversely, stomach cancer rates among these groups have declined. In both cases, environmental pathways explain the differential patterns. To the extent that fat- and sugar-rich, fiber- and nutrient-poor diets in the United States replace traditional Asian diets, thus increasing obesity, breast, and prostate cancer rates also increase. Stomach cancer, however, declines in such an instance as *Helicobacter pylori* bacterial infections, although they are linked to higher colorectal cancer rates and are less common in the United States.

Carcinogens

Tobacco

Approximately one-third of all U.S. cancer deaths yearly are caused by direct and indirect environmental exposure to tobacco smoke. Tobacco exposure increases the risk of a wide range of cancers: lung, mouth, bladder, colon, kidney, throat, nasal cavity, voice box, esophagus, lip, stomach, cervix, liver, and pancreas, as well as leukemia. Cigarette smoke releases over 100 carcinogens into the environment. A synergistic effect is observed whereby mouth, voice box, and esophagus cancer incidence is exacerbated among smokers who also consume at least two drinks/day; similar synergies are documented for smoking and radiation exposure in exacerbating lung cancer incidence.

Diet/Weight/Physical Inactivity

Numerous studies have corroborated that high consumption of red and processed meats increases colorectal and stomach cancer incidence. Conversely, higher and more diverse consumption of fruits and vegetables decreases a host of cancers, particularly esophageal, stomach, and colorectal cancers. Obesity is highly correlated with cancers of the breast, kidney, esophagus, and colon. Independent of obesity, sedentary lifestyles increase colon and breast cancer risk. Ensemble, inactivity and obesity are linked to one-quarter or more of breast and colon cancers.

Alcohol Consumption

The consumption of more than two drinks daily is linked to an increased risk of cancer, particularly among smokers. Cancers particularly connected to heavy drinking are mouth, throat, voice box, liver, and esophageal. However, emerging evidence suggests that the type of alcohol matters. Some studies, for example, have shown a notably higher risk of certain lymphomas among high drinkers of spirits than among red wine drinkers. Naturally occurring compounds present at relatively high levels in red wine, such as quercetin and resveratrol, may be responsible for the relatively lower cancer incidence rate among red wine drinkers.

Viruses, Bacteria, and Fungi

Viruses and bacteria promote the emergence of several cancers. Among the clearest links researched is the sexually transmitted virus, human papillomavirus (HPV), as the main cause of cervical and anal cancer. Other known virus–cancer links are hepatitis B

(HBV) and hepatitis C (HCV) as major determinants of liver cancer. A virus for which most U.S. adults are carriers, Epstein-Barr virus (EBV), known commonly as mononucleosis, is linked to certain types of lymphoma, including a several-fold increased incidence of Hodgkin's lymphoma within several years of suffering EBV symptoms. Similarly, Kaposi's sarcoma–associated herpesvirus (KSHV), or human herpesvirus 8 (HHV-8), is linked to Kaposi's sarcoma. Lastly, *Helicobacter pylori* bacteria infection, the primary cause of gastritis and peptic ulcers, is a precursor to stomach cancer.

Some species of fungi foment aflatoxin growth on food, which can lead to liver cancer. Grains and peanuts house aflatoxin-producing fungi. Meat and dairy products can also be contaminated through aflatoxin-contaminated feed given to livestock and dairy animals. Agricultural laborers, particularly in the developing world where screening is lax, are particularly at risk through inhaled airborne dust particles.

Prescription Drugs

Research is increasingly encountering links between prescription drugs and cancer, including, perversely, drugs used to treat cancers. Some research suggests an increased risk of breast and endometrial cancers among users of estrogen and synthetic estrogen-like substances, consumed to treat postmenopausal conditions. When used as a contraceptive, synthetic estrogen-like products have also been implicated in cervical and vaginal cancers. However, estrogen use is also associated with a reduced incidence of colon cancer. It is therefore unsurprising that Tamoxifen, a synthetic hormone that blocks estrogen, is used to treat breast cancer recurrence or as a preventative measure for women with high risk for contracting breast cancer.

A host of drugs employed on the front line against cancer increase the risk of second cancers. Synergistic effects leading to multiplicative increased risks for secondary neoplasms are observed with certain combinations of chemotherapy and radiotherapy. Drugs used as immunosuppressants, such ascyclosporin and azathioprine for patients having organ transplants, are related to increased cancer incidence, especially lymphomas. Conversely, some over the counter medicines have been associated with reduced risk of certain cancers, such as aspirin and colon cancer.

Radiation

Atmospheric, invisible, and odorless, radiation is constantly damaging our DNA. Most of these damaging effects are repaired by complex intracellular mechanisms. When they are not, cancers sometimes result. Ultraviolet (UV) radiation from the sun causes premature skin aging and DNA damage that can lead to melanoma and other skin cancers. Skin cancer incidence is increasing quickly. However, recent research suggests that the overall incidence of cancer may be lower among people with regular exposure to solar radiation than among those with very low exposure. The latter is associated with low levels of vitamin D, which is increasingly linked to higher rates of a host of cancers. Research on links between vitamin D deficiency and breast cancer incidence is particularly convincing and has led to promising results with vitamin D supplementation in clinical trials.

While we are all exposed to small amounts of cosmic radiation (which may account for up to 1 percent of our total cancer risk), radioactive gas from soil sources such as radon (resulting from uranium decomposition) naturally emits ionizing radiation. Radon gas enters homes through underlying soil and bedrock through cracks and fissures in the home's foundation. Approximately 5 percent of all U.S. homes are exposed to measurable levels of radon. Approximately 20,000 lung cancer deaths yearly are linked to radon exposure. Another source of radiation is from radioactive substances released by medical procedures, nuclear weapons, or energy sources, leading to a wide range of cancers, including breast, thyroid, lung, and stomach. While few people today are exposed to most of these sources, irradiative medical treatment is becoming more common. Since radiation effects are multiplicative over time, people treated for acne, ringworm, or cancer with radiation as children suffer particularly high rates of diverse cancers as older adults.

Industrial Chemicals

Benzidine was among the first chemicals found to cause cancer in humans. Higher bladder cancer rates were observed in the first half of the 19th century among textile and paper dye production workers. Dyes metabolize to benzidine in vivo, but are hazards primarily in close proximity to dye and pigment plants, particularly where waste is discharged.

A handful of solvents used as paint thinners, grease removers, and in the industrial dry cleaning industry have been associated with animal cancers. One of them, benzene, causes leukemia in people. Used widely as a solvent and a component of gasoline, it was banned in 1997 in the U.S. for use in pesticides. However, the chemical remains ubiquitous in a host of industrial products where people are principally exposed through inhalation. Present in high quantities around gas stations and in cigarette smoke (half of all benzene exposure in the United States is estimated to be through cigarette smoke), about half of the U.S. population is exposed to benzene in industrial solvents and virtually all of us are exposed through gasoline. Diesel exhaust particles are also likely carcinogens; elevated lung cancer rates are found among people exposed to unusually high levels of diesel exhaust, such as train attendants, miners, mechanics, bus drivers, and truckers.

Exposure to certain fibers and dust is omnipresent in industrial settings and is linked to elevated cancer risk. Elevated rates of mesothelioma, a rare lung and abdominal cancer, have been definitively linked to asbestos exposure, which accounts for the largest portion of occupation-related cancers. Further, ceramic fibers used as insulation materials in lieu of asbestos also cause lung cancer in animals. Wood and silica dust, linked to lung cancer and cancers of the nasal cavities and sinuses, are known carcinogens for unprotected workers who are exposed regularly from sanding operations and furniture manufacturing.

Dioxins emerge in the environment as unintended by-products of chemical processes in paper and pulp bleaching. They are also present in some insecticides, herbicides, and wood preservatives. Dioxins accumulate in human fat tissue. Increased cancer-related death rates are observed in people working in related industries. Similarly, lung, skin, and urinary cancers are higher in people exposed to polycyclic aromatic hydrocarbons (PAHs). PAHs are used in the plastics industry to produce containers, film wrap, sundry housing products, and credit cards. PAHs are produced by burning wood and fuel in homes and are also present in gasoline and diesel exhaust, cigarette smoke, and charcoal-broiled foods. Vinyl chloride is a colorless gas related to liver, brain, and lung cancers, and angiosarcomas (tumors of the blood vessels). While proximity to a plastics plant increases airborne exposure, risk to people as end users of the product appears low to absent.

Metals

Arsenic, found in various levels in drinking water, is related to skin, lung, bladder, kidney, and liver cancers; cancers are particularly linked to higher levels of exposure, such as found in mining and copper smelting and in herbicide production. Beryllium compounds, used in a wide range of industries including fiber optics and cellular networks, aerospace, and defense industries as an ingredient in glass and plastics, dental applications, and sports equipment, causes lung cancer as observed in significantly higher morbidity among workers in beryllium production industries. People outside beryllium-related industries are exposed mainly through coal and fuel oil combustion and by inhaling or consuming trace beryllium residues in air and food. Another cancer-causing metal is cadmium, used as a coating to prevent corrosion, in batteries, and in fungicides, although food remains the primary source of human exposure to cadmium. Smoking, likely vis-à-vis contaminated topsoil, represents the primary nonoccupational exposure to cadmium. Other cancer-causing metals include chromium compounds, used to prevent corrosion in steel and popular in stainless steel products; lead acetates and phosphates used in coating metals and in inks, paints, hair dyes, and poison ivy treatments; and nickel, related to cancers of the nasal cavity and lungs, and ubiquitous in the air, water, and soil due to its wide use in steel, dental additives, magnets, and batteries.

Pesticides

While only a tiny fraction (approximately 20 of the nearly 1,000 tested) of the ingredients used in pesticides within the United States are proved to cause cancer in animals, many remained untested. As a result of pesticide links to cancer, many pesticides have been outlawed or highly regulated in the United States but remain legal in much of the world. Where particularly noxious pesticides are unregulated and farmers have frequent contact with them, skin cancers and cancer of the lip, stomach, lung, brain, and prostate are notably elevated.

Green Prevention

With so many natural and human-made carcinogens in our environment, it seems a daunting task to prevent exposure. Living a green lifestyle, however, can go a long way in preventing the majority of cancers. Since at least two-thirds of cancer cases are exogenous, that is, environmental, there is much we can do. Behavioral modification can drastically reduce the higher cancer rates associated with smoking, obesity, and excessive alcohol consumption. Most cancers can be prevented through behavioral change, limiting risk through these three avenues. To live green, reduce environmental cancer risk, and increase longevity and vitality, a handful of simple guidelines conclude this article:

- Do not smoke and avoid secondhand smoke.
- Exercise regularly.
- Shun calorie- and fat-dense and processed food in favor of the rainbow spectrum of naturally colored fruits, nuts, and vegetables.
- Drink alcohol moderately (e.g., one or two alcoholic drinks a day).
- Avoid sunlight at midday and sunburns in favor of more consistent and mild exposure in the morning and evening.
- Avoid viral and bacterial infections by engaging in safe sex, avoiding injection drugs, and acquiring vaccinations when appropriate, such as to HPV and hepatitis B.
- Check your home for radon levels.
- Avoid contact with pesticides, solvents, and industrial chemicals.

See Also: Arsenic Pollution; Chemical Pesticides; Computers and Printers (Ink); Dental Mercury Amalgams; Dry Cleaning; Hormone Therapy; Mercury Sources and Health; Organic Produce; Pest Control; Radon and Basements; Smoking; Stomach Ulcers and *Helicobacter Pylori*.

Further Readings

American Cancer Society. "Nutrition and Diet Guidelines." http://www.cancer.org/eprise/main/docroot/PED/ped_3_1x_ACS_Guidelines?sitearea=PED (Accessed January 2010).

Centers for Disease Control and Prevention, National Center for Health Statistics. http://www.cdc.gov/nchs (Accessed May 2010).

Doll, R. "Epidemiological Evidence of the Effects of Behavior and the Environment on the Risk of Human Cancer." *Recent Results in Cancer Research*, 154 (1998).

National Cancer Institute. "Cancer Library." http://cancer.gov/cancer_information/cancer_literature (Accessed May 2010).

National Cancer Institute. "Cancer Mortality Maps and Graphs." http://www3.cancer.gov/atlasplus (Accessed May 2010).

U.S. Department of Health and Human Services. "Household Products Database." http://householdproducts.nlm.nih.gov (Accessed May 2010).

U.S. Department of Health and Human Services, National Institutes of Health, National Cancer Institute, National Institute of Environmental Health Sciences. "Cancer and the Environment: What You Need to Know. What You Can Do." http://www.cancer.gov/images/Documents/5d17e03e-b39f-4b40-a214-e9e9099c4220/Cancer%20and%20the%20Environment.pdf (Accessed May 2010).

U.S. National Library of Health, National Institutes of Health. http://www.nlm.nih.gov (Accessed May 2010).

World Health Organization. "Diet, Nutrition and the Prevention of Chronic Diseases." http://www.who.int/hpr/nph/docs/who_fao_expert_report.pdf (Accessed May 2010).

David L. Carr
Independent Scholar

CARBON FILTERS

Carbon filters or charcoal filters have been used for several centuries and are still considered one of the most efficient and oldest methods of water purification. Historians have shown evidence that carbon filters may have been used in ancient Egyptian cultures for both air and water decontamination.

Carbon filters are being widely used in the healthcare and industry fields to control the level of contamination in liquid and gaseous materials. Carbon filters remove impurities and contaminant particles or clusters using the chemical adsorption method. An activated carbon or activated charcoal filter uses carbon that has been processed to improve adsorption by creating a number of large internal compartments that increase the surface area and create a maze-like structure that improves the probability of attracting the contaminants.

It has been shown that one kilogram of activated carbon can contain a surface area of nearly 2 square kilometers.

Activated carbon is superheated in a controlled environment; in the absence of oxygen, which alters the material by creating tiny nanometer-wide cracks, the result is an immense surface area that attracts gaseous contaminants through a process called adsorption, which is the key in the filtration process. Carbon has a natural affinity for organic pollutants like benzene, which bind to its binding sites, distributed in the surface of its internal compartments. When a material adsorbs something, it attaches to it by chemical attraction, and the huge surface area of activated carbon provides countless bonding sites in which chemicals may be attached to the surface and are trapped. The major force in the trapping interaction is the van der Waals forces, which are relatively weak bonds compared to chemical reactions.

Activated carbons (charcoals) are suitable for trapping other carbon-based impurities, as well as other compounds like chlorine. Chemicals such as sodium, nitrates, and so on, which are not attracted to the trapping sites of the activated carbons, will pass through the carbon filters with no adsorption, so an activated carbon filter acts selectively by removing certain impurities while screening others. It should be noted that once all of the trapping sites are filled, the activated carbon filter becomes saturated and stops working. At this time, the filter must be replaced with a fresh activated filter. The speed of saturation highly depends on the type and concentration of contaminants, as well as on the flow of the carbon filter.

Carbon is usually activated with a positive charge and is designed to attract negatively charged contaminant particles or clusters. Since carbon filters usually have a positive charge, they are suitable for removing negatively charged particles such as chlorine, sediment, and volatile organic compounds (VOCs) from water. However, activated carbons are poor for removing minerals, salts, and dissolved inorganic compounds. Carbon filters can filter particles or clusters ranging from 0.5 to 50 micrometers and are categorized and described based on the range of the particle size that they are able to filter. The efficacy of a carbon filter is based on the flow-rate regulation. When water or air is allowed to flow through the carbon filter at a slower rate, the contaminants are exposed to the internal compartments for a longer time and the chance of trapping increases as the passage of time increases.

Some manufacturers produce as many as 150 types of activated carbon filters, which differ in characteristics such as density and pore size. Carbon filters are selected according to their usage: a carbon filter with large holes would be best at picking up heavy organic chemicals, while smaller pores would catch the lighter pollutants. Different types of carbon filters can be constructed by different methods of activations, or they can come from different sources of materials. Some are made from coal, wood, or sawdust, while others are made from peach pits, olive pits, or coconut shells.

Carbon filters are commonly used for water purification, but they are also used in air purifiers and are mainly used for filtration of hydrocarbons, organic vapors, some acids, esters, sulphur compounds, hydrogen compounds, solvents, and odors, as well as in air decontamination and radiation measurement and detection.

There are two major types of carbon filters used in the filtration industry: powdered block filters and granular activated filters. In general, carbon block filters are more efficient in filtering the highly concentrated contaminants. These filters rely mainly on the increased surface area of their internal compartments. In addition, some carbon filters use a secondary medium, such as silver or KDF-55, to prevent the growth of bacteria and molds within the filter compartments.

Carbon filters are used in individual homes as point-of-use water filters and, sporadically, in municipal water treatment facilities. In medicine, they are also used as pretreatment devices for reverse-osmosis systems and as specialized filters designed to remove chlorine-resistant cysts, such as giardia and cryptosporidium.

See Also: Supplying Water; Tap Water/Fluoride; Water-Borne Diseases.

Further Readings

Bansal, Roop Chand and Meenakshi Goyal. *Activated Carbon Adsorption.* Oxfordshire, UK: Taylor & Francis, 2005.
Environmental Protection Agency. "A Citizen's Guide to Activated Carbon Treatment." (December 2001). http://www.epa.gov/tio/download/citizens/activatedcarbon.pdf (Accessed August 2010).
Sutherland, Ken. *Filters and Filtration Handbook*, 5th ed. Oxford, UK: Elsevier Science, 2008.

Hassan Bagher-Ebadian
Independent Scholar

CARDIOVASCULAR DISEASES

Cardiovascular disease, including coronary heart disease (CHD), rheumatic heart disease, and hypertensive heart disease, are public health concerns throughout the world. Globally, more than 17 million people die of cardiovascular disease every year, with over 80 percent of those deaths occurring in low- and middle-income countries. Heart disease is the most common cause of death in the United States. In 2006, of the 631,636 total deaths recorded in the country, 26 percent were caused by heart disease. Many of the risk factors for cardiovascular disease are behavioral and include diet, exercise, and tobacco use. There is great interest in the scientific and medical communities to identify ways the environment relates to the risk factors for, and incidence of, heart disease. For instance, quality of diet may be affected by the availability of healthy versus unhealthy foods in a person's neighborhood, and people's activity levels may be related to how easy it is for them to integrate exercise into their daily lives.

The Epidemiology of Coronary Heart Disease

CHD, the most common type of heart disease, is caused by the failure of the blood vessels supplying the heart muscle due to fatty acid deposition, calcification, and subsequent plaque formation within the linings of the coronary arteries. This plaque leads to narrowing of the artery opening, thus reducing the amount of blood flow to the myocardium, followed by symptoms of chest pain and angina.

Culture determines the prevalence of individuals with elevated blood lipids, the total outcome of risk frequency, and the potential for prevention of coronary heart disease. People with diets composed of significant fatty acids, vegetable protein, cholesterol, complex carbohydrates, caloric excess, and a high intake of sodium are at the greatest risk for heart disease. Tobacco use also is associated with the increased risk of heart disease. Excessive caloric intake also can influence human health by metabolic maladaptations of obesity, hyperlipidemia, hyperinsulinemia, and hyperuricemia (gout).

Atherosclerosis (colloquially called "hardening of the arteries") is four times more frequent in males than in females. The risk of sudden death will vary with the extent of the disease present along with the degree of impairment of ventricular function. Death will probably result from ventricular fibrillation. The most common complication of coronary disease is myocardial infarction. Within the United States, 3 million people will have an infarction, with most deaths occurring within 30 to 60 minutes after the start of symptoms.

Coronary artery disease (CAD) is responsible for nearly half of all deaths, and one-third of all deaths in people between the ages of 35 and 65 within the United States. The coronary heart, or ischemic heart disease, patient suffers from what is called pump failure, which is a decline in the contraction force of the ventricles. Associated risk factors include hypertension, hyperlipidemia, diabetes mellitus type 2, smoking, alcohol consumption (very heavy), obesity, sedentary lifestyle, hormone replacement therapy, and psychosocial factors. Typical risk factors for coronary heart disease that are nonmodifiable are genetics, race (blacks more than whites), age, male gender, and diabetes type 2.

Chronic Heart Failure

The initial loss of function of the myocardial cells places an abnormal burden on the left ventricle, causing an overload that may, over time, not be tolerated by the remaining normal myocardial cells. In these patients, this scenario can lead to acute or chronic heart failure (CHF).

Chronic heart failure is the loss in the heart's ability to properly replenish the systemic or pulmonary blood volume. Right-sided heart failure leads to overflow in the systemic circulation leading to systemic venous congestion. With left-sided heart failure there is an increase in pulmonary venous congestion. The most common cause of CHF is a myocardial infarction (MI, commonly known as a "heart attack"), which results in ischemic changes of the myocardium. Certain diseases and conditions can cause cardiomyopathy (deterioration of the heart muscle) that may lead to CHF, including substance abuse and hemochromatosis (increased iron overload). Intrinsic cardiopulmonary diseases such as pulmonary hypertension and aortic valve problems (regurgitation) can lead to chronic heart failure.

More common problems include high output failure, which is the inability of the heart to meet the body's needs for oxygen and nutrients, when those needs have increased due to conditions such as anemia, beriberi, A-V fistulas, Paget's bone disease, and thyrotoxicosis, and thus are greater than the heart can supply. Dyslipidemia (abnormal levels of lipids such as fat or cholesterol in the blood) is a cause of nearly half of CHF cases, with most having a familial component. There are five types of dyslipidemias: IIA familial hypercholesterolemia; under 200–240 hypertriglyceridemia (may be familial); VLDL and chylomicron levels; HDL greater than 60, which is said to be cardioprotective; and LDL under 100 with HDL greater than 160–190, which is considered to be very high. Also, very high triglyceride levels can lead to severe pancreatitis and become a possible complication.

Hypertension

Hypertension, also called high blood pressure, is a major risk factor for heart disease. The primary definition of hypertension is a diastolic (lower) pressure of above 90 mm Hg and a systolic (upper) pressure greater than 140 mm Hg. Elevations in systolic pressure are generally thought to be less significant than the elevations in generally diastolic pressure. Ninety percent of all hypertension is of uncertain origin and is labeled essential or primary hypertension. The other 10 percent is called secondary hypertension, which is associated with various renal, endocrine, neurologic, or vascular disorders.

Primary or essential hypertension is present in about 5 percent of the adult population in the United States. Females are much more affected than males. This occurs usually within the fourth and fifth decades of life. Hypertension incidence increases with age, and as much as half of the general population over the age of 50 suffers from primary hypertension that can eventually lead to moderate to severe cardiovascular and cerebrovascular disorders. This is primarily due to the elevation of the vascular blood pressure.

The milder form of this disease is called benign hypertension, which can lead to atherosclerosis. A thickening of the inner walls of the artery and the smaller arterioles in the kidney leads to systemic hypertension.

Essential hypertension is caused by a combination of environmental and genetic factors. The genetic factor is prevalent with African Americans and is a leading cause of cardiovascular and cerebrovascular diseases within the United States.

Behavior patterns, obesity, stress, and the use of oral contraceptives are common causes of hypertension. The overuse of dietary sodium intake has a relationship to the development of essential hypertension. Americans consume eight or more grams of sodium every day; this correlates to an 8 to 30 percent incidence in hypertension.

Hypertension can cause serious disorders, such as an enlarged left ventricle due to the pressure of increased blood volume, which over time creates an elevated pressure. This volume and pressure increase is called hypertrophy. There is an increased incidence in arterioscleroses with myocardial infarcts with hypertension, even when left ventricular hypertrophy is not present. Finally, hypertensive individuals also are predisposed to thromboses or blood clots of the cerebral vessels that can lead to cerebral hemorrhage (stroke) or a cerebral vascular accident.

The formation of a hemostatic plug is started by damage to the inner lining of the vascular wall, called the endothelial lining. This leads to contraction of the vessel (vasoconstriction), contact activation of platelets with subsequent platelet aggregation, and activation of the coagulation pathways. Platelets serve two different functions. They protect the vascular integrity of the endothelium, and they initiate repair when blood vessel walls are damaged. This platelet–vessel wall interaction is called primary hemostasis. The clotted mass itself is called a thrombus; if some part of a thrombus breaks loose, it is called an embolus. This embolus may flow downstream and lodge at a distant site. The potential consequence of both thrombosis and embolism is ischemic necrosis (death due to restriction of the blood supply) of cells and tissue known as an infarction. Thromboembolic infarctions of heart and brain are dominating causes of morbidity and mortality in industrialized nations and account collectively for more deaths than those caused by all forms of cancer and infectious diseases combined.

Obesity

Obesity is an excessive amount of body fat, with a body weight of more than 20 percent of the standard weight for an individual's height and body frame. This excess fat deposit occurs because energy intake exceeds energy output; that is, a person consumes more calories than are burned in activity. Child obesity tends to run in families with either parent being overweight. These children sometimes are more overweight than their parents. Obesity is mild at 20 to 40 percent over ideal weight; moderate levels are between 41 and

100 percent over ideal weight; and severe if the person who is above 100 percent over ideal weight. Obesity is associated with many health risks, including cardiovascular disorders, diabetes mellitus, many cancers, cholelithiasis, fatty liver and cirrhosis, osteoarthritis, and premature death. In women, social factors are often associated with obesity, which occurs six times more frequently in the lower socioeconomic classes than in upper-income groups. Ethnic and religious factors also are closely linked to obesity, partly due to differences in diet and exercise behaviors.

Many factors are involved in regulating food intake, beginning with pathways regulated by preabsorptive and postabsorptive signals from the gastrointestinal (GI) tract, and changes in plasma nutrient levels provide short-term feedback to regulate food intake. GI hormones such as the peptide hormone, cholecystokinin reduces food intake, while Ghrelin increases food intake. The hypothalamus integrates various signals involved in the regulation of energy balance and activates pathways that increase or decrease food intake.

Hormonal factors, as seen in hyperadrenocorticism where corticosteroid excess leads to an increase in gluconeogenesis and a greater demand for insulin leading to an increase in lipogenesis, also are elements that must be taken into consideration in obesity. Certain metabolic abnormalities may occur in the hypothalamus of the brain, causing an increased level of fullness in the satiety center, leading to metabolic obesity.

Fat cells create a reservoir of energy that expands or contracts according to the energy balance of the organism. These fat cells come from preadipocytes to take on excess calories with the continuing increase in food intake, which increases new energy balance. The new adipocytes from the precursor cells will increase the total number of new fat cells. Throughout weight loss, fat cell numbers will remain fixed. Free fatty acids can enter the adipocytes and can again be esterified to triglycerides and stored.

Fat is distributed differently in men and women. In men, fat cells are distributed mostly in the upper body above the waist; in women, fat is distributed predominately within the lower body. This is due to the presence of testosterone and estrogen hormones and may act differently on upper and lower fat cells. The distribution as well as the amount of body fat can have health consequences: excessive body fat in the upper body is associated with an increase in the morbidity and mortality rates more than when excess fat is distributed in the lower body. Increased fat distribution in the abdominal region is associated with increased risk of cardiovascular disease, moderate to severe hypertension, and type 1 to type 2 diabetes mellitus. These conditions are more common in males than in women.

People with a blood pressure of greater than 160/95 mm Hg are three times more common in overweight individuals than those who maintain normal weight. Associated cardiovascular disease is seen with obese people, which can lead to an increased blood volume, stoke volume, and filling pressure. All of these lead to a high output failure. This failure can lead to a left ventricular hypertrophy (an enlarged left ventricle) and increase the risk of congestive heart failure.

Obesity can lead to chronic hypoxia with cyanosis and hypercapnia. These people have an increased demand for ventilation and suffer from breathing overload. Severely obese individuals often will have varicose veins and venous stasis, which can lead to thrombophlebitis and thromboemolism, which are associated with pulmonary embolism of the lungs.

Diabetes

Insulin release by the beta cells of the pancreas promotes the deposit of intracellular triglycerides by stimulating the body's glucose into fatty acids and glyceride to glycerol by maintaining adipose tissue lipoprotein lipase levels and inhibiting intracellular lipolysis.

The hormone estrogen can operate at the level of the adipocyte and can lead to an increased development of subcutaneous adipose tissue in women.

Diabetes mellitus is a disease marked by impaired insulin secretion and variable degrees of peripheral insulin resistance leading to hyperglycemia. The early symptoms related to hyperglycemia include polydipsia, polyphagia, and polyuria; later complications include cardiovascular disease, peripheral neuropathy, and predisposition to severe infections. The prognosis varies with the degree of glucose control.

Within large vessels, atherosclerosis results from the hyperinsulinemia, dyslipidemias, and hyperglycemia. These are all characteristics of diabetes mellitus. Microvascular disease leads to the three most common devastating manifestations of diabetes mellitus—retinopathy, nephropathy, and neuropathy. Impaired skin healing is also common, so that even minor breaks in the skin integrity can develop into deeper ulcers and can lead to severe infection.

Diabetic cardiomyopathy is thought to result from many factors, including epicardial atherosclerosis, hypertension and left ventricular hypertrophy, microvascular disease, endothelial and autonomic dysfunction, obesity, and metabolic disturbances. Most will develop heart failure due to impairment in left ventricular systolic and diastolic function and are more likely to develop heart failure after a myocardial infraction.

Some obese individuals may go on to develop noninsulin-dependent diabetes mellitus. The prevalence of diabetes is three times higher in overweight persons versus nonoverweight persons, and in the United States, 85 percent of those with noninsulin-dependent diabetes mellitus are obese.

Rheumatic Fever

Rheumatic fever often includes carditis (inflammation of the heart muscle), which may involve the myocardium, endocardium, and pericardium. In later years it will lead to chronic rheumatic heart disease (regurgitation, arrhythmias). Generally, fibrinous nonspecific pericarditis sometimes can cause an effusion only in people with endocardial inflammation. It usually subsides without any permanent damage within the heart. Potentially dangerous valve changes may occur (aortic, mitral). When valvitis occurs, there is an acute interstitial valvular edema. If this is left untreated, the valve may thicken, causing a fusion of the valve itself. Destruction of the leaflets can lead to a long-standing stenosis (narrowing) and/or an insufficiency (failing). Similarly, the tendons that hold the valves will shorten or thicken, leading to a regurgitation (leaking) of the valves. The valves most commonly affected by rheumatic fever are, in order, mitral, aortic, tricuspid, and pulmonic.

Although rheumatic fever may occur at any age, 90 percent of patients will have their first attack between the ages of 5 and 15 years. The most common pathogenesis is an infection by streptococcal pharyngitis (sore throat). Many people are unaware of the antecedent infection and have a negative throat culture at the time rheumatic fever is discovered. Males are more affected than females, and the incidence is higher among the poor. Most deaths from rheumatic fever occur long after the acute disease has subsided and result from endocardial involvement (the valves).

Congenital Heart Disease

Congenital heart disease is generally present in about 1 percent of North American and British populations, making this the most common category of congenital structural malformations. There are two divisions of congenital heart disease: noncyanotic and cyanotic. Patients with right to left shunts fall into the cyanotic category, whether they have readily recognizable cyanosis or not. Patients who do not have right to left shunts, even if they are not cyanotic for other reasons, such as low cardiac output, are generally placed in the noncyanotic category.

Eight percent of all congenital heart defects are known to be associated with a single gene mutation or chromosome abnormalities. The remainder are due to various other causes. Environmental factors that cause congenital heart disease are diabetes, alcohol intake, progesterone usage, certain viruses, and many teratogens. The effect of the rubella virus is probably independent of hereditary factors and can lead to patent ductus arteriosus and pulmonary artery branch stenosis. Acquired heart disease such as rheumatic fever has a much stronger environmental influence. There are 13 common congenital heart disease disorders, with ventricular septal defect as the most common and pulmonary atresia as the least.

Myocarditis

Myocarditis involves inflammatory changes of the myocardium. The most common cause is microbiologic infections and can be induced by hypersensitivity reactions, radiation therapy, and any chemical or physical agents or drugs that can induce acute myocardial fibernecrosis and a secondary inflammatory change.

Viral myocarditis is by far the most common clinical pattern of myocarditis and the one that most often presents as a primary infection. The coxsackie group of viruses, principally coxsackie B, affects males twice as often as females. In adults, the disease is usually associated with pericarditis and is often benign, but in debilitated and immunosuppressive adults the infection can be far more severe. It may even cause sudden death. Other viruses, like poliomyelitis and German measles, can cause myocarditis. Bacterial myocarditis is rare compared to viral myocarditis, such as diphtheria, that can lead to a toxic myocarditis by its exotoxin.

See Also: Health Disparities; Obesity; Physical Activity and Health.

Further Readings

American Heart Association. "Conditions." http://www.heart.org/HEARTORG/Conditions/ Conditions_UCM_001087_SubHomePage.jsp (Accessed August 2010).

Centers for Disease Control and Prevention. "Heart Disease." http://www.cdc.gov/ heartdisease (Accessed August 2010).

Durstine, J. Larry, ed. *Pollock's Textbook of Cardiovascular Disease and Rehabilitation.* Champaign, IL: Human Kinetics, 2008.

Mayo Clinic. "Heart Disease." http://www.mayoclinic.com/health/heart-disease/DS01120 (Accessed August 2010).

Phibbs, Brendan. *The Human Heart: A Basic Guide to Heart Disease.* 2nd ed. Philadelphia, PA: Lippincott, Williams & Wilkins, 2007.

Shepard, Donald S. *Lifestyle Modification to Control Heart Disease: Evidence and Policy.* Sudbury, MA: Jones and Bartlett, 2010.

World Health Organization. "Cardiovascular Diseases." http://www.who.int/topics/ cardiovascular_diseases/en (accessed August 2010).

Richard Wills
Independent Scholar

CELL PHONES

Cell phones are mobile communication devices that have grown in popularity as technology has advanced. As usage of cell phones has increased globally, so has concern over their potential harm to environmental health and to human health. Future trends in research for cell phone manufacturing and usage are starting to address the issues of impending harm. Current mobile telephonic communication evolved from radio transmission dated from the 1920s to two-way radios and walkie-talkies, to mobile telephone systems (MTA and MTB), to analogue and digital. During the 1980s and 1990s, rapid changes in cellular communication included physical characteristics of cell phones, functionality of cell phones, and radio/cellular network communication. Decades of cellular technology are referred to as first generation, second generation, third generation, and fourth generation (1G, 2G, 3G, and 4G). The 1980s generation of analog devices is represented by 1G, 2G denotes the movement during the 1990s into digital network signaling, and 3G is correlated with the technology of increased data transfer seen in the 2000s. The forward movement into research and development for the 2010s introduced the 4G network, which focuses mostly on speed of data transfer.

The sheer volume of cellular phones in global circulation contributes to negative human health effects and damages the environment. Toxic metals used in the manufacture of these phones include arsenic, antimony, beryllium, cadmium, copper, lead, mercury, nickel, selenium, and thallium.

Source: iStockphoto.com

When mobile phones were first introduced into the market, they were quite costly to purchase and to service. However, during the later 1980s and through the 1990s into the 2000s, cell phone usage and availability became market competitive and pervasive. In 2008, there were over 1.1 billion cellular phones and cellular devices sold in the global market.

From a green health standpoint, cellular phones have become a concern due to the sheer volume of phones in global circulation, the materials involved with cellular communication, their short life span, and their end-of-life product disposal. Concerns include both human health effects from usage and damage to the environment from manufacturing and discarding.

Toxic metals identified in the process of making these phones include arsenic, antimony, beryllium, cadmium, copper, lead, mercury, nickel, selenium, and thallium. These metals are mined almost exclusively in a manner that is devastating to the environment. Additionally, most of these metals have toxic properties and are suspected carcinogens. Plastic parts of a mobile phone initiate apprehension when addressing negative impacts on the environment. Cell phones are disposed of by rudimentary recycling methods in third world countries with the practice of incineration of plastics and other phone components.

Incineration can introduce hazardous volatile organic compounds (VOCs) into the atmosphere. These compounds include hundreds of congeners of polychlorinated/brominated dioxins, furans (PCDDs/PBDDs, PCDFs/PBDFs), and biphenyls (PCBs/PBBs). The congener species variation depends on how many chlorine or bromine atoms are attached to the molecule and the location of their attachment.

Dioxins, furans, and biphenyls are characterized as persistent organic pollutants (POPs). They are toxic, can travel long distances, bioaccumulate in the food chain, and persist in the environment. Through both epidemiological and animal studies, these compounds are thought to be endocrine disruptors, hepatotoxic, teratogenic, immunotoxic, and carcinogenic. The level of toxicity and damage as a result of exposure to these compounds varies greatly, depending on which species or congener was exposed; 2,3,7,8-tetrachlorodibenzo-p-dioxin (TCDD), which is emitted during electronic waste (eWaste) incineration, is considered the most toxic organic compound and is classified by the U.S. Environmental Protection Agency as a group 1 carcinogen.

Aside from disposal method exposure, there is a concern about electromagnetic radiation exposure from the usage of cell phones. The scientific community is not in agreement as to whether or not cell phone usage with close proximity to the ear is an exclusive cause for an increase in brain tumors or other health risks. Neurosurgeons have stated publicly that they will only use an ear device to speak on a cell phone in order to minimize the radiation exposure to the brain, and some epidemiological and animal studies have concluded that there may be health risks from mobile phone usage. Many scientific studies, however, have shown no evidence of physical damage from cell phone usage. The rapid and global public health concern to better understand the field of electromagnetic radiation and its implicative capacity has generated multicountry collaboration of research and results. Current regulations are set for the peak amount of watts that can be absorbed into the human head from the handset. This is known as specific absorbed radiation (SAR). The SAR regulations are set by individual country government agencies and may vary by region. There are several research projects under way that will yield more evidence about possible health concerns, if any.

Engineers, toxicologists, environmentalists, chemists, and economists are initiating an interdisciplinary approach when considering alternative components used in the manufacturing of electronic devices and the creation of infrastructures for handling electronic waste.

See Also: Industrial Ecology; International Policies; Lead Sources and Health; Radiation Sources.

Further Readings

"International Chemical Safety Cards: 2,3,7,8-TETRACHLORODIBENZO-p-DIOXIN" (2003). National Institute for Occupational Safety and Health. http://www.cdc.gov/niosh/icps/icstart.html (Accessed April 2010).

Lincoln, J. D., O. A. Ogunseitan, J. D. Saphores, and A. A. Shapiro. "Leaching Assessments of Hazardous Materials in Cellular Telephones." *Environmental Science & Technology,* 41 (2007).

Safe, S. H. "Comparative Toxicology and Mechanism of Action of Polychlorinated Dibenzo-p-dioxins and Dibenzofurans." *Annual Review of Pharmacology and Toxicology,* 26/1 (1986).

Stewart, E. S. "Emissions From the Incineration of Electronics Industry Waste." IEEE International Symposium on Electronics and the Environment 2003 ISEE-03 (2003).

Kathleen Hibbert
University of California, Irvine

Centers for Disease Control and Prevention (U.S.)

The U.S. Centers for Disease Control and Prevention (CDC), based in Atlanta, Georgia, operates under the U.S. Department of Health and Human Services. Established in 1946, the CDC was initially known as the Communicable Disease Center. It was founded by Dr. Joseph Mountin and evolved from World War II's Malaria Control in War Areas program. During the 1950s, two major health crises helped solidify the CDC's position as an instrument for protecting public health: tracing a contaminated vaccine to a California lab and tracing the 1957 influenza epidemic; interestingly, the CDC's handling of vaccines and influenza today has given rise to intense criticism of the agency's scientific rigor and credibility.

The CDC is credited with eradicating smallpox worldwide by 1977 by embracing a vigorous surveillance program coupled with refined vaccination techniques. The agency is also credited with helping to identify Legionnaires' disease and AIDS, among other diseases. In its efforts to prevent disease outbreaks and implement disease prevention strategies, the CDC collaborates with states and other partners and also has personnel stationed in more than 25 countries to help guard against international disease transmission. The CDC celebrated its 60th anniversary in 2006 and has stated that it is intent on becoming more efficient and effective by focusing on five strategic areas:

- Supporting state and local health departments
- Improving global health
- Implementing measures to decrease causes of death
- Strengthening surveillance and epidemiology
- Reforming health policies

The CDC's focus on public health is credited with adding 25 years to the life expectancy of U.S. citizens; the agency counts the following as its "Ten Great Public Health Achievements, 1900–1999":

- Vaccination
- Motor vehicle safety
- Safer workplaces
- Control of infectious diseases
- Decline in deaths from coronary heart disease and stroke
- Safer and healthier foods
- Healthier mothers and babies
- Family planning
- Fluoridation of drinking water
- Recognition of tobacco use as a health hazard

Despite its successes, criticism of the CDC reaches back for decades, to the Tuskegee study, which it took over from the Public Health Service in 1957; the study involved researching the effects of untreated syphilis. Although by the late 1940s penicillin was known to be an effective treatment for syphilis, the black men in the study were not given penicillin until the CDC's action was brought to public attention. Another incident that raised questions about the CDC's credibility was the agency's efforts to vaccinate the U.S. population against swine flu in 1976; some who received vaccinations developed Guillain-Barre syndrome, which stopped the vaccination campaign immediately. The epidemic never occurred.

Criticism of the CDC has continued, particularly while under the direction of Dr. Julie Gerberding from July 2002 until January 2009, when she and several other senior-level CDC officials submitted their resignations at the behest of the Obama administration.

In 2004, critics charged the CDC with allegations including wasting money on dubious research, altering or disregarding their own scientists' work, and retaliating against those who objected to their manipulation and censorship of scientific research. By 2005, low morale among CDC employees prompted five former CDC directors to send Gerberding a letter regarding what was seen as inappropriate politicization of the CDC—a politicization that was considered to be detrimental to the CDC's national and international reputation.

The major contemporary complaints against the CDC are that the agency ignores scientific findings—even those from its own scientists—and refuses to link industrial pollution and chronic disease. Critics point to the CDC's 2002 investigation of cancer clusters in Nevada and Arizona, in which the CDC denied any relationship between industrial pollution and cancer, as evidence of the agency's lack of integrity and scientific rigor. Another CDC controversy involved the delayed disclosure of a study connected to the Great Lakes region and the demotion of the study's lead scientist. The study found that exposures to toxins including lead, mercury, and pesticides may cause problems, including low birth weights and elevated rates of certain kinds of cancer. Critics argued that because of the close relationship between the U.S. government and the chemical industry, the CDC was compelled to downplay the link between illness and industrial pollution. In 2008, Michigan Representatives John Dingell and Bart Stupak launched a congressional investigation charging the CDC with covering up information that put the public health at risk.

The CDC has also drawn criticism for downplaying the cancer risks associated with formaldehyde exposure in the 144,000 trailers purchased by the Federal Emergency Management Agency (FEMA) for Hurricane Katrina victims. The CDC was alleged to have suppressed vital information and to have retaliated against the study's lead scientist—the same scientist who led the Great Lakes study, Dr. Christopher De Rosa. In all these cases, the CDC has been accused of being more concerned with assigning liability to chemical manufacturers than with protecting the health of the American people.

The CDC has also been accused of hyping certain health threats, such as tuberculosis, in order to raise funds by scaring the public; the CDC has refused to release documents related to its funding strategies or the tuberculosis scare. It has also been accused of overstating influenza deaths, which it claims kill nearly 40,000 Americans annually. When pushed to explain how it arrived at this number, the CDC admitted that it was a computer-generated guess; investigative journalist Kelly O'Meara found, using the CDC's own data, that the greatest number of influenza deaths since 1979 was 3,006 in 1981. CDC critics charge that the CDC hypes certain health threats like influenza and bird flu while ignoring threats such as cancer and autism. Critics point to the CDC's handling of questions regarding vaccine safety as more evidence of its refusal to ascribe liability to corporations, particularly those who make vaccines. Many believe that some cases of childhood autism are tied to vaccines, though the CDC continues to vehemently deny this link, even though the U.S. government has awarded compensation to at least 10 families whose children developed autism after being vaccinated. There are more than 5,000 other claims of vaccination-injured children awaiting review in vaccine court.

Critics are suspicious of the relationship between the CDC and the vaccine industry; the CDC and vaccine manufacturers are linked in various, unsettling ways, such as sharing vaccine patents; further, CDC owns stock in vaccine companies and receives money from vaccine manufacturers for monitoring vaccine tests. It is difficult to see how the CDC could investigate the efficacy of vaccines in an unbiased way when it is linked ideologically and financially to the vaccine industry.

In May 2009, President Barack Obama tapped Dr. Thomas R. Frieden, an infectious disease specialist, to direct the CDC. Frieden had previously served under Mayor Michael Bloomberg as New York City's top health official. As director of the CDC, Frieden faces administrative and organizational problems related to Gerberding's reign, which was criticized for creating too many bureaucratic layers. Another issue on Frieden's plate is deciding which functions are best fulfilled by contract employees and which should be handled in-agency; under the George W. Bush administration, the CDC added thousands of contract workers. Finally, healthcare reform will profoundly impact how the CDC functions to improve public health, and Frieden is likely to be faced with decisions that could create controversy. Further, it remains to be seen if, under Frieden's direction, the CDC will continue drawing allegations of favoring corporations and industry over the health of U.S. citizens.

See Also: Cancers; Children's Health; Environmental Illness and Chemical Sensitivity; Tuberculosis; Vaccination/Herd Immunity.

Further Readings

Centers for Disease Control and Prevention. "About CDC." http://www.cdc.gov/about (Accessed June 2010).
The Huffington Post. "CDC Under Siege." http://www.huffingtonpost.com/deirdre-imus/cdc-under-siege_b_94720.html?view=screen (Accessed June 2010).
Tetrahedron Publishing Group. "UPI Investigates: The Vaccine Conflict." http://www.tetrahedron.org/articles/vaccine_awareness/UPI_Investigates.html (Accessed June 2010).

Tani Bellestri
Independent Scholar

CHEMICAL PESTICIDES

Pesticides are agents used to prevent, destroy, or control populations of animals, plants, insects, and microbial pests and pathogens including fungi. Chemical pesticides are often classified by structure or usage, with usage categories including insecticides, herbicides, fungicides, and fumigants. Some major chemical classes include the organochlorine, organophosphate, and carbamate insecticides and the inorganic, triazine, phenoxyacid, and thiocarbamate herbicides.

Over time, there has been a general trend toward use of less toxic and more specific-acting pesticides, which confer lower risk of unintended harms to humans and other species. In Illinois, grower Joe Zumwalt applies a low-insecticide bait to target western corn rootworms, which both feed on and lay eggs in soybeans.

Source: U.S. Department of Agriculture Agricultural Research Service/Ken Hammond

While there are benefits of pesticide use, such as increased crop yields and improved control of vector-borne illnesses, there are also concerns about potential harms to human health and the environment. Over time, there has been a general trend toward use of less toxic and more specific-acting pesticides, which confer lower risk of unintended harms to humans and other species. Most recently, a new category of pesticides called biopesticides was developed, which shows promise in the specific control of target pest populations and limited risks for other species. However, the toxicity of many pesticides is still not fully understood, particularly the effects of chronic (long-term) exposure on health outcomes such as cancers. Ongoing studies will help improve understanding of these effects, which will, in turn, better inform guidelines and regulations regarding pesticide use.

The inherent nature of a given chemical to harm living things, which characterizes a successful pesticide, may result in adverse effects on other species than the target pest. For example, use of the organochlorine insecticide dichlorodiphenyltrichloroethane (DDT) in the western Pacific in the 1950s and 1960s was highly effective at reducing mosquito populations and the spread of malaria. However, the insecticide also conferred unintended harm to a number of additional species, such as parasites of caterpillars that fed off thatched roofs, which led to a rise in caterpillar populations and roof damage, as well as to populations of predatory birds (e.g., the bald eagle) and possibly domestic cats as well. These unintended harms, combined with the substantial persistence of the chemical in the environment, influenced the decisions to ban DDT in the United States and a number of other areas in the world beginning in the 1970s. However, the subsequent increasing burden of malaria led the World Health Organization (WHO) in 2006 to recommend more widespread indoor spraying with DDT.

Given the large amounts of chemical pesticides manufactured and applied—the most recent U.S. Environmental Protection Agency estimates, for 2000 and 2001, show over 5 billion pounds/year applied worldwide, including over 1.2 billion pounds/year in the United States—the potential for harms to human health warrants thorough investigation. People may be exposed to pesticides through their occupations, such as jobs in agriculture or in industry manufacturing pesticides, or through environmental sources, such as pesticide-contaminated drinking water, air, and food. Environmental exposures include exposure from residential use, which was reported in 82 percent of U.S. households around 1990. While occupational exposures are generally higher than environmental exposures, U.S. studies that have tested human tissues/bodily fluids for markers of exposure to pesticides have confirmed that pesticide exposures are common in the general population.

In acute (short-term), high-dose exposures, many chemical pesticides are highly toxic to humans. A 1990 WHO report estimated that 3 million human pesticide poisoning cases, involving about 220,000 deaths, occur each year. Since about the mid-20th century, there has been a general trend toward development and use of chemicals with lower acute toxicity to humans and other nontarget species. Although DDT, first used in the 1940s, is not known to possess acute human toxicity at usual usage levels, it confers substantial harms to other species as discussed above. Reductions in the use of DDT and other harmful organochlorines starting in the 1970s were followed by increased use of organophosphate and carbamate chemicals, which were developed in the 1940s and 1950s and which exert toxicity by inhibiting the enzyme acetylcholinesterase. Acetylcholinesterase inhibition allows the neurotransmitter acetylcholine to accumulate between nervous system cells and, when pronounced, can cause neurotoxicity in humans and other species. While organophosphate and carbamate chemicals continue to be used due to their effectiveness against target pests, the 1970s and 1980s brought a wave of chemicals that were predicted to possess less severe adverse effects to humans, including the widely used pyrethroids, whose effects at most common levels of acute exposure include irritation to the skin, nose, or airways. In the 1990s, a new category of pesticides called biopesticides was developed that is believed to pose limited human health risks. An example is the genetic engineering of crops to produce a bacterial chemical that has relatively specific toxicity to unwanted pests.

Human health effects of chronic chemical pesticide exposures are generally less well understood than those of acute, high-dose exposures. This is in part because effects of the latter exposures usually manifest soon after the exposure occurs, which helps establish a link between the pesticide and the health outcome. In contrast, possible effects of chronic exposures, such as cancers and other chronic diseases, might only manifest due to a complex constellation of genetic susceptibility and environmental and lifestyle exposures that occurred many years or decades in the past, given the delay in developing such diseases. Additionally, exposures over a long period of time tend to be more difficult to measure. Thus, the role of chronic exposure to a particular pesticide in disease development is more challenging to identify.

Data are beginning to accumulate that can speak to the effects of chronic exposure. One such data source is the Agricultural Health Study, a prospective study of about 57,000 licensed pesticide applicators and 32,000 spouses in North Carolina and Iowa that has been gathering extensive information on exposures to some of the most widely used chemical pesticides, as well as cancers and other diseases, since 1993. Thus far, considering a number of studies, there are generally consistent findings of increased risk for prostate cancer and some immune system cancers, such as non-Hodgkin's lymphoma and leukemia, in agricultural populations, which suggest possible roles of pesticides in these cancers. Additionally, parental exposure to pesticides has been linked with a variety of birth defects, as well as cancers, asthma, and neurodevelopmental effects among children, who are considered a more susceptible population. Ongoing studies aim to further investigate these associations and to more fully explore the effects of individual pesticides on health and the factors that influence these effects (e.g., genetic background). Some pesticides of interest to study include chemicals of the organochlorine, organophosphate, carbamate, and pyrethroid families due to current or historical widespread use and accumulating data regarding metabolism/storage or action in the body that have generated concern. For example, organochlorines are known to collect in body fat and there is increasing evidence that some organophosphates might lead to oxidative stress and/or DNA damage in humans. Continued study of these and other chemicals will inform assessments of risks associated with pesticides that can be weighed against their benefits in the generation of enlightened guidelines and regulations regarding pesticide use.

See Also: Biological Control of Pests; Cancers; Genetically Engineered Crops; Malaria; Pest Control.

Further Readings

Alavanja, Michael C. R., Mary H. Ward, and Peggy Reynolds. "Carcinogenicity of Agricultural Pesticides in Adults and Children." *Journal of Agromedicine*, 12/1 (2007).

Karr, Catherine J., Gina M. Solomon, and Alice C. Brock-Utne. "Health Effects of Common Home, Lawn and Garden Pesticides." *Pediatric Clinics of North America*, 54 (2007).

Kumar, Suresh, Amaresh Chandra, and K. C. Pandey. "*Bacillus thuringiensis (Bt)* Transgenic Crop: An Environment Friendly Insect-Pest Management Strategy." *Journal of Environmental Biology*, 29/5 (2008).

O'Shaughnessy, Patrick T. "Parachuting Cats and Crushed Eggs—The Controversy Over the Use of DDT to Control Malaria." *American Journal of Public Health*, 98/11 (2008).

World Health Organization in collaboration with the United Nations Environment Programme. *Public Health Impact of Pesticides Used in Agriculture*. Geneva, Switzerland: World Health Organization, 1990.

Kathryn J. Hughes
Yale University

Children's Health

Children's health encompasses the physical, mental, emotional, and social well-being of children from birth through adolescence and is determined by the interaction of a multitude of influences and complex, dynamic processes, which can be divided into biological, behavioral, and environmental (physical and social) categories. Measuring the effects of these factors on health is difficult, as they are highly intertwined and difficult to isolate. In children, measurement is complicated further by their rapidly changing nature, as they grow and undergo developmental processes that play an important role in shaping and determining their health. Traditionally, health surveillance and monitoring systems have been based on adult matrices of health, focusing on prevention and treatment of morbidity and mortality that resulted primarily from infectious diseases, as well as monitoring chronic diseases among adults. Only recently, perhaps due to increases in scientific information about the development of health, the role of prenatal and early childhood health on adult health outcomes, and the importance of predisease pathways that begin in childhood, have efforts been made to track more detailed, systematic, and longitudinal data on the internal and external influences on children's health.

Despite an effort to recognize children as a population with unique health issues, children's health is still often lumped together with maternal health, as the two are closely linked, especially during the first few years of a child's life. A critical indicator used to determine the well-being of children within a country is the Under-Five Mortality Rate (U5MR). Other indices may include the proportion of newborns born small or too early, neonatal or infant mortality, disease-specific incidence and mortality rates, and proxy-reported ratings of health or activity limitations. Available data come from various sources: vital statistics, surveys, and clinical and administrative datasets.

In the United States, the Maternal and Child Health Bureau of the U.S. Department of Health and Human Services, Health Resources and Services Administration coordinates the collection of large portions of the nation's data on children's health. A consortium of other federal organizations, such as the Eunice Kennedy Shriver National Institute of Child Health and Human Development, the National Institute of Environmental Health Sciences of the National Institutes of Health and the Centers for Disease Control and Prevention under the umbrella of the U.S. Department of Health and Human Services, the U.S. Environmental Protection Agency, and the U.S. Department of Education, have come together for projects such as The National Children's Study. This study will examine the effects of environmental influences on the health and development of 100,000 children across the United States, following them from before birth until age 21.

Common measures of biological health of children, particularly in more industrialized countries such as the United States, include rates of preterm birth and low birth weight, infant mortality, emotional and behavioral difficulties, adolescent depression, activity limitation, diet quality, obesity, and asthma. In the United States, 12.3 percent of infants were born before 37 weeks of gestation and 8.2 percent of infants weighed less than 5 pounds 8 ounces at birth in 2008. Approximately 5 percent of children aged 4 to 17 reported a serious difficulty with emotions, concentration, behavior, or getting along with other people in 2008. Also in 2008, 8 percent of youth aged 12 to 17 were reported to suffer from a major depressive episode. Approximately 19 percent of children aged 6 to 17 were considered obese, and 9 percent of children aged 0 to 17 were diagnosed with asthma in the United States in 2008.

Common behavioral effects on child health, particularly adolescent health, include substance abuse, such as use of cigarettes, alcohol, or illicit drugs; early sexual activity; and participation in violent crime. In the United States in 2009, among children in the 10th grade, 6 percent reported smoking cigarettes daily in the past 30 days, 18 percent reported having five or more alcoholic beverages in a row in the past two weeks, and 18 percent reported illicit drug use in the past 30 days. Early sexual activity has been associated with both emotional and physical health risks. Among high school students surveyed in the United States in 2007, 48 percent reported having had sexual intercourse.

Physical safety greatly affects the development of children. Indoor air pollutants, such as secondhand smoke, are associated with lower-respiratory infections, asthma, and other respiratory conditions. Within the United States, 60 percent of children aged 0 to 17 live in counties where levels of one or more air pollutants were above allowable levels in 2008. Contaminated drinking water is associated with gastrointestinal illness, development delays and learning disorders, and chronic disease, such as cancer. In 2008, 6 percent of children were living in communities with water systems that did not meet all applicable health-based drinking water standards. Exposure to lead, particularly through lead paint, is associated with cognitive disability. Violent crime is also considered a physical hazard affecting children's health. Within the United States, 12 out of 1,000 children aged 12 to 17 were reported to have been victims of serious violent crimes.

The social environment of a child also has a great influence on his or her health. Characteristics such as family composition and structure can affect the economic security of a child, thereby also possibly affecting a child's health. In 2009, 67 percent of children in the United States, ages 0 to 17, were living with two married parents. Poverty, associated with food insecurity, housing issues, and reduced access to health and dental care, affects child health and well-being. Approximately 19 percent of children aged 0 to 17 were reported to live in poverty in the United States in 2008. Approximately 22 percent of children aged 0 to 17 lived in households classified by the U.S. Department of Agriculture as "food insecure." Maltreatment, including neglect and medical neglect and physical, sexual, and psychological abuse, also affects the health of a child.

International data on children's health are complicated by resource limitations and the infrastructure capacity of each country. Many turn to global organizations like the World Health Organization (WHO) and the United Nations Children's Fund (UNICEF) as the leading source for data on the state of children's health globally.

In November 2009, UNICEF released its report "The State of the World's Children: Celebrating 20 Years of the Convention on the Rights of the Child," which ranked the U.S. 149th out of 193 countries and territories listed in descending order by their estimated 2008 under-5 mortality rate (U5MR). The report estimated the U.S. U5MR at 8 per 1,000 live births and an infant mortality rate of 7 per 1,000 live births, as compared to a global average U5MR of 65 per 1,000 live births and a global infant mortality rate of 45 per 1,000 live births in 2008. The WHO contends that about two-thirds of child deaths are preventable through practical, low-cost interventions. To assist in implementing measures to improve children's health globally, WHO and UNICEF have launched projects such as the Integrated Management of Childhood Illness (IMCI), which began in 1997 as a strategy for reducing mortality among children under 5 years old. The IMCI strategy includes three components: improving health worker skills, the healthcare system, and family and community practices in the more than 100 countries that have adopted this strategy to date. The evidence-based integrated approach has been shown to improve health workers' performance and quality of service.

Major Causes of Death in Neonates and Children Under 5

United States

Reports indicate that 10,780 children between the ages of 1 and 14 years died of various causes in 2006 (based on Centers for Disease Control and Prevention, National Center for Health Statistics, and National Vital Statistics System, 2006).

Overall U5MR = 28.4 per 100,000 children in that age group occurs from the following:

- Unintentional injury (35 percent)
- Congenital anomalies (birth defects) (11 percent)
- Malignant neoplasms (cancer) (8 percent)
- Homicide (8 percent)
- Diseases of the heart (4 percent)

Overall mortality rate among 5- to 14-year-old children = 15.2 per 100,000 children in that age group occurs from the following:

- Unintentional injury (37 percent)
- Malignant neoplasms (15 percent)

- Homicide (7 percent)
- Congenital anomalies (birth defects) (6 percent)
- Diseases of the heart (4 percent)

International

Based on WHO's *World Health Statistics 2008*, global health indicators, and U5MR for 2006, the overall U5MR was 71 per 100,000 children in that age group. Thirty-five percent of under-5 deaths were due to the presence of undernutrition.

Based on WHO's *The Global Burden of Disease: 2004 Update*, the causes of under-5 deaths were the following:

- Neonatal Deaths (36 percent) → Caused by:
  - o Prematurity and low birth weight (31 percent)
  - o Birth asphyxia and birth trauma (23 percent)
  - o Neonatal infections (26 percent)
  - o Congenital anomalies (7 percent)
  - o Diarrheal diseases (3 percent)
  - o Neonatal tetanus (3 percent)
  - o Other (7 percent)
- Acute respiratory infections (post-neonatal) (17 percent)
- Diarrheal diseases (post-neonatal) (16 percent)
- Other infectious and parasitic diseases (9 percent)
- Malaria (7 percent)
- Measles (4 percent)
- Noncommunicable diseases (post-neonatal) (4 percent)
- Injuries (post-neonatal) (4 percent)
- HIV/AIDS (2 percent)

Approximately 2.6 million deaths occurred among 10- to 24-year-olds in 2004. Ninety-seven percent of these deaths occurred in middle- and low-income countries, and almost two-thirds of all deaths occurred in Africa and Southeast Asia.

The top-10 major killers can be divided into three overall categories:

- Communicable diseases and maternal causes (36 percent)
  - o Maternal conditions
  - o Lower-respiratory infection
  - o TB
  - o HIV
  - o Meningitis
- Noncommunicable diseases (22 percent)
- Injury (42 percent)
  - o Road traffic
  - o Self-inflicted
  - o Violence
  - o Drowning
  - o Fires

According to the WHO, from 1 month to 5 years of age, the main causes of death are pneumonia, diarrhea, malaria, measles, and HIV. Pneumonia is the primary cause of death in children under 5 years of age. Nearly three-quarters of all cases occur in just 15 countries.

Addressing the major risk factors—including malnutrition and indoor air pollution—is essential to preventing pneumonia, as are vaccination and breastfeeding. Antibiotics and oxygen are vital tools for effectively managing illnesses. Diarrheal diseases are a leading cause of sickness and death among children in developing countries, and access to clean water is instrumental in prevention. The spread of vector-borne diseases, such as malaria, can be prevented by use of insecticide-treated nets, especially in Africa, where one African child dies every 30 seconds from malaria. Over 90 percent of children with HIV are infected through mother-to-child transmission, which can be prevented with antiretroviral drugs, as well as safer delivery and feeding practices. About 20 million children under 5 worldwide are severely malnourished, which leaves them more vulnerable to illness and early death. Malnutrition is estimated to contribute to more than one third of all child deaths.

The WHO Millennium Development Goal (MDG) 4 is to reduce child mortality by two-thirds by the year 2015. Interventions with proven success in approaching the goal include expanded immunization programs, promotion of breastfeeding, and provision of mosquito nets in endemic regions. The United Nations, with leaders of governments, foundations, nongovernmental organizations, and businesses, has developed a Global Strategy for Women's and Children's Health. The strategy focuses on enhancing financing, strengthening policy, improving service delivery, and implementing global reporting of progress. UNICEF is also furthering this MDG through purchase of vaccines and education and skills transfer to families.

In June 2009, 600 participants from 60 countries and 223 national and international organizations met at Busan, the Republic of Korea, to draw renewed and urgent attention to children's environmental health issues, reposition children's environmental health (CEH) in the global public health agenda, and improve and promote practical protective policies and actions at all levels. The resulting Busan Pledge for Action on Children's Environmental Health is as follows:

> We pledge to develop a global plan of action to improve CEH, monitor and report on progress, and we urge WHO and its partners to facilitate the development of this plan in collaboration with all relevant agencies. We will implement activities in close interactive partnerships with governmental and nongovernmental organizations, centres of excellence, academia, professional bodies, educators and other sectors. We commit to take CEH issues to the consideration of higher authorities in our respective countries and to the attention of the international agencies concerned about children's health and the environment and the needs for green growth and sustainability.

A draft of the Global Plan of Action for Children's Health and the Environment (2010–2015) resulted from the Busan meeting, outlining a strategy including data collection and analysis, collaborative research, advocacy, clinical capacity building to increase service delivery, and awareness raising and education. Each strategy also includes an associated plan of action and list of expected outcomes.

In the United States, a consortium of federal partners, including the U.S. Department of Health and Human Services and the United States Environmental Protection Agency, are leading the National Children's Study. The study was authorized by the U.S. Congress through the Children's Health Act of 2000. The study plans to examine the effects of environmental influences on the health and development of 100,000 children across the United States from birth to 21 years of age. The environment encompasses national and manmade environmental factors, biological and chemical factors, physical surroundings, social factors, behavioral influences and outcomes, genetics, cultural and family influences, and geographic locations. Researchers will analyze the interaction between the environmental factors and identify if they are associated with healthcare access and disease occurrence. The goal of the study is to improve the health and well-being of children.

See Also: Centers for Disease Control and Prevention (U.S.); Healthcare Delivery, Health Disparities; Malaria; Water-Borne Diseases; World Health Organization's Environmental Burden of Disease.

Further Readings

National Research Council and Institute of Medicine. "Children's Health, the Nation's Wealth: Assessing and Improving Child Health." Committee on Evaluation of Children's Health. Board on Children, Youth, and Families, Division of Behavioral and Social Sciences and Education. Washington, DC: National Academies Press, 2004.
United Nations Children's Fund. "The State of the World's Children: Celebrating 20 Years of the Convention on the Rights of the Child—Special Edition." New York: UNICEF, 2009. http://www.unicef.org/rightsite/sowc/pdfs/statistics/SOWC_Spec_Ed_CRC_Statistical_Tables_EN_111809.pdf (Accessed April 2010).
U.S. Department of Health and Human Services, Health Resources and Services Administration, Maternal and Child Health Bureau. "Child Health USA 2008–2009." Rockville, MD: U.S. Department of Health and Human Services, 2009. http://mchb.hrsa.gov/chusa08 (Accessed April 2010).
U.S. Department of Health and Human Services, Health Resources and Services Administration, Maternal and Child Health Bureau. "The National Survey of Children's Health, 2007." Rockville, MD: U.S. Department of Health and Human Services, 2009. http://mchb.hrsa.gov/nsch07 (Accessed April 2010).
World Health Organization (WHO). "CAH Progress Report Highlights, 2008." Geneva: WHO, 2009. http://www.who.int/child_adolescent_health/documents/9789241597968/en (Accessed April 2010).
World Health Organization (WHO). "Global Plan of Action for Children's Health and the Environment, 2010–2015." Geneva, Switzerland: WHO, 2010. http://www.who.int/ceh/cehplanaction_10_15.pdf (Accessed September 2010).

Rebecca A. Malouin
Sarah Merten
Michigan State University

CHLORINATION BY-PRODUCTS

Drinking water comes from natural sources in the environment, with naturally occurring disease-causing microorganisms. Because of this, water must be disinfected in order to be potable and safe for human consumption. However, common disinfectants may also react with naturally occurring components of the water to create by-products that can have cancerous or other health effects.

Drinking water sources typically derive from either surface or ground waters. Surface water has naturally occurring pathogens, which are biological agents, such as bacteria, that cause human diseases. Some of the most dangerous of these pathogens originate from the gastrointestinal (GI) tracts of mammals. When animals or humans defecate near surface water, the water may become contaminated with these pathogens, which causes illness when humans drink the water. In countries with adequate drinking water treatment, waterborne illnesses have become a rare event. But in the developing world, unsafe water is still a major cause of illness.

According to Water for People, a charity dedicated to providing adequate drinking water and wastewater sanitation throughout the world, globally 884 million people lack safe drinking water and 2.5 billion people lack adequate sanitation, resulting in the deaths of approximately 6,000 people daily, mostly children. Many common illnesses in the developing world have been almost eradicated in societies with adequate water disinfection, for example, cholera, dysentery, Legionnaires' disease, Pontiac fever, typhoid fever, and polio.

Within the United States, one of the most famous outbreaks of waterborne illness was the 1993 *Cryptosporidium* outbreak in Milwaukee, Wisconsin, causing illness in over 400,000 people, resulting in over 100 deaths, and creating an estimated $96.2 million in costs associated with illness. Although the cause of this outbreak was believed to be an ineffective filtration process, publicly supplied drinking water is meant to have an adequate chlorine residual remaining in the water when it leaves the plant to ensure adequate disinfection when it arrives at a consumer's tap.

Although chlorine is used in drinking water treatment to disinfect and protect humans, it also reacts with organics and inorganics in the water to form chlorination by-products, many of which have been shown to cause cancer or to cause adverse reproductive or developmental effects (such as certain trihalomethanes and certain haloacetic acids) in laboratory animals. Other methods of disinfection exist, which also have disinfection by-products such as chlorine dioxide and ozone. Chlorine dioxide is chlorine containing a compound used to disinfect water. It reacts in water to form chlorite, which has been shown to cause adverse effects in laboratory animals.

The U.S. Environmental Protection Agency (EPA) does not regulate chlorination by-products alone, but regulates all disinfection by-products, including trihalomethanes, haloacetic acids, chlorite, and bromate through the stage 1 Disinfectants/Disinfection Byproducts Rule, as follows:

- Trihalomethanes (THMs) include chloroform, bromodichloromethane, dibromochloromethane, and bromoform. THMs form when chlorine and other disinfectants react with organic and inorganic matter in water. The maximum allowable annual average of THMs is 80 parts per billion (µg/l).
- Haloacetic acids (HAA5) include monochloroacetic acid, dichloroacetic acid, trichloroacetic acid, monobromoacetic acid, and dibromoacetic acid. HAA5s form when chlorine and other disinfectants react with organic and inorganic matter in water. The maximum allowable annual average of HAA5 is 60 parts per billion (µg/l).
- Chlorite is formed when chlorine dioxide is used to disinfect water. The maximum allowable annual average of chlorite is 1 part per million (mg/l).

- Bromate forms when ozone used for disinfection reacts with bromide present in water. The maximum allowable annual average of bromate is 10 parts per billion (µg/l).

Another form of chlorine may also be used for disinfection of drinking water. Chlorine can be combined with ammonia to form chloramines. Chloramines, on the whole, are more stable than chlorine. Because of this, they are less likely to react with organic matter, reducing the formation of THMs and HAA5s. This characteristic also makes chloramines more stable in the distribution system and weaker as disinfectants when compared to chlorine. Because of the lower production of by-products and greater stability in the distribution system, chloramines are gaining in popularity as disinfectants within the United States. Within humans, chloramines are neutralized within the digestive system, but chloramines can be toxic to fish. Tap water treated with chloramines should be treated prior to use in aquariums.

See Also: Bottled Water; Groundwater; Reverse Osmosis; Waterborne Diseases.

Further Readings

Corso, P. S., M. H. Kramer, K. A. Blair, D. G. Addiss, J. P. Davis, and A. C. Haddix. "Cost of Illness in the 1993 Waterborne *Cryptosporidium* Outbreak, Milwaukee, Wisconsin." *Emerging Infectious Diseases* (April 2003).
"National Primary Drinking Water Regulations: Disinfectants and Disinfection Byproducts." *Federal Register,* 63/241 (December 16, 1998).
Water for People. http://www.waterforpeople.org (Accessed February 2010).

Michelle Edith Jarvie
Independent Scholar

CITIES

The U.S. Census Bureau estimates that approximately 80 percent of the country's population reside in urban and suburban settings. As the largest consumer of energy, American cities are responsible for a great portion of greenhouse gases and pollution through energy consumption to support industrial, commercial, residential, and transportation needs. On a human scale, they are places of social, environmental, and physical patterns and behaviors affecting population health.

As cities grow in size and population, more attention is being paid to infrastructure needs as well as to the environmental impacts of urbanization. Within cities, much of the conversation and action centers on the concept of sustainability. Urban planners are increasingly engaged in the conversation about population health. They are working with organizations and industry to create programs and interventions to reduce the negative effects of urbanization.

A concise definition of sustainability has not been developed, and a variety of principles have been adopted by cities and organizations. It is not a matter of disagreement about what needs to happen with the built and social environments; it is a matter of disagreement about what sustainability will ultimately create. For example, Chicago, which touts itself as America's greenest city (among others that do likewise), emphasizes economy in its discussion of sustainability. Seattle aims for environmental protection. Portland seeks to limit growth. Sonoma County recognizes the link between the economy, society, and the environment and the need to manage resources for current and future generations. This last sample may best summarize the concept of sustainability. Basically, the driving force is being responsible to the Earth's ecosystems and creating places where practices reduce the negative effects of consumption. These principles include designing streets for increased physical activity, development that reduces environmental impact, neighborhood economies to reduce travel, urban gardening, and influencing stewardship through the creation and purchase of renewable energy sources. Environmental efforts are inextricably linked to health, such as reducing pollution outputs, building green space, and creating economies and specialized workforces to remediate harmful environmental factors to reduce exposure to hazards.

It is impossible to measure the number of efforts under way for creating sustainable cities, and the long-term effects on population health are yet known. Cities are increasingly involved in public health initiatives, including banning smoking in public places, promoting farmers markets, and restricting fast food establishments near schools. Cities are instrumental in these efforts through community and economic development programs, revitalization efforts, urban development practices, and program design. Some efforts can potentially have multiple positive effects on population health. For example, creating safe parks and streets can reduce violence and injury, promote physical activity, mitigate urban heat, and combat carbon output of surrounding buildings.

The role of the city in green health is one that is emerging as initiatives are implemented and as programs expand. Many cities are making efforts to remediate what currently exists; other cities are passing regulations to proactively intervene with future development. For example, some cities have passed ordinances or created tax incentives for new building standards to incorporate green principles that aim to increase the efficiency of resource use. Remediation might include redevelopment of brownfields (cleaning up contaminated property), reuse of abandoned or historic buildings, and removal of lead from older housing stock. Other efforts might serve community and individual needs, such as partnering with neighborhood groups and schools to turn empty lots into community gardens.

These efforts are often associated with the needs, design, and culture of a place. Trends are emerging, but what is also increasingly clear is that what is embraced in one place will not necessarily work in another. Also, many cities are promoting themselves as "green," but standards vary and are often contested by activist organizations. The term *greenwashing* has gained popularity to describe seemingly green practices that are often canceled out by other negative environmental impacts. For example, cities have encouraged citizens to reduce energy consumption and participate in Earth Hour (during which homeowners are asked to turn off their lights during a particular hour or a particular date). This is canceled out by the purchase of downtown office lighting to promote sports teams.

Practices vary according to a city's civic culture, branding, socioeconomics, and other urban variables such as crime. Urban health initiatives include building sidewalks to reduce automobile dependence and encourage walking. This has not had much success in most suburban communities where street corridors have heavy car traffic and where homes are separated from commercial places. Within higher-density urban areas, physical activity is dependent upon streetscape (e.g., cleanliness, lighting, lower crime risk, and accessibility of commercial amenities and resources). Healthcare disparities among minority and low-income populations can also be linked to some of these same factors. Spatial patterns of poor health within urban spaces are being examined to determine causal factors and interventions. Private foundations, such as the Robert Wood Johnson Foundation, fund programs to increase access to healthy, affordable food; promote physical activity in communities; and distribute health resources to populations living in underserved places. These efforts are often concentrated in older, poorer, urban neighborhoods.

Culture also plays a role in citizenry behavior within urban environments. Biking as a form of transportation gained in popularity during the late 2000s due to gasoline prices, vehicle emissions awareness, and public health social marketing campaigns. The Chicago region has developed a bike transit network composed of paths, street bike lanes, bike racks and parking areas, allowance on trains, and a downtown bike commuter center with lockers and showers. This has created a year-round bike commuter culture in spite of the harsh winter climate. Other cities may promote biking to work but without the infrastructure to support this and perhaps because of safety issues for bikes; these communities struggle to create a bike culture. For example, New Orleans is a much more compact city, where citizen health and safety are closely linked to the surrounding environment. However, crime, street conditions, and intense heat and humidity serve as barriers to outdoor physical activity and bike commuting. The city has failed to develop a program and resources to mitigate these conditions.

Some green city movements are extreme and seek to design cities that will have minimal environmental impact and that will be inhabited by people who will engage in practices to reduce their environmental impact. Some consider this image utopian in nature. The ideal is a place where the buildings will meet green architectural standards, people will organically farm on their green rooftops and shun motor vehicles, and waste outputs will be recycled into new consumption forms. Each of these efforts is a component to creating healthy, viable, and sustainable cities. However, cities are preexisting places that were created over decades and centuries. Incorporating green principles into a city requires a more realistic image of how people behave, what exists structurally, and how services are designed, funded, and delivered.

See Also: Automobiles (Emissions); Climate Change; Fast Food; Health Disparities; Home-Grown Food; Lead Sources and Health; Physical Activity and Health.

Further Readings

Fitzpatrick, Kevin and Mark LaGory. *Unhealthy Places: The Ecology of Risk in the Urban Landscape.* New York: Routledge, 2000.

Marks, James. "Why Your Zip Code May Be More Important to Your Health Than Your Genetic Code." *The Huffington Post* (April 23, 2009). http://www.rwjf.org/pr/product .jsp?id=42029 (Accessed December 2009).

McKinnon, R. A., J. Reedy, S. L. Handy, and A. B. Rodgers. "Introduction: Measuring the Food and Physical Activity Environments: Shaping the Research Agenda" http://www .rwjf.org/pr/product.jsp?id=42370 (Accessed April 2009).

Register, Richard. *Ecocities: Building Cities in Balance With Nature.* Gabriola Island, BC, Canada: New Society Publishers, 2006.

Denese M. Neu
HHS Planning & Consulting, Inc.

CLEAN COAL

In the first decade of the 21st century, clean coal emerged as an incredibly contentious and politicized term. The roots of this contentiousness lie in the larger context of the mostly

partisan debates in the United States over climate change and its root causes and solutions. In the most general terms, those on the American Left see "clean coal" as a bit of marketing deception, a term put forth by the coal lobby to convince the American public that burning coal can be made "clean" through technological know-how. For many liberals and others, clean coal symbolizes the great lengths energy companies will go to in order to convince the American public that global warming is either not a problem or is easily solved with technology. For conservatives, clean coal symbolizes the power of the market to develop newer, more efficient technologies, not through regulation but through competition. Given this politicized context, it can be difficult to determine what clean coal is and whether or not it is a viable solution to the problem of coal pollution.

A coal pile in front of a so-called clean coal plant in the American Midwest. Clean coal encompasses a number of technologies intended to make burning coal less polluting but has been contested in the cultural and political realms.

Source: iStockphoto.com

Clean coal, which encompasses a number of technologies intended to make burning coal less polluting, has been most contested in the cultural and political realms. Some particularly incendiary historical examples illustrate this. In December 2008, an organization called America's Power released a Christmas-themed music video on its website, www.americaspower.org (and subsequently reposted on YouTube, where it has been viewed an additional 10,000 times). The music video featured the Clean Coal

Carolers, a group of animated coal lumps singing about the merits of clean coal to the tune of "Frosty the Snowman."

America's Power is an organization that is funded almost exclusively by the coal industry, and the coal carolers' video both delighted and angered many, depending on their position on the political spectrum. For example, Joseph Romm, a prominent blogger for the liberal think tank Center for American Progress, argued, "Note to industry . . . Clean Coal is magical in the way Harry Potter is magical—neither of them exist in the real world. . . . In the twisted minds of the industry madmen who put this together, it makes perfect sense to turn songs about the birth of Jesus into songs about 'clean coal.'"

Romm's position is frequently echoed by most on the Left, who believe clean coal is an industry-propagated myth. One popular rebuttal to the Clean Coal Carolers campaign appeared in the form of a spoof on a public-service-style commercial, directed by prominent filmmakers Ethan and Joel Coen, titled "Get Clean Coal Clean!" (http://www.youtube.com/watch?v=uFJVbdiMgfM).

Clean Coal and Politics

Clean coal also played a role in the 2008 presidential elections in the United States. Both candidates (Barack Obama and John McCain) pledged to support the research and development of clean coal technologies. The election proved to be a particularly contentious one when looked at from the vantage point of energy-related issues. McCain's supporters could frequently be found chanting "Drill, baby, drill," while Obama's supporters often wanted him to take stances against offshore drilling, nuclear power, and clean coal. One could argue that it would be difficult for any candidate to win election to the presidency without making overtures to large energy companies and that the candidates' endorsements of clean coal might also be read as endorsements of business-as-usual energy policies. A more generous interpretation could also allow for the possibility that both candidates believed that coal-scrubbing or storage technologies were worthwhile investments, given the country's reliance on coal for electricity production.

Speaking in the most general terms, therefore, we could say that those who advocate for clean coal are typically Republicans who support traditional forms of resource extraction (petroleum, coal, and natural gas) and who typically do not believe global warming poses a significant threat to humankind. They believe in the power of technology and the markets to regulate themselves, even when it comes to pollution. Those who oppose clean coal are typically Democrats who are suspicious of big business. They are especially suspicious of large energy companies, which have been known to fund disinformation campaigns about the harmful effects of fossil fuel burning in the past. Clean coal represents a commitment to business-as-usual energy and environmental policies rather than to progressive changes that challenge the status quo.

But this representation oversimplifies, concealing a more complex understanding of clean coal and its potential and limitations. Clean coal refers to a number of technologies, all of which are largely in the research and development phases. This can make it difficult to track clean coal's progress. However, there are some generalizations about clean coal that we can make with confidence.

First, we do know that at this time coal-fired power plants are not clean. The United States relies on coal-fired power plants to produce approximately 50 percent of its electricity needs. Although the coal industry has successfully used some chemical additives to make coal produce less sulfur dioxide pollution, therefore making it technically "cleaner," the big picture of coal production is a very dark and polluted one. Coal-fired power plants produce massive amounts of carbon dioxide, a potent greenhouse gas, and "black carbon," which consists of fine particulates that enter the atmosphere and can pose significant health and environmental risks. Furthermore, a growing percentage of U.S. coal production comes from the process of mountaintop removal, wherein the tops of mountains (mostly in the American south) are blown off, exposing coal seams below. For some companies, this is an easier and more cost-effective process for accessing coal than underground mining. However, this process is devastating not only for the mountaintops but also to the many miles of streams below, which are effectively buried by the fallen debris. The death of these streams has been disastrous for the fish that live in them, for surrounding wildlife and plant populations, and for people who have relied on those streams for drinking water and livelihoods for generations. It makes perfect sense, then, that many of these people—and those in sympathy with them—would be suspicious of the term *clean coal*.

Clean Coal and CCS

However, instead of talking about clean coal, it may be more accurate and productive to talk about particular technologies, such as carbon capture and sequestration (CCS). Carbon capture and sequestration is what many scientists, engineers, and politicians mean when they reference clean coal. In fact, CCS relies on an understanding that coal is not clean, that it is in fact quite polluting, and that it needs to be sequestered. There are a number of technologies that might make CCS possible, but the general idea behind CCS is that the carbon dioxide produced by a coal-fired power plant might be somehow separated from the other by-products of coal burning, and this carbon dioxide could then be sequestered, most likely by pumping it underground. The general philosophy behind CCS is that coal is an integral part of our economy (via electricity production) and it will be technologically, economically, and politically impossible to phase it out overnight, or even within decades. However, carbon dioxide is an undesirable pollutant that must be prevented from entering the atmosphere and contributing to global warming. So it is worthwhile to consider the possibility of continuing to burn coal for power while sequestering its most harmful pollutant. Some argue that CCS could be a useful bridge technology, preventing carbon dioxide from entering the atmosphere until our energy production systems can fully decarbonize by transitioning to nuclear, wind, solar, or other forms of renewable power generation.

CCS is not wholly undesirable, therefore, nor is it mythical. The same might be said of other clean coal technologies, such as gasification. CCS is, in fact, a theoretically sound concept. Perhaps ironically, petroleum engineers have often used carbon dioxide, pumped underground, to extract difficult-to-obtain petroleum. The carbon dioxide then remains sequestered in those underground veins. Furthermore, there are a number of CCS demonstration projects—most notably in Germany but also in the United States, at the Mountaineer Plant in West Virginia—that may yield important results. Although some oppose CCS for philosophical reasons (they do not want to support the continued burning of coal, no matter what), a better approach might be to advocate both for the decreased use of coal and for technologies that could make coal-fired power plants less polluting in the near term.

Finally, we must acknowledge that CCS is not yet proved, at the commercial level, to be economically or technologically viable. This does not make it a myth, nor does it make clean coal a fact. The most we might say about the technology is that it shows promise but is not yet proven. One could argue that, given the enormity of the energy and environmental challenges we face in the 21st century and our massive reliance on coal for energy production, it may be worthwhile to approach CCS as we would any other technology— not as a silver bullet that will save us from the problems of rising energy demands and global warming but as one technology among many deserving of resources and careful attention.

See Also: Asthma; Carbon Filters; Climate Change; Electricity.

Further Readings

Center for Science and Policy Outcomes and Clean Air Task Force. "Innovation Policy for Climate Change: A Report to the Nation." Washington, DC: Center for Science and Policy Outcomes and Clean Air Task Force, September 2009. http://www.cspo.org/projects/eisbu/report.pdf (Accessed March 2010).
"The Future of Coal: An Interdisciplinary MIT Study." Cambridge: Massachusetts Institute of Technology, 2007. http://web.mit.edu/coal (Accessed March 2010).
National Energy Technology Laboratory, Department of Energy. "Clean Coal Technology Compendium." http://www.netl.doe.gov/technologies/coalpower/cctc/index.html (Accessed March 2010).

Jen Schneider
Colorado School of Mines

CLIMATE CHANGE

The United Nations Framework Convention on Climate Change (UNFCC), located in Bonn, Germany, defines climate change as "a change of climate which is attributed directly or indirectly to human activity that alters the composition of global atmosphere and which is in addition to natural climate variability observed over comparable time periods." Global climate change poses a serious threat that can generate social upheaval, population displacement, economic hardships, and environmental degradation. In order to achieve a green world, according to the new ecological trends in society, mitigation of global climate change should be a priority for the world and its governments.

Climate System

A climate system is defined as a highly complex system consisting of five major components: the atmosphere, the hydrosphere, the cryosphere, the land surface, and the biosphere, with high dynamics interactions between them. The climate system evolves in time under the influence of its own internal dynamics and because of external and human forcings.

Effects of Climate Change

Climate change, as well climate variability, can exert multiple and different effects on human population and, generally, on life or on the biosphere. These phenomena may affect

No matter what the degree of preparedness, projections suggest that future extreme weather events will be catastrophic because of their unexpected intensity, as seen during Hurricane Katrina in 2005. Here, a National Guard helicopter works to close the breach in the 17th Street Canal in New Orleans after that storm.

Source: U.S. Army Corps of Engineers/Alan Dooley

life direct or indirectly, usually in a dynamic interplay with multiple factors interacting between them as a consequence of the significant changes of environmental elements such as temperature, rainfall patterns, storm severity, frequency of flooding or droughts, and rising sea levels.

Such changes impact different elements of the biosphere, including the balances between health and disease. Most populations will see the effects of climate change on health in the next decades, as billions of people face a greater risk of illnesses and uncomfortable environments, which could put their lives and well-being in danger. Today and in the recent past, some studies and well-done scientific research have demonstrated that climate change and climate variability produced significant shifts in the incidence and prevalence of different diseases—both communicable and noncommunicable.

Status of Climate Change and Predictions

During this century, the Earth's average surface temperature increase is likely to exceed the safe threshold of two degrees Celsius per century, which is the maximum increment above preindustrial average temperature. Again, this increment may produce direct or indirect effects on human health. Rises will be greater at higher latitudes, with medium-risk scenarios predicting two to three degrees Celsius rises by 2090 and four to five degrees Celsius rises in northern Canada, Greenland, and Siberia. Currently, different academic groups are outlining the major threats—both direct and indirect—to global health from climate change through changing patterns of disease, water and food insecurity, vulnerable shelter and human settlements, extreme climatic events, and population growth and migration in order to improve the understanding of them, as well to propose strategies to reduce them and to develop plans to mitigate their potential effects.

Most scientific studies indicate that health will be significantly affected by climate change and climate variability, generating increases of endemic diseases and emergence and reemergence of communicable or transmissible diseases, particularly vector-borne diseases such as malaria, dengue, yellow fever, some vector-borne encephalitis, and leishmaniasis, among others known as regional tropical diseases. Based on predictions from the 2007 Intergovernmental Panel on Climate Change (IPCC) report, global changes in temperature and precipitation patterns in different regions may impact the incidence and range of several infectious diseases within endemic areas and their introduction to free areas.

Climate Change and Human Health

Although vector-borne diseases will expand their reach and death tolls, especially among elderly people, will increase because of heat waves, the indirect effects of climate change on water, food security, and extreme climatic events are likely to have the biggest effect on global health.

Much of the impact of climate on vectorborne diseases can be explained by the fact that the arthropod vectors of these diseases are ectothermic (cold-blooded) and, therefore, subject to the effects of fluctuating temperatures on their development, reproduction, behavior, and population dynamics. Temperature also can affect pathogen development within vectors and interact with humidity to influence vector survival and, hence, vectorial capacity. The seasonality and amounts of precipitation in an area also can strongly influence the availability of breeding sites for mosquitoes and other species that have aquatic immature stages.

Despite that, in general, very little research on climate and disease published in English in peer-reviewed journals has been conducted in the countries with the highest numbers of child deaths (India, Nigeria, China, Pakistan, and the Democratic Republic of Congo), especially under-5 mortality rates (Sierra Leone, Niger, Angola, Afghanistan, and Liberia). If the United Nations Millennium Development Goals (MDGs) are to be met, then these countries should be the focus of future research, interventions, healthcare worker training, and infrastructure development. For those countries in which natural or political disasters have created a state of emergency or regions emerging from periods of political instability where rapid progress must be made to reestablish health services, then predictive studies in neighboring countries and extrapolations may help to identify high-risk locations and times and facilitate the appropriate healthcare, via nongovernmental organizations (NGOs), for example. Some studies today elucidate many disease-specific gaps and help to define areas in need of future research and how climate analysis or tools may aid this endeavor.

For example, very little information is documented on cholera–climate interactions outside of Asia, although recent studies from western Africa indicate that regional climate variability may be a driver. This disease is of particular concern on that continent, since over 90 percent of the cases reported to the World Health Organization (WHO) occur in Africa. While it is difficult to collect accurate data, especially when most cases are from epidemics occurring from natural disasters and/or political instability, retrospective analysis of selected outbreaks and meteorological data may highlight putative climatic risk factors. Furthermore, some authors have restricted their analyses to those most climate-sensitive diseases, but other diseases such as Chagas disease (American trypanosomiasis), Crimean-Congo hemorrhagic fever, tick-borne encephalitis, shigellosis, typhoid, and influenza should also be considered in future climate-related studies of disease.

For practical purposes, many climate-change indicators and variables have been identified and in consequence are assessed, evaluated, and even forecasted. Indicators that are regularly sensed by satellites are the El Niño Southern Oscillation (ENSO) indexes, which are currently measured in order to follow the impacts of ENSO on human health, and particularly vector-borne diseases, such as malaria in Africa. The identification of biological impacts of ENSO may be elusive and demands an important effort to sort out putative from real effects. In coastal marine ecosystems, where dynamic changes are the rule, the availability of long-term data seems crucial to define what is normal and what constitutes a deviation. In this sense, a careful selection of variables should also allow inference of structural and functional properties of the study system in order to minimize confounding effects. In some cases, however, a wrong or inaccurate assessment of ENSO effects may compromise more than just the quality of scientific knowledge in a theoretical sense. This is highly relevant when we deal with the management and conservation of natural resources, but it is especially critical when we use that knowledge to make decisions involving public health. In fact, researchers are increasingly incorporating ENSO in the dynamic study of marine biotoxins and diseases such as cutaneous leishmaniasis, dengue, malaria, or cholera. In such a context, thus, the quality of the analyses and information on ENSO effects has become a matter of co-responsibility with a clear bioethical dimension.

Considerable uncertainty will remain about projected climate change at geographical and temporal scales of relevance to decision makers, increasing the importance of risk management approaches to climate risks. However, no matter what the degree of preparedness is, projections suggest that some future extreme events will be catastrophic because of the unexpected intensity of the event and the underlying vulnerability of the affected population. The European heat wave in 2003 and Hurricane Katrina in 2005 are examples. The consequences of particularly severe extreme events will be greater in low-income countries. A better understanding is needed of the factors that convey vulnerability and, more importantly, the changes that need to be made in healthcare, emergency services, land use, urban design, and settlement patterns to protect populations against heat waves, floods, and storms.

Research Priorities

Key research priorities include addressing the major challenges for research on climate change and health in the following ways: development of methods to quantify the current impacts of climate and weather on a range of health outcomes, particularly in low- and middle-income countries; development of health-impact models for projecting climate-change–related impacts under different climate and socioeconomic scenarios; investigations of the costs of the projected health impacts of climate change; effectiveness of adaptation; and the limiting forces, major drivers, and costs of adaptation.

Low-income countries face additional challenges, including limited capacity to identify key issues; collect and analyze data; and design, implement, and monitor adaptation options. There is a need to strengthen institutions and mechanisms that can more systematically promote interactions among researchers, policy makers, and other stakeholders to facilitate the appropriate incorporation of research findings into policy decisions in order to protect population health no matter what the climate brings in the near or distant future.

See Also: Airborne Diseases; Arsenic Pollution; Asthma; Cities; Education and Green Health; Environmental Illness and Chemical Sensitivity; Environmental Protection Agency (U.S.); Industrial Ecology; Malaria; Smog; Suburbs; United Nations Environment Programme; Waterborne Diseases; World Health Organization's Environmental Burden of Disease.

Further Readings

Cabaniel, Gilberto, Liliana Rada, Juan J. Blanco, Alfonso J. Rodriguez-Morales, and Juan P. Escalera. "Impact of the El Niño Southern Oscillation (ENSO) Events on Cutaneous Leishmaniasis in Sucre, Venezuela, Using Satellite Information, 1994–2003." *Peruvian Journal of Experimental Medicine and Public Health*, 22 (2005).

Camus, Patricio A. "Understanding Biological Impacts of ENSO on the Eastern Pacific: An Evolving Scenario." *International Journal of Environment and Health*, 2 (2008).

Cardenas, Rocio, Claudia M. Sandoval, Alfonso J. Rodriguez-Morales, and Carlos Franco-Paredes. "Impact of Climate Variability in the Occurrence of Leishmaniasis in Northeastern Colombia." *American Journal of Tropical Medicine & Hygiene*, 75 (2006).

Cardenas, Rocio, Claudia M. Sandoval, Alfonso J. Rodriguez-Morales, and Paul Vivas. "Zoonoses and Climate Variability: The Example of Leishmaniasis in Southern Departments of Colombia." *Annals of the New York Academy of Sciences*, 1149 (2008).

Confalonieri, Ulisses and Bettina Menne. *Human Health. Climate Change 2007: Impacts, Adaptation and Vulnerability. Contribution of Working Group II to the Fourth Assessment Report of the Intergovernmental Panel on Climate Change*, M. L. Parry, O. F. Canziani, J. P. Palutikof, P. J. van der Linden, and C. E. Hanson, eds. Cambridge, MA: Cambridge University Press, 2007.

Costello, Anthony, Mustafa Abbas, Adriana Allen, Sarah Ball, Sarah Bell, Richard Bellamy, Sharon Friel, Nora Groce, Anne Johnson, Maria Kett, Maria Lee, Caren Levy, Mark Maslin, David McCoy, Bill McGuire, Hugh Montgomery, David Napier, Christina Pagel, Jinesh Patel, Jose Antonio Puppim de Oliveira, Nanneke Redclift, Hannah Rees, Daniel Rogger, Joanne Scott, Judith Stephenson, John Twigg, Jonathan Wolff, and Craig Patterson. "Managing the Health Effects of Climate Change." *The Lancet*, 373 (2009).

Intergovernmental Panel on Climate Change (IPCC). *Climate Change 2007: Impacts, Adaptation and Vulnerability. Contribution of Working Group II to the Fourth Assessment Report of the Intergovernmental Panel on Climate Change*, M. L. Parry, O. F. Canziani, J. P. Palutikof, P. J. van der Linden, and C. E. Hanson, eds. Cambridge, UK: Cambridge University Press, 2007.

Kelly-Hope, Louise and Madeleine C. Thomson. "Climate and Infectious Diseases." In *Seasonal Forecasts, Climatic Change and Human Health and Climate*, M. C. Thomson, R. Garcia-Herrera, and M. Beniston. Book Series, Advances in Global Change Research. Dordrecht, Netherlands: Springer, 2008.

Pinto, J., C. Bonacic, C. Hamilton-West, J. Romero, and J. Lubroth. "Climate Change and Animal Diseases in South America." *Revue Scientifique et Technique* (International Office of Epizootics), 27 (2008).

Rodriguez-Morales, Alfonso J. "Ecoepidemiology and Satellite Epidemiology: New Tools in the Management of Problems in Public Health." *Peruvian Journal of Experimental Medicine and Public Health*, 22 (2005).

World Health Organization, World Meteorological Organization, and United Nations Environment Programme. *Climate Change and Human Health: Risks and Responses*. Geneva: WHO, 2003.

Alfonso J. Rodriguez-Morales
Universidad de Los Andes, Universidad Central de Venezuela

COMPUTERS AND PRINTERS (INK)

As information technology has become more important to all aspects of life, ways to make computers and printers more environmentally friendly have become popular. To

The major environmental concern related to ink-jet printers is linked to their disposable cartridges, which generate waste and harm during both the manufacturing and disposal processes.

Source: iStockphoto.com

this end, efficient and effective components, such as monitors, printers, central processing units (CPUs), and the like, have become more advanced, thereby minimizing their impact on the environment. Both governmental regulation and private initiatives have played a role in the push to more efficient computer and printing technologies. Research continues in ways to make computers and printers more sustainable.

Computers

In 1992, the U.S. government began its Energy Star program in an effort to encourage energy-efficient monitors, CPUs, and related components. The Energy Star program was created by the U.S. Environmental Protection Agency (EPA) in an attempt to encourage a reduction in energy consumption, with a resulting decrease in greenhouse gas emissions by power plants. A voluntary labeling initiative, the Energy Star program allows manufacturers whose products use 20 to 30 percent less energy than required by federal standards to affix a label bearing the program's logo to their goods. The Energy Star program led to awareness among consumers of the relative energy efficiency of various options and pressured manufacturers to produce more environmentally friendly equipment. Interest in green computing soared shortly after the Energy Star program was initiated.

Although initially focused solely on computers and related peripherals, the Energy Star program has grown to include thousands of products, including home appliances, office equipment, lighting, and other energy-using devices. The program has critics who question its standards and procedures, asserting that energy savings were unreliable and often manipulated by manufacturers and hindered by outdated testing. Despite this, the EPA estimates that Energy Star saves consumers over $15 billion in energy costs each year.

Manufacturers have also initiated their own attempts to make computers greener. Efforts to make computing more sustainable focus on four distinct aspects: design, manufacturing, use, and disposal. The manufacturing of a computer accounts for 70 percent of the natural resources it uses. As a result, design that extends the life of that computer by enabling inexpensive and easy upgrades has a significant environmental impact. During the manufacturing process, computer makers are using more environmentally friendly materials to build their products. Materials that are renewable or produced using less energy are favored, and ecofriendly processes are used when possible, such as when lead soldering is eliminated. Incorporating better power-saving mechanisms assists in making computers greener during use, as does using programming algorithms. Interest in algorithmic efficiency, which waned as the cost of hardware declined, has increased as a result of the desire to make computers more energy efficient. Finally, disposal of electronic waste often threatens to add lead, hexavalent chromium, and mercury to landfills and the environment. Recycling computer equipment can prevent this, and such practices also can eliminate the need to purchase equipment that would otherwise need to be manufactured, thus saving further energy and preventing emissions. Some computer equipment can be repurposed for other uses or donated to schools, charities, or other nonprofit organizations that have a need for certain computing processes that do not require the most up-to-date equipment.

Printers (Ink)

Because printers benefit from many of the innovations to make computers more environmentally friendly, attempts to make them greener have centered on ink and other substances used for printing. Laser and ink-jet printers are the most commonly used methods of printing in the home and workplace. Laser printers use dry ink (toner) to be affixed to paper through a combination of direct contact and heat. Although economical and reliable, laser printers—because of their use of toner—present several health risks that have caused other methods of printing to be investigated. Specifically, laser printers can emit microparticles that can cause respiratory diseases. Laser printers can emit ozone, styrene, and xylenes. Ink-jet printers reproduce digital images by propelling liquid ink onto the page. Ink-jet printers present fewer health risks to the public than do laser printers, although they can emit pentanol. The major environmental concern caused by ink-jet printers deals with waste caused by disposable cartridges that generate waste and harm incurred during the manufacturing process. Concerns with laser and ink-jet printers have led to the investigation of alternative methods, including green ink and solid ink.

Ink that is environmentally friendly is sometimes referred to as green ink. Experiments have been made with soy ink, an ink substitute that is more environmentally friendly than petroleum-based ink, in an attempt to use it with desktop printers. Soy ink is made from soy oil produced from nonfood soybeans. Soy ink produces brighter colors than petroleum-based ink and makes recycling paper easier. Soy ink does, however, take more time to dry than petroleum-based ink, which has not permitted its use with personal printers.

Solid-ink printers use solid ink sticks instead of the toner or liquid ink used by laser or ink-jet printers. When used for reproduction, the solid ink is melted and used to produce images on paper in a manner similar to offset printing, which transfers an image to a plate or rubber blanket and then to a printing surface. Solid-ink printing produces more vibrant colors than other methods and is environmentally friendly due to reduced waste output and use of nontoxic materials. Solid-ink printers do require a lengthy warm-up time and use increased power to keep a portion of the solid ink heated for use between copying jobs.

See Also: Advertising and Marketing; Environmental Protection Agency (U.S.); Government Role in Green Health; Private Industry Role in Green Health.

Further Readings

Espejo, Roman, ed. *What Is the Impact of E-Waste?* Farmington Hills, MI: Greenhaven, 2008.
Kuehr, Ruediger and Eric Williams. *Computers and the Environment: Understanding and Managing Their Impacts (Eco-Efficiency in Industry and Science)*. Boston: Kluwer Academic, 2003.
Schultz, G. *The Green and Virtual Data Center*. Boca Raton, FL: CRC Press, 2009.

Stephen T. Schroth
Adam J. Mize
Knox College

COST-BENEFIT ANALYSIS FOR ALTERNATIVE PRODUCTS

Cost-benefit analysis is a type of economic analysis that requires the quantification of the costs and benefits of any action and that may be used to choose among various actions, depending on which is projected to yield the greatest benefits in relation to costs. It is often used in fields such as healthcare and environmental studies because it provides a means to assign numeric value (which may be expressed in monetary terms such as dollars or in other units) to entities that are not ordinarily thought of in those terms (e.g., the value of preserving one human life or of maintaining a wilderness area in unspoiled condition). There are many controversies involved with cost-benefit analysis, and different analysts may come up with different results for the same problem, depending on their assumptions and methodology. However, it has proven to be a useful technique for decision making and has the advantage of allowing a broad consideration of societal costs and benefits within the decision-making process.

Economic Analysis

Cost-benefit analysis is one among a family of economic analyses that may be applied in particular circumstances to guide the choice among alternatives. All require that the scope of costs and benefits be specified: for instance, is cost simply the dollar amount paid, or does it include societal considerations such as whether a choice will increase pollution in some part of the world? And do benefits refer only to the immediate benefits to an individual or clearly specified group, or are more diffused social benefits also considered?

A *cost-minimization analysis* is the simplest, because it assumes that the outputs, or benefits, would be the same under several different courses of action and therefore seeks to identify which would have the lowest cost. For instance, a university might choose between a number of suppliers for a given product according to which offered it at the lowest price (this assumes that the products in question are identical or equivalent and that there are no other considerations such as the green policies of the suppliers). A *cost-effectiveness analysis* may be used when the courses of action have not only different costs but also different effectiveness in producing the output: for instance, choosing to implement one diabetes intervention plan rather than another depending on which saves the greatest number of disability days per dollar invested. A *cost-utility analysis* adds the opportunity to incorporate a consideration of the value of the output: for instance, measuring the benefits of a program in terms of quality adjusted life years (QALYs), which incorporates societal valuations of the quality of life in different states of health (rather than simply counting the days).

Cost-benefit analysis allows for consideration of outputs that are not common to all alternatives (e.g., increasing employment in one town versus preserving a wilderness area near another), to place those outputs on a common scale, and to determine which choice of action has the greatest net social benefit for the least net social cost. The ideas of cost-benefit analysis are not new: they date back to the 19th century, and a well-known early application dates from the 1930s and the U.S Army Corps of Engineers. They developed a system for measuring the benefits and costs of their projects in order to evaluate projects, because in the Federal Navigation Act of 1936, they were charged with carrying out projects to improve the waterway system if the benefits (broadly defined) exceeded the costs. The principles of cost-benefit analysis were further refined by economists in the 1950s, and there are still areas of disagreement today due to the complex nature of the questions that must be considered.

In order to perform a cost-benefit analysis, a common unit of measurement must be established for all costs and benefits. Often it is the local monetary unit (e.g., the dollar in the United States) because this has a value anyone can understand. The costs and benefits must also be expressed in terms of value at a particular point in time, and benefits that are expected in the future are generally valued less than if the same benefits were available today. This is based on common sense as well as economic theory: you probably value the promise of a dollar five years from now less than receiving an actual dollar today, and economic theory allows you to specify that difference by assigning a present value or a discounted value to the dollar you expect to receive in the future. A third specification is that costs and benefits are determined within a specified location, whether that is a single company or the entire world: the choices made in choosing which area or location to consider can radically affect the results of the cost-benefit analysis. Another way to look at this is that the analysis must specify whose costs and whose benefits will count in your calculations.

Performing a cost-benefit analysis often requires establishing the value in the chosen units for things that may not ordinarily carry such a price tag. There are various ways economists have tried to establish these values: for instance, asking a varied group of individuals how much money they would accept as equivalent to the benefit in question. Because people often find these questions difficult to answer, some economists try to find real-life examples of people making choices that can be interpreted as an indication of how much they value something intangible. For example, if people have the choice of parking in a lot close to their office for $10 or in a lot a 30-minute walk away for free, you could say they valued their time at $10/hr (assuming a round-trip walk). Of course, choices seldom present themselves so clearly, not everyone makes the same choice (people place different values on their time: some may enjoy the exercise provided by the walk), and there may be other alternatives as well (carpooling to split the parking fee or using public transportation, for instance). Often values have to be calculated using less-satisfactory examples: for instance, you could calculate the value of easy access to good schools by considering the price of equivalent homes in a good versus an inferior school district.

As can be seen, arriving at satisfactory valuations for the costs of intangibles is as much art as it is science, and if the question is something like the value of enjoying a year of life in good health, then it becomes even trickier. Yet these types of calculations are often required in the decision-making process, and the alternative is to fail to account for them in the model. Philosophically, a single human life may hold infinite value, but in terms of economic analysis, such an approach precludes including the value of lives saved in your model at all and for the purposes of the analysis has the same result as if human life were assumed to hold no value.

Cost-Benefit Analysis for Alternative Products

In the case of alternative products, the question is often whether some goal should be achieved using conventional or alternative means: for instance, should a planned building be constructed following conventional building practices or by incorporation of environmentally friendly techniques? This type of evaluation may go beyond simple considerations of dollar cost to include factors such as aesthetics, worker health, and public image, but the technique of cost-benefit analysis allows such considerations to be assigned a value and included in the calculations. Of course, arriving at such valuations may be more problematic than, for instance, determining the cost of building materials; a reasonable estimate can often be found for intangible factors, and because the green movement places a value on such things (e.g., quality of life and reductions in pollution), it is logical to include them in calculations.

Some cost-benefit analyses also consider the effects that a new product or building may have on the current environment with the acknowledgment that the effect may not be simply additive. For instance, if an electrical grid is already overburdened and the electrical supply is subject to shortages (or to the need to purchase additional electricity at high prices), then those factors might be included in the cost of electricity for the new product. By the same reasoning, if there is already a water shortage in an area, then adding a project that makes further water demands may significantly increase the cost or reduce the available supply of water for everyone. To take a third example, if a new building will bring large numbers of new people to an area and thus require improvement in the transportation infrastructure, then the projected costs of this upgrade could also be included in the analysis. In all these cases it is important to specify whose costs and benefits are included in the analysis. For instance, building new roads to accommodate increased traffic costs someone (perhaps the local government) money even if the company that built the new building does not have to pay for them directly—will that cost be included in the analysis?

Green Building

A good example of a cost-benefit analysis for an alternative product is the 2003 report (Greg Kats) on the financial benefits and costs of green buildings, which evaluated the success of a green building program for public buildings in California. The major conclusion of this report was that although green design and building presented slightly higher initial costs (about 2 percent on average), over the life cycle of the average building, lower operating costs and other benefits would save about 20 percent of the initial construction costs, returning 10 times the initial increase in expenditure.

This report considered many different types of benefits, including not only those on which it is relatively easy to place a dollar valuation (lower costs for energy and water use, waste disposal, and maintenance) but also more intangible factors like increased productivity and health. To find valuations for these intangibles, the authors reviewed the available literature and used estimates from previous studies to arrive at valuations for the different benefits. For instance, one of the tenets of green building is that natural lighting or daylighting (i.e., having indoor spaces lit by sunlight whenever possible) is preferred by most people to using artificial lighting and will result in improved school and work performance. Kats and his colleagues located studies supporting this, which they included in their calculations: for instance, one study determined that students in classrooms with natural light performed up to 20 percent better than those in classrooms with little daylight, and another found that worker productivity increased 7 percent after a company moved to a daylit facility.

The final calculations in this report made it clear that green buildings, despite the higher initial cost of construction, more than paid for that initial investment over time. The actual savings in energy use were enough to offset the higher initial cost (net present value per square foot of $5.79 versus green cost premium of $4), but the greatest savings came in terms of improved productivity and health. Those savings increased for buildings that incorporated more green features as measured by their status in terms of LEED certification (a rating by the U.S. Green Building Council). Worker health and productivity was esti-mated to have a net present value of $48.87 per square foot for buildings that were LEED-certified or had LEED Silver status and $67.31 per square foot for buildings that had the higher levels of LEED Gold or Platinum status. This result emphasizes that intangibles such as worker health and productivity, if they are considered valuable, should be included in cost-benefit calculations (this gets back to the question of whose costs and benefits will be included in the analysis) because they may represent a major factor in the evaluation of alternative products.

The report by Kats and his colleagues used a life cycle costing (LCC) approach that looked at the costs and benefits over the expected life cycle of a product (in this case, a building, which was estimated at 20 years). A more rigorous approach is that of life cycle analysis (LCA), which looks at all the upstream and downstream costs of an activity, pro-ducing a "cradle-to-grave" impact and assessment. Although LCA is well developed in some industries, this is not the case for building, and data were not available to complete the model, so the more limited LCC approach was used instead.

Ethanol

The use of ethanol, a fuel source made from crops such as sugar or corn, is sometimes touted as an environmentally friendly alternative to petroleum-based fuels, as well as a means for countries such as the United States to reduce their dependence on foreign oil and to bring economic benefits to economically depressed regions. However, the net financial and environmental benefit of ethanol is a contested issue, and scholars have come to dif-fering conclusions as to whether the benefits outweigh the costs. Michelle Isenhouer reviewed various studies that have looked at this issue and concluded that the benefits of ethanol production do not outweigh the costs in the United States in the short term. She provides an interesting review of how political and economic local interests may influence evaluation of the costs and benefits of any changes and emphasizes the importance for any cost-benefit study of deciding at the start exactly what the geographic scope of the study will be and whose costs and benefits will be considered.

In the first place, Isenhouer distinguishes between ethanol produced from sugar (com-mon in Brazil), ethanol produced from corn (common in the United States), and ethanol produced from cellulose (e.g., switchgrass, corn stalks, wood chips), which is currently rare in the United States but could be expanded. Although the example of Brazil's success in adopting sugar ethanol as a fuel is often cited as an argument in favor of increasing ethanol production in the United States, the two cases are not comparable. Sugar ethanol can be produced more cheaply and easily than corn ethanol, and the process is also more environ-mentally friendly. The United States lacks an appropriate climate for large-scale sugar production, so any discussions of the United States must be limited to the practicalities of corn and cellulosic ethanol.

Corn ethanol offers no environmental benefits due to the large amounts of fertilizer and pesticides used in cultivation, but it does produce financial benefits: ethanol production has increased wages and employment in corn-growing states and also provides industrial employment in states both within and outside the Corn Belt. People involved in the ethanol industry purchase products from other businesses and pay taxes on their profits or wages, and corn prices have increased since the introduction of ethanol.

The production of cellulosic ethanol does produce environmental benefits because it burns more cleanly than gasoline and produces fewer emissions. It can be made from plant waste products such as corn stalks left after harvesting and wood chips from logging operations and from plants such as switchgrass, which can be grown on land not otherwise in use and which is minimally destructive to the environment. Cellulosic ethanol can also be grown in much of the United States (unlike corn, which is mainly produced in only a few states) and could provide the financial benefits of this industry to many parts of the country.

The financial and environmental costs of ethanol production differ depending on the type of fuel used. In some states such as Kentucky, there is political pressure to use coal to power ethanol plants rather than more environmentally friendly fuels such as livestock manure (which actually reduces the amount of pollutants in the air by consuming methanol created from the manure). Costs associated with subsidies (corn production is heavily subsidized by the federal government) and tax credits should also be included in the calculation, as well as any enhancements to the infrastructure that is required to support the industry.

Ethanol production has raised the price of both corn and meat, and this must also be considered a cost to society as well as a benefit to the farmer. Opportunity costs should also be considered if productive land is shifted due to the ethanol industry (e.g., from growing food crops to growing corn for ethanol); this is less likely to apply to cellulosic ethanol, which is often created from waste products or crops such as switchgrass, which can grow on land otherwise unsuited for cultivation.

Substitution of corn production for rotated crops may result in environmental costs and greater erosion and runoff. Finally, the environmental costs from the pesticides and fertilizers used in corn production should include the fact that they may end up in streams and rivers that empty into the Mississippi River and eventually the Gulf of Mexico, in some cases seriously damaging the ecosystem.

Criticisms of Cost-Benefit Analysis

Although cost-benefit analysis is a common technique used in economic evaluation and decision making, it also has its critics. One of the most common arguments against cost-benefit analysis is that it requires placing a numeric value (often a monetary value) on things such as human life, whose value is essentially infinite. In some cases people have argued that placing a value on intangibles simply produces license to destroy those things. For instance, requiring a company to pay a financial penalty for polluting a river might be interpreted as facilitating that pollution because paying the compensation gets the company off the hook, morally speaking.

A second problem is that there is no universally accepted method to set the value of intangibles and that any such calculation will require making many assumptions, which leaves the process open to manipulation. This means someone with an interest in the outcome can select the method and assumptions that will produce the most favorable result from their point of view. To take a recent example in the United States, the value of a human life has been calculated as both $6.1 million (by the Environmental Protection Agency, during the presidency of Bill Clinton) and $2.6 million (during the presidency of George W. Bush).

One must also consider whether use value or existence value will be considered in the analysis. For instance, after an oil spill it is possible to estimate the economic losses to people who live and work in the area (e.g., due to lost employment), but individuals may place a much higher value on avoiding an oil spill (e.g., having a pristine environment) than simply the loss of business. In the case of the *Exxon Valdez* oil spill, the use value loss was estimated at $300 million, while the existence value loss was calculated at $9 billion. A final problem is that the use of discounting makes sense from the point of view of people living at a particular moment of time, but decisions made about the environment should consider future generations as well. For instance, the costs of climate change in the future may be very small to someone living today (in fact, there may be no major changes during that person's lifetime), but to someone living 100 years from now, those costs may be very high.

See Also: Antibiotic Resistance; Biological Control of Pests; Chemical Pesticides; Genetically Engineered Crops; Light Bulbs; Organic Produce; Pest Control; Phaseout of Toxic Chemicals; Recycled Water.

Further Readings

Ackerman, Frank. *Priceless: On Knowing the Price of Everything and the Value of Nothing.* New York: Free Press, 2004.

Environmental Literacy Council. "Cost Benefit Analysis." http://www.enviroliteracy.org/article.php/1322.html (Accessed August 2010).

Heinzerling, Lisa and Frank Ackerman. "Pricing the Priceless: Cost-Benefit Analysis of Environmental Protection." (2002). http://www.ase.tufts.edu/gdae/publications (Accessed August 2010).

Isenhouer, Michelle. "Is Going Yellow Really Going Green? A Cost-Benefit Analysis of Ethanol Production in America." *Pepperdine Policy Review*, 1 (2008). http://publicpolicy.pepperdine.edu/policy-review/2008v1/ethanol-production.htm (Accessed August 2010).

Kats, Greg. "The Costs and Financial Benefits of Green Buildings: A Report to California's Sustainable Task Force." (October 2003). http://www.usgbc.org/Docs/News/News477.pdf (Accessed August 2010).

Pearce, David W., Giles Atkinson, and Susana Mourato. *Cost-Benefit Analysis and the Environment: Recent Developments.* Paris: Organisation for Economic Co-operation and Development, 2006.

Revesz, Richard L. and Michael A. Livermore. *Rethinking Rationality: How Cost-Benefit Analysis Can Better Protect the Environment and Our Health.* New York: Oxford University Press, 2008.

Smith, V. Kerry, ed. *Environmental Policy Under Reagan's Executive Order: The Role of Cost-Benefit Analysis.* Chapel Hill: University of North Carolina Press, 1984.

Watkins, Thayer. "An Introduction to Cost Benefit Analysis." http://www.applet-magic.com/cba.htm (Accessed August 2010).

World Health Organization. "Cost Benefit Analysis of Interventions." http://www.who.int/indoorair/interventions/cost_benefit/en/index.html (Accessed August 2010).

Sarah Boslaugh
Washington University in St. Louis

DEGENERATIVE DISEASES

Degenerative diseases are characterized by a progressive degenerative change in tissue. The result of the degeneration is a reduction of structure or function of the associated tissue. These diseases usually follow a clinical course that includes an asymptomatic stage with a gradual progression in the manifestation of symptoms. Progression can continue from mild symptoms all the way to lethality. The incidence of degenerative diseases correlates well with aging and will more likely be symptomatic in the later years.

Degenerative diseases can strike any body organ or system. Most are not communicable and do not have an infectious etiology, although there are some exceptions to this. Acquired immune deficiency syndrome (AIDS) is caused by the human immunodeficiency virus (HIV); stomach ulcers and cancer are associated with *Helicobacter pylori*; liver cancer can be caused by the hepatitis B virus; and cancers of the cervix are linked to the human papilloma virus (HPV).

The Nervous System

One of the more well-known groupings of degenerative diseases strikes the nervous system. Degenerative nerve diseases can affect many of the body's activities, such as movement and balance, heart function, talking, and breathing. Certain medical conditions, toxins, chemical exposures, and viruses can cause degenerative changes, such as alcohol abuse, tobacco abuse, tumors, and strokes. However, a significant number of these conditions are genetically based, meaning they occur in families or are caused by a genetic mutation.

Alzheimer's disease is the most common form of dementia among the elderly. Onset usually occurs after the age of 60 and begins slowly by affecting a person's ability to control thought, memory, and language. Symptoms worsen over time and sufferers become unable to read, write, and perform activities of daily living. They may lose the ability to recognize family members and become anxious or aggressive or possibly wander away from their residence. In the end, people with Alzheimer's disease require total care. Some medications are available to delay the progression of the disease, but none have proved effective in stopping it entirely.

Amyotrophic lateral sclerosis (ALS), commonly known as Lou Gehrig's disease, attacks nerve cells, called motor neurons, in the brain and spinal cord. These neurons lose the ability to voluntarily control muscles. Initial symptoms include trouble with walking or running, writing, and speech. As the disease progresses, the patient loses strength and is unable to move. When the muscles of respiration fail, the individual is no longer able to breathe and dies from respiratory arrest. Onset of symptoms is usually between the ages of 40 and 60 years, and occurs more often in men than women. The exact cause of ALS is unknown, but some evidence suggests the immune system may be involved. Approximately 90 to 95 percent of cases are sporadic. There are no cures for ALS, but medicines are available to help relieve symptoms of the disease and prolong the patient's lifespan.

Parkinson's disease is a disorder affecting neurons in an area of the brain called the substantia nigra. As these neurons degenerate, the brain's ability to generate body movements becomes disrupted and symptoms of Parkinson's disease occur. These symptoms include tremors, rigidity, loss of spontaneous movement, slowness of movement, and problems with walking and posture. There are no medications to cure the disease, but many treatments can slow the progression substantially. Many people with Parkinson's disease do not reach advanced stages of the disease because they continue to receive significant benefit from their anti-Parkinson medications.

The Musculoskeletal System

The musculoskeletal system can also be affected by degenerative diseases. It is composed of the systems of the body that provides support, form, stability, and locomotion, among other tasks. The musculoskeletal system consists of the body's bones and the surrounding soft tissues, which are muscles, tendons, ligaments, cartilage, and other connective tissue. The term *connective tissue* is used to describe the tissue that supports and binds tissues and organs together. There are actually only a few degenerative diseases of the bones, as most degenerative diseases of the musculoskeletal system attack the soft tissues.

Osteoporosis is by far the leading form of degenerative disease of the bone. In osteoporosis, the bones lose density at an accelerated pace and can become weak, increasing the individual's risk of fracture. Bones are in a constant state of change. Bone mineral is broken down and replaced by new mineral deposits. Bone density builds during childhood and the early adult years. During this critical period early in life, diet and exercise are of vital importance to build as much bone mass as possible. From around 30 years of age onward, bones slowly lose density because the body's ability to replace bone that has broken down with new bone mineral is compromised. This bone loss can become accelerated in women after menopause due to the loss of production of estrogen, the hormone that helps protect against bone loss. It is due to this fact that women are four times as likely to suffer from osteoporosis as men. One in every two women over the age of 50 will have an osteoporosis-related fracture in her lifetime.

There are several medications from the class bisphosphonates that are available to help prevent further bone loss in individuals who exhibit early signs of increased bone loss. However, the best treatment of osteoporosis is aggressive prevention. Education on the importance of diet and exercise in building a higher peak bone mass early in life is the most successful measure to prevent osteoporosis.

A second type of degenerative disease of the bone is osteitis deformans, also known as Paget's disease. This disease is characterized by a progressive focal disorder of bone remodeling, where normal bone is destroyed and replaced with abnormal bone. In Paget's disease, there are exceptionally large osteoclasts, cells that reabsorb bone, at locally affected sites. When areas of bone are affected by these osteoclasts, there is often a marked reaction to form new bone, but the bone is formed in a very disorganized and defective manner. Symptoms of Paget's disease include pain in the area affected, enlarged yet brittle bones, and damage to cartilage in the joints. The exact cause of Paget's disease is still unknown, but research has looked at a genetic link as well as an infectious etiology. Bisphosphonates are used to treat Paget's disease and can prevent worsening of the disease, but do not improve or reverse damage that has already occurred.

There are several degenerative diseases of the muscles, with the most well-known being the muscular dystrophies. Muscular dystrophy refers to a group of genetic, hereditary muscle diseases that weaken the muscles that move the human body. There are more than 100 diseases with similarities to muscular dystrophy. The hallmark characteristics include progressive weakness in skeletal muscles, defects in muscle proteins, and death of muscle cells and tissue. Duchenne muscular dystrophy is the most well-known type of the disease. It is an X-linked recessive inherited disorder and as such affects males much more than females because males only have one X chromosome. For a female to have Duchenne muscular dystrophy, both X chromosomes must have mutations to cause the disorder. Diagnosis is based on biopsy of muscle tissue. In some types of muscular dystrophy, muscle wasting can be difficult to assess without a biopsy, as muscle tissue can be replaced with connective tissue and fat, making the muscles appear larger than normal. The prognosis for patients with muscular dystrophy varies depending upon the type. Some people with more mild forms of the disease have few symptoms until late in life, while more severe forms of the disease can lead to functional disability and death in infancy. There is no known cure for muscular dystrophy. Since the disease seems to follow a poor course with inactivity, the mainstay of treatment is physical and occupational therapy, orthopedic appliances for support when needed, and occasionally surgery.

Far more commonly, degenerative diseases of the musculoskeletal system strike the joints. Arthritis is the most well-known degenerative disease affecting the joints of the body. The word *arthritis* means joint inflammation. There are more than 100 types of arthritis, but two types are much more commonly known than the others. Osteoarthritis is a degenerative process of articular cartilage on the surface of a joint due to wear and tear, and is the most common of all arthroses. Wear and tear can occur due to many factors, including heredity, obesity, joint injury, repeated overuse, and aging. Onset of symptoms usually occurs slowly and can include steady or intermittent joint pain, stiffness after periods of inactivity, and swelling in the affected joints. As the process advances, destruction of the joint can occur. This destruction can lead to overgrowth of bone called osteophytes, narrowing of joint space, deformity of the joint, and sclerosis or hardening of bone at the joint surface.

Rheumatoid arthritis is an autoimmune disease that causes chronic inflammation of the affected joints. Individuals with rheumatoid arthritis have antibodies in their blood that attack and destroy their own tissue. Not only joint tissue can be affected in rheumatoid arthritis; many other tissues within the body can be attacked as well. A newer term used to describe the condition is *rheumatoid disease* instead of rheumatoid arthritis. Even though it is a prolonged and progressive disorder, patients may often have periods without symptoms.

Arthritis treatment begins with conservative and over-the-counter (OTC) remedies at home. Minimizing risk factors when applicable, such as weight loss for the obese patient with osteoarthritis, should be attempted. Topical pain medications can be applied over affected areas, as can heat applications. This includes dipping the affected joint in melted paraffin wax. Other available OTC pain medications include acetaminophen and nonsteroidal anti-inflammatory drugs (NSAIDS). When conservative measures fail, a healthcare provider may move to stronger dosages of NSAID or corticosteroid medication to control the pain and inflammation associated with arthritis. Immune system–based arthritis may require medications designed to suppress the immune system. Medications such as methotrexate and sulfasalazine are examples. Surgery is a last resort treatment option for individuals with severe joint destruction from arthritis. The goals of arthritis surgery are to relieve pain, improve function, and correct deformity.

Diabetes

Diabetes mellitus is a multisystem disease caused by lack of production or underutilization of insulin, resulting in a state of elevated blood glucose (sugar) levels. In type 1 diabetes, the body does not make insulin. In type 2 diabetes, the most common type, the body does not make or use insulin well. Over time, elevated blood glucose can lead to degenerative changes in most organ systems within the body. Diabetes is a leading cause of blindness due to retinopathy from damage to the tiny blood vessels on the inside of the eye. For this reason, diabetics should have a complete eye examination yearly.

Elevated blood glucose also can also damage the covering on nerve cells. The damaged nerves then cannot properly send messages through the areas of damage. This condition, called diabetic neuropathy, occurs in about half of all diabetics and can lead to numbness in the hands or feet, urinary problems, dizziness, shooting pains, burning, or tingling. Of particular concern is when damage occurs to the nerves innervating the bottom of the feet. Lack of sensation in the feet can lead to blisters, cuts, or sores the patient cannot feel. If left untreated, they can become infected and form ulcers and lead to potential amputation of the affected section of the extremity. Kidney damage, termed diabetic nephropathy, can begin long before any symptoms develop. Diabetics should be monitored regularly for protein in the urine, a sign that kidney damage is occurring. If the damage is left unchecked, the end result is kidney failure. People with kidney failure need either a kidney transplant or dialysis.

Diabetes is treated by aggressively managing blood glucose levels. Early on or in less severe cases, diet modification and exercise regimes are attempted to control the disease. When those measures fail, medication is needed to help control blood glucose levels. In Type 1 diabetes, since the pancreas does not make insulin, patients need to supplement with insulin preparations. In type 2 diabetes, medications are used to assist the body in the use of insulin. Several types of medications are available, and often patients need more than one type of oral diabetic medication to control their blood glucose levels. With all types of diabetes, the patient should receive extensive education on the disease and monitoring of blood glucose levels.

A final area of note is a small group of infectious diseases with degenerative sequela. Creutzfeldt-Jakob disease (CJD) and bovine spongiform encephalopathy (BSE, also known as mad cow disease) are infectious diseases that cause degenerative damage to the brain and other nerve tissues within the body. These diseases are caused by infectious protein particles called prions, which are misfolded protein particles. Prions are unable to replicate themselves; instead, they induce normal proteins within the body to convert to the misfolded form. These new misfolded forms can then repeat the process, starting a chain reaction that causes the prions to accumulate in previously healthy tissue, causing cell damage and death.

See Also: Cardiovascular Diseases; Kidney Diseases; Men's Health; Mental Exercises; Musculoskeletal Diseases; Neurobehavioral Diseases; Obesity; Physical Activity and Health; Women's Health.

Further Readings

eMedicineHealth. "Arthritis." http://www.emedicinehealth.com/arthritis (Accessed January 2010).

Harrison's Principles of Internal Medicine. New York: McGraw-Hill Professional, 2005.

MedlinePlus. http://www.nlm.nih.gov/medlineplus/diabetesmedicines.html (Accessed January 2010).

National Institute of Arthritis and Musculoskeletal and Skin Diseases. http://www.niams.nih .gov/Health_Info/Bone/Pagets/default.asp (Accessed January 2010).

National Institute of Neurological Disorders and Stroke. http://www.ninds.nih.gov/disorders/ amyotrophiclateralsclerosis/detail_amyotrophiclateralsclerosis.htm (Accessed January 2010).

National Institute on Aging. "Alzheimer's Disease Education and Referral (ADEAR) Center." http://www.nia.nih.gov/Alzheimers (Accessed January 2010).

National Osteoporosis Foundation. "Osteoporosis." http://www.nof.org/osteoporosis/ diseasefacts.htm (Accessed January 2010).

University of Maryland Medical Center. "Maryland Parkinson's Disease and Movement Disorders Center." http://www.umm.edu/parkinsons (Accessed January 2010).

Rance McClain
Independent Scholar

DEHYDRATION

Dehydration means an abnormal depletion of body fluids, which occurs when fluid intake does not match body fluid loss, giving rise to imbalances. Dehydration is often the result of a scarcity of drinking water, which in turn is linked to population growth and global warming.

Dehydration Effects on Health

Fluid intake occurs principally during drinking (about 60 percent) and eating (about 30 percent); 10 percent of fluid intake results from metabolic processes within the body.

Fluid loss occurs through excretion from the kidneys (which in resting conditions amounts to approximately 60 percent of the total fluid loss), excretion from the large intestine (approximately 5 percent), and evaporation of water from the respiratory tract and sweat secretion (about 35 percent). Excessive fluid loss can be fatal, because water is essential for human life: it constitutes nearly 60 percent of body weight and about 72 percent of lean body mass and forms the basis for all body fluids, among them blood. As water is the principal constituent of the body, even a 9 to 12 percent fluid loss can have serious consequences.

The body fluids have vital functions: they contribute to the transportation and absorption of nutrients, allow muscle contraction, and help in eliminating waste. For this reason, it is essential to maintain the correct parameters of these fluids, especially blood parameters, such as blood volume and blood pressure. In effect, the body tries to defend blood volume as much as possible also in hypohydration situations: when there are deficits in body water content, water passes from inside the cells to the bloodstream. Through this fluid exchange, dehydration causes a redistribution of body water, which largely derives from a depletion of the water content of muscles and skin. Thus, depletion of body levels of fluid can significantly stress the body, impairing physical performance.

Hence, dehydration should be limited, or better still, prevented by drinking more fluids. Yet if fluid intake cannot balance fluid loss, dehydration becomes more severe. The body tissues start to dry out to such an extent that cells may shrivel and malfunction.

Symptoms of dehydration can be thirst; dry mouth, skin, and mucous membranes; decreased sweating and urine output; headache; muscle weakness; excessive fatigue and tiredness; dizziness; cramps; fever; low blood pressure; and in the most severe cases, damage to internal organs, confusion, delirium, or even unconsciousness. Fluid loss rates can vary greatly depending on many factors. Some of them concern specific characteristics of each individual (e.g., gender, age, health status); others may result from climatic conditions, such as humidity and temperature.

Therefore, dehydration should be prevented by consuming drinks and foods high in water content, such as fruit and vegetables.

Global Warming

Research has shown that the average temperature of the Earth's near-surface air and oceans has been increasing since the 1950s. Many scholars claim that this general rise is due to the presence of specific gases in the atmosphere. These gases, which include water vapor, carbon dioxide, methane, nitrous oxide, and chlorofluorocarbons (CFCs), cause the so-called greenhouse effect—absorbing infrared radiation from the sun and trapping it in the troposphere, the lowest layer of the atmosphere, directly above the Earth's surface.

Some of the greenhouse gases are human produced: for instance, the increase of carbon dioxide results principally from fossil fuel burning, industrial activity, and deforestation. Trees are logged and/or burned in order to extend pasture, obtain charcoal, and establish new human settlements. Thus, deforestation can contribute to global warming. In turn, global warming can cause a decrease in rainfall; this lack of available water threatens forests, which can no longer exert their climate protection effect. This in turn provokes the expansion of subtropical deserts.

In the last 60 years, due to a lack of effective environmental management, heat waves have been more frequent, intense, and longer lasting. Because the amount of water resources greatly depends on climate, the quantity of available water in heat-stressed regions has gradually lessened. Since water is of primary importance in all forms of life, heat-related impact often includes dehydration.

Water Scarcity

Since the end of the 19th century, total water demand has greatly increased all over the world because water is used for drinking purposes and in many production processes.

About 95 percent of the Earth's water is seawater; the available freshwater is only 5 percent, and amounts are rapidly becoming depleted in populous countries such as India and China. The 2002 United Nations Environment Programme predicted that about half of the world's population will suffer water shortage by the year 2032.

Although improved agricultural practices have recently decreased the percentage of water used in agriculture, the amount of available freshwater has decreased because of agriculture-related groundwater pollution. Groundwater represents the source of two-thirds of available freshwater, but its quality is threatened by agricultural pollutants such as herbicides and pesticides.

Measures to Be Adopted

For all these reasons, our currently unsustainable production practices and lifestyles cannot be maintained. In order to avoid a scarcity of water and the subsequent risk of dehydration, it is necessary to improve environmental management and conservation of resources, for instance, reducing the dispersion of drinking water through leaking sewers and distribution systems, and limiting the use of drinking water as a transport medium for human and animal waste.

Furthermore, it is necessary to enhance pollution prevention, promote the use of non-pollutant agroadditives, and decrease emissions of gases that can be considered co-responsible for the greenhouse effect.

See Also: Chemical Pesticides; Climate Change; Groundwater; Supplying Water; Water Scarcity.

Further Readings

Hunt, Constance Elizabeth. *Thirsty Planet: Strategies for Sustainable Water Management.* London: Zed Books, 2004.

Jury, William A. and Henry Vaux. "The Role of Science in Solving the World's Emerging Water Problems." *Proceedings of the National Academy of Sciences of the United States of America*, 102 (2005).

United Nations World Water Assessment Programme. *United Nations World Water Development Report: Water, A Shared Responsibility.* Oxford, UK: Berghahn Books, 2006.

Alessandra Padula
University of L'Aquila

Dental Mercury Amalgams

Mercury amalgams are alloys of mercury (a heavy metal that can produce severe organ damage) with other metals. They are used in dentistry as filling materials for treatment of dental caries. In some countries, dental mercury amalgams are classified as medicines and therefore subjected to rigorous regulations. In other countries (among them the United

States), they are considered devices; therefore, they have not been submitted to the strict official government testing required for medicines.

Although mercury toxicity is well known, scientific studies on the potential health effects of mercury amalgams used in dentistry have come to opposite conclusions.

Mercury

Mercury is, after plutonium, the most toxic substance on the Earth. If mercury is ingested or inhaled, it is quickly assimilated and accumulates in various tissues, affecting primarily the nervous, immune, and urinary systems. In fact, in past centuries, because mercury was widely used in the felt hat industry, hatters often drooled, trembled, and had bouts of paranoia: this gave rise to the saying "mad as a hatter."

Mercury can increase production of free radicals, diminish availability of antioxidants, and increase retention of other toxins, causing rapid oxidation, dysfunction, and cell death.

The U.S. Environmental Protection Agency (EPA), aware of mercury's toxicity and concerned about mercury release in the environment (e.g., from incinerators, crematoria, and coal-fired power plants), designated this substance as constituting a waste disposal hazard—in fact, mercury can pollute the environment, imposing health risks upon the population. For this reason, in order to limit damage to the environment, some countries arrange for the recovery of mercury from wastewater, and other countries ban the release of mercury into the public sewer system.

However, in regard to mercury emissions, some studies reported controversial results: whereas a 1992 study showed that 86 percent of discarded mercury was released from batteries, and only 0.56 percent from dental amalgam, in 2005 the World Health Organization (WHO) reported that mercury from dental offices and laboratories accounted for one-half of total mercury emissions. Despite this, many countries (including the United States) do not regulate the use of mercury in dentistry.

Use of Mercury in Dentistry

Mercury has been used as a component of amalgam fillings in Europe since 1603 and in the United States since 1833. Compared with other dental restorative materials, this compound has several advantages: strength, durability, ability to limit growth and reproduction of bacteria, easy application, and low cost. For these reasons mercury-containing amalgams have been widely employed in the United States since the mid-1800s; yet a controversy over their use in dentistry dates back to the 1840s. In fact, by 1845 all the members of the American Society of Dental Surgeons pledged not to use amalgam fillings, fearing mercury poisoning in both patients and dentists. Yet the American Dental Association (ADA), founded in 1859, did not ban the use of amalgams and still holds this position.

Amalgam fillings are frequently called "silver fillings" but contain much more mercury than silver: on average, 50 percent mercury, 35 percent silver, 13 percent tin, and minimal quantities of copper and zinc. However, a 2006 poll conducted in the United States found that only 28 percent of interviewees were aware of the mercury percentage in silver fillings.

Furthermore, WHO reported in 2005 that, because many countries do not impose specific waste-disposal procedures for mercury-containing waste products, these are nearly always flushed down the drain, accounting for one-third of the mercury present in the sewage system. Already in 2002, the Association of Metropolitan Sewerage Agencies (AMSA) found that dentistry mercury represented 40 percent of the total mercury content in wastewater treatment plants, and dental offices and laboratories were by far the heaviest contributors to this load.

Health Effects From Amalgam Fillings

In the last 75 years, several studies have reported that changes in health may result from exposure to amalgam fillings, among them are the following:

- Fatigue, weakness, tremors
- Lack of concentration and memory, irritability, insomnia
- Taste, appetite, and digestion problems
- Hoarse voice
- Skin alterations, destructive changes in the mucous membranes
- Oral lesions, inflammatory periodontal tissue reactions
- Hearing loss, tinnitus, vertigo
- Changes in visual system, double vision, graying of the lens of the eye
- Cardiovascular, chest, joint, mouth, and throat pain, painful swallowing and menstruation, headaches
- Fine-motor capability deterioration, neurobehavioral changes

These symptoms led researchers to hypothesize a connection between mercury poisoning from amalgam fillings and various disorders:

- Chronic viral illnesses: Epstein-Barr virus (EBV), cytomegalovirus (CMV), human immunodeficiency virus (HIV), herpes zoster and genital herpes, chronic fatigue, and immune dysfunction syndrome (CFIDS)
- Chronic fungal illnesses: thrush and others
- Recurrent episodes of bacterial infections: chronic sinusitis, tonsillitis, bronchitis, bladder/prostate infections, HIV-related infections
- Autoimmune disorders: multiple sclerosis, lupus erythematosus, thyroiditis, and eczema
- Mental disorders: dementia, anorexia
- Disorders of the central nervous system: Alzheimer's disease, amyotrophic lateral sclerosis (ALS), autism, epilepsy and seizures, migraines, paralysis, Parkinson's disease
- Cancer
- Emotional diseases: mood depression, timidity, erethism
- Hematopoietic diseases: leukemia, Hodgkin's disease
- Cardiovascular system diseases
- Disorders of the respiratory system: asthma
- Diseases of the genitourinary system
- Skin disorders: acne
- Allergies

Actually, the ADA maintains that there is no consistent evidence of adverse health effects attributed to mercury amalgams used in dentistry.

However, in 2006, considering the question of the safety of silver fillings, a panel of doctors and dentists recommended special caution in using amalgam as restorative material with specific populations: children under the age of 6, girls and women of reproductive age, pregnant women, subjects with mercury allergy or hypersensitivity, and dental personnel.

Controversial Positions

Many countries, such as Canada, Austria, Germany, Denmark, and Norway, have already banned the use of silver fillings, and other countries such as Japan, Finland, and Sweden have severely restricted it.

On the contrary, in July 2009, the U.S. Food and Drug Administration (FDA) issued a new regulation on dental fillings, classifying mercury amalgams as class II medical devices, like other filling materials (e.g., composite and gold). In this way, the FDA affirmed that according to the present scientific standpoint, dental mercury amalgams are effective and safe. Antiamalgamists hypothesized, however, that one of the reasons for this resolution was to avoid a massive class-action lawsuit for mercury amalgams.

See Also: Environmental Illness and Chemical Sensitivity; Environmental Protection Agency (U.S.); Mercury Sources and Health; Oral Diseases; Solid Waste Management.

Further Readings

Mercury Policy Project, et al. "What Patients Don't Know: Dentists' Sweet Tooth for Mercury." http://www.nrcm.org/documents/dental_mercury_report.pdf (Accessed February 2010).

Public Health Service, Department of Health and Human Services. "Dental Amalgam: A Scientific Review and Recommended Public Health Service Strategy for Research, Education and Regulation." http://web.health.gov/environment/amalgam1/ct.htm (Accessed February 2010).

U.S. Food and Drug Administration. "Medical Devices." http://www.fda.gov/MedicalDevices/ProductsandMedicalProcedures/DentalProducts/DentalAmalgam/ucm171120.htm (Accessed February 2010).

Alessandra Padula
University of L'Aquila

Dry Cleaning

Dry cleaning, the process by which clothing is cleaned using organic chemical solvents rather than water, is a major source of hazardous waste and air pollution, at least when traditional methods are used. Since the 1980s, those concerned with environmental risks of dry cleaning have sought to increasingly regulate individual operations and to create green alternatives to toxic solvents traditionally used in the process. Efforts to create green substitutes to traditional dry cleaning practices have involved the development of strategies to balance environmentally friendly practices with effectiveness and efficiency for a diversified and decentralized industry.

Dry cleaning dates to the mid-19th century, when French industrialist Jean Baptiste Jolly began using petroleum-based solvents such as kerosene and gasoline to clean clothes and other fabrics. The dry cleaning process was especially popular with garments that otherwise required hand laundering, and the practice became very popular. The use of flammable solvents caused many fires and explosions in dry cleaning establishments, and local governments soon began regulating the industry. By the 1920s, dry cleaners had begun using chlorinated solvents instead of petroleum-based alternatives. Chlorinated solvents were much less likely to cause fires or explosions and provided superior cleaning capacity. As early as the 1930s, the solvent perchloroethylene (perc) had been adopted as the industry standard, and perc is still commonly used today. Although dry cleaning traditionally had garments dropped off at local shops and then taken to a central facility for processing, after World War II, in-store equipment became the norm. Dry cleaning machines are similar to washing machines, and garments are placed in a chamber that is then filled to about one-third capacity with perc. The perc is forced into the chamber and then passed through a filtration chamber and reused. The typical dry cleaning "wash" cycle lasts between 8 and 16 minutes, depending on the type, quantity, and soiling of the garments. After the wash cycle, perc is recovered and distilled for reuse. The garments are air dried, deodorized, and pressed.

Traditional methods of dry cleaning are a major source of hazardous waste and air pollution, because clothes are cleaned using chemical solvents rather than water. Since the 1980s, there has been a push to regulate the industry and to create greener alternatives to standard dry cleaning.

Source: iStockphoto.com

The use of perc in dry cleaning has proven problematic from an environmental perspective. During the 1970s, the U.S. Consumer Product Safety Commission (CPSC) classified perc as a carcinogen. Although this classification was later withdrawn, perc's status as the first chemical so designated caused it to receive a great deal of negative attention. The International Agency for Research on Cancer, a subagency of the United Nations' World Health Organization (WHO), has designated perc as a member of its Group 2A category, meaning perc is probably carcinogenic to humans. Perc also is a common soil contaminant and is toxic at low levels when found in groundwater. Perc from dry cleaning operations is also a major source of airborne particles and classified by the U.S. Environmental Protection Agency (EPA) as a hazardous air contaminant. State and local authorities also have taken action to reduce the amount of perc used in dry cleaning operations. For example, concerns about perc led the California Air Resources Board (CARB) to identify perc as a toxic air contaminant in 1991. As a result of this identification, regulations regarding the equipment, operations, and maintenance of dry cleaning machines, record keeping, and reporting requirements for perc were put into place in California. As of 2003, perc emissions had been reduced by over 70 percent, but CARB determined that further action was needed. In 2007, CARB instituted new regulations that would eliminate the use of perc by 2023.

Concerns about perc have led to a variety of alternative solvents being investigated for use by dry cleaners. Alternative solvents used include water-based cleaning, carbon dioxide (CO_2), hydrocarbon solvent, GreenEarth solvent, propylene glycol ether, Stoddard solvent, and PureDry solvent. Water-based cleaning uses water and biodegradable soap in conjunction with special dryers and stretching machines that allow fabric to retain its shape and size. Carbon dioxide is effective but requires machines that each cost nearly $100,000 more than perc machines. Hydrocarbon cleaning uses solvents produced by petroleum companies such as ExxonMobil and Chevron Phillips; these solvents are flammable and contribute to smog. GreenEarth solvent is a proprietary process that uses liquid silicone in a closed-loop system. Propylene glycol ether is formed by the base-catalyzed reaction of propylene oxide with alcohols like methanol, ethanol, propanol, butanol, or phenol. Stoddard solvent, also known as white spirit, is a paraffin-based clear liquid that predates perc. PureDry is a patented process that uses enhanced hydrocarbon. All of these processes hold promise and are being used increasingly as alternatives to perc. As environmental concerns lead to additional regulation and control of the dry cleaning industry, all of these alternatives will continue to be explored. Consumers who seek green alternatives have begun to patronize shops that use these substitutes to perc.

See Also: Air Filters/Scrubbers; Cost-Benefit Analysis for Alternative Products; Green Chemistry; Occupational Hazards; Phaseout of Toxic Chemicals.

Further Readings

Green, J. D. and J. K. Hartwell, eds. *The Greening of Industry: A Risk Management Approach*. Cambridge, MA: Harvard University Press, 1997.

Howle, S. J. and L. A. Elfers. *Ambient Perchloroethylene Levels Inside Coin-Operated Laundries With Drycleaning Machines on the Premises*. Research Triangle Park, NC: Environmental Monitoring Systems Laboratory; and Cincinnati, OH: Center for Environmental Research Information, 1982.

McKernan, L. T., A. M. Ruder, M. J. Heim, C. L. Forrester, W. T., Sanderson, D. L. Ashley, and M. A. Butler. "Biological Exposure Assessment to Tetrachloroethylene for Workers in the Dry Cleaning Industry." *Environmental Health: A Global Access Science Source*, 7/12 (2008).

Stephen T. Schroth
Jason A. Helfer
Maurice J. McDavid
Knox College

Ecosyndemics

The notion of ecosyndemic derives from the broader perspective of ecohealth, which recognizes inextricable linkages between humans and their biosocial environments that are reflected in situations in the continuum from healthy to disease states. Specifically, the term refers to environmental conditions that promote the clustering and adverse interaction of two or more diseases in a population. Research shows that when individuals suffer from multiple diseases, there are biological or other pathways through which interactions occur that can significantly impact health. An example is that damage to the skin caused by one disease may create vulnerability to other diseases, such as infection by opportunistic pathogens. This kind of disease synergy, known as a syndemic, occurs, for example, when a person suffers from co-infection of human immunodeficiency virus (HIV) and herpes. The result is a speeding up of the onset and pace of the damaging effects of both diseases. A person suffering from herpes is five times more likely to become infected from a single sexual contact with a person with HIV than someone who does not have herpes. Syndemics like this play a significant role in human health and mortality. A syndemic interaction between viral and bacterial species, for instance, appears to have driven the influenza pandemic of 1918–1919, which caused as many as 100 million deaths worldwide.

Syndemics are more likely to occur under unstable environmental conditions. Eco-crises commonly cause social disruptions that fray personal and communal support networks, increase individual emotional stress, and reduce the ability of the human body to initiate effective immune response to a microbial threat. Environmental instability can threaten subsistence while ushering in malnutrition with its damaging effects on the immune system. Syndemics that are mediated by environmental factors, including the movement or elimination of species, the development local eco-crises, changes in weather patterns, and other degradations to the environment caused by human activities, are known as ecosyndemics.

A growing factor in the onset of ecosyndemics is global warming caused by the buildup of greenhouse gases. Global warming is disrupting natural ecosystems and contributing to the geographic diffusion of disease-causing viruses, bacteria, fungi, and other pathogens. This pathogenic migration, involving species that are restricted in their distribution by seasonal temperatures, is occurring at a time when human population growth, much of which is now concentrated in overcrowded megalopolises around the world, places an unprecedented number of people at risk for overlapping exposure to multiple infectious diseases and resulting syndemics. Diseases that have increased their range because of global warming include mosquito-borne maladies such as malaria, dengue, Rift Valley fever, West Nile disease, Chikungunya fever, and yellow fever. Increased planetary warming is also a factor in the spread of disease-bearing rodents like those responsible for the outbreak of hanta virus in the U.S. southwest in 1993 and pneumonic plague in India the following year, as well as disease-causing ticks that carry encephalitis and Lyme disease and sand flies that transmit visceral leishmaniasis. Additionally, waterborne diseases like cholera are spreading because of ever-warmer oceans. From an ecosyndemic perspective, questions of great concern include the following:

- What happens to diseases as they spread to new environments?
- To what degree do they encounter and cluster with other diseases in new host populations?
- Precisely which other infectious and noninfectious diseases do they interact with in the new environment?
- What are the pathways of disease interaction?
- What are the health consequences for human populations?
- Which populations are most at risk for a particular ecosyndemic?
- What are the best methods of preventing and treating identified ecosyndemic disease outbreaks?

In Kenya, for example, as malaria has diffused to new areas, it has begun interacting with HIV infection, a synergy that may be causing both diseases to spread more quickly. Once a person with HIV is infected through a mosquito bite and begins to suffer malaria fever episodes, the level of the HIV virus skyrockets. This significantly increases the risk of the person infecting a sexual partner with HIV. At the same time, a person with HIV appears to be more susceptible to malaria infection.

Beyond rapid pathogen migration, global warming is having a direct effect on the metabolism of disease vector species, resulting in increased rates of growth and cell division. Research suggests that an increase in planetary temperature of only 4 degrees Fahrenheit would more than double the metabolism of mosquitoes, leading to more frequent feedings and resultant increased opportunities for the transmission of infections.

Warmer seawater is associated with increases in extreme weather. Warmer oceans have higher rates of evaporation, fueling heavier rains, and more intense storms can cause flooding and the spread of waterborne diseases. Heat exhaustion brought on by global warming is also increasing, with resulting adverse impact on the human immune system. The absolute virulence of many pathogens is kept in check by the body's immune system. If the body's natural defenses are compromised, previously harmless pathogenic strains can become lethal.

While the study of ecosyndemics is new, and work to date on this emerging critical health issue is limited, it is evident that the development of ecosyndemics adds to the rationale for significantly enhancing and accelerating efforts to mitigate the harmful effects of global warming.

See Also: Asthma; Climate Change; Malaria; Sexually Transmitted Diseases.

Further Readings

Baer, Hans and Merrill Singer. *Global Warming and the Political Ecology of Health: Emerging Crises and Systemic Solutions.* Walnut Creek, CA: Left Coast Press, 2009.

Epstein, Paul. "Climate Change and Human Health." *New England Journal of Medicine*, 353 (2005).

Singer, Merrill. "Ecosyndemics: Global Warming and the Coming Plagues of the 21st Century." In *Plagues, Epidemics and Ideas*, Alan Swedlund and Ann Herring, eds. London: Berg, 2009.

Merrill Singer
University of Connecticut

Education and Green Health

It is well recognized among experts that environmental conditions have important effects on human health. The general public is only recently beginning to understand this relationship, learning about the links between our personal health, the health of the local environment, and the larger ecology of the Earth. Green health is the understanding of the interrelationship between sustainability of personal health and the sustainability of planetary resources and processes. The understanding is occurring on the following two fronts:

- Individuals, communities, and the healthcare industry are broadening their understanding of health by enhancing our understanding of the determinants of health, providing more holistic approaches to health and healthcare beyond the typically narrow focus of traditional Western medicine.
- Conscientious individuals, nongovernmental organizations, and governmental agencies are acknowledging the need for protecting and promoting sustainability of life forms on the Earth, thus our responsibility to promote ecological health is an emerging trend in agriculture, energy, and other industries that directly affect the planet's ecology.

Essential to this growing movement of green health is the way in which these multiple stakeholders are learning how to participate in this changing global worldview.

Emergence of Ecohealth Literacy

Individual understanding of green health occurs with improved literacy about ecology and health. Ecohealth literacy is the ability of individuals to recognize broader environmental conditions that produce good health. The way individuals experience health and the environment is reframing how people think about their own bodies. Key trends toward this change include, for example, accepting the reality of global climate change, the epidemiologic transition shifting of disease burden from infections disease to chronic disease, the growing understanding of the costs and resource constraints on caring for an aging population, the shifting of regulation from national to both international and local theaters, and the changing of personal values to include greater emphasis on personal health. Each of these five drivers represents incentives for increased appreciation of green health.

Early signs of global change to the climate system were recorded during the 1980s. The first report from the Intergovernmental Panel on Climate Change (IPCC) was issued in 1990, though real acceptance was not until after the release of the movie, *An Inconvenient Truth* by former vice president Al Gore. As of 2003, the World Health Organization

(WHO) recommended a transition to behaviors, technologies, and practices that promote sustainability by governments, society, and individuals. Though at times conflicting, interests continue to debate the consequences of global climate change, or question the human role in creating the phenomenon. Possibly more than any other ecological challenge, global climate change has brought together both personal and ecological health.

A legacy of the 20th-century science and technological development was the success of pharmaceutical medicines in reducing infectious disease. Beginning in the mid-1930s, with the successful release of penicillin, the 20th century witnessed a lasting reduction in mortality from bacterial and viral infections. With these successes came significantly longer life spans, with a gain of almost 30 years since 1950. However, a new challenge is emerging in the early years of the 21st century, rising chronic disease, which coincides with an aging population. A recent report by the Milken Institute predicts a 42 percent increase in the incidents of the seven most common ailments in America: diabetes, heart disease, hypertension, cancer, stroke, pulmonary conditions, and mental disorders, sometimes called "lifestyle diseases." A driver to green health is the understanding of the causes of chronic illness, which are not simply microbial but rather multifactorial. The current healthcare system emphasis is on acute infectious diseases. Responses to lifestyle illness include policy changes promoting healthier urban development, along with changes in agricultural practices such as limiting overuse of antibiotics and limiting improper disposal of waste, and developing healthier workplaces, among others.

Today's healthcare system continues to require more resources. As of 2009, healthcare costs consumed 16 percent of the U.S. gross domestic product, without any signs of slowing. Coupled with an aging population, which will surpass 20 percent over the age of 65 by 2020, this provides ominous signs about our ability to manage dwindling medical resources. With the projected rise in chronic illness, the current healthcare delivery system will not be able to manage as currently organized. Alternative health systems exist in several countries, providing basic services for all members of society. The most promising is primary healthcare, which places an emphasis on promotion, prevention, treatment, and rehabilitation. Promotion and prevention involve setting health priorities specific to local communities. In this way, communities are better able to allocate resources, learning the value of health-promoting behaviors.

Much of the early policy for the health of the environment was shaped in the United States by federal policy, including the Clean Air Act (1963), the Clean Water Act (1977), Toxic Substances Control Act (1976), and the Endangered Species Act (1973). However, recent policy shifts occur either within industry or internationally. Examples from within industry include the development of LEED Certification by the U.S. Green Building Council and the emergence of organic food certification in California and Maine. Recent examples of changes in international policy include Registration, Evaluation and Authorization of Chemicals (REACH), and the European Union (EU) Integrated Product Policy (IPP), which take a holistic approach to stewardship in the development, use, and disposition of products and services. Policy changes for green health are occurring on many levels, as self-organizing systems find necessary ways to regulate new standards that reflect the commitment to health and the environment.

Personal values about health and healthcare have evolved significantly in the past 30 years. Consumers emphasizing personal health and environmentally friendly practices are driving changes in consumer products. This translates into a growing demand for products that are healthier for consumers and that have been manufactured for a healthy planet, and for people who are willing to pay for these items. The growing number of ecofriendly consumer goods are in many domains, including food and nutrition, transportation and energy, home care, finances, and healthcare. Thus, increasing ecohealth literacy is supporting a growing culture of green health.

Place Matters for Green Health

Given the current drivers of green health, citizens, communities, business, and government are increasingly learning aspects that grow green health. More domains become part of the effort to protect body, home, communities, and the planet. As more people try to live sustainably, more organizations and communities try to operate sustainably. The essential ingredient will continue to be place, which comes to matter more in individual health. The demand for information about place continues to grow. Citizens increasingly monitor the environment using emerging technologies that capture ever-finer qualities of data. These data will continue to inform the consumer in several ways, including increasing health-driven product development with a broader expectation for healthy production, distribution, health, and environmental concerns. Local actions grow as green health moves out into the community. These include community gardening, community supported agriculture (CSA), and local data mapping for tracking environmental risk.

The link between health and place fundamentally challenges the understanding of risk and disease. Along with learning more about environmental contamination, location and living conditions are becoming a greater factor in overall health outcomes. Improvements in the ability to measure risk improve as new technologies of measurement continue to enter the market. Increasing knowledge is revealing serious inequities in the current distribution of health opportunities defined by local environmental features. Inequities in access to clean air, healthy food, recreation space, educational opportunities, living wages, and decent housing are all factored into overall risk. Green health begins with participation by local citizens and leaders in promoting public health measures designed to lower risk in these disadvantaged communities. An increasing understanding of what good health means emphasizes green public spaces and green workplaces.

New technologies continue to inform green health trackers, not only mapping risk but also identifying and mapping resources. Sharing strategies that promote health across electronic frontiers help communities develop and promote best practices. Land-use policies are greatly affected by the growing information clusters. While previous land-use policies were concerned with open space, new green health initiatives promote land-use policies focusing on local disease burdens, environmental services, or health risks associated with different land-use opportunities.

Green health changes the understanding of risk by increasing knowledge of the environmental burden on human health and the financial risks of this burden. The emerging field of biomonitoring is accumulating scientific knowledge about true concentrations of chemicals in human bodies. National and local databases are growing as policy efforts begin to emphasize greater precaution. The precautionary principles promote alternatives before putting humans at risk for environmental contamination. Cradle-to-cradle product design and stewardship offers manufacturing processes of consumer goods that reduce chemical loads on the environment by manufacturing products that are completely reusable. Reducing chemical inputs in the environment reduces potential harm caused by unknown agents entering the human life cycle.

Schools are engaging in green health programs by teaching ecoliteracy through school lunch initiatives that teach children the value of locally grow produce. To be ecoliterate means understanding the principles of organization of ecological communities (i.e., ecosystems) and using those principles for creating sustainable human communities. As more children relearn where and how food is grown through school gardening programs, the shift away from industrial agricultural will continue.

In the workplace, programs are encouraging personal sustainability. Companies are creating personal sustainability programs that encourage health and wellness of workers. Food sources are shifting to more organic and sustainable sources providing workers with increased energy and well-being. Ergonomic assessments encourage healthy workplace habits, reducing sick days and improving physical conditions.

Urban planning processes are becoming green by fostering healthier environments. This drives a proliferation of community gardens, urban farming, and therapeutic programs that foster ecological interactions. Health clinics and hospitals are greening through the reduction of resource use and promotion of environmental behaviors. Hospitals are offering farmers markets. Healthcare providers are shifting to promote more sustainable services that promote local environmental health.

The Future of Green Health

Modern society is becoming more ecohealth literate, shifting toward green health. These shifts include greener manufactured products both on the outside and in production, including increased research and service opportunities for healthier product design, and improved safety for workers in all aspects of industry and government. Green health opportunities grow through improved communication systems, through increased participation by citizens locally and regionally shaping the "health commons," and by fostering coalitions and collectives. Finer technological monitoring will increase mapping, which will continue to promote individual engagement.

Healthcare will continue to reflect the shifting values toward green health by creating healthier environments in both hospital and clinical settings. The entire healthcare sector, including hospice, long-term care facilities, pharmacies, and primary care clinics, will continue to emphasize better resource management, waste reduction, and pollution prevention. Clinical encounters will teach the importance of the environment. Clinical interventions will utilize the natural world more sustainably to help patients gain the most from natural resources including sunlight, water, air, and the outdoors. Promoting healthy environments will take center stage for all human resource programs, promoting healthier lifestyles, disease prevention, and wellness.

Hospitals have already made the largest commitment to green health. The green hospital movement is spearheaded by Practice Greenhealth, a member organization that provides knowledge and services for shifting to more ecofriendly practices. New hospitals that meet LEED certification standards are currently under construction. Food service is including organic food and foodstuffs from local farmers markets. Waste and energy consumption are evaluated and minimized.

Ecofriendly dentistry is gaining popularity, reducing exposure to mercury and lowering the footprint of dental offices. Green clinics are serving as resource centers for local communities, providing information and expertise on local environmental health issues. Health professionals, including physicians and nurses, are shifting clinical practices to meet the demands of dwindling resources and promoting cleaner environments. As patients and businesses interact with this growing green community, knowledge of ecohealth literacy will continue to grow. Individuals and communities will further take advantage of opportunities to better care for the environment by placing more emphasis on personal health behaviors.

See Also: Climate Change; Government Role in Green Health; Health Disparities; Hospitals (Carbon Footprints); Personal Consumer Role in Green Health; World Health Organization's Environmental Burden of Disease.

Further Readings

Almeda County Public Health Department. "Life and Death From Unnatural Causes: Health and Social Inequality in Alameda County" (April 2008). http://www.acphd.org/AXBYCZ/Admin/DataReports/unnatural_causes_exec_summ.pdf (Accessed August 2009).

DeVol, R., A. Bedroussian, et al. "An Unhealthy America: The Economic Burden of Chronic Disease—Charting a New Course to Save Lives and Increase Productivity and Economic Growth." Milken Institute, October 2007. http://www.milkeninstitute.org/publications/publications.taf?function=detail&ID=38801018&cat=ResRep (Accessed August 2009).

Environmental Working Group. "Chemical Index." http://www.ewg.org/chemindex (Accessed August 2009).

Myers, N. "The Rise of the Precautionary Principle: A Social Movement Gathers Strength." *The MultiNational Monitor*, 25/9 (2004). http://multinationalmonitor.org/mm2004/090 12004/september04corp1.html (Accessed August 2009).

Practice Greenhealth. http://www.practicegreenhealth.org (Accessed April 2010).

World Health Organization. "Climate Change and Human Health." http://www.who.int/globalchange/climate/summary/en/index12.html (Accessed August 2009).

World Health Organization. "Primary Health Care." http://www.who.int/topics/primary_health_care/en (Accessed August 2009).

Joel H. Kreisberg
Teleosis Institute

ELECTRICITY

The use of electricity by civilized society has produced a plethora of benefits for human health. With the advent of electric lighting replacing gas and oil lamps, indoor air quality has improved. Where electricity has replaced locally burned fuels, local environmental quality improves; however, it is important to consider the health impacts of different methods of electricity production. Electricity as an energy vector produces no emissions at the point of use and is relatively easy to transport. Historically, in many cases electricity has been substituted for consuming local fuels—which, when burned to produce energy, resulted in local pollution and environmental degradation.

Use of electricity as an energy vector is expected to increase as part of global efforts to decarbonize our society; however, while substituting electrically based solutions for fuel-based solutions often results in some carbon savings, it is important for true decarbonization to examine how the electricity is being produced in the first place and transition to renewable methods of production.

Using Services Versus Using Electricity

In appraising how best to deliver human well-being, we need to differentiate between the use of electricity and the use of the services that electricity delivers. Electricity is an energy vector; it is a means of easily moving energy around; but of itself it has no inherent value or use. It is only when we use energy in a device that we realize its potential usefulness. People consume "services."

Electricity does not produce any emissions or wastes at the point of use; however, its production, if produced from nonrenewable sources, can have manifold health impacts for not only those at all stages of the energy supply chain, but also for the greater society.

Health Impacts of Producing Electricity

In the conventional thermal electricity generating plant, heat is produced by burning fossil fuels, renewable alternatives such as biomass, or by using the process of fission to produce heat. This heat is used to raise steam by boiling water. The steam then passes through a turbine, which converts the motion of the steam into rotary motion that can be used to turn a generating set to generate electricity. How this steam is produced determines the major portion of the health impacts that arise from a traditional thermal electricity generating plant.

Nonrenewable Energy

Human health impacts arise in the extractive industries through workers' exposure to different risks as they extract nonrenewable fuels from the ground. In thermal power plants, it is the fuels that are used to produce heat that produce different pollutants that have the potential to harm human health in different ways.

Fossil Fuel Energy

Natural gas is the cleanest-burning fossil fuel; methane, whose chemical formula is CH_4, burns to produce only carbon dioxide and water. The carbon emissions from burning methane are lower than those of other fossil fuels. Oil and emulsions of heavy oils and bitumens such as orimulsion can be burned for electricity production. Oil, however, is a dirtier fuel than gas and results in greater carbon emissions in addition to a range of other atmospheric pollutants. Coal is the most carbon-intensive fuel for electricity generation as it has a low hydrogen-to-carbon ratio. It should also be understood that coal-fired generation increases the public's exposure to radiation as vast quantities of coal, some containing traces of radioactive elements, are burned, releasing these traces into the environment.

Carbon Capture and Sequestration is one technology that has been proposed by the fossil-fuel energy industry to ameliorate concerns over the carbon emissions resulting from fossil-fuel energy production. This technology would involve capturing the emissions produced by the fossil fuel–burning plant, compressing these emissions, and sequestering them underground in geologically stable formations of rock. There is still a long way to go before this technology is proved; at the moment, captured carbon—where it has been used in demonstration projects—has been used in "CCS EOR" or "Enhanced Oil Recovery"; this is a process whereby carbon dioxide is forced underground into oil wells that are at the end of their life to displace oil and gas and increase the oil production of the well.

Ironically, when considered overall with a systems perspective, this ultimately results in greater carbon emissions and environmental degradation (as the oil also produces emissions when burned) than the carbon dioxide released into the atmosphere in the first place.

Nuclear Energy

The human health impacts of uranium are well documented; however, this should be offset by the fact that a relatively small quantity of uranium can be used to produce a significant amount of energy. While producing nuclear power does not produce any emissions directly, the supporting infrastructure for fabricating, transporting, and disposing of nuclear fuel inevitably has some environmental and therefore human health impacts. Nuclear power plants do contribute measurably to an increase in local radiation levels; however, this increase is relatively modest considered against natural background radiation levels. Nuclear waste is highly radioactive and can have significantly deleterious effects on genetic material. There is still no long-term consensus globally over the safe disposal of nuclear waste.

Renewable Energy

Renewable energy, by its nature, produces no emissions in use and, therefore, it does not contribute to the problems of anthropogenic climate change or pollution of the natural environment; however, there are some impacts to different renewable technologies.

Large-scale hydro developments often require the flooding of vast valleys to create a reservoir behind the dam to generate power and can restrict the flow of water to areas downstream of the dam. This can have adverse effects on the local population, which is often forcibly relocated.

The manufacture of some photovoltaic technologies requires the use of chemicals that are toxic. This presents a hazard to human health at the manufacture stage; however, as these chemicals are encapsulated in the product, there is little concern at the point of use. Solar "concentrating" technology is a method of thermal electricity generation, substituting the polluting methods of heating water with benign solar energy concentration using an array of mirrors onto a central focus. The heat energy can also be stored (e.g., in molten salt) to allow electricity generation to continue overnight.

In generating power from wind, there is a certain amount of noise that is produced that has the potential, if poorly sited, to serve as an irritation, causing stress and unease. Furthermore, light passing between fast-rotating turbine blades can cause what is known as "shadow flicker," which can cause problems in individuals who are sensitive to flickering light. Human exposure to both of these can be reduced through sensible design, and many software packages are available to wind developers that can be used to simulate noise and shadow flicker before a wind farm is built. Noise and shadow flicker are not considerations for off-shore developments.

Using Electricity-Efficient Devices

In the drive toward sustainability, consumers are encouraged to conserve energy, use it wisely, and switch to devices and appliances that are more frugal in their use of energy. In some cases, greener devices may have different health implications from traditional means of consuming power, and these should be borne in mind. Compact fluorescent lamps, which offer a higher efficacy than traditional incandescent bulbs, contain small quantities of mercury; while not a problem in normal use, consideration should be given to what happens in the event that a bulb breaks or needs to be disposed of.

See Also: Climate Change; Light Bulbs; Personal Consumer Role in Green Health; Radiation Sources.

Further Readings

Comar, C. L. and L. A. Sagan. "Health Effects of Energy Production and Conversion." *Annual Review of Energy* (1976).

Lave, L. B. and L. C. Freeburg. "Health Effects of Electricity Generation From Coal, Oil, and Nuclear Fuel." *Nuclear Safety*, 14/5 (1973).

Markandya, A. and P. Wilkinson. "Energy and Health 2: Electricity Generation and Health." *The Lancet* (2007).

Rashad, S. M. and F. H. Hammad. "Nuclear Power and the Environment: Comparative Assessment of Environmental and Health Impacts of Electricity-Generating Systems." *Applied Energy* (2000).

Gavin D. J. Harper
Cardiff University

EMERGENCY ROOMS

The Federal Emergency Medical Treatment and Active Labor Act (1986) mandates that hospitals must provide care to anyone needing emergency healthcare treatment regardless of ability to pay. About one out of every five Americans visited an emergency room (ER) in 2007, resulting in roughly 110 million emergency room visits, or nearly four visits every second. Emergency rooms are straining to serve their communities, and, in many ways, are failing. About once every minute, an ambulance carrying a sick person is turned away from an overburdened ER and diverted to another, wasting precious minutes that could mean the difference between life and death for many people. Even more troubling are the horror stories about patients who died in the waiting rooms of emergency departments, sometimes having waited hours to be seen, even after exhibiting classic symptoms of conditions that require immediate treatment. ERs are burdened beyond capacity, and yet, even as the number of those seeking emergency care rises, ERs across the country are disappearing as more and more hospitals are closed. Further, there is deep disagreement regarding just what type of patients visit ERs, seeking what type of care.

Despite the common myth that it is the uninsured who make up the majority of ER patients, a 2010 report by the National Center for Health Statistics (NCHS) reveals that the uninsured were no more likely than the insured to visit an ER. The report also found that most ER patients have insurance and that 90 percent of those who visited an ER were seeking care for an urgent medical condition, refuting the notion that ERs are overtaxed by those seeking nonurgent care.

The NCHS report also found that Medicaid recipients make up the largest proportion of those seeking ER care, likely due to the difficulty of finding primary care doctors who accept Medicaid. Faced with the inability to find a primary care doctor, many Medicaid recipients put off seeking medical attention until their condition demands emergency care. Making Medicaid more attractive to doctors could improve the overall health of a significant portion of the population while easing the burden on ERs.

Recent healthcare reform legislation may further overwhelm ERs, as around 30 million newly insured people will enter into a paradox that finds them with healthcare insurance but no primary healthcare provider. There is growing consensus around the notion that the United States lacks the primary care infrastructure necessary to cover the 30 million newly insured patients that healthcare reform legislation will create. While some argue that previously uninsured people who gain health insurance will go to a doctor rather than put off care until faced with an emergency, others claim that newly insured people will still seek primary care at ERs because they do not have doctors. Pointing to the 2006 Massachusetts bill that created near-universal coverage for its residents, many note that despite claims that insurance coverage would ease ER traffic for Massachusetts's hospitals, wait times there have either increased or stayed the same.

It has been three decades since emergency medicine was recognized as a specialty by the American Board of Medical Specialties, and today, board-certified emergency physicians are available at more than half of all ERs. Still, a 2004 survey by the American College of Emergency Physicians found that 66 percent of emergency department directors stated that inadequate on-call specialist coverage was a problem in their ERs. A nationwide nursing shortage further chips away at the ability of ERs to meet patient needs. Further, hospitals often reserve in-patient beds for those entering hospitals for elective surgeries who have established their ability to pay for care; conversely, the payment status of the next patient to visit the ER is unknown. This establishes a cycle of reserving in-patient beds for elective surgeries, boarding patients in the ER until acute care beds become available, and diverting ambulances carrying patients who may not be able to pay for service to another hospital. This overcrowding is also difficult for the caretakers themselves to manage, as they struggle to meet the needs of too many patients seeking emergency care while witnessing terrible human suffering, which creates a heavy psychological burden. Overrun with patients and lacking proper resources and specialists, America's ERs are themselves sick, and those who staff them and those who seek out care in them each pay a different yet exacting price for what some call a systemic failure in healthcare.

Many claim that a major catastrophic event, such as a pandemic flu or a major natural or man-made disaster, could overload an already taxed system to the breaking point. Healing America's ailing ERs, easing the burden placed on the physicians and nurses who staff them, and providing care to the millions of Americans who enter ERs each year is a daunting task, and it remains to be seen what changes will be made to remedy a bad situation balanced on the edge of becoming much, much worse.

See Also: Biomedicine; Healthcare Delivery; Health Insurance Reform; Nursing, Lack of.

Further Readings

Consumer Affairs. "U.S. Emergency Rooms Are Getting More Crowded." http://www
.consumeraffairs.com/news04/2010/05/crowded_emergency_rooms.html (Accessed June
2010).

Heritage Foundation. "The Crisis in America's Emergency Rooms and What Can Be Done." http://www.heritage.org/Research/Reports/2007/12/The-Crisis-in-Americas-Emergency-Rooms-and-What-Can-Be-Done (Accessed June 2010).

Physicians for a National Health Program. "Our Ailing Emergency Rooms." http://pnhp.org/blog/2008/07/21/our-ailing-emergency-rooms (Accessed June 2010).

Tani Bellestri
Independent Scholar

ENVIRONMENTAL ILLNESS AND CHEMICAL SENSITIVITY

Environmental illness (EI) and *chemical sensitivity* (CS) may be considered synonymous terms. The older term *multiple chemical sensitivity* (MCS) is no longer used by the American Academy of Environmental Medicine, which is responsible for training physicians to diagnose and treat the conditions. CS is defined as "a multisystem disorder, usually polysymptomatic, caused by adverse reactions to environmental incitants present in air, food, water, and habitats."

Chemical sensitivity is common. Numerous surveys by government agencies have shown that the prevalence of disabling chemical sensitivity is between 3 to 5 percent in the U.S. population. Further, it has been shown that 15 percent of the general population is at least mildly intolerant of chemicals, whereas the prevalence among the elderly, deployed Gulf War veterans, and patients with chronic fatigue syndrome and fibromyalgia is double this rate, at 30 percent. It has been shown that the cerebrospinal fluid of the patients with Gulf War syndrome, CFS, and fibromyalgia all have a unique group of abnormal proteins that show they have a common pathophysiologic mechanism, but it is unknown whether the cause is viral, chemical, or related to toxic mold. Chemical sensitivity is a condition usually precipitated by exposure to mold toxins, pesticides, chemicals, or toxic metals at home or work that then leads to intolerance, in some way, to other subsequent environmental exposures.

The history of exposure to potential risk factors is the key to discovering the possible environmental influences on health. Risk factors for CS include exposure to new carpeting or cabinetry, musty basements, and exposure to a chemical spill. The preliminary screening questions used to determine whether a person is chemically sensitive are: do you find perfume, diesel exhaust, or the detergent aisle of the grocery store offensive or disabling? Does red wine or newsprint give you a headache or make you tired? Do clothing tags itch your neck? Are you sensitive to medications, to many foods, even to vitamins? Do you have trouble reading or remembering things, fatigue, muscle weakness, anxiety? Has your face become reddened like an alcoholic? Do you get a sore throat in the mall or do you become very tired? Do tire stores smell awful to you? Signs of electrical sensitivity are: Does the cell phone heat up in your hand or do fluorescent lights give you headache?

Electrical sensitivity occurs in 15 to 30 percent of people with chemical sensitivity and is poorly understood. It may be a result of damage to the autonomic nervous system and brain stem that leads to an intolerance of cell phones, televisions, refrigerators, fans, and computers. Patients with this problem may not be able to find housing because of an intolerance of wireless networks (Wi-Fi) and wiring methods used in the average home or apartment. Recommended therapy begins with treatment of the chemical sensitivity and avoidance of magnetic and/or electric fields, with wearing leather-soled shoes, and taking magnesium or potassium supplementation. Checking the mouth for oral galvanism—where two disparate metals touch and create current—is an essential part of the workup in these patients as well. Current produced in the mouth, so close to the brain stem, can lead to incapacitating symptoms of electrical and chemical sensitivity.

Common Conditions of Chemically Sensitive People

Common conditions that develop in a chemically sensitive patient include adrenal insufficiency, especially in women. Weakening of this stress gland causes low blood pressure (systolic less than 100 mm Hg), dizziness on standing, headaches, allergies, tearfulness, and anxiety. The Centers for Disease Control determined that women with chronic fatigue syndrome are more likely to have adrenal insufficiency than men, and U.S. Army research found that only female rats exposed to mold toxins in the air develop adrenal necrosis and that, if given, testosterone can be protective.

Dysautonomia, or a difficulty in maintaining an upright posture without increasing heart rate, also occurs as the disease progresses. Symptoms of this disorder are the "pretzeling" of one's legs, fast heartbeat, or the desire to lie down in stores, after eating large meals, upon standing up, or in the heat. It frequently occurs in people with chronic fatigue syndrome.

Cognitive damage and neurologic disease, autoimmunity, arrhythmias, rashes, gastrointestinal problems, and allergies often develop. Treatment is aimed at identifying any recent exposure precipitating the event and halting it. Depending on the level of illness and toxicity of the exposure, more drastic steps will need to be taken. If exposure to neurotoxigenic mold is diagnosed, it means leaving one's premises and clothing behind, and moving to a clean room, preferably without carpeting (the oasis bedroom). The patient should strive for charcoal-filtered clean air, pesticide-free food, and water bottled in glass in an effort to lower the toxic load. After five days in this environment, many patients will become more sensitive to everyday exposures, so the use of a charcoal mask may be helpful when near traffic exhaust. Limited exposure to stores, chemicals, and of course, pesticides, is recommended.

At a clinic, the treatment will include oxygen for two hours a day by ceramic mask, if indicated by venous oxygen measurement. The patient will have provocation and neutralization (P and N) allergy testing to decrease the response to all incitants in the environment. Measuring hormones, nutrients, and neurotransmitters, and replacing those that are low is paramount. Genetic assessment of the ability to detoxify certain chemicals may prove helpful. Immune modulation with a lymphocyte factor created from the patient's own cells appears to be very promising. After patient stabilization and the administration of intravenous vitamins B, C, and other nutrients like glutathione, the use of a low-temperature sauna (140 degrees Fahrenheit) is started to enhance detoxification, and long-term sauna use is recommended. Continuation of P and N antigen shots is sometimes required for years to prevent relapse. Return to functionality is possible for those disabled by allergies, although permanent disability is common among people who do not receive treatment.

See Also: Air Filters/Scrubbers; Bottled Water; Carbon Filters; Dental Mercury Amalgams; Fungi and Sick Building Syndrome; Indoor Air Quality.

Further Readings

American Academy of Environmental Medicine. http://www.aaemonline.org (Accessed August 2010).

Bell, Iris. *Clinical Ecology: A New Medical Approach to Environmental Illness.* Bolinas, CA: Common Knowledge Press, 1982.

Dickey, Lawrence D. *Clinical Ecology.* Springfield, IL: Charles C Thomas, 1976.

Gorman, Carolyn. *Less Toxic Alternatives.* Hot Springs National Park, AR: Optimum Publishing, 2004.

Rapp, Doris J. *Our Toxic World—A Wake Up Call.* Buffalo, NY: Environmental Medicine Research Foundation, 2004.

Rea, William J. *Chemical Sensitivity.* Boca Raton, FL: CRC Press, 1996.

Lisa Lavine Nagy
The Preventive and Environmental Health Alliance, Inc.

ENVIRONMENTAL PROTECTION AGENCY (U.S.)

Upon its creation, the Environmental Protection Agency was tasked with repairing degradation and harm in the environment. This restored waterfront area and former Superfund site in New Bedford, Massachusetts, was once contaminated by polychlorinated biphenyls (PCBs) from two local manufacturers of electrical capacitors.

Source: National Oceanic and Atmospheric Administration

The U.S. Environmental Protection Agency (EPA, or occasionally referred to as USEPA) is a U.S. federal regulatory agency charged with protecting public health and the environment through the oversight of water quality, air quality, and fuel quality standards, as well as chemical use. In essence, the agency works toward a cleaner, healthier environment for the American people by addressing and confronting the environmental challenges before us.

Having been established in 1970 under the Richard Nixon administration, EPA is one of the newer governmental regulatory agencies. The White House and Congress created EPA in response to public concern about the environment for cleaner land, water, and air. Prior to its creation, the U.S. government's structure did not allow for an orchestrated means to combat pollutants that degraded the environment and were harmful to public health. The establishment of EPA is an example of a shift in American cultural values, altering political landscapes and the allocation of resources.

EPA strives to lead the nation's environmental science, research, education, and assessment efforts and is responsible for much of the environmental policy in the United States.

It is committed to three core values: science-based policies and programs, complete transparency, and adherence to the rule of law.

In terms of organization, EPA is composed of approximately 17,000 full-time employees across the United States. Over 8,000 of these individuals are scientists, engineers, and policy analysts. Washington, D.C., houses the headquarters, but 10 regional offices as well as more than a dozen labs are located across the United States.

The head of EPA is the administrator, appointed by the president of the United States. While EPA is not an agency within the cabinet, the administrator is typically given cabinet rank. The first head of the Environmental Protection Agency in 1970 was William Ruckelshaus. At the time of this writing, Lisa P. Jackson is the administrator; she was appointed to the position in December 2008 to serve in the Obama administration.

Examples of EPA legislation include the Clean Air Act, the Endangered Species Preservation Act, Superfund, and the Federal Food, Drug, and Cosmetic Act. As the name suggests, the Clean Air Act enforces clean air standards, reducing smog and air pollution. The Endangered Species Preservation Act outlaws any action or behavior that can harm a threatened or endangered species, including its habitat.

Superfund is the widespread name for the Comprehensive Environmental Response, Compensation, and Liability Act: this federal law was created to clean up abandoned sites of toxic hazardous waste and enforce property damage liabilities; in other words, the law sanctioned EPA to identify parties responsible for contamination of sites and require them to clean them. Finally, the Federal Food, Drug, and Cosmetic Act authorizes the EPA to set maximum tolerances for pesticides put on or in foods or animal feed.

Examples of EPA programs include Energy Star, Safe Drinking Water, and WaterSense. Energy Star exists as a voluntary program that encourages energy efficiency. The Safe Drinking Water program enforces quality standards of drinking water for the public by setting regulations for over 160,000 water systems across the United States in order to protect public health. WaterSense is a voluntary program developed to foster water efficiency by using a specific label on consumer products. Such products include irrigation equipment and high-efficiency toilets (HETs) and sink faucets.

Traditionally, EPA officials have followed a prescriptive and orderly approach to implement federal legislative requirements that protect human health and the environment. Initially, environmental legislation seemed to be focused on environmental problems that were easiest to tangibly see. For example, legislation was created to address lakes experiencing immense amounts of algae. Today, legislators who work with EPA recognize that environmental problems are more systemic and obstinate than they had previously realized in the early 1970s. For instance, the Clean Water Act's strategy of controlling point sources of pollution has improved the water quality of the nation's lakes and rivers, but legislators now realize nonpoint sources, which contribute substantially to surface water pollution, are harder to control both politically and technologically.

EPA has certainly been charged with challenging tasks. Upon its creation, EPA was tasked with repairing degradation and harm in the environment and delineating new criteria to guide the American public in moving forward to produce a cleaner and healthier environment. Since its inception, the amount of released toxins has fallen, the air and water are cleaner, and methods of waste disposal have improved.

Yet, despite the daunting nature of the EPA's tasks and its accomplishments thus far, many people believe that the agency is not responding sufficiently to its charge. While the environmental regulatory process was initially characterized by great deference to EPA expertise and decision making, some now question the agency's reliance on science. Arguments have been made that EPA often relies more on policy makers than on scientific proof; in other words, EPA scientists may be subject to political pressure that may alter the nature of their work. Others contend that EPA only enacts regulations to correct problems that were previously identified by congressional legislation.

Increasingly since its creation, EPA has delegated significant authority over environmental protection to the states. U.S. states are expected to mirror EPA standards. Nevertheless, the states often complain that they are still bound by excessive federal control. These complaints increase when the government decreases funding for environmental programs. Yet, the federal government is hesitant to release too much federal control if it cannot trust the states to protect the environment and human health.

Like most large agencies, EPA has been involved with several federal lawsuits. For example, in *Massachusetts v. Environmental Protection Agency* (2007), the U.S. Supreme Court ruled in a 5–4 decision that EPA holds the authority to regulate the emission of heat-trapping gases in automobile emissions. The Supreme Court further stated that EPA must regulate the emission of greenhouse gases that contribute to climate change unless the agency can provide a scientific reason for its refusal.

Despite the criticism that EPA has received, this federal regulatory agency is committed to its mission of protecting human health and the environment. EPA is critical to maintaining and enhancing the public health of the nation. Without a federal regulatory agency like EPA to coordinate environmental protection efforts, the health of the American people would suffer.

See Also: Chemical Pesticides; Climate Change; Environmental Illness and Chemical Sensitivity; Government Role in Green Health; Indoor Air Quality.

Further Readings

Collin, Robert W. *The Environmental Protection Agency: Cleaning Up America's Act (Understanding Our Government)*. Westport, CT: Greenwood Press, 2005.

Ondich, Gregory G. "New Role Orientations for U.S. EPA Officials in the Next Generation System of Environmental Protection." Ph.D. dissertation. Blacksburg: Virginia Polytechnic Institute and State University, 2001.

Raloff, Janet. "EPA's Strategic Revolution." *Science News*, 138 (1990).

U.S. Environmental Protection Agency. "About EPA." http://www.epa.gov/aboutepa.htm (Accessed February 2010).

Annie W. Bezbatchenko
New York University

Ergonomics

Ergonomics is an interdisciplinary field that designs objects, systems, and environments to improve human capability and interaction. It is largely synonymous with the term *human factors*, which covers the same technical areas but originated and developed in Europe. Although ergonomics has a broad scope, it commonly refers to workplace health, performance, and safety. An emerging approach is the application of green ergonomic principles, particularly the design of environmentally friendly equipment and systems. This article outlines the ergonomics field and the position of green approaches within it.

A central theme of ergonomics is the environment–behavior relationship, in which *environment* is defined loosely to include built environments (buildings, workplaces) such as the material environment (lighting, temperature, noise), technical environment (equipment, workstations, parameters of machines), organizational environment (work processes), and social environment. All of these can be designed and optimized in environmentally sustainable ways; therefore, ergonomists hold potential key roles as environmental stewards to society.

Ergonomics is divided into several broad domains: physical, cognitive, organizational, and environmental. Within these domains there is wide diversity. For example, the Human Factors and Ergonomics Society organizes more than 20 specific application areas. They include the application of ergonomics to aviation and space environments; computer systems; education and training; environmental design; health and quality of life; human communication; Internet technology; manufacturing and industry; product design; and transport. Practitioners come from a variety of backgrounds, drawing upon anthropology, biomechanics, biology, computing, engineering and design, biology, occupational therapy, and psychology. The aim is to provide a complete picture of human interaction with products and systems. A recent trend has been the integration of a green agenda into projects. Rather than an alternative to conventional approaches and subfields, a green approach to ergonomics provides a unifying framework, integrating environmental stewardship into the various domain sub-branches.

Reference to green approaches in the ergonomics literature is relatively recent, but many long-standing goals of the field encourage prudent use of resources. Ergonomists optimize conditions by assessing and designing best fits between people and their environment. This usually involves, for example, reductions in operating and resource costs of buildings and workplaces to maximize efficiency and productivity, which also achieves energy efficiency. Green ergonomics takes this one step farther. A good example is sustainable building design, which can include energy-efficiency improvements; the use of renewable energy sources; energy audits to measure energy consumption; compliance with environmental laws and regulations; environmental implications of infrastructure; and social justice issues.

Furthermore, ergonomics ensures workability through preventative occupational health and safety, largely by designing workplaces to promote well-being and to prevent illness and accidents. Regulating workplace health has long been standard. Healthy indoor environments, work–life balance, and other goals pursued by ergonomists reconcile business goals with socially sustainable development, a requisite for green health. Outside the workplace, there is also a role to be played by ergonomists in the management and planning of natural environments to achieve concordance between ecological processes and social milieu.

In addition to developing technological solutions to ecological problems, ergonomics helps to understand social responses to the environment. As human behavior is shaped through product and system design, users' capabilities and lifestyles are taken into account in projects. Good designs are more likely to be adopted by consumers. Indeed, the term *human factor*—a key feature of the field—refers to physiological, psychological, and social attributes that influence interaction with the environment. This represents a broad spectrum of opportunities to promote environmentally prudent behavior and lifestyles at the individual and society level. This is important because research shows that consumers and householders have a poor knowledge and understanding of how their behavior impacts the environment.

Despite these opportunities, green ergonomics is a recent development, and it is currently not a priority consideration. There is slow adoption of sustainable design practices by both ergonomists and the public. Work environments, for example, are likely to be designed in terms of operational performance and cost effectiveness, rather than sustainability. Also, the construction of sustainable programs can be difficult, expensive, and time consuming. This is hampered by a paucity of empirical research linking sustainable strategies to improved occupant health and work performance, compared to buildings designed around standard practice. Green ergonomics has yet to be fully understood and embraced, and shifts in policy and practice have been slow to materialize.

Altogether, ergonomics has come a long way since its origins in aviation design and safety in World War II. The recent development of green approaches like sustainability offer a meaningful application of the field to pressing world problems. The broad scope of ergonomics means that it has potential to put sustainability at the forefront of people's work, leisure, and other aspects of daily lives. Although green approaches are relatively new and preliminary, the potential impact on green health is significant. It also offers a complementary approach to numerous cognate academic disciplines involved in advancing the green movement.

See Also: Alternative Energy Resources (Solar); Automobiles (Emissions); Indoor Air Quality; Lighting; Recycled Water.

Further Readings

Dul, Jan and Bernard Weerdmeester. *Ergonomics for Beginners: A Quick Reference Guide.* Washington, DC: CRC Press, 2001.
Karwowski, Waldemar. *International Encyclopedia of Ergonomics and Human Factors.* Washington, DC: CRC Press, 2006.
Norman, Donald. *The Design of Everyday Things.* Cambridge: MIT Press, 1998.
Wickens, Christopher, John Lee, Yili Liu, and Sallie Gordon-Becker. *Introduction to Human Factors Engineering.* Newark, NJ: Prentice Hall, 2003.

Gareth Davey
University of Chester

Fabrics

The word *green* is an adjective used to describe something that is perceived to be beneficial to the environment. So, green textile materials should be fibers, fabrics, clothing, and related products that would be friendly or beneficial to the environment. At the same time, green materials should be also sustainable products, which are defined as having no negative impact on natural ecosystems or resources. Most green materials should be recyclable, renewable, and reusable. Recycled products are made in whole or part from materials recovered from the waste stream.

Renewable products are those that can be replaced by natural ecological cycles or sound management practices, while reusable products can be repeatedly used after salvaging, or special treatment or processing. Textile manufacturing processes may involve the use of raw materials from either nonrenewable or renewable resources such as oil, coal, agricultural products, energy, and water. The production of textiles should also follow the principles of green chemistry, including 12 principles generally recognized by scientists. The principles include the following:

1. It is better to prevent waste than to treat or clean up waste after it is formed.

2. Synthetic methods should be designed to maximize the incorporation of all materials used in the process into the final product.

3. Wherever practicable, synthetic methodologies should be designed to use and generate substances that possess little or no toxicity to human health and the environment.

4. Chemical products should be designed to preserve efficacy of function while reducing toxicity.

5. The use of auxiliary substances (e.g., solvents, separation agents) should be made unnecessary wherever possible and innocuous when used.

6. Energy requirements should be recognized for their environmental and economic impacts and should be minimized. Synthetic methods should be conducted at ambient temperature and pressure.

7. A raw material or feedstock should be renewable rather than depleting wherever technically and economically practicable.

8. Reduce derivatives—unnecessary derivatization (blocking group, protection/deprotection, temporary modification) should be avoided whenever possible.

9. Catalytic reagents (as selective as possible) are superior to stoichiometric reagents.

10. Chemical products should be designed so that at the end of their function they do not persist in the environment but break down into innocuous degradation products.

11. Analytical methodologies need to be further developed to allow for real-time, in-process monitoring and control prior to the formation of hazardous substances.

12. Substances and the form of a substance used in a chemical process should be chosen to minimize potential for chemical accidents, including releases, explosions, and fires.

In general, green textile materials should be renewable, reusable, recyclable, and eco-friendly at the grave; also, they should provide improved health and aesthetic benefits to humans and should be produced in green chemistry processes.

Green Fibers and Chemicals

Textile fibers include natural fibers such as silk, wool, cotton, and flax, and manufactured fibers like rayon, polyester, polyamide (nylon), acrylics, and polypropylene. Rayon is regenerated cellulosic fiber and can be produced from woods and many renewable cellulose resources. All natural fibers and regenerated cellulose fibers are renewable materials. However, the production of silk, wool, cotton, and flax are restricted by the limited farmable lands and scales of agriculture production and cannot be increased without limit to meet the future increase of the world population. Since rayon and regenerated cellulose fibers can be made from almost any cellulose resources and crops produce million of tons of biomass (which is a major renewable cellulose resource), regenerated cellulose fibers are a focus of development of sustainable textiles fibers. Traditional rayon (xanthate) production process causes significant pollution in the form of emission of carbon disulfide and large quantities of wastewater, as well as sodium sulfate as a by-product. Most of the current research focuses on development of novel and environmentally friendly solvents or solvent systems for regeneration of cellulose. Lyocell fiber is a great example of regenerated cellulose fibers produced by using an environmentally friendly process and a new solvent, and ionic liquids have attracted a lot of attention recently.

Polyester, nylon, acrylics, and polypropylene fibers are the most widely used fibers in apparels and clothing materials; they are all petroleum-based synthetic materials and are traditionally considered as nonrenewable if the world supply of petroleum oil runs out. However, novel biobased synthetic polymers such as polylactic acid (PLA), polyhydroxy-alkanoates (PHAs), and partially biobased polytrimethylene terephthalate (PTT) can be totally or at least partially made from agricultural crops, making some synthetic materials sustainable. PLA and PHAs are biodegradable polymers, and disposal of these polymers would have reduced impact on the environment. In recent years, polyester fibers made from recycled polyester bottles have been widely adopted in textile applications, resulting in products with greener images.

Textiles also use various chemicals and additives, including colorants, softeners, and functional finishing agents, which are mostly made from nonrenewable resources. Although colorants could come from natural resources, the traditional production processes are nonsustainable due to the low yields of natural productions. This is similar to the supply of natural fibers due to limited lands and production yields. However, natural resources could be more efficiently utilized to produce textile chemicals. Examples include utilization of agricultural wastes (biomasses) in production of colorants and textile chemicals and efficient production of colorants by fermentation processes. Biomasses contain a significant amount of lignin, a major natural polymer with a lot of aromatic phenyl rings, which could be a building-block component for colorants. On the other hand, the use of some synthetic cationic softeners and flame retardants on textiles has raised concerns about human health and environmental safety. There is an urgent need to develop green functional textile chemicals.

Textile dyeing and finishing operations consume large amounts of water and energy and also produce large volumes of wastewater, which is a key factor affecting the energy footprints of textiles, or the green factor. Any novel technology that can reduce the use of water and energy or cut production of wastewater is considered an improvement for the textile industry. The use of supercritical carbon dioxide in textile finishing and dyeing has been considered as an option in the reduction of water usage. Research and development on processes and machines that can consume less energy and less water is important.

Green Fabrics

Textile fabrics are in the forms of knit, woven, and nonwoven structures. Knit and woven fabrics, also labeled in general as woven, are made of yarns, while nonwoven fabrics can be made of fibers directly. Woven and nonwoven fabrics are not only different in structure but are more unlike in fabric performance and applications of their products. Due to the differences in manufacturing processes and fabric structures, the woven fabrics, both knit and woven, are more durable and washable but may consume more energy and cost more to produce than nonwovens. As a result, almost all reusable and washable textiles are made of woven fabrics, and all single-use products are made of nonwovens. Due to the low cost of production, nonwoven fabrics are economically more advantageous.

Reusable and single-use products can be found everywhere, particularly in healthcare, hygienic, and institutional products. Single-use products are only used once and then disposed of, producing more wastes and lowering the value of the materials. Thus they are not as environmentally friendly as reusable ones. However, reusable products need to be washed, cleaned, and restored to the original performance for reuse, which may also produce more wastewater and consume more energy. The key advantages of reusable textiles are generating less solid wastes, consuming less fiber resources, and providing lower costs for overall usage. Overall, green textiles should be reusable, regardless of whether woven or nonwoven.

See Also: Green Chemistry; Personal Consumer Role in Green Health; Phaseout of Toxic Chemicals.

Further Readings

Alihosseini, F., J. Lango, K. S. Ju, B. D. Hammock, and G. Sun. "Mutation of Bacterium Vibrio Gazogenes for Selective Preparation of Colorants." *Biotechnology Progress*, 26 (2010).

Bowman, L. E., C. Caley, R. Hallen, and J. L. Fulton. "Sizing and Desizing Polyester/Cotton Blend Yarns Using Liquid Carbon Dioxide." *Textile Research Journal*, 66 (1996).

Darnerud, P. O. "Toxic Effects of Brominated Flame Retardants in Man and in Wildlife." *Environment International*, 29 (2003).

Libkind, D. and M. van Broock. "Biomass and Carotenoid Pigment Production by Patagonian Native Yeasts." *World Journal of Microbiology & Biotechnology*, 22 (2006).

Liebert, T. "Chapter 1. Cellulose Solvents: For Analysis, Shaping and Chemical Modification." In *Cellulose Solvents—Remarkable History, Bright Future*, T. Liebert, T. J. Heinze, and K. J. Edgar, eds. ACS Symposium, Vol. 1033. Washington, DC: American Chemical Society, 2010.

Moncrieff, R. W. *Man-Made Fibers*, 6th ed. Hoboken, NJ: Wiley, Newnes-Butterworth, 1974.

Gang Sun
University of California, Davis

Fast Food

America's love affair with fast food reaches back to 1921 when Walter Anderson and Billy Ingram opened the first White Castle restaurant. By the 1950s, Carl's Jr. and McDonald's were established, and in 1958, McDonald's sold its 100 millionth hamburger. While other fast food restaurants existed—they were usually drive-ins with carhops—none changed the natural and dietary landscape so thoroughly as did the three mentioned here. They were later joined by other fast food restaurants, including Wendy's, Burger King, Pizza Hut, and Taco Bell, and today there are nearly 200,000 fast food restaurants in the United States. While early love of fast food was tied to the novelty of the automobile and the accompanying growth of an entire infrastructure for a newly mobile society, today fast food has simply become a ubiquitous fact of our existence.

Not only do fast food restaurants themselves pervade our surroundings, fast food can be found everywhere, from airports to grocery stores to zoos. Further, billboards, television, and radio advertisements bombard us with pictures of fast food, visions of happy people eating fast food, and jingles and slogans about having it "our way." Fast food is also heavily marketed to children, and in addition to simply offering Happy Meals augmented with toys, billions are spent yearly on product and movie tie-ins designed to attract children. Further, McDonald's is the largest private operator of playgrounds and one of the country's top toy distributors. Research has shown that rising obesity rates in children are linked to the fast food marketing efforts directed at them, particularly since the 1980s.

About a quarter of the American population visits a fast food restaurant each day, and nearly half the money used by families to buy food is spent on fast food. McDonald's hires about one million people each year—more than any other U.S. organization, public or private. Fast food workers are the largest group of minimum-wage earners in the United States, the vast majority of whom work part time, lack benefits, and leave after a few months of employment. Fast food, then, impacts us in ways far beyond the 15 minutes it takes to buy and consume a burger and fries. Books, movies, and documentaries have been produced detailing our relationship with fast food, like Morgan Spurlock's *Super Size Me*, a 2004 documentary that follows Spurlock as he eats nothing but McDonald's for 28 days and gains 25 pounds, fat accumulation in his liver, and soaring cholesterol numbers, as well as experiences depression, mood swings, and a plummeting sex drive. Eric Schlosser's 2001 book, *Fast Food Nation*, examines various aspects of the fast food industry, from slaughterhouse to advertising. Schlosser provides a sobering account of the revolutionary force that fast food has unleashed on us dietetically, economically, and culturally. This focus on fast food's impact is part of a larger recent movement to reexamine our relationship with food; authors like Schlosser, Michael Pollan, and Peter Pringle are helping to direct attention to methods of food production and to eating habits in general, and fast food restaurants are responding in various ways, from offering healthy alternatives to standard menu items to scaling back on portion sizes. Despite these and other changes, fast food remains, for the most part, incredibly unhealthy and far more costly than its inexpensive price tag suggests.

Americans spend more than $115 billion a year on fast food—more than they spend on higher education or new cars. In return for these billions of dollars, they are served food that is loaded with calories, sugar, salt, and fat; in fact, one fast food meal can easily deliver a total day's intake of fat. In a country that has seen, over the past few decades, obesity in adults and children soar, it is difficult to not think about the role of fast food in this transformation, particularly as people increasingly claim that hectic schedules combined with low-priced fast food provide the driving force behind their desire for fast food. Some researchers think there may be another factor at play: addiction.

Scientists have theorized that eating foods that are high in fat and sugar, like fast food, stimulates the brain's natural opioids, producing a high similar, though less intense, to that produced by drugs like cocaine and heroin. So, for some, saying "no" to fast food can be as difficult as it is for addicts to say "no" to heroin and cocaine. Thus far, the research has been limited to rats, and the preliminary conclusions have been met with skepticism by some; still, others say that it has been known for years that eating fast foods and other foods high in sugar and fat stimulates a cycle of instant satiation, followed by rapidly plunging blood sugar levels, leading to cravings for another "fix." Researchers have also discovered that hormones that regulate appetite are profoundly affected by portion size and by the level of calories, fat, and sugar contained in what we consume. The hormone leptin, which works to keep the body's fat reserves stable, is particularly affected by high-fat diets, which cause the body to make more and more leptin and to lose its ability to regulate fat levels; drops in leptin brought on by losing weight actually cause the body to send out warning signals of starvation. For most people, this signal, of course, stimulates the desire to eat more.

Desire to eat more is reinforced by fast food portion sizes, which have been steadily increasing since the early 1980s, when market researchers found that fast food restaurants are sustained by a core of "heavy users." In an attempt to capture more of these "heavy users," Taco Bell dropped their prices; in response, Taco Bell saw a huge increase in customers and found that dropping prices encouraged people to buy more food. Taking their cue from Taco Bell, McDonald's, Burger King, and Wendy's all embraced lowering prices in order to gain bigger market shares. Soon, the increase in revenue leveled off, and fast food franchises needed another way to attract more customers and increase profits: "supersizing" and "up-selling." Increasing portion size hardly increases the cost of producing, for example, a bag of fries, yet it can nearly double the profit made on the sale. And anyone who has ever visited a fast food restaurant has been asked something like "Would you like to supersize that for 75 cents more?" Most Americans are unable to pass up such a "good" deal. Thus, price cutting, increasing portions, and up-selling have all combined to create fast food meals that are grossly laden with sugar, salt, fat, and calories. Rising obesity rates in the United States, the fact that poor diet combined with inactivity is one of the leading causes of preventable death in the United States, and the knowledge that fast food is both unhealthy and possibly addictive suggest that Americans would do well to think twice before including fast food in their diets.

See Also: Advertising and Marketing; Calorie Labeling for Restaurants; Children's Health; Fast Food Warnings; Obesity.

Further Readings

Organic Consumers' Association. "Fast Food Is a Major Public Health Hazard." http://www .organicconsumers.org/foodsafety/fastfood032103.cfm (Accessed August 2010).
Schlosser, Eric. *Fast Food Nation: The Dark Side of the All-American Meal.* New York: HarperPerennial, 2002.
WebMD. "Fast Food Creates Fat Kids." http://www.webmd.com/parenting/news/20040105/ fast-food-creates-fat-kids (Accessed August 2010).

Tani E. Bellestri
Independent Scholar

Fast Food Warnings

Rising environmental consciousness has raised awareness of potential threats to health in a variety of areas, including the fast food industry. Many have looked to fast food companies as one of the principal causes for the increase in obesity in the United States and have demanded action to help reverse this trend. In an effort similar to the battle against tobacco, groups have called upon fast food companies to post warnings and to provide information regarding the risks associated with the consumption of their foods.

Since 1990, with the passage of the Nutrition Labeling and Education Act, fast food companies have been required to provide nutritional information regarding their meals. This legislation, however, does not specify where the information is to be made available, leading many providers to limit access to websites or brochures kept behind counters. Several additional warnings have been recommended for fast food items. An early concern involved french-fried potatoes and acrylamide. Acrylamide, a white, odorless chemical that is produced when starches such as potatoes are introduced to high temperatures, had been shown to cause cancer and reproductive problems when given to laboratory mice in high doses. While the amount of acrylamide in french fries reported in studies investigating potential harm have been found to be below hazardous levels, some have sought warning labels for french fries. Under California's Proposition 65, approved in 1986, any material containing any substance known to have an association with cancer risks is required to carry a warning label. Acrylamide has been listed under Proposition 65 since 1990, and several lawsuits have been filed to demand that the public be made aware of the trace acrylamide levels in french fries.

Trans fats, or unsaturated fat with trans-isomer fatty acid, have also generated calls for fast food warnings. Trans fats are associated with greater probabilities for heart disease, diabetes, and other health risks and were common in many fast foods. In 2006, the U.S. Food and Drug Administration (FDA) requested that manufacturers state the amount of trans fats in their respective products; this has caused fast food companies to change recipes and processing to reduce or eliminate trans fats.

Aside from specific chemicals or fat additives, many groups, such as the Center for Science in the Public Interest, have pushed for more information regarding the health consequences of certain fast foods. Recommended warnings include the posting of caloric content and macronutrient breakdowns of food items in fast food restaurants solely for the customers' benefit, as consumers often underestimate the calories in meals consumed in fast food restaurants. Menu labeling proponents argue that providing more information to consumers will allow for better choices regarding proper nutrition, which will in turn reduce the number of nutrition-related health risks occurring in the United States. Several lawsuits have been filed against fast food companies regarding the deceptively high calories and levels of saturated fat. In response to these claims, fast food companies have taken measures to promote the nutritional quality of their foods including removing larger-sized beverage and french fry options, adding lighter fare such as fresh fruits and salads to menus, and promoting healthy activities and lifestyle changes through national marketing campaigns.

Despite proactive measures taken by fast food companies, multiple legislative efforts have been made to create laws to warn consumers of the nutritional content of fast food. California was the first state to pass a menu-labeling law in 2008 and stipulated that restaurants with 20 or more locations in California must inform customers of the caloric and nutritional values of foods served. However, the law does not specify that the information must accompany the food listings on the menu but may be published in a brochure. Maine, Massachusetts, Oregon, and New Jersey have followed suit and enacted statewide legislation as well, and 29 other states have introduced similar bills in their respective legislatures. At the federal level, legislation such as the Menu Education and Labeling (MEAL) Act seeks to expand menu-labeling requirements to restaurants with more than 20 locations nationwide. The MEAL Act would require information regarding calories, carbohydrates, protein, fat, and other nutrients on all menus. In response, the restaurant industry initially supported a competing bill, the Labeling Education and Nutrition (LEAN) Act, which would have required calories to be posted, but only near menus. The restaurant industry has since given support to the MEAL Act in hope of avoiding coordination of national campaigns with differing individual state measures regarding menu labeling. While many fast food restaurants have waited for legislation to mandate their actions, Yum Brands, Inc., which owns KFC, Taco Bell, and Pizza Hut, has announced that it will voluntarily add information to its menus regarding calorie content in all of its restaurants by the beginning of 2011.

Fast food warnings aim to provide consumers with more complete and easily accessible information regarding the foods they eat. It is hoped that providing nutritional information directly next to menu listings will help consumers make more healthy choices regarding what they eat. Although fast food companies have created healthier alternatives and reduced portion sizes, momentum continues to build for increased information about and warnings regarding fast foods.

See Also: Calorie Labeling for Restaurants; Children's Health; Fast Food; Obesity.

Further Readings

Kincheloe, J. L. *The Sign of the Burger: McDonald's and the Culture of Power*. Philadelphia, PA: Temple University Press, 2002.
Simon, Michele. *Appetite for Profit: How the Food Industry Undermines Our Health and How to Fight Back*. New York: Nation Books, 2006.
Simon, P., C. J. Jarosz, T. Kuo, and J. E. Fielding. *Menu Labeling as a Potential Strategy for Combating the Obesity Epidemic: A Health Impact Assessment*. Los Angeles: Los Angeles County Department of Public Health, 2008.

Stephen T. Schroth
Jason A. Helfer
Michael A. Wipper
Knox College

FERTILIZERS

Agriculture is a major industry in almost all parts of the world, important not only for its financial benefits but also because of its necessity in feeding the population. Raising crops takes valuable nutrients out of the soil, and farmers frequently use fertilizers to replenish these to promote plant growth. The chief nutrients provided by fertilizers are nitrogen, phosphorus, and potassium, although other nutrients are sometimes added as well. Both inorganic and organic fertilizers are available, and both have relative advantages and disadvantages. As a larger number of consumers become interested in purchasing and using food that is environmentally sustainable, discussions regarding the relative merits of inorganic and organic fertilizers have intensified. Additionally, the use of any sort of fertilizer may impact the surrounding environment, which has led to concerns related to the uses of certain types of fertilizer.

Uses of Fertilizer

Fertilizers began to become common during the 19th century. Based upon research conducted independently and concurrently in England, France, Germany, the United States, and other countries, understanding of the need for plant nutrition grew, along with understanding that different types of soil were more beneficial for plant growth than others. The finding that certain substances, such as nitrogen, could assist plant growth when added to the soil as well as the discovery of substances such as coprolite (fossilized animal dung), guano (feces and urine of bats, seabirds, and seals), and bone meal that could serve as fertilizer spurred its use. Developments in chemistry also allowed for the creation of inorganic fertilizers, which were less expensive than natural alternatives and increased the market for such products.

Fertilizers provide nutrients to the soil that encourage plant growth. Fertilizers are usually applied directly to the soil, although they can also be spread through foliar feeding, where they are sprayed on leaves of plants. Fertilizers are sometimes devised to address specific soil or crop needs but generally contain both macronutrients and micronutrients. Fertilizers usually contain the following:

- Nitrogen, phosphorus, and potassium, collectively known as the three primary macronutrients
- Calcium, magnesium, and sulfur, together known as the three secondary macronutrients
- Boron, chlorine, copper, iron, manganese, molybdenum, selenium, and zinc, which are micronutrients or trace minerals

Macronutrients are consumed in larger quantities and are found in plant tissues, when tested, to a percentage as great as 4 percent of the dry-matter weight basis. Micronutrients are also present in plant tissues but in a much smaller proportion when compared to macronutrients.

The use of fertilizers greatly increased food production during the 20th century, a development that was met with an exponential increase in the Earth's population. The period between the end of World War II and the 1970s is sometimes referred to as the "green revolution," an era that dramatically increased food production in the developing world. The green revolution stemmed from the expansion of irrigation and the introduction of hybridized seeds, inorganic fertilizers, and pesticides to farmers in developing nations. Yields of crops such as rice, corn, and wheat increased greatly during this era. Critics of the green revolution note that reliance on certain crops undermined food security and that the actual yield of crops to energy declined during this period, making its continued sustainability questionable.

Inorganic Fertilizers

Inorganic fertilizers are those composed of synthetic chemicals, minerals, or a combination thereof. Inorganic fertilizer may come in a variety of forms, the most popular of which is granular fertilizer, which comes in a powder and is often shipped in bags or boxes. Liquid fertilizer is also popular and is easily applied. Recently, slow-release fertilizer has also become available for agricultural and home use, a form that has the advantage of not exposing plants to excessive nutrients. Many forms of inorganic fertilizer are synthesized using the Haber-Bosch method, which introduces an enriched iron or ruthenium catalyst to nitrogen and hydrogen gases, resulting in ammonia. Ammonia is necessary for the nitrogen in fertilizer and can easily be combined with rock phosphate and potassium to form a compound fertilizer. Well over a third of the Earth's food supply is produced using inorganic chemicals. Inorganic fertilizers are frequently used to treat fields growing barley, corn, rapeseed, sorghum, soy, and sunflowers because they are inexpensive and easy to obtain.

Inorganic fertilizers do present certain problems. Many inorganic fertilizers lack certain trace minerals, which are depleted by crops over time. This has been traced to the 75 percent reduction of the quantities of minerals present in fruits and vegetables grown using inorganic fertilizers. Overfertilization can also be harmful because fertilizer burn may result. Inorganic fertilizers also demand a great deal of energy to produce. The production of synthetic ammonia, for example, consumes 5 percent of the world's natural gas production annually, or 2 percent of the Earth's energy production. The ways in which they are currently produced means that inorganic fertilizers are problematic in terms of long-term sustainability.

Organic Fertilizers

As problems with inorganic fertilizers have become known, interest has grown in organic alternatives. Organic fertilizers include naturally occurring materials such as compost, guano, manure, peat moss, seaweed, and worm castings, or naturally occurring mineral deposits such as saltpeter. Organic fertilizers can, in addition to boosting crop yield, improve the biodiversity and long-term productivity of soil. Organic fertilizers have the benefits of mobilizing existing soil nutrients, releasing nutrients at a slower rate, assisting in retaining soil moisture, and helping prevent topsoil erosion. If available locally, organic fertilizers are also less expensive than inorganic alternatives, and are more sustainable. Disadvantages also exist with organic fertilizers, however. Because their composition is more complex and varied than inorganic fertilizers, their effect on soil is less standardized. Improperly processed organic fertilizers may also contain plant or animal pathogens that are harmful to humans. Additionally, because of the greater mass, organic fertilizers can require more energy to transport than their inorganic alternatives if transportation over a distance is necessary.

Environmental Concerns

All fertilizers present certain environmental concerns. Because fertilizers provide nitrogen to enrich the soil, nitrogen-rich runoffs that enter oceans, rivers, and lakes can deplete their oxygen levels and diminish their ability to sustain aquatic life, such as fish. Visually, water with depleted oxygen becomes discolored or cloudy. Nitrogen-containing fertilizers can also lead to the acidification of soil, which limits available nutrients and must be combated by applying lime or calcium hydroxide. For fertilizers containing phosphates, a risk of heavy metal accumulation also accrues. Heavy metals such as cadmium and uranium are common in soil that has been treated with phosphate fertilizers. Soil that has been treated with steel-industry wastes, a popular source for zinc, often contains increased levels of arsenic, chromium, lead, and nickel. As a result of these risks, nutrient budgeting is recommended, as this balances nutrients coming into the farming system with those leaving. Nutrient budgeting not only improves the ecological balance of farms and the surrounding land and water but also reduces costs as overfertilizing is reduced.

See Also: Animal Products; Biological Control of Pests; Genetically Engineered Crops; Organic Produce; Pest Control.

Further Readings

DeGregori, T. R. "Green Revolution Myth and Agricultural Reality?" *Journal of Economic Issues*, 38/2 (2004).
Eilittä, M., J. Mureithi, and R. Derpsch. *Green Manure/Cover Crop Systems of Smallholder Farmers: Experiences From Tropical and Subtropical Regions*. Dordrecht, Netherlands: Kluwer Academic, 2004.

Shiva, V. *The Violence of the Green Revolution: Third World Agriculture, Ecology and Politics.* London: Zed Books, 1992.

Stephen T. Schroth
Jason A. Helfer
Knox College

FIREWOOD AND CHARCOAL

Firewood and charcoal have been used for thousands of years to provide heat and light to humans. Both are popular today, although in more prosperous parts of the world this choice is based more on aesthetics than on need. Although firewood and charcoal are both renewable sources of energy, demand for both can outpace forests' ability to regenerate on a local or national basis. Although good forestry practices can ensure better local supplies, overharvesting can lead in extreme cases to desertification. Both firewood and charcoal have environmental consequences as a result of being burned, and green alternatives have been developed that have a less adverse affect on the ecosystem.

Firewood and charcoal have been used for thousands of years to provide heat and light to humans. Both are popular today, although in more prosperous parts of the world, this choice is based more on aesthetics than on need.

Source: iStockphoto.com

Use and Problems

Both firewood and charcoal have been used for heat, light, and fuel for thousands of years. Firewood is classified as either seasoned (dried) or unseasoned (green) and can derive from either hardwood or softwood. Different types of wood provide different burning experiences, as hardwoods (beech, oak, and other deciduous trees) take longer to burn, while softwood (pine, spruce, and other conifers) are easier to harvest and cut. Although the harvesting of firewood varies by region or culture, most firewood in the United States comes from forests or woodlots cultivated specifically for that purpose. Standing dead timber and recent deadfall are preferred because this wood is already partially seasoned. Harvesting dead timber or deadfall reduces the intensity of brush or forest fires, as it removes a potential fuel source from such disasters. Most firewood is cut during the winter, when trees contain less sap and will season more quickly. Firewood is generally cut within a distance of 50 miles from its point of sale; this is believed to be the longest distance it should be moved to reduce the spread of disease.

Charcoal is produced through a process of slow pyrolysis, which occurs when wood, sugar, bone char, or other substances are heated in the absence of oxygen. The result of pyrolysis is charcoal, which is soft, black, brittle, lightweight, and porous. Charcoal in many ways resembles coal, and is composed of 50 percent to 95 percent carbon, with the remainder being ash and volatile chemicals. Charcoal is preferable to wood because it produces a more consistent source of heat, and is lighter and more compact. The production of charcoal also produces a variety of by-products such as acetic acid, acetone, and methyl alcohol that have valuable commercial uses. Charcoal can cause severe health problems, especially when burned indoors, as it causes carbon monoxide production. Even when burned outdoors, charcoal is a major contributor to air pollution.

Environmental Solutions

Demand for firewood can lead to deforestation if trees are overharvested. Deforestation is damaging and can also lead to soil erosion. If forest ecosystems are properly managed, more firewood can be produced and in a manner that is beneficial to the environment. When the planting, protecting, thinning, controlled burning, felling, harvesting, and processing of lumber is managed properly, competing ecological, economical, and social needs may all be met. Those interested in ensuring that proper forestry practices take place often support third-party certification, which has become common since the 1990s. Certification programs were developed after the forestry industry had come under intense criticism for failing to manage forests in an environmentally sensitive manner.

Firewood combustion contributes significantly to air pollution, especially in terms of particulates. Large amounts of particulates in the atmosphere can contribute to a variety of human health conditions, including asthma and heart disease. The burning of firewood releases no more carbon dioxide than the natural biodegradation of the wood itself, making it a carbon neutral exercise. The harvesting and transporting of firewood, however, does contribute significantly to greenhouse gas emissions. Using a pellet stove or other slow-feeding device makes the most efficient use of firewood.

In developing nations, charcoal is becoming one of the primary contributors to greenhouse gas emissions and environmental degradation. In an effort to stem this tide, biomass briquettes have been suggested as a greener alternative to charcoal. Biomass briquettes are composed of various natural materials, such as rice husks, fibrous residue such as the stalks from sugarcane or sorghum, groundnut shells, and the like. Biomass briquettes provide a higher level of caloric value per dollar expended than do charcoal or coal when used for industrial boilers or other heating sources. Biomass briquettes can only limit the use of charcoal to a certain extent, but as industries and governments understand the benefits of limiting pollution, they are becoming increasingly popular. Through a process of biomass pyrolysis, biochar can be produced. Biochar is a substitute for wood charcoal that is made, like biomass, from agricultural waste. Biochar generally releases fewer greenhouse gases than does charcoal and assists greatly in preventing deforestation. Deforestation is especially problematic in developing nations, as government regulation of forestry is more difficult to enforce.

Alternative fuel sources do present some problems, however. Some critics suggest that firewood and charcoal alternatives such as biomass and biochar will create agricultural, economic, and social problems of their own. As the demand increases for agricultural products that can be used for alternative fuels, the price of feedstock will increase as well. This could cause livestock production costs to increase, harming many small producers as well as consumers of meat. Additionally, as demand for certain crops that can be used in biofuel production increases, farmers will shift away from growing other less profitable grains. This shift will cause an alteration in crop production diversity, ultimately diverting crops away from the human food chain. Such shifts could have repercussions for consumers because those who already have difficulty obtaining sufficient food would most likely be severely affected by price increases for food.

See Also: Air Filters/Scrubbers; Arsenic Pollution; Clean Coal; Environmental Illness and Chemical Sensitivity; Environmental Protection Agency (U.S.).

Further Readings

Bruges, James. *The Biochar Debate: Charcoal's Potential to Reverse Climate Change and Build Soil Fertility.* Totnes Devon, UK: Green Books, 2009.
Houghton, J. T. *Global Warning: The Complete Briefing*, 3rd ed. Cambridge, UK: Cambridge University Press, 2004.
Schmidt, G. and J. Wolfe. *Climate Change: Picturing the Science.* New York: W. W. Norton, 2009.

Stephen T. Schroth
Jason A. Helfer
Keely E. Campbell
Knox College

FOOD ALLERGIES

A food allergy is an exaggerated, adverse immune reaction to food affecting 6 to 8 percent of children under 3 to 6 years old and 3 to 4 percent of adults, and the incidence is rising. A food allergy is different from the more common food intolerance. There are nine common food allergies and several lesser ones. Most allergies are IgE mediated, involve histamine release, and can be diagnosed by one of several tests. The only treatment is strict avoidance.

A food allergy is a type I immediate hypersensitive reaction mediated by the immune system involving an antibody immunoglobulin, IgE, and mast cells. When the individual is first exposed to the food, his or her white blood cell lymphocytes produce the IgE antibody to the food allergen, a protein in the food. Upon reexposure, the IgE antibody on the mast cell prompts the release of histamine. This can occur anywhere from within seconds to after two hours. The location of the tissue from which the histamine is released and to which it is carried determines the symptoms that an individual develops.

The classic IgE-mediated reaction occurs from oral exposure and can produce oral symptoms. Second, the IgE and/or non-IgE-mediated reaction can produce esophagitis, gastritis, and gastroenteritis. Third, the non-IgE-mediated response can produce protein-induced enterocolitis syndrome; food-protein proctocolitis or proctitis; protein-induced enteropathy (e.g., Celiac disease); milk–soy protein intolerance (especially during infancy or childhood); and Heiner syndrome, a serious lung disease due to milk-protein IgG antibody-immune complexes. The classic type I mediated IgE response can include angioedema, with swelling of the eyelids, face, lips, tongue, and voice box. The patient may also develop hives and itching of the mouth, throat, eyes, and skin. Nausea, vomiting, diarrhea, and abdominal cramping pain can also occur. Nasal congestion is possible, as is wheezing, a scratchy throat, shortness of breath, and difficulty swallowing. Anaphylaxis is a severe systemic reaction leading to a sudden drop in blood pressure and edema, causing bronchoconstriction that can be fatal within minutes. This is why avoidance of the offending food is essential.

Diagnosing a Food Allergy

A physician can perform one of several allergy tests to confirm the diagnosis of a food allergy. The skin prick test introduces a small amount of the food allergen under the skin. A hive will form confirming the diagnosis. Non-IgE-mediated allergies cannot be detected with this method. An immunoCAP-RAST (radioallergosorbent test) blood test can detect the level of IgE antibodies present for a particular food. This method is valid for dairy, egg, peanut, fish, soy, wheat, and other allergies. A blood serum test can determine the level of IgE and IgG4 antibodies present for a particular food. This method has been used for several decades and will test about 300 foods. An antigen leukocyte cellular antibody test (ALCAT) uses a whole blood assay to detect blood cell changes in response to food. Approximately 100 foods can be tested. It is more appropriate when testing for food intolerance than for food allergy.

The most common food allergies are dairy, egg, peanut, tree nuts (e.g., almond, walnut, pecan, Brazil nut, hazelnut, and macadamia), seafood, shellfish (e.g., shrimp, crab, lobster, snails, and clams), soy, wheat, and sesame. Peanut and tree nut allergies are the least likely to be outgrown.

Source: U.S. Department of Agriculture, Agricultural Research Service/Scott Bauer

A food challenge test, preferably double blind and placebo controlled, is the gold standard for diagnosing non-IgE-mediated food allergies. The placebo and/or food allergen is given in a capsule, and the patient is monitored for symptoms. Because of the risk of anaphylaxis, this must be conducted in a hospital in the presence of a physician.

The food elimination diet pioneered by Drs. A. H. Rowe and A. Rowe was the only method of testing prior to the above laboratory tests. The patient was placed on a hypoallergenic diet, a possible allergic food was introduced one at a time, the patient's response was monitored, and the food was recorded. The patient was returned to the hypoallergenic diet, and a new food was introduced when the patient no longer had symptoms. This method can also detect food intolerances. Finally, an endoscopy, colonoscopy, and biopsy may be ordered to determine the effect of the allergy on gastric tissue or to measure allergic response.

The differential diagnosis of a food allergy includes several conditions as well as the very common food intolerance. Food intolerance is a nonimmunoglobulin-mediated reaction that can present with a variety of non–life-threatening symptoms, such as flushing, abdominal upset and pain, diarrhea or constipation, and mild mucosal or skin reactions. Lactose intolerance can present at any age due to a deficiency of lactase enzyme. While it can occur among people of European descent, it is very common among non-Europeans. Celiac disease is a non-IgE-mediated food allergy that triggers an autoimmune response to gluten in wheat, oats, rye, barley, spelt, kamut, triticale, and related grains. Irritable bowel syndrome presents with constipation and/or diarrhea, abdominal pain, and a sensitivity reaction to various foods or food groups. C1 esterase inhibitor deficiency (hereditary angioedema) is rare and may present with angioedema; swelling of the face, lips, tongue, and throat; abdominal cramping and pain; diarrhea; and occasionally shock.

The most common food allergies are dairy, egg, peanut, tree nuts (e.g., almond, walnut, pecan, Brazil nut, hazelnut, macadamia), seafood, shellfish (e.g., shrimp, crab, lobster, snails, clams), soy, wheat, and sesame. In east Asia, rice allergies are common; celery allergies are found in eastern Europe, buckwheat allergies in Japan, and beef allergies in Australia. Corn allergies are common in the United States and Canada. Fruit allergies are common to members of the rose family (e.g., apples, pears, cherries, peaches, nectarines, and strawberries) and to jackfruit in the mulberry family.

One Treatment—No Cure

The treatment for a food allergy is avoidance. There is no cure. Even trace amounts of the food allergen can cause a hypersensitivity reaction. Approximately half of all children will outgrow their dairy, egg, soy, and wheat allergy by age 6. If the allergy is still present at age 12, they have less than an 8 percent chance of outgrowing it. Peanut and tree nut allergies are the least likely to be outgrown.

Allergy desensitization shots have not been clinically effective in research trials compared to the more successful pollen and dander desensitization shots. Allergic persons should avoid all forms of the food to which they are allergic. They should carry an autoinjector of epinephrine in case of exposure to the offending food(s). They should also wear some form of medical alert jewelry and inform their family and friends of this life-threatening allergy, that they are carrying an autoinjector of epinephrine, and to contact emergency services if they begin to develop any allergic symptoms or have to use their epinephrine.

Allergic patients need to read labels carefully before making any food or snack purchases. Not all countries have stringent labeling laws, and not all foods and beverages (e.g., alcoholic beverages) are labeled or adequately labeled. Contact a knowledgeable person at the manufacturing company if you have a question before making a purchase. When eating out, it is important to choose restaurants carefully. Inform the wait staff of your allergy; ask questions about the food choices and how they are prepared; do not hesitate to question the chef before making a food selection; and if you are in any doubt, do not consume the food or beverage. Not all restaurants use the same cooking method or recipes or ingredients, even within the same franchised restaurant chain. These same precautions must be followed for all nutritional supplements or anything that will be ingested. Patients should also avoid skin exposure to their allergen. Read all cosmetic and topical medication labels carefully. Remember that even trace exposure to the food allergen can cause a severe allergic reaction. Each repeated exposure to the food allergen will heighten the patient's allergic response.

The development of a food allergy is more common in individuals with atopic syndrome. These individuals often produce IgE antibodies to dust mite feces, animal dander (e.g., cat, dog, horse, bird), pollens, and other environmental elements. They are more prone to allergic rhinitis or hay fever and asthma. Symptoms can include watery eyes, itchy eyes, runny nose, sneezing, cough, sinus congestion, headache, swollen throat, shortness of breath, wheezing, acute asthma attack, and anaphylaxis.

There are several theories for the development of hypersensitivity. Influenza and yellow fever vaccines are egg based, and the measles-mumps-rubella vaccine was egg based until 1994. Eggs are a very common food allergen, and vaccines given to children at too early a developmental stage may trigger a hypersensitivity reaction. Introduction of protein foods early in an infant's diet before their digestive tract is ready for complex proteins may cause an immune reaction in the infant's gut. The hygiene hypothesis is based on not only the reduced exposure to dirt and germs but also the increased use of antibiotics and antibiotic cleaners in the developed world. The result could be an overreaction to foods as a result of a shift in the individual's Th1 and Th2 immune balance. Food allergy and allergies in general are much less common in the developing world, where hygiene standards are lower. Leaky gut syndrome, perhaps due to antibiotic use, may lead to food allergies. It is important to note that the gut-associated lymphoid tissue (GALT) comprises about 70 percent of our immune system. Damage to this system or early triggering of the GALT to foods or other ingested proteins may affect our immune system reaction to food and many other substances. In the developed world, baby skin-care products are often based on peanut oil, a cheaper alternative to nonallergenic products. Studies have shown that breastfeeding for at least four to six months compared to cow- and soy-based formula can prevent or delay the onset of food allergy and wheezing. Recent studies show that exposure in infancy to omega-3-rich fish may also reduce the risk of food allergy and atopic conditions such as eczema and asthma.

Food allergy is a serious, life-threatening condition. It requires medical diagnosis and intervention. It should be considered a life-long condition that requires hypervigilance to the offending foods and their various by-products. There are a number of dietary alternatives to the allergenic food(s). There are numerous cookbooks with recipes to help one make safe, tasty, nonallergic foods for the allergic person.

See Also: Asthma; Children's Health; Fast Food Warnings; Home-Grown Food; Wine and Other Alcohols.

Further Readings

Emerton, V. *Food Allergy and Intolerance*. London: Royal Society of Chemistry, 2002.

Oh, C. K. and C. Kennedy. *How to Live With a Peanut Allergy*. New York: McGraw-Hill, 2005.

Rowe, A. H. and A. Rowe, Jr. *Food Allergy: Its Manifestations and Control and the Elimination Diets. A Compendium*. Springfield, IL: Charles C Thomas, 1972.

Sicherer, S. H. *Understanding and Managing Your Child's Food Allergies*. Baltimore, MD: Johns Hopkins University Press, 2006.

Wood, R. A. and J. Kraynak. *Food Allergies for Dummies*. Hoboken, NJ: John Wiley, 2009.

Paul Richard Saunders
Canadian College of Naturopathic Medicine

FUNGI AND SICK BUILDING SYNDROME

Any discussion of green health should include a discussion of the illnesses related to toxic exposure, called *environmental illnesses,* or EI. The field of environmental medicine focuses, in part, on the treatment of individuals made ill by their environment. *Sick building syndrome* (SBS) is a term coined in the 1970s to describe otherwise unexplained illness in individuals working or living in the same nonindustrial building. This vague term refers to a building inclusive of its population rather than to a clinical state in an individual. Sick building syndrome is really a form of basic environmental illness and is not a specific disease on its own. Indoor air contamination by chemicals or mold within the building as well as inadequate air exchange and contamination by outdoor pollutants in air or soil have been described as causes. Molds growing in damp buildings can produce mycotoxins and volatile organic compounds as well as allergenic spores. These mycotoxins act like manmade chemicals in that they can also cause environmental illness and often are carcinogenic. The health effects of indoor mold exposure are varied and substantial, as well as controversial, and are not yet significantly recognized in traditional medicine circles. A detailed description of environmental illness/chemical sensitivity and its diagnosis and treatment is suggested for further reading to complement this topic.

Symptoms of SBS characteristically improve upon leaving the building and upon breathing fresh air at the end of the day and over the weekend—only to recur upon reentry. As the exposure goes on, symptoms persist even when the workers are at home. Eventually, permanent symptoms develop and often disable the inhabitants. In fact, in a Swedish study, 45 percent of victims of SBS no longer had the capacity to work. Thus, this can be a devastating yet preventable health problem that has been reported to affect up to 20 percent of workers in the United States. In fact, 25 percent of American workers surveyed felt that the poor quality of air at work was adversely affecting their health.

Since the 1990s, mold contamination has been found to occur in 35 to 50 percent of sick buildings with air quality issues. Indoor air pollution can also be caused by formaldehyde-filled building materials, cabinetry, pressed board, and any chemical spill, as well as from carpeting and the adhesives used to lay carpeting. New carpeting was determined to have made 200 people ill in the 1980s at the Environmental Protection Agency (EPA) in Washington, D.C. Elevated levels of carbon dioxide (CO_2) and inadequate air exchange has also been a culprit. Lowered productivity and increased sick leave are associated with poor air quality. Air conditioning systems with improper maintenance, design, and operation also contribute to the development of symptoms of SBS or EI.

SBS is often acknowledged because dozens of people all complain of becoming ill. Initially, officials and physicians hired by management name this "mass hysteria" until a lawsuit proves otherwise. When only one or two people are ill in a private home, there is even less acknowledgment by doctors of a possible environmental exposure. The average physician receives no training in medical school in the signs and symptoms of environmental illness, though it affects tens of millions of people to some degree. Often these patients behave strangely—due to toxic brain effects. They struggle to be believed, and they are often referred erroneously to a psychiatrist.

Chemical Overload

Environmental illness or chemical sensitivity is acknowledged in the Americans with Disabilities Act as a potentially disabling condition. According to the EPA, people suffering from sick building syndrome usually experience symptoms such as headaches (70 percent); sleepiness (60 percent); fatigue; dry eyes, nose, and lips; sore throat; dry cough; dry skin; dizziness and nausea; difficulty concentrating; and "extra sensitivity to odors." In addition, nosebleeds, skin rashes and burning sensation, irritability, insomnia, and many other symptoms have been described from mold exposure. Characteristically, symptoms occur in many organ systems and with different manifestations among individuals. Some people may have confusion or problems reading, and others experience depression or anxiety. Persons in one part of the building may feel worse than others depending on proximity to the offending problem. In fact, over time, significant environmental exposures within buildings may result in full-blown chemical sensitivity. This syndrome can be thought of as occurring because of an overloading of the body's ability to detoxify contaminants.

One result of this overload is that the person can no longer tolerate other chemicals, natural smells, or even sounds, lights, and electrical appliances or electromagnetic sensitivity. This phenomenon occurs because of damage to the autonomic and central nervous system, immune system, and endocrine system, as well as hepatic detoxification system. A preventive strategy to counter the occurrence of SBS is to lower the "total environmental load." This means to minimize exposure to chemicals and molds and to supply ample fresh, unpolluted air exchanges.

Mold exposure inside buildings can lead to a variety of mental and physical symptoms and conditions—some life threatening. Classically, waking up with a headache can indicate mold in the bedroom. Symptoms depend on the specific toxins produced by particular molds. Not all molds produce these mycotoxins, but one can presume molds that grow due to water intrusion or leaks to be largely toxic. It is essential to identify the genus and species of a mold by doing cultures/mold plates in order to surmise what toxins may be in a building—and what health effects to look for in the future. Simply entering the name of the mold and the word *mycotoxin* into a search engine will quickly provide a short list of chemicals produced by the mold. These toxins can be absorbed by clothing, furniture, books, and paper as well as stored in the body for decades. Treatment is involved, often only partially successful, and costly.

If the inhabitants of a building exhibit neurologic symptoms like headaches, tremors, memory loss, impaired balance, confusion, and difficulty reading, then the toxins are obviously neurotoxic, whatever they may be. Therefore, it is wisest to leave the workplace and take no items from the building, perhaps even discarding clothing at home and the vehicle used to go to and from work. Once the contaminated clothing has been washed at home, even the laundry machines have been contaminated and therefore the rest of the clothing at home as well. There is no scientific evidence that mycotoxins can be cleaned by washing or dry cleaning—they are extremely heat stabile and have been researched extensively by the army because they have been used as agents of bioterrorism.

SBS and Psychiatric Problems

Significant psychiatric problems can occur with mold exposure. Anxiety, bipolar illness, attention deficit, and depression are common. In a Brown University study of 5,800 Europeans. mold exposure in the home was found to be associated with a 40 percent incidence of depression. This illustrates that some psychiatric disease is, in fact, environmentally induced and may be effectively treated by doctors using the environmental medicine approach.

Mold exposure has been reported to induce autoimmunity, especially leading to antibodies to components of the nervous system such as myelin basic protein, myelin-associated glycoprotein, tubulin, and so on. This in part explains how exposed individuals could develop multiple sclerosis from working in a moldy building, for example. Nerve conduction velocities have also been reported to be abnormal, and this leads to peripheral neuropathies (tingling, numbness, muscular weakness). Studies also have found that there is a change in the EEG, demonstrating lowered activation of the frontal cortex stemming from brain stem involvement. SPECT scanning of the brain has also shown dramatic changes in perfusion in the brains of mold-exposed patients.

A host of other medical problems can occur in people living in damp buildings: bronchitis, pneumonia, asthma, and chronic rhinosinusitis and nasal polyp formation. The Mayo Clinic published that 93 percent of patients with chronic sinusitis (there are 37 million of them) have positive fungal cultures (in addition to bacteria) and may benefit from nasal or oral antifungals. Significant changes in T and B lymphocyte cells occur and mediate the development of infections, cancer, autoimmune disease, and allergy. Dysautonomia causes difficulty standing without heart rate elevation, and mitochondrial damage can cause severe muscle weakness and even difficulty breathing akin to Lou Gehrig's disease.

Many biologic markers of mold exposure have been described as well. Dr. Ritchie Shoemaker described these parameters as abnormal in patients from water-damaged buildings: matrix metalloproteinases 9 (MMP 9), leptin, alpha melanocyte stimulating hormone (MSH), vascular endothelial growth factor (VEGF), and pulmonary function. High levels of MMP 9 indicate a proinflammatory cytokine response, leptin and MSH abnormalities indicate disruption of the proopiomelanocortin pathway in the hypothalamus, and VEGF reflects the level of tissue hypoxia. Additionally, this researcher found that the administration of the medication cholestyramine improved these markers and reduced the levels of mycotoxins as measured in urine and helped to reduce symptoms in some patients.

The toxins produced by certain molds are numerous. The trichothecenes, ochratoxin, and aflaxtoxin B are now measurable in urine. The trichothecenes can be measured affordably in house dust. The trichothecenes, the most lethal group of toxins, have been studied by the U.S. Army. One study of rats by the army showed that inhaled trichothecenes led to adrenal necrosis. They showed this damage to the stress gland is prevented with the administration of testosterone first, hence proving the difference between men's and women's responses to living in a moldy home. Women develop adrenal insufficiency more often than men, who have 10 times more testosterone from which they can steal to make the adrenal hormones survive. Along the same lines, the Centers for Disease Control and Prevention (CDC) found that women with chronic fatigue (which can be caused by mold exposure) develop low adrenal function much more often than men. This loss of adrenal function can explain dizziness, depression, allergy, weight loss, intolerance of stress, increased sense of smell, and immune problems, as well as anxiety in women exposed to mold. Hormone replacement gives substantial relief to the sufferers.

The stigma of becoming ill from perhaps undetermined exposures is slowly lifting. Successful treatment of individuals will start with acknowledgment that they are ill and not just malingering for financial gain. Those scientists who claim that no illness comes from mold or chemical exposure in buildings should be queried as to their financial ties to industry and benefits from expert-witness fees in litigation. Recognition of the environmental illnesses by medical societies, medical schools, and the government will determine the speed of acceptance of the science of environmental medicine.

See Also: Air Filters/Scrubbers; Carbon Filters; Environmental Illness and Chemical Sensitivity; Environmental Protection Agency (U.S.); Men's Health; Women's Health.

Further Readings

Environmental Health Center of Dallas. http://www.ehcd.com (Accessed February 2010).

Kilburn, Kaye H., ed. *Molds and Mycotoxins: Papers From an International Symposium.* Washington, DC: Heldref Publications, 2004.

Murphy, Michelle. *Sick Building Syndrome and the Problem of Uncertainty: Environmental Politics, Technoscience, and Women Workers.* Durham, NC: Duke University Press. 2006.

Rea, William. *Chemical Sensitivity*, vols. 1–4. Boca Raton, FL: CRC Press, 1997.

Lisa Lavine Nagy
The Preventive and Environmental Health Alliance, Inc.

GASTROENTERITIS

Gastroenteritis is defined as an infection of the gastrointestinal tract (stomach, small intestine, and/or large intestine). Gastroenteritis is common, and the resulting vomiting, dysentery, and/or diarrhea is the most common cause of death in developing countries, accounting for approximately 2.5 million fatalities per year. In the United States alone, it is estimated that there are over 10,000 deaths annually due to gastroenteritis. Bacteria, viruses, and parasites that cause gastroenteritis are most commonly transmitted to humans through contaminated food or water but in some cases can be directly transmitted from an infected individual or other object. Some infections are self-limiting and do not require treatment except for rehydration, while others are extremely serious and require immediate medical treatment.

In general, gastroenteritis can be broadly divided into three main groups based on the pathogenesis of the causative agent: (1) bacterial intoxication, (2) viral gastroenteritis, and (3) gastroenteritis infection, which can be noninflammatory or inflammatory.

Bacterial Intoxication

Bacterial intoxication (also known as food poisoning or toxemia) occurs when a preformed bacterial toxin is ingested. The most common causative agents of bacterial intoxication gastroenteritis are *Staphylococcus aureus*, *Bacillus cereus*, and *Clostridium perfringens*. The bacterium that produces the toxin need not be ingested, and therefore this type of gastroenteritis is not technically considered to be an infection. Symptoms of bacterial intoxication normally occur within two to 12 hours after ingestion of the toxin. Normally, a fever is absent, and fecal leukocytes are not present. In general, the symptoms of bacterial intoxication are self-limiting, and treatment other than rehydration is not required. Because of the absence of a bacterial infection, the use of antibiotics is not warranted.

Bacterial intoxication caused by *Staphylococcus aureus* (gram-positive coccus) is due to the *Staphylococcus* enterotoxin. These toxins are resistant to boiling and act by binding to the emetic reflex center of the brain, resulting in nausea and vomiting. *Bacillus cereus* (gram-positive rod) can produce two different enterotoxins depending on the type of food contaminated. Type 1 enterotoxin (emetic form) is produced in high-carbohydrate food (rice, pasta) and causes nausea and vomiting. Type 2 enterotoxin (diarrheal form) is produced in high-protein food (meats) and causes diarrhea. A third bacterium, *Clostridium perfringens* (gram-positive rod), produces an enterotoxin that disrupts intestinal ion transport, altering membrane permeability and resulting in watery diarrhea.

Viral Gastroenteritis

A common cause of acute diarrhea in the United States is viral gastroenteritis. The U.S. Centers for Disease Control and Prevention (CDC) estimates that viruses cause 9.2 million of 13.8 million total cases of food-related illness each year. The most common causative agents of viral gastroenteritis are transmitted by the fecal–oral route and include rotaviruses, Norwalk virus, noroviruses, and adenoviruses. Typical symptoms are a low fever, nausea, vomiting, abdominal pain, and watery diarrhea. Viral gastroenteritis is normally self-limiting, but rehydration to replenish fluids and electrolytes is necessary, especially in infants.

Rotaviruses and Norwalk virus are found in the intestinal tract of many domestic and wild animals. Gastroenteritis caused by rotavirus is most common in infants and young children, and diarrheal infections are more commonly seen in winter. Gastroenteritis caused by Norwalk virus, however, is more common in older children and adults, and outbreaks are typically seen in the summer. It is estimated that Norwalk virus causes about one-third of cases of viral gastroenteritis in individuals over 2 years of age.

Norovirus is also called Norwalk-like virus and can be transmitted by contaminated food or water or by contact with contaminated surfaces or objects. These viruses are thought to be the most common cause of outbreaks of gastroenteritis in developed countries and in crowded areas such as cruise ships. Norovirus infections are more common in the winter and typically cause vomiting and diarrhea. Adenoviruses, meanwhile, are thought to be the second-most-common cause of viral gastroenteritis and are most commonly associated with diarrhea in infants.

Gastroenteritis Infections

Gastroenteritis infections can lead to either noninflammatory or inflammatory diarrhea, depending on the causative agent. Gastroenteritis infections are normally transmitted by the fecal–oral route and through contaminated foods. The most common causes of noninflammatory gastroenteritis include enterotoxigenic *Escherichia coli* (ETEC), enteropathogenic *E. coli* (EPEC), and *Vibrio cholerae*.

ETEC, like other *E. coli*, are gram-negative rods. Once ingested, ETEC multiplies in the small intestine and produces toxins that stimulate the secretion of chloride by intestinal cells, resulting in watery diarrhea and abdominal cramps typically lasting less than 24 hours. ETEC is a common cause of diarrhea in infants and also "traveler's diarrhea" in adults. EPEC strains, on the other hand, do not produce a toxin and instead bind to cells of the gastrointestinal tract upon ingestion. The binding to the enteric cells results in death of those cells and a mild diarrhea. EPEC commonly causes diarrhea in infants less than 6 months of age. Finally, *Vibrio cholerae* (a gram-negative curved rod) causes cholera and is transmitted via contaminated food or water. The bacteria produce an enterotoxin (cholera toxin) and other bacterial factors once they reach the small intestine, which stimulate a very rapid and extensive loss of fluids. In severe cases, several liters of liquid from the intestinal tract may be lost within a few hours, resulting in death if untreated.

Inflammatory gastroenteritis is also known as dysentery, and the causative agents are invasive and result in a large inflammatory response by the patient's immune system.

Symptoms of inflammatory gastroenteritis usually include a fever and abdominal pain, and the three most common bacterial causes are the gram-negative rods *Salmonella* spp., *Shigella* spp., and *Campylobacter jejuni*.

Salmonella spp. cause approximately 3 million cases of bacterial gastroenteritis per year in the United States. Most *Salmonella* spp. colonize the intestines of animals, and transmission is from animals to humans or through ingestion of contaminated animal products. Once ingested, *Salmonella* attaches to and kills cells of the intestine to gain entry into the bloodstream, causing fever, abdominal pain, and diarrhea. *Salmonella* infections are more common in the summer, when warm outdoor weather allows bacterial growth in contaminated foods, such as those found at picnics. Symptoms of *Salmonella* gastroenteritis are usually self-limiting. *Shigella* spp. are closely related to *E. coli* and are transmitted by the fecal–oral route and by flies. *Shigella* spp. are typically ingested, where they penetrate the cells of the intestine. These bacteria multiply in the intestinal mucosa and produce toxins (enterotoxin and *Shiga* toxin) that cause intestinal cell death. *Shigella* infections are more common in children ages 1 to 4, and symptoms include fever, abdominal pain, and bloody diarrhea. Infections are normally self-limiting, but infected individuals are highly contagious. Lastly, *Campylobacter jejuni* is believed to be the most common cause of infectious bacterial diarrhea in the United States. The organisms are transmitted by the fecal–oral route in contaminated food and water. After ingestion, the bacteria invade the cells of the intestine and produce a toxin, resulting in intestinal cell death and watery diarrhea, although symptoms are usually mild and self-limiting.

See Also: International Travel; Nosocomial Infections; Oral Diseases; Personal Hygiene; Waterborne Diseases.

Further Readings

Centers for Disease Control and Prevention (CDC). http://www.cdc.gov (Accessed August 2009).
Chamberlain, N. R. *The Big Picture: Medical Microbiology*. New York: McGraw-Hill, 2008.
Thielman, N. M., et al. "Clinical Practice. Acute Infectious Diarrhea." *New England Journal of Medicine*, 350/38 (2004).

Tracey A. H. Taylor
Kansas City University of Medicine and Biosciences

GENETICALLY ENGINEERED CROPS

Throughout human history, plants have been bred selectively to promote the accumulation of desirable traits, such as increased crop yields, increased nutrition, and resistance against pests and disease. Food crops were first domesticated from naturally occurring plants over 10,000 years ago when early hunter-gatherers shifted toward agricultural cultivation. Today, the process of genetic engineering allows for the selective transfer of genetic material between unrelated organisms to produce novel, desirable phenotypes in all types of organisms. Genetically engineered crops are just one type of genetically modified organism (GMO), where

In the United States, the regulatory process of genetically engineered crops is dictated by three government agencies, including the U.S. Department of Agriculture. This plant physiologist at a government laboratory in Maryland examines genetically engineered tomatoes.

Source: U.S. Department of Agriculture Agricultural Research Service/Scott Bauer

the organism's original genetic code has been modified through the process of biotechnology. Most often, genetic engineering involves the use of molecular cloning and genetic transformation. The combination of genes from different organisms is known collectively as recombinant DNA technology, and the outcome is a resultant transgenic organism that has been genetically modified from its original state. The process of genetically engineering food crops is a rapidly developing and somewhat controversial technology.

The first genetically engineered food plant was described in 1983, when a gene for antibiotic resistance was successfully inserted into a tobacco plant. The next major advancement was the approval by the Food and Drug Administration in the early 1990s of the first genetically engineered food crop, the Flavr Savr tomato, engineered by the California company Calgene to have a longer shelf life than traditional tomatoes. Genetic engineering techniques have since been broadly applied to crop production, and the technology is on the rise in both developed and developing nations. As of 2008, over 250 million acres of transgenic crops were planted in a total of 22 countries. In 2008, an estimated 13 million farmers globally were growing some type of genetically engineered crops, including an estimated 12 million in developing countries. In 2008, global transgenic crop production came from the United States (53 percent), Argentina (17 percent), Brazil (11 percent), Canada (6 percent), India (4 percent), China (3 percent), Paraguay (2 percent), and South Africa (1 percent). As of 2009, 23 countries globally were cultivating transgenic crops, including 12 developing and 11 industrial countries. The number of transgenic traits and planted acreage of genetically engineered crops are currently predicted to double by 2015.

Most of the engineered crops being produced are herbicide- and pest-resistant soybeans, cotton, corn, canola, and alfalfa. Other crops that are currently being field-tested for commercial production include a virus-resistant sweet potato and a type of super rice containing increased amounts of iron and vitamins. Governments around the world are working to establish regulatory processes to monitor and study the impacts of farming with genetically engineered crops. In the United States, the regulatory process is dictated by the three government agencies that have jurisdiction over genetically modified foods: the Department of Agriculture (USDA) evaluates whether plants are safe to grow; the Food and Drug Administration (FDA) evaluates whether plants are safe to eat; and the Environmental

Protection Agency (EPA) is responsible for regulating substances that might cause harm to the environment.

The most common application of genetically engineered crop technology is the production of patent-protected food crops that demonstrate a significant level of resistance against commercial herbicides, or that are alternatively able to produce pesticide-resistant toxins within the plant itself. Recent technologies have combined both of these traits to produce so-called stacked trait seeds that protect the plant widely against damage from a number of potential sources. The American corporation Monsanto is at the forefront of this technology and currently owns the largest global share of genetically engineered crops being produced around the world. It was estimated that in 2007, stacked trait seeds produced by Monsanto were planted over 246 million acres (1 million square kilometers [km^2]) around the world. Within the United States, Monsanto's genetically engineered triple-stack corn, which combines traits for weed and insect resistance, is by far the most highly produced variety. In fact, in 2007, corn farmers planted an estimated 17 million acres (69,000 km^2) of Monsanto's triple-stack corn in the United States. It was predicted that Monsanto's genetically engineered corn alone would be planted on over 50 million acres (200,000 km^2) of farmland in the United States by 2010.

While crops have been genetically engineered most effectively for resistance against pests and herbicides, some efforts have modified food crops to contain increased nutritional levels as well. Genetically modified sweet potatoes have been enhanced with protein and other nutrients; and Golden Rice, developed by the International Rice Research Institute, has been discussed as a possible supplement to directly address vitamin A deficiency in third world countries. Technologies for genetically modifying foods offer dramatic promise for meeting some of the 21st century's greatest environmental challenges in the poorest countries of the world. As the human population continues to grow exponentially, agricultural output will have to keep pace in a manner that will not severely impact environmental and human health. Genetic modification offers one highly targeted mechanism to potentially increase food yield, as well as food quality and nutrition, in an environmentally green way.

Risks

However, like all new technologies, there are significant risks that have been associated with the production of genetically modified organisms, especially in the realm of food crop production. Controversies surrounding genetically modified foods and crops most commonly focus on the issue of human and environmental safety. Extensive use of genetically modified crops engineered against insect and herbicide damage has raised legitimate concerns about their prolonged impact on the natural environment and on the genetic health of naturally occurring organisms.

One of the biggest concerns is the unintended harm that genetically engineered crops could introduce to other organisms in the natural environment. Although the majority of research indicates little risk associated with genetic engineering, one concern involves the potential horizontal transfer of antibiotic genes to nontarget organisms. Horizontal gene transfer, also known as lateral gene transfer, is the process by which an organism transfers its genetic material to another organism other than its offspring. This genetic material is then incorporated into and expressed in the recipient organism. Gene transfer into nontarget species could result in the development of superweed varieties that are herbicide-resistant and difficult to control. Introduced genes also have the potential to cross over into natural organic crops through cross-pollination. Genes from engineered crops have been shown to be able to spread widely by pollen, even to other plants that are great distances away.

Targeted insects can also become resistant to engineered toxins very rapidly, resulting in the widespread susceptibility of crops to pests again within only a few generations. Additionally, insects that interact with genetically engineered chemicals from engineered crop pollen could experience detrimental effects from these nontargeted pesticides. The most commonly engineered crop trait comes from a gene derived from the naturally occurring widespread soil bacteria, *Bacillus thuringiensis* (Bt). *Bacillus thuringiensis* produces insecticidal proteins called Bt toxins that target insect larvae that feed upon the plant. Chemical spraying of Bt for controlling pest insects (e.g., moths, butterflies, beetles, mosquitoes) has been implemented across the United States since the 1920s, but has been linked to the widespread decline of a number of nontarget insects. The alternative of genetically engineering plants to express the Bt toxin on their own reduces the need for aerial spraying and is a rapidly growing trend in agriculture. As of July 2008, the U.S. government had approved the production of 13 different strains of Bt corn (*Zea mays*), five strains of Bt cotton (*Gossypium hirsutum*), five strains of Bt potato (*Solanum tuberosum*), and one strain of Bt tomato (*Lycopersicum esculentum*). While the use of Bt crops reduces the need for widespread chemical application, a number of published studies have documented the effects of Bt crops on nontarget insect species. Two well-known studies examined the effects of Bt crops on monarch and black swallowtail butterflies. Monarch caterpillars were shown to die at a higher frequency when feeding on pollen produced from Bt corn in comparison to pollen from conventional corn, while no significantly negative effects of Bt pollen were found in the swallowtail caterpillars. Studies examining the effects of transgenic crops on nontarget insects continue to try to tease apart their effects.

Another growing concern is that foods produced from genetically engineered crops may have a number of unexpected and negative impacts on human health. The most common concerns relating to human health involve questions about the inherent toxicity of introduced genes and their products, the potential for novel allergens to develop in engineered food, unanticipated toxicity of engineered food, and potential nutritional alterations. Because genetically engineered crops are still a relatively new technology, the long-term effects or benefits from consumption are still not fully known. However, the general scientific consensus seems to accept genetically engineered foods as low risk to human health. Scientific evidence indicates that genetically engineered foods are as safe and nutritious as natural food.

Benefits

While there are legitimate concerns about how genetically engineered crops may affect the environment and human health, one major benefit of this technology could be a drastic reduction in the amount of chemicals used to combat crop pests and disease. With traits for pest and disease resistance engineered directly into crop DNA, the amount of chemical pesticides and herbicides applied externally by farmers could be dramatically reduced worldwide. It was estimated that in 1998 approximately 8 million fewer pounds of active pesticide ingredients were used on corn, cotton, and soybeans in the United States than in the previous year (1997) as a direct result of the increase in genetically engineered crops. The most recent evidence suggests that farmers who have adopted the use of genetically engineered crops have received both environmental and economic benefits. Although genetically engineered seeds often cost more, farmers require less labor and fewer chemicals to produce their crops. They can also reduce soil erosion caused by extensive tilling by using genetically engineered herbicides to control weeds. The increased efficiency of engineered crop plants to uptake and utilize fertilizer offers economic benefits to farmers, while reducing the amount of chemical runoff polluting streams and other natural waterways.

Genetic engineering is a very recent and rapidly developing technology that continues to attract a lot of attention. The potential beneficial outcomes of genetically engineering crops are unquestionable; however, with any new technology, the potential risks must continue to be evaluated. Overall, genetically engineered crops offer tremendous potential to address a number of food-related crises in both developed and developing countries in an environmentally green way.

Timeline of Genetic Engineering of Crops

Prehistoric times to 1900: Human hunter-gatherers collect food from plants they find in nature. Humans begin to plant seeds and domesticate crops around 10,000 years ago.

Early 1900s: Scientists begin to implement Gregor Mendel's genetic theory to selectively breed plants for agriculture.

1953: James Watson and Francis Crick publish the three-dimensional double helix structure of DNA.

1973: Herbert Boyer and Stanley Cohen create the first successful recombinant DNA organism.

1980: The U.S. Supreme Court rules that genetically altered life forms can be patented, allowing the Exxon Oil Corporation to patent the first oil-ingesting microorganism.

1982: The U.S. Food and Drug Administration approves the first genetically engineered drug, Humulin, by Genentech, a form of human insulin produced by bacteria.

1984: The U.S. Environmental Protection Agency approves the release of genetically engineered tobacco in the United States. The first field tests of genetically engineered tobacco plants are conducted in Europe.

1992: The U.S. Department of Agriculture approves the commercial production of the first genetically engineered crop, the Flavr Savr tomato by Calgene, designed to resist rotting for a longer period of time, giving it a longer shelf life in stores.

1992: The U.S. Food and Drug Administration declares that genetically engineered foods are not inherently dangerous and, therefore, do not require special regulation.

1994: Tobacco, the European Union's first genetically engineered crop, is approved in France.

2000: International Biosafety Protocol is approved by 130 countries at the Convention on Biological Diversity in Montreal, Canada. The protocol agrees on labeling of genetically engineered crops, but still needs to be ratified by 50 additional nations before it goes into effect.

2003: Zambia cuts off the flow of genetically modified food (mostly maize) from the United Nations World Food Programme, leaving a famine-stricken population without food.

2004: Venezuela announces total ban on genetically modified seeds.

2005: Hungary announces ban on importation and planting of genetically modified maize seeds.

2006: American exports of rice to Europe are interrupted when the U.S. crop is found to contain biologically engineered genes, most likely due to accidental cross-pollination with conventional organic crops.

2009: The U.S. Food and Drug Administration issues final regulations governing the approval of genetically engineered animals. The rules do not require consumer labeling for foods produced from these animals.

2010: The European Union Commission officially allows the cultivation of the first genetically modified plants in Europe since 1998, the antibiotic-resistant potato Amflora.

See Also: Chemical Pesticides; Environmental Protection Agency (U.S.); Government Role in Green Health; Home-Grown Food; Organic Produce; Personal Consumer Role in Green Health; Rural Areas.

Further Readings

Batista, R. and M. Oliveira. "Facts and Fiction of Genetically Engineered Food." *Trends in Biotechnology*, 5 (2009).

Conner, Anthony J., Travis R. Glare, and Jan-Peter Nap. "The Release of Genetically Modified Crops Into the Environment." *The Plant Journal*, 33 (2003).

Gerngross, Tillman U. "Can Biotechnology Move Us Toward a Sustainable Society?" *Nature Biotechnology*, 17 (1999).

Lemaux, Peggy G. "Genetically Engineered Plants and Foods: A Scientist's Analysis of the Issues." *Annual Review of Plant Biology*, 60 (2009).

Malarkey, Trish. "Human Health Concerns With GM Crops." *Mutation Research*, 544 (2003).

Redenbaugh, K., W. Hiati, B. Martineau, et al. *Safety Assessment of Genetically Engineered Fruits and Vegetables: A Case Study of the FLAVR SAVR Tomato. 1992.* Boca Raton, FL: CRC Press, 1997.

U.S. Department of Agriculture. "Adoption of Genetically Engineered Crops in the U.S." http://www.ers.usda.gov/data/biotechcrops (Accessed March 2010).

U.S. Department of Energy. "What Are Genetically Modified (GM) Foods?" http://www.ornl.gov/sci/techresources/Human_Genome/elsi/gmfood.shtml (Accessed April 2010).

Carrie Nicole Wells
Clemson University

Government Role in Green Health

Although much of the green movement is being led by private organizations, government organizations are funding many programs and informational materials. With regard to health, these run the gamut from community-based health initiatives to improve outcomes to the publication of guidelines to help healthcare providers incorporate green principles into buildings and operations.

The U.S. Environmental Protection Agency (EPA) is the federal driver behind pollution prevention. One of the challenges of greening healthcare is that waste and toxicity are inherent to the industry. Medical waste is biohazardous and must be disposed of in a way to protect the public; disposal of pharmaceuticals creates toxins; and personal protective equipment necessary for the workforce generates additional waste during service delivery.

The American Hospital Association has a voluntary agreement with the EPA to reduce waste volume and toxicity. The program, Hospitals for a Healthy Environment (H2E), is now administered by Practice Greenhealth, a national membership and networking organization for institutions in the healthcare community that have made a commitment to sustainable, ecofriendly practices. This is an example of how a government entity leads the development of a program with performance goals but then provides the resources for implementation by nongovernmental organizations (NGOs). The performance goals of mercury elimination, waste management, and chemical management are currently not regulatory requirements but are driving industrial innovations and spawning new green services. Some of these innovations and services also receive federal funding for development, testing, implementation, and assistance to take them to market.

Federal regulations that are directly associated with the greening of the healthcare sector include the Clean Air Act (CAA); Clean Water Act (CWA); Health Insurance Portability and Accountability Act (HIPAA); Federal Insecticide, Fungicide and Rodenticide Act (FIFRA); Resource Conservation and Recovery Act (RCRA); Toxic Substances Control Act (TSCA); and the Occupational Safety and Health Administration Act (OSHA).

OSHA regulations aim to ensure the safety and health of employees. While extremely necessary in the healthcare industry due to risk factors, this act creates an exceptional challenge for greening of the industry. For example, many healthcare products are not reusable; employees generate waste by changing gloves between patients and procedures, and paper products (such as gowns) have replaced washable items. The need for these precautions has resulted in businesses developing new ecofriendly (or ecofriendlier) products to mitigate the negative environmental effects. The federal government has helped this process by providing competitive grants through the Small Business Innovation Research/Small Business Technology Transfer program (SBIR/STTR). This program provides funding for research and development and, in recent years, has created economic development of the green product industry.

Another set of regulations that is not specific to healthcare but that has both negative and positive environmental consequences is from the Department of Transportation (DoT). These regulations govern packaging and transporting to ensure that healthcare facilities properly manage hazardous materials, such as chemicals and medical waste, harmful to the environment and to people. Although not always associated with the concept of greening the industry, these DoT regulations have helped reduce infectious and chemical waste contamination and damage.

Compliance with the Clean Air Act and the Clean Water Act is an essential component of the greening of the healthcare industry. While many new facilities are being built according to standards that require compliance, older facilities remain in operation. Within buildings that were built merely for function, the CAA and CWA force healthcare providers to make changes in their physical plant and operational procedures. The CAA regulates air emissions and is designed to protect and enhance the nation's air resources for the public health and welfare. The CWA is intended to protect the quality of surface water resources. These acts made it unlawful to discharge pollutants without permits from the EPA. Considering that the healthcare industry comprises 17 percent of the Gross Domestic Product, these acts have made substantial greening impact ranging from factory production of products to reducing asbestos debris during renovations and keeping hazardous chemicals from entering municipal water supplies.

The FIFRA governs antimicrobials for public health purposes. The TSCA gives the EPA authority over the manufacture of over 75,000 industrial chemicals produced or imported into the Unites States. Healthcare facilities are subject to lead, PCB, and asbestos reduction and abatement rules. Use of new ecofriendly building products is encouraged in new construction and building rehabilitation.

The last act that is generalized across industries but that has specific agency in the healthcare industry is the Resource Conservation and Recovery Act. The RCRA oversees solid and hazardous waste. These controls are considered cradle-to-grave and establish management requirements on generators and transporters of hazardous wastes and on owners and operators of hazardous waste treatment, storage, and disposal facilities. This requires a healthcare facility to identify all of its hazardous waste (defined as ignitable, corrosive, reactive, or toxic) to ensure proper storage and disposal procedures are implemented. These practices help reduce the threat to both human and environmental health.

The final government regulations and standards with regard to greening of the industry are specific to healthcare. The Health Insurance Portability and Accountability Act of 1996, widely known as HIPAA, requires healthcare facilities to protect patient confidentiality. Although it was not drafted for environmental purposes, the rules provided the opportunity for recycling patient records. According to Practice Greenhealth's website, it is estimated that the U.S. healthcare industry produces 2 billion pounds of paper and cardboard waste each year. Healthcare regulations require patient records be maintained for a certain length of time after discharge, but each year records are sent to be destroyed. Waste management policies and procedures can allow compliance while supporting environmental responsibility. Efforts are also under way to reduce healthcare paper waste by creating electronic records that will ease record transfer while protecting patient privacy. The U.S. Department of Health and Human Services is leading this effort, which is also designed to improve quality of care and cost effectiveness.

In addition to federal programs, state and local environmental requirements apply. These differ according to location and the healthcare industry must develop practices for compliance.

Beyond federal, state, and local rules, there are other incentives for the industry to adopt green practices. For new or older facilities, operations are encouraged to be integrated. If designed and implemented well, these practices can reduce budgetary waste and provide additional environmental benefits. Operations that eliminate duplication across departmental lines, reduce transport activity, and manage resources/inventory can be viewed as green practices.

More obvious green practices are those of sustainable site management. These responsible practices can be conducted for both interior and exterior environments. As the green movement becomes integrated into economics, more businesses are landscaping to control for erosion and in accordance with the natural environment. For example, hospitals are encouraged to landscape using indigenous plants to reduce water usage and to plant for heat reduction. Some municipalities, such as Chicago, are providing incentives or using public funds for large buildings to install green roofs that will reduce heat as well as provide peaceful areas and help manage storm water. Retrofitting rooftops and using green construction products is growing in popularity as capital projects are completed.

There are no required standards for incorporating green products and principles in the construction of healthcare facilities. At first look, the prices of these products might lead a healthcare business development team to reject greening initiatives. The government has funded a widespread social marketing campaign to educate business about the long-term cost benefits of the initially higher construction costs. Through organizations such as Enterprise Community Partners, the U.S. Green Building Council, and Global Green, industrial decision makers are becoming increasingly aware of the value-added bottom line of green building. The reduction of energy, water, and other operating costs will mitigate the initial capital outlay. Additionally, some community development financing and tax credits provide additional incentives for green construction, particularly to for-profit providers to serve low and moderate income populations that carry a larger healthcare burden that is exemplified by disparities data.

As the confluence of environmental and health initiatives continues to grow, new programs are being developed and implemented at each level of government and across the for-profit and nonprofit sectors, often with government funding. Thus, it is impossible to measure the full scale and scope of the government role in green health. Likewise, many of the programs are not immediately recognized as green but have significant impact on controlling environmental health impacts of the industry.

See Also: Cost-Benefit Analysis for Alternative Products; Environmental Protection Agency (U.S.); Healthcare Delivery; Hospitals (Carbon Footprints); Infectious Waste; Occupational Hazards.

Further Readings

GHC. Green *Guide for Health Care: Best Practices for Creating High Performance Health Environments*, Version 2.2 (2007). http://www.gghc.org (Accessed July 2009).

Practice Greenhealth Resources. http://www.practicegreenhealth.org (Accessed July 2009).

Wall, Derek. *The No-Nonsense Guide to Green Politics*. Oxford, UK: New Internationalist, 2010.

Denese M. Neu
HHS Planning & Consulting, Inc.

GREEN CHEMISTRY

Green chemistry presents an attractive alternative to traditional industrial and manufacturing practices because of traditional methods' far-reaching effects on the environment and on human health. Industry's use of chemistry in manufacturing creates hazardous by-products as a result of the production process as well as products that, after use, may also be harmful. Because of industry's mass production, the hazardous results created in common industrial processes have attracted the attention of concerned individuals and groups calling for a rethinking of the chemistry involved in such production.

Green chemistry seeks to reduce or eliminate hazardous chemicals in the production, use, and disposal of goods with the goal of minimizing, or at least reducing, negative environmental and human impact. Green chemistry advocates go about achieving these ends through the reduction of produced waste in chemical reactions, the use of energy-efficient production, and the use and production of chemicals that degrade into substances that are not environmentally harmful. Green chemistry aims at finding solutions that eliminate the problem of chemical waste rather than simply keeping toxicity of chemicals within an acceptable range. The reduction of waste is superior to the reuse, treatment, or disposal of chemical wastes.

Paul T. Anastas and John C. Warner released a widely cited set of main principles that guide green chemistry efforts to achieve the ends of reduction or elimination of hazardous chemicals. The principles ask users of chemicals to do the following:

- Prevent waste rather than clean it up after it is formed
- Incorporate as many materials used in the process as possible into the final product
- Use synthetic substances that present minimal hazard to humans and the environment whenever practicable
- Design chemicals to reduce toxicity and maximize efficacy
- Use auxiliary substances such as solvents or separation agents as sparingly as possible
- Consider energy requirements with regard to environmental and economic impacts and minimize them when possible
- Renew raw materials or feedstock whenever practicable
- Avoid derivatives
- Consider catalytic reagents superior to stoichiometric reagents
- Design chemical products so as not to persist in the environment at the end of their function
- Further develop analytical methodologies
- Choose substances used in chemical processes to minimize chemical accidents

These principles lay a foundation for the implementation of practices that work toward the aims of green chemistry. Various advances have been made that incorporate these principles into practice, including the creation of new methods for chemical production that avoid the use of hazardous chemicals, new energy-efficient ways to produce chemicals, and advances in the pharmaceutical industry's creation of medicines, such as ibuprofen, that reduce the amount of waste created.

Green chemistry has grown into an international concern in which actions are being taken to promote and assist in the attainment of the aims of green chemistry. In the United States, the Environmental Protection Agency (EPA) promotes green chemistry through its Presidential Green Chemistry Challenge Awards Program, which provides awards to individuals or groups who have made significant progress in some aspect of green chemistry. In Europe, the European Union (EU) has promulgated legislation that concerns itself with the registration, evaluation, authorization, and restrictions of chemicals (REACH). The European Chemicals Agency (ECHA) manages the chemical information REACH is concerned with, coordinating the relations between industry and this information.

Further demonstrating international interest regarding green chemistry, the 2005 Nobel Prize was awarded to a group of researchers in green chemistry. The award was given for a process that makes various syntheses of chemicals more efficient, easier to carry out, and with less impact on the environment. This international recognition of a process devoted to the aims of green chemistry exemplifies the importance of such a process to the betterment of the state of the world as a whole. Such important recognition and attention draws the public eye toward the goals and achievements thus far of the green chemistry movement.

Green chemistry can also be economically advantageous to those industries that adopt it. In utilizing means of production in correspondence with green chemistry aims, industrial production creates less waste material and thus reduces the costs associated with the treatment of chemical waste. In addition to these savings, by using renewable feedstock, a common starting material for chemical reactions, industrial companies can avoid using other nonrenewable resources. Furthermore, an increased chemical yield resulting from green chemistry practices provides industry economic incentives.

With the growing concerns about environmental and human impacts, a number of groups have emerged to assist in education concerning green chemistry. The American Chemical Society provides various resources for exploring green chemistry in the classroom for students in elementary through graduate school. Such efforts aimed at increased awareness are significant because, in comparison to the importance given to the utility of such practices, little is being done to teach students about green chemistry. The aims of education in green chemistry include both an increased awareness of the environmental and human impacts of chemistry and a contribution of new thoughts and viewpoints on the creatively demanding problem of the reduction of such impacts.

While there appear to be significant benefits in the research and implementation of green chemistry practices, unfortunately the available funding for such advancement appears to be rather limited. While groups such as the National Science Foundation provide some funding, in comparison to other funding that the foundation provides, the amount applied toward green chemistry is extremely small. Further, while the EPA supports green chemistry through the Presidential Green Chemistry Challenge Awards Program, the agency does not provide grants toward the research of green chemistry practices.

Green chemistry promises to better the state of industrial production through the reduction or elimination of hazardous chemicals that affect the environment and human health. While this promise appears to be beneficial both for industry itself through means of economic benefits and to the average population through the improvement of the environment, there is still much progress that needs to be made. The implementation of green chemistry in schools and the attempts made by various agencies all help to bring the issue of green chemistry to the forefront, although funding, awareness, and lack of action hinder the initiative.

See Also: California's Green Chemistry Initiative; Education and Green Health; Pharmaceutical Industry.

Further Readings

Anastas, P. T. and J. C. Warner. *Green Chemistry: Theory and Practice.* New York: Oxford University Press, 2000.

Dunn, Peter, Andrew Wells, and Michael T. Williams. *Green Chemistry in the Pharmaceutical Industry.* Weinheim, Germany: Wiley/VCH, 2010.

Grossman, G. *Chasing Molecules: Poisonous Products, Human Health, and the Promise of Green Chemistry.* Washington, DC: Shearwater, 2009.

Stephen T. Schroth
Jason A. Helfer
Michael J. Kaminski
Knox College

GROUNDWATER

Groundwater is the water located below the ground's surface. This reservoir of freshwater supports multiple domestic, agricultural, and industrial sectors of society. Groundwater is a renewable resource if properly managed, but it can also be mined faster than it is replenished. Groundwater is susceptible to contamination from pollutants and saltwater intrusion. As the human population continues to grow and intensely develop the Earth, management of groundwater reserves is essential to the survival of the human species.

The cycle of groundwater replenishment includes snowfall and rainfall, and movement of some surface stream and lake water into or out of underground reservoirs. Groundwater removal occurs through the roots of plants, evaporation, and artesian springs. Here, the U.S. Department of Agriculture measures water levels among skunk cabbage.

Source: U.S. Department of Agriculture Agricultural Research Service/Stephen Ausmus

Water found in soil pore spaces, sedimentary deposits, and fractures in rock formations is called groundwater. The depth at which the soil and sediments become saturated with groundwater is called the water table. The entire volume of groundwater in a geographical region that is held in the soil and porous substrate under the soil is called the groundwater reservoir, or aquifer. The science of groundwater is called hydrogeology, or groundwater hydrology, and experts are known as hydrologists. Groundwater can be liquid and flowing through the aquifer, or it can be virtually immobile such as permafrost or water locked into bedrock formations. Groundwater is 20 percent of the world's freshwater supply, but only 0.61 percent of the Earth's total water resource. The Earth's freshwater is approximately one-third groundwater, less than 1 percent surface water, and two-thirds frozen in glaciers and ice caps. The natural cycle of groundwater replenishment includes snowfall and rainfall infiltration and runoff, and movement of some surface stream and lake water into or out of underground reservoirs. The natural cycle of groundwater removal is through the roots of plants, evaporation, and artesian springs where hydrostatic pressure and the level of the water table force it onto the surface. This water is then evaporated and returns to the Earth's surface as rain or snow to complete the hydrologic cycle. Unnatural groundwater removal occurs through wells for domestic, agricultural, and industrial uses.

Occasionally, surface water is pumped back into the ground in an effort to replenish the groundwater.

Groundwater aquifers can be small or extensive and often cross political boundaries. Australia's Great Artesian Basin covers nearly 2 million square kilometers. The water extracted from the western portion of this aquifer may be over 1 million years old. Water travels from its eastern surface recharge sources at about 1 meter per year. Groundwater withdrawal for agricultural and domestic use from the Ogallala aquifer extending from Texas and New Mexico to Wyoming and South Dakota can exceed 230 million liters per square kilometer per year, a rate far greater than the natural infiltration in a short-grass prairie climate. This mining of water has dropped the water table hundreds of feet. As the water table drops, hydraulic pressure is removed, soil and rock particles come closer together, and land subsidence occurs. Land subsidence in San Jose, California, in the early 1900s was almost 4 meters due to overwithdrawal of water. New Orleans, Louisiana, is below sea level in part due to groundwater removal that exceeds its natural replenishment. In some regions of the world, land subsidence has exceeded 6 meters as the aquifer is drawn down. If subsidence occurs near an ocean, the hydraulic pressure of the salt water can infiltrate the freshwater aquifer. This process is called saltwater intrusion, and it permanently contaminates the freshwater aquifer.

Pollutants can enter the soil and sedimentary layers and contaminate groundwater. These pollutants can spread over a wide area within the aquifer and be returned to the surface by well pumping or artesian springs, contaminating lakes and rivers. The health emergency at Love Canal in western New York State occurred when homes were built over and near an industrial landfill. Cancer rates and birth defects in this neighborhood rose at an alarming rate as chemicals seeped from the landfill into local water sources and basements. Eight hundred families had to be compensated and moved. The Ganges Plain of northern India suffers from arsenic pollution in about one-quarter of its wells due to a chemical reaction in the aquifer between iron and organic compounds resulting in the release of arsenic into the groundwater.

Nitrogen fertilizers added to the soil to promote agricultural crop growth can move through the soil and into the groundwater as nitrate. The U.S. Environmental Protection Agency (EPA) has set 10 ppm or 10 mg/L nitrates as the upper limit of safety for drinking water. The incidence of blue-baby syndrome is significantly higher in nitrate-contaminated groundwater regions. Nitrate and bacteria can also come from human and animal waste that seeps into the groundwater. This type of contamination caused severe illness and death in Walkerton, Ontario, Canada. Other agricultural contaminates of groundwater include pesticides that move through the soil into the groundwater, or run off into surface water that eventually flows into the groundwater.

Groundwater protection requires knowledge of the local geology and hydrology, careful design and placement of the well field, protection of the infiltration zones, and continuous monitoring of the water quality and quantity withdrawn from the aquifer. When problems are detected they need to be addressed immediately to protect this valuable, renewable resource.

See Also: Arsenic Pollution; Bottled Water; Fertilizers; Tap Water/Fluoride; Water Scarcity; Waterborne Diseases.

Further Readings

Freeze, A. R. and J. A. Cherry. *Groundwater*. Englewood Cliffs, NJ: Prentice Hall. 1979.

Job, C. A. *Groundwater Economics*. Boca Raton, FL: CRC Press. 2009.

Journal of the American Water Resources Association. http://www.awra.org/jawra (Accessed September 2010).

National Groundwater Association. *Groundwater Monitoring and Remediation Journal*. http://www.ngwa.org/publication/gwmr/index.aspx (Accessed September 2010).

Paul Richard Saunders
Canadian College of Naturopathic Medicine

Healthcare Delivery

Healthcare delivery refers to the system by which services are delivered to allow people to live healthy lives. It includes not only medical care but related care as well, such as treatment by a dentist or physical therapist. Healthcare may be delivered in many different settings, including hospitals, freestanding clinics, individual physician offices, community and workplace clinics, and the patient's home, and may be funded publicly, privately, or some combination of the two.

In most developed countries, healthcare costs constitute a significant portion of national expenditures, and healthcare institutions such as hospitals use large quantities of natural resources and produce large quantities of waste. Recently, many hospitals and other healthcare organizations have become interested in ways to make their operations greener by means such as reducing energy consumption and waste and instituting environmentally preferable purchasing plans. The impact of individual healthcare consumers on the environment has also become a focus of attention, and many cities and states have instituted plans to allow patients to safely dispose of hazardous materials such as sharps and unused drugs; the exact system used differs by country and historical period. This article concentrates on the current healthcare delivery system in the United States, although many of the principles also apply to other countries.

Environmentally Preferable Purchasing

Healthcare facilities use large quantities of many types of products, and the adoption of an environmentally preferable purchasing plan can significantly lower their environmental impact without necessarily increasing costs or reducing quality. *Environmentally preferable* refers to products that have a decreased impact on the environment and/or human health when compared to comparable items used for the same purpose. This evaluation encompasses all aspects of the product over its lifetime, including use of raw materials, manufacturing, packaging, distribution, maintenance, and disposal.

An example of environmentally preferable purchasing in the healthcare industry is the decision by many hospitals to avoid purchasing products containing PVC (polyvinylchloride) when good alternatives are available. PVC is the most commonly used polymer in plastic hospital products and is used in many medical devices, including intravenous bags and tubing, nasogastric tubes, umbilical artery catheters, and blood bags and infusion tubing. However, the use of PVC poses concerns both because incineration of PVC products can release dioxin (identified by the EPA as a probable carcinogen) into the environment and because many PVC products use DEHP (di-[2-ethylhexyl]phthalate), also identified by the EPA as a probable carcinogen, as a plasticizer. Lists of substitute products that do not contain PVC are available from several organizations, including the international coalition Health Care Without Harm.

Hospitals also purchase large quantities of nonmedical items like cleaning products, disinfectants, and pesticides that may contain toxic chemicals. Indoor air pollution has become an increasing concern worldwide and is particularly relevant to hospitals because of the increased risk to patients who may have chemical sensitivities and compromised immune systems. The effects of harmful chemicals on hospital staff is also a concern because their exposure is much greater than the average patients'. For these reasons, many hospitals have moved to integrated pest management (IPM), a system that emphasizes nontoxic methods of controlling pests such as improved sanitation and structural maintenance and, if a chemical pesticide is necessary, uses the least toxic product available.

In terms of disinfectant and cleaning products, many contain hazardous materials like gluraraldehyde and ethylene oxide, which have been shown to have serious health effects, while the common antibacterial product triclosan is considered a threat in increasing the ability of bacteria to resist antibiotics. Unscented rather than scented products are preferred in most cases because fragrance chemicals (volatile organic compounds that vaporize into the air) are associated with many adverse health effects, including headaches, eye irritation, nausea, asthma attacks and other respiratory distress, hormonal disruption, and neurotoxic systems. Many hospitals have adopted a fragrance-free policy for these reasons, which applies to staff (e.g., no perfume or scented shampoos) as well as cleaners and other maintenance products. The nonprofit organization Green Seal produces a list of cleaners that are considered environmentally friendly (as evaluated by publicly available standards), and many institutions, including hospitals, use the list as a guideline when deciding which products to purchase.

Food Service

Many U.S. hospitals run large food service operations that provide meals and snacks not only to patients but also to staff and visitors. Although hospital food has often been the butt of jokes in the past (and, in fairness, provision of meals meeting many different types of nutritional demands and for people who may be on restricted diets is a daunting task, to say nothing of providing staff meals around the clock), currently many hospitals take an active approach in ensuring that the food they provide will be tasty and environmentally friendly, as well as meet dietary guidelines.

The international organization Health Care Without Harm has numerous suggestions for making hospital food service more environmentally friendly and notes that, due to the quantity of food produced by hospitals, their choices can have a significant influence on the market. It recommends purchasing meat, poultry, and seafood from sustainable sources that raise animals without the use of nontherapeutic antibiotics and avoiding poultry that consumed feed containing arsenic (used to kill parasites, improve pigmentation, and promote growth). In addition, it suggests lowering consumption of meat through programs such as one voluntary meat-free day per week and recommends purchasing milk from cows that have not been treated with recombinant bovine growth hormone (rBGH).

This hormone is given to cows in order to increase milk production, but it already is prohibited in many countries, including all members of the European Union, Canada, Australia, New Zealand, and Japan. It also promotes purchasing local, organic, and Fair Trade products and avoiding genetically engineered foods.

Health Care Without Harm advocates removing fast food operations from hospitals (a 2006 report from the American Medical Association found that 42 percent of 234 hospitals surveyed sold brand-name fast food on their campus) due to the well-known poor health qualities of much fast food (high in fat, sugar, and salt and low in nutrition), as well as the environmentally unfriendly practices of many fast food corporations. They also suggest limiting the number of vending machines available and requiring that healthy foods be available in the machines, including whole fruit, juice, and low-fat and low-sugar snacks.

Hospital food service operations typically use large amounts of disposable products (such as paper cups and plastic cutlery) and produce large amounts of waste in terms of uneaten food (10 percent of the typical hospital's waste stream, according to Health Care Without Harm). Some hospitals are beginning to reduce their environmental impact by diverting or composting their food waste rather than sending it to a landfill or incinerator, purchasing bio-based and/or compostable containers and cutlery for takeout use, using reusable food service items for in-cafeteria use, using recycled paper products, and recycling glass, metal, and paper products.

Medical Waste Reduction and Safe Disposal

The healthcare system in the United States produces large amounts of trash, in part because of the heavy use of single-use devices (SUDs); that is, disposable items intended to be used only once or on a single patient, then discarded, which makes it easier to ensure sterility. The most recent estimate available, from the 1990s, is that the United States produces about two million tons of medical trash per year. The most common means of disposal are depositing the waste in landfills (by one estimate, healthcare facilities are the second-largest contributor to landfills, following the food industry) or incinerating it, the latter being a particular concern because it may release toxic chemicals into the air. Until recently, most hospitals and other healthcare delivery systems did not give much thought to the issue, but the need to cut costs as well as growing environmental awareness has encouraged them to consider ways to reduce both the amount of waste produced and the ways it is disposed. One of the most important tools to cut waste is the use of reprocessed medical devices, discussed below.

In 1998, the American Hospital Association and the U.S. Environmental Protection Agency signed a memorandum of understanding that established goals for waste reduction. This project also led to the establishment of the Hospitals for a Healthy Environment Initiative, now known as Practice Greenhealth, which provides information and resources to help hospitals reduce their environmental impact. Specific goals in the memorandum of agreement (renewed in 2001) include reduction of the total volume of hospital waste by 50 percent by 2010 and virtual elimination of mercury (commonly used in thermometers, sphygmomanometers, and esophageal dilators) from the healthcare industry waste stream.

Recently, greater attention has been placed on providing consumers with environmentally friendly ways to discard medical items used in home healthcare. In years past, patients were often told to flush unused medications down the toilet or (in the case of liquid medications) pour them down the drain in order to prevent them from being used improperly. However, these methods of disposal mean that eventually they will turn up in the nation's waterways: a U.S. Geological Survey study in 1999–2000 revealed that 80 percent of U.S. rivers and streams had measurable levels of drugs such as antibiotics, hormones, contraceptives, and steroids, presumably due to discarded human drugs.

Because these drugs can have an adverse effect on aquatic life and may lead to the development of drug-resistant bacteria, the current recommendation is to discard unused medications with the trash (i.e., as solid waste) after securely wrapping them in plastic and sealing the container with tape. Some cities and states also offer special collection centers where unused prescription and over-the-counter drugs may be disposed of. The disposal of sharps (such as hypodermic needles) used at home also presents a safety concern, and many states have instituted programs that allow households to dispose of sharps at hospitals free of charge.

Reprocessing Medical Devices

One method to reduce medical waste is through reprocessing; that is, following a procedure that renders a used device safe to be reused. This is not a new idea: before the introduction of SUDs, most medical equipment was designed to allow cleaning and sterilization so it could safely be used many times on many different patients. SUDs were introduced because they simplified medical care: the device was sterile until its package was opened, and it was discarded after one use, saving the need for staff to clean and resterilize it. The human immunodeficiency virus (HIV) epidemic in the 1980s heightened fears about transmitting infections and encouraged adoption of more SUDs.

As of 2010, over 100 types of SUDs can be reprocessed, generally by third-party reprocessing companies. As of 2002, about one-quarter of U.S. hospitals used at least one kind of reprocessed SUD, with large hospitals more likely than small to use reprocessed SUDs. Two large companies, Ascent Healthcare Solutions and SterilMed, provide about 95 percent of reprocessing of medical devices in the United States. Reprocessing saves both in terms of costs (on average the cost of reprocessed SUDs is about half that of purchasing new devices) and garbage (Ascent estimates that in 2009, its 1,800 hospital clients diverted 2,650 tons of garbage from landfills by using reprocessed devices).

The Food and Drug Administration (FDA) regulates reprocessing in the United States and classifies three types of devices that may be reprocessed. Class I includes devices such as elastic bandages and tourniquet cuffs, which pose low risk to patients and are exempt from premarket submission requirements (meaning reprocessing firms do not have to provide safety data to the FDA). Most reprocessed SUDs fall into Class II, or medium-risk devices (e.g., ultrasound catheters, drills, laparoscopic equipment), which require the reprocessing company to provide evidence that the reprocessed device is equivalent in terms of safety, effectiveness, and intended use to devices already on the market. Class III devices pose the greatest patient risk (e.g., balloon angioplasty catheters, implanted infusion pumps) and must go through a stringent approval process (including scientific data on safety and effectiveness as well as an inspection of the reprocessing facility), and so reprocessing of these items is less common.

One major concern with the use of reprocessed medical devices is their effect on patient safety, with particular concerns being the sterility of the reprocessed device and whether it can be counted on to function as effectively as a new device. There have been efforts by physicians and politicians to require patient consent before reprocessed devices are used, but they have been unsuccessful (and would pose interesting ethical dilemmas if enacted: for instance, would a different fee be charged if a new device were used?). By law, reprocessed devices must be labeled as such, and a 2008 report from the U.S. Government

Accountability Office concluded that reprocessed devices do not pose a significant health risk when compared to new devices.

Even reprocessed devices must eventually be discarded, and some medical experts believe that a more effective solution would be to return to permanently reusable equipment. Other measures to reduce waste include streamlining the items included in surgical kits (operating rooms produce 20 to 30 percent of a typical hospital's waste) so seldom-used items are not automatically included and donating unused supplies and reusable equipment that would otherwise be discarded to charitable causes such as health clinics in developing countries.

See Also: Animal Products; Antibiotic Resistance; Antibiotics; Biological Control of Pests; Emergency Rooms; Indoor Air Quality; Infectious Waste; Organic Produce.

Further Readings

Chen, Ingfei. "In a World of Throwaways, Making a Dent in Medical Waste." *New York Times* (July 5, 2010). http://www.nytimes.com/2010/07/06/health/06waste.html?_r=1&ref=health (Accessed July 2010).

Food and Drug Administration, U.S. Department of Health & Human Services. "Guidance for Industry and FDA Reviewers—Reprocessing and Reuse of Single-Use Devices." http://www.fda.gov/MedicalDevices/DeviceRegulationandGuidance/GuidanceDocuments/ucm073758.htm (Accessed July 2010).

Green Guide for Health Care. "Best Practices for Creating High Performance Healing Environments." http://www.gghc.org (Accessed July 2010).

Green Seal. http://www.greenseal.org (Accessed July 2010).

Health Care Without Harm. http://www.noharm.org (Accessed July 2010).

Kwakye, Gifty, Peter J. Provonost, and Martin A. Makary. "Commentary: A Call to Go Green in Health Care by Reprocessing Medical Equipment." *Academic Medicine*, 85/3 (2010). http://journals.lww.com/academicmedicine/Fulltext/2010/03000/Commentary__A_Call_to_Go_Green_in_Health_Care_by.10.aspx# (Accessed July 2010).

Practice Greenhealth. http://practicegreenhealth.org (Accessed July 2010).

Sustainable Hospitals. http://www.sustainablehospitals.org/cgi-bin/DB_Index.cgi (Accessed July 2010).

U.S. Government Accountability Office. "Reprocessed Single-Use Medical Devices: FDDA Oversight Has Increased, and Available Information Does Not Indicate That Use Presents an Elevated Health Risk" (January 2008, GAO-08-147). http://www.gao.gov/new.items/d08147.pdf (Accessed July 2010).

Sarah Boslaugh
Washington University in St. Louis

HEALTH DISPARITIES

There are varying definitions of health disparities, or healthcare inequality, within the healthcare community. Generally, a health disparity is defined as a substantial difference in health between one population and another. The National Institutes of Health (NIH) defines health disparities as "the differences between groups of people that can affect how frequently a disease affects a group, how many people get sick, or how often the disease causes death." The NIH identifies the different populations influenced by health disparities as "racial and ethnic minorities; residents of rural areas; women, children, the elderly; and persons with disabilities." Similarly, the Centers for Disease Control and Prevention (CDC) defines health disparities as "differences in the incidence, prevalence, and mortality of a disease and the related adverse health conditions that exist among specific population groups." The CDC characterizes these groups by "gender, age, ethnicity, education, income, social class, disability, geographic location, or sexual orientation." Some of these populations are characterized by a greater rate of avoidable disabilities, disease, and death compared to other populations and nonminorities.

Causes

There are several factors that contribute to health disparities. Most notably, these disparities are attributed by the CDC to result from the "complex interaction among genetic variations, environmental factors, and specific health behaviors." Specific factors affecting such disparities include personal, socioeconomic, and environmental characteristics or social determinants; access to healthcare system delivery; and quality of healthcare received. The disparities may arise from increased risk of disease from occupational exposure or from underlying genetic, ethnic, or familial factors as well. Examples include evidence that low-income and racial and ethnic minorities are affected by disparities in rates of insurance, lack of access to a primary care provider, and ineffective or miscommunication with a healthcare provider. Because of these factors, it is more difficult for certain populations to achieve health compared to others. The well-being of these groups is significantly associated with their socioeconomic status and race. For example, low-income Americans and racial and ethnic minorities have relatively limited access to care and also higher rates of disease, especially obesity, cancer, diabetes, and acquired immune deficiency syndrome (AIDS). The CDC has found the most common and prevalent diseases correlated to health disparities within each of these particular groups.

Examples

According to the 2000 U.S. census, African Americans account for 13 percent of the U.S. population, or 36.4 million individuals. Major health disparities affecting African Americans include heart disease and stroke, cancer, adult immunization rates, diabetes, and infant mortality. In 2001, the African American age-adjusted death rate for heart disease (316.9 per 100,000) was 30.1 percent higher than for white Americans (243.5 per 100,000) and 41.2 percent higher than for white Americans for stroke (78.8 per 100,000 vs. 55.8). The age-adjusted death rate for cancer was 25.4 percent higher for African Americans (243.1 per 100,000) than for white Americans (193.9 per 100,000) in 2001. Breast cancer death rates are higher for African American women despite having a mammography-screening rate that is nearly the same as the rate for white women. Influenza vaccination coverage among adults 65 years of age and older was 70.2 percent for whites and 52 percent for African Americans in 2001. The gap for pneumococcal vaccination coverage among older adults was even wider, with 60.6 percent for whites and 36.1 percent for African Americans. The age-adjusted death rate for African Americans in 2001 was more than twice that for white Americans (49.2 per 100,000 versus 23 per 100,000). The infant death rate among African Americans is more than double that among whites.

According to the 2000 U.S. census, Hispanics/Latinos represent 13.3 percent of the U.S. population, or 38.8 million individuals. Examples of major health disparities affecting Hispanics and Latinos include diabetes, asthma, tuberculosis, and low birth weight. The diabetes death rate in 2000 was highest among Puerto Ricans (172 per 100,000), Mexican Americans (122 per 100,000), and Cuban Americans (47 per 100,000) for Hispanics/Latinos. Hispanics living in the United States are almost twice as likely to die from diabetes as are non-Hispanic whites. In the northeast United States, from 1993 to 1995, Hispanics/Latinos had an asthma death rate of 34 per million, more than twice the rate for white Americans (15.1 per million). Although comprising only 11 percent of the total population in 1996, Hispanics accounted for 20 percent of new cases of tuberculosis. The rate of low-birth-weight infants is lower for the total Hispanic population compared with that of whites, but Puerto Ricans have a low-birth-weight rate that is 50 percent higher than the rate for whites.

According to the 2000 U.S. census, American Indians and Alaska Natives comprise 0.9 percent to 1.5 percent of the U.S. population and have the highest poverty rates of all Americans. Examples of health disparities affecting American Indians and Alaska Natives include diabetes, liver disease, infant mortality, and sexually transmitted diseases. The 2002 age-adjusted prevalence rate of diabetes was over twice that for all U.S. adults, and the American Indian and Alaska Native mortality rate from chronic liver disease was nearly three times higher. The Pima tribe in Arizona have one of the highest rates of diabetes in the world. The American Indian and Alaska Native rates were 1.6 times higher than non-Hispanic white rates. The American Indian and Alaska Native sudden infant death syndrome (SIDS) rate was the highest of any population group, more than double that of whites in 1999. The syphilis rate among American Indians and Alaska Natives was six times higher than the syphilis rate among the non-Hispanic white population, the chlamydia rate was 5.5 times higher, the gonorrhea rate was four times higher, and the AIDS rate was 1.5 times higher in 2001.

According to the 2000 U.S. census, Asian Americans represent 4.2 percent of the U.S. population, or 11.9 million individuals. Examples of health disparities affecting Asian Americans include cancer, tuberculosis, and hepatitis B virus (HBV). During 1988–1992, the highest age-adjusted incidence rate of cervical cancer occurred among Vietnamese American women (43 per 100,000), almost five times higher than the rate among non-Hispanic white women (7.5 per 100,000). Asian Americans and Pacific Islanders had the highest tuberculosis (TB) case rates (33 per 100,000) of any racial and ethnic population in 2001 (14 per 100,000 for non-Hispanic blacks, 12 per 100,000 for Hispanics/Latinos, 11 per 100,000 for American Indians/Alaska Natives, and 2 per 100,000 for non-Hispanic whites). The rate of acute hepatitis B (HBV) among Asian Americans and Pacific Islanders has been decreasing, but the reported rate in 2001 was more than twice as high among Asian Americans and Pacific Islanders (2.95 per 100,000) as among white Americans (1.31 per 100,000).

According to the 2000 U.S. census, Native Hawaiians and Other Pacific Islanders represent 0.3 percent of the U.S. population, or 874,000 individuals. Examples of health disparities affecting Native Hawaiians and Other Pacific Islanders include diabetes, infant mortality, and asthma. From 1996 to 2000, Native Hawaiians were 2.5 times more likely to be diagnosed with diabetes than non-Hispanic white residents of Hawaii of similar age. In 2000, infant mortality among Native Hawaiians was 9.1 per 1,000, almost 60 percent higher than among whites (5.7). Native Hawaiians in Hawaii had an asthma rate of 139.5 per 1,000 in 2000, almost twice the rate for all other races in Hawaii (71.5 per 1,000).

Organizations Addressing Health Disparities

The diseases associated with health disparities are not explained by current biologic and genetic characteristic information about racial and ethnic minorities, but are believed to be a result of varying healthcare factors. Because of this, there are many public, private, nationwide, and statewide projects that are looking to reform and to prevent health disparities so that all Americans will be able to receive affordable and valuable healthcare. There have been and will be many organizations and projects in the past, present, and future working toward a change in health disparities including the Office of Minority Health (OMH) created by the CDC in 1988; the Disadvantaged Minority Act passed by Congress in 1990; the Healthy People Initiative 2010; and Racial and Ethnic Approaches to Community Health (REACH) 2010 created by the CDC in 1999. One of the major factors contributing to the importance of these reforms is that the racial and ethnic group is expected to grow significantly in size and number within the next decade. Therefore, the goal of these plans is to reduce and eliminate heath disparities within the American population and to yield quality care and wellness.

Several national governmental agencies are tasked with addressing health disparities within the United States. The U.S. Department of Health and Human Services (HHS) supports the Office of Minority Health (OMH) with the mission "to improve and protect the health of racial and ethnic minority populations through the development of health policies and programs that will eliminate health disparities." The OMH provides advice to the secretary of HHS and the Office of Public Health and Science on public health program activities affecting American Indians and Alaska Natives, Asian Americans, Blacks/African Americans, Hispanics/Latinos, Native Hawaiians, and Other Pacific Islanders.

Within HHS, the CDC created the Office of Minority Health (OMH) in 1988. The NIH created the National Center on Minority Health and Health Disparities in 1990 as the Office of Research on Minority Health (ORMH). In 1993, ORMH was formally established through Public Law 103-43, the Health Revitalization Act of 1993. The ORMH became a Center in 2000 after the president signed Public Law 106-525. The mission of the center is to "promote minority health and to lead, coordinate, support, and assess the NIH effort to reduce and ultimately eliminate health disparities." The Agency for Healthcare Research and Quality (AHRQ) produces a report titled "National Healthcare Disparities Report" and ensures that research includes all minority populations with a special emphasis on women, children, older adults, residents of rural areas, and individuals with disabilities or special healthcare needs.

Suggested Solutions

The Commonwealth Fund, in a 2004 report to states, recommended the following strategies to address health disparities: (1) improved and consistent data collection about disparities, (2) effective evaluation of state programs to reduce disparities, (3) an emphasis on cultural and linguistic competence in all health services, (4) a workforce development program to increase minority representation within health services, (5) expanded access to health services (insurance) for all ethnic and racial groups, (6) expanded government offices for minority health, and (7) involvement of minority representation in minority health improvement efforts.

See Also: Centers for Disease Control and Prevention (U.S.); Emergency Rooms; Health Insurance Reform.

Further Readings

Agency for Healthcare Research and Quality (AHRQ). "National Healthcare Disparities Report, 2008." http://www.ahrq.gov/qual/nhdr08/Key.htm (Accessed November 2009).

Centers for Disease Control and Prevention. "OMHD." http://www.cdc.gov/omhd (Accessed November 2009).

Health Reform. "Higher Rates of Disease." http://www.healthreform.gov/reports/health disparities (Accessed November 2009).

McDonough, John E., Brian K. Gibbs, Janet L. Scott-Harris, et al. *A State Policy Agenda to Eliminate Racial and Ethnic Health Disparities*. New York and Washington, DC: The Commonwealth Fund, June 2004.

National Center on Minority Health and Health Disparities. "Health Disparities: Closing the Gap." http://ncmhd.nih.gov/hdFactSheet_gap.asp (Accessed November 2009).

Rebecca A. Malouin
Samantha Martens
Michigan State University

HEALTH INSURANCE INDUSTRY

Health insurance is a system by which a group of people pool their risk of medical expenses by paying a certain amount of money (a premium) on a regular basis in return for the assurance that the insurance company or fund will cover some or all of the costs of medical treatment, should they need it. There are many different types of health insurance, and not all countries have a private health insurance industry, as some provide healthcare on a national basis (every citizen is enrolled in a system that is funded by tax dollars). In the United States, there is no national system of either healthcare or health insurance. Healthcare is provided by a number of independent providers (some of which have voluntarily joined group practices or other associations) and institutions such as hospitals; health insurance is provided by a number of competing companies as well as several large government-funded programs, including Medicare and the Veterans Health Administration. The implications of such a fragmented system are a major contributor to the industry's environmental impact in terms of resource use (e.g., reliance on paper rather than electronic records) and lack of incentive for companies to invest in broad-reaching programs that would be environmentally friendly and also promote good health in the long term.

Environmental Impact of the Industry

In terms of environmental impact, health insurance companies resemble many other types of knowledge-based businesses: they have offices, computers, records storage locations, telecommunications systems, and so on. One way health insurance companies differ from many similar businesses is their continuing heavy dependence on paper records. In a typical healthcare transaction, an individual seeks some kind of treatment, then the provider applies to the health insurance company for reimbursement; if the claim is approved, the provider receives payment. Often the person seeking service also pays a fixed fee (a co-pay) or a percentage of the cost (coinsurance). All this business is usually conducted through hard copy; that is, claims and other documents are printed on paper forms and mailed or faxed. If there is an adjustment or dispute over the amount or whether a procedure is covered, that will very likely also require printing out more sheets of paper and mailing or faxing them. Some companies also send printed "explanation of benefits" statements to individuals once a particular transaction is concluded, generating even more hard copies.

Because of privacy concerns and lack of national integration, health insurance has lagged behind comparable industries in adopting electronic records (see below), although internal company records are likely to be electronic, and paper records are often transcribed or digested and stored electronically. At the level of the individual physician or hospital, paper records remain common (the patient "chart" is typically a manila folder holding numerous pieces of paper), and if information is required in a different location—for example, to support a submitted insurance claim—records are often simply photocopied and mailed or faxed to the insurer. An additional complication is that most physicians treat patients covered by a variety of insurance companies, each of which may require different paperwork on the company's own individual form, which creates yet another barrier to adopting electronic records.

Electronic Medical Records

Standardized electronic healthcare records would be environmentally preferable to the current system of paper records, not only because of the reduced environmental impact (paper records have to be printed, stored, duplicated, and eventually discarded or destroyed) but because it can improve healthcare efficiency by making all of an individual's medical information readily available to any provider who is treating him or her. However, lack of standardization and privacy concerns have impeded the development of a standard electronic medical record. Privacy concerns are amplified by the fact that patients have an incentive to conceal medical conditions that could cause them to be refused private insurance or charged a higher rate for a policy.

One exception to the rule of paper records is the Department of Veterans Affairs (VA), the largest integrated healthcare system in the United States. In 2009, 8.1 million patients were enrolled in the VA healthcare system, and 5.7 million patients were treated. The VA has been a leader in streamlining procedures and implementing programs that make the provision of healthcare more efficient and frequently reduce the environmental burden as well. For instance, the VA has had a standardized electronic medical record system for years, and in 2003 the VA system was proposed by Tommy Thompson, then U.S. Secretary of Health and Human Services, as a model for development of similar systems across the United States.

The VA electronic medical record system, called VistA, provides a single location for healthcare providers to access, view, and update the medical records of an individual. This record includes a variety of information, such as the patient's allergies, reported health problems, hospital and outpatient clinic visits, current medications, X-rays, nursing orders, and laboratory results. Use of an integrated medical record system not only improves healthcare by providing all of a patient's information in one place but also eliminates the need for paper copies of many records and for multiple copies of the same information if an individual sees numerous providers.

Medicare, a federally funded program that provides health insurance for people over 65, the disabled, and those with end-stage renal disease, is not an integrated healthcare system like the VA but a method of paying for healthcare received from individual practitioners and hospitals. For this reason Medicare cannot mandate the use of a standard medical record (also known as an EHR or electronic health record) but has proposed a plan (as of 2010) to encourage physicians and hospitals to switch to electronic medical records by providing bonus payments. Under the proposed system, bonuses would be provided to those who use electronic records for at least 80 percent of their patients (for physicians) or complete 10 percent of their orders electronically (for hospitals). A prototype regional database for physicians in southwest Michigan was in use by mid-2010.

A Green Insurance Company

In 2009, *Newsweek* magazine produced the Green Rankings report, which names the top 500 U.S. companies in terms of their environmental friendliness. Each company's score had three components: an environmental impact score (based on global environmental impact, as normalized by annual revenues to allow comparison of large and small companies), a green policies score (based on the company's environmental policies and performance), and a reputation score (based on an opinion survey of academics and other professionals). No health insurance companies were selected as particularly outstanding in terms of environmental friendliness: in fact, the highest ranked was UnitedHealth Group at number 253. UnitedHealth Group, the largest insurance company in the United States (ranked 25th among Fortune 500 companies in 2008), received an overall score of 70.62 (out of 100) and was commended for requiring regular audits and environmental impact assessments of company operations and services.

UnitedHealth Group provides a good example of ways health insurers can become more environmentally friendly. In 2009, UnitedHealth moved 2.6 million customers to online health statements, reducing the amount of paper used as well as the resources required to deliver hard copies while also improving client access to online sources of information about health and well-being. UnitedHealth securely destroyed and then recycled 8.2 million pounds of paper in 2009, which is equivalent to saving 69,000 trees. UnitedHealth also funded a number of local programs to encourage healthy eating and physical activity and collaborated on a research project with MIT and Columbia University to study American food systems (the network of food production, processing, distribution, sale, and consumption) and their impact on health and obesity.

Complementary and Alternative Medicine

The use of complementary and alternative medicine (CAM), which includes many different practices—such as the use of nutritional supplements, yoga, hypnotherapy, and acupuncture—is becoming increasingly popular in the United States. Because CAM practices often place less of a burden on the environment compared to standard Western medicine (e.g., taking pharmaceutical products, undergoing surgery), increasing the use of effective CAM treatments could decrease the ecological impact of the healthcare system. However, health insurers have been reluctant to include these types of programs as covered benefits (meaning that the insurer would pay all or part of the cost), and in 2007, Americans spent almost $34 billion out-of-pocket on CAM, accounting for 1.5 percent of total healthcare expenditures in the United States for that year and 11.2 percent of total out-of-pocket healthcare expenditures.

Some CAM treatments are covered by some insurers (e.g., chiropractic treatment), but insurers are often reluctant to cover CAM therapies that have not been scientifically evaluated for effectiveness. Even if covered, CAM treatments may be considered an "extra" for which there is a higher co-payment or a limit on the annual amount that will be reimbursed. If not covered by health insurance, some CAM treatments are deductible from income taxes as a medical expense: for instance, in 2009, the United States allowed the cost of acupuncture, chiropractic, and osteopathic care to be deducted on federal income tax returns.

The National Center for Complementary and Alternative Medicine (NCCAM), located within the National Institutes of Health, was established in 1991 as the Office of Alternative Medicine for the purpose of promoting research into the effectiveness of CAM and to provide research training and outreach to the medical profession. Among other activities, the NCCAM supports clinical trials, considered to provide the highest quality information about the effectiveness of medical treatments.

Some health insurance companies have voluntarily instituted programs that encourage policyholders to engage in environmentally friendly behavior that also is believed to improve health and therefore lower the costs of covering that individual. For example, many provide websites or newsletters with information about choosing a healthy diet and avoiding processed and fast food, beginning and sustaining an exercise program, and quitting smoking. Some provide discounts for individuals who do not smoke or who agree to participate in a program to quit smoking, and some provide payment for services such as consultation with a nutritionist or enrollment in exercise classes.

Health Insurance and the Quality of Life

The preamble to the constitution of the World Health Organization (WHO), adopted in 1946, includes an often-cited definition of health as "a state of complete physical, mental and social well-being and not merely the absence of disease or infirmity." The preamble also states that it is a fundamental right of every individual to enjoy "the highest attainable standard of health," which was a radical notion at the time. This statement was updated in 2005 when WHO passed a resolution officially recognizing the increased amount of medical care available and the subsequent increases in cost as well as the deleterious effects of requiring individuals to pay most of their health costs directly out of pocket. This situation is typical of that among developing countries and also exists to a degree in the United States because uninsured individuals may find themselves facing very high costs for medical treatment. This system of individual, out-of-pocket payment, WHO declared, is incompatible with the concept of health as a human right, because those who need care may not be able to afford it and necessary care may be delayed, allowing a condition to worsen.

The WHO resolution states the importance of sustainable and equitable financing for healthcare (meaning that households contribute according to their ability to pay) as a necessary part of the concept for health for all as well as a means of reducing poverty and encouraging development. The resolution points out the need for health-financing systems in order to develop equal access to healthcare and cites the need to protect individuals from financial risk while providing access to necessary services.

Even in a wealthy country like the United States, which provides various means for people without health insurance to get healthcare, numerous studies have suggested that these efforts are not sufficient and the lack of health insurance can be detrimental to a person's health. For instance, a 2009 study by a research group from Harvard University (led by Andrew Wilper) found that about 45,000 deaths annually in the United States are associated with a lack of health insurance, exceeding the number due to common killers such as kidney disease. Wilper's group found that even after controlling for baseline health, health behaviors (including smoking, drinking, and obesity), income, and education, working-age American adults under age 65 without insurance have a 40 percent higher risk of death as compared to their peers who are privately insured.

See Also: Fast Food; Health Disparities; Health Insurance Reform; Healthcare Delivery; Obesity; Paper Products.

Further Readings

Centers for Medicare and Medicaid Services. "EHR Incentive Programs." http://www.cms .gov/EHRIncentiveprograms (Accessed July 2010).

National Center for Complementary and Alternative Medicine, National Institutes of Health. "Paying for CAM Treatment." http://nccam.nih.gov/health/financial (Accessed July 2010).

Newsweek. "Green Rankings: The 2009 List." http://greenrankings2009.newsweek.com/ (Accessed July 2010).

UnitedHealth Group. "2009 Social Responsibility Report." http://www.unitedhealthgroup .com/main/SocialResponsibility.aspx (Accessed July 2010).

Wilper, Andrew P., Steffe Woolhandler, Karen E. Lasser, Danny McCornick, David H. Bor, and David U. Himmelstein. "Health Insurance and Mortality in U.S. Adults." *American Journal of Public Health*, 99/12 (2009). http://www.ncpa.org/pdfs/2009_harvard_health_ study.pdf (Accessed July 2010).

World Health Organization. "Constitution." http://www.who.int/about/en (Accessed July 2010).

World Health Organization. "Sustainable Health Financing, Universal Coverage and Social Health Insurance." World Health Assembly Resolution 58.33 (2005). http://www.who.int/ health_financing/documents/cov-wharesolution5833/en/index.html (Accessed July 2010).

Sarah Boslaugh
Washington University in St. Louis

HEALTH INSURANCE REFORM

On March 30, 2010, U.S. President Barack Obama signed into law the complete package of the Health Care Reform Act, representing a watershed in the protracted debate on public healthcare in the United States. The debate in the United States regarding health insurance reform reaches back to the New Deal policies of the 1930s, when President Franklin D. Roosevelt instigated a national conversation on the subject by considering adding comprehensive health coverage to Social Security legislation. Almost six decades later, in 1993, this debate raged again when President Bill Clinton appointed his wife,

Hillary Rodham-Clinton, to chair a task force charged with creating a universal healthcare plan for all Americans. Neither President Roosevelt nor President Clinton was successful in instituting health insurance reforms, and in the 2008 presidential campaign, the question of health insurance reform was again considered. Two years later, the debate raged on, with the Obama administration's proposal for health insurance reform dividing the country along mostly partisan lines: most Democrats and those who define themselves as progressive support some form of universal healthcare, and most Republicans and those who define themselves as conservative argue that the Obama administration's proposal for health insurance reform not only would be costly but would also impinge on personal freedoms.

At the heart of the debate is the question of whether or not healthcare in the United States should be considered a right or a privilege, and, further, if there are flaws in the system, how best to fix them—through government-mandated actions or through free-market solutions? These differences of opinion have sparked a fiery debate, with those on each side citing often-contradictory facts and figures. For example, while those who support the Obama administration's plan put the number of uninsured Americans at 46 million, opponents argue that this figure is flawed, as it includes a portion of illegal immigrants and does not take into account those among the 46 million who can afford health insurance but do not purchase it or those who qualify for government-funded programs but have chosen not to apply for them or are unaware that they qualify for them. Thus, when considering the number of those in the United States without health insurance, the figure fluctuates between 46 million and 21 million. The fact that there is so much debate over the seemingly simple question of how many Americans are uninsured highlights the fact that information about health insurance reform can be difficult to navigate and evaluate. While it would be easy to focus on the cultural war that health insurance reform has sparked, the thornier debate is between those who recognize that the United States needs health insurance reform but are at odds with how such reform should manifest.

Beyond a concern for those who are either uninsured or underinsured, when citing the major reasons that the U.S. health insurance industry needs an overhaul, health insurance reform supporters from all sides of the political spectrum cite the following:

- Profound health disparities between those with health insurance and those without it
- Skyrocketing insurance costs
- Discriminatory insurance company practices
- A bloated, under-regulated pharmaceutical industry

Further, reform supporters note that the per capita cost of health insurance in the United States is higher than that of any other industrialized nation. They also point out that although the United States spends more on healthcare than any other country, the country ranks 23rd in infant mortality and 38th in longevity when compared to other nations.

The Obama administration's health insurance reform proposal, which the administration claims will be paid for upfront without increasing the deficit, includes the following major features:

- Ending discrimination against those with preexisting conditions
- Protecting Medicare for seniors and eliminating the gap in prescription drug coverage
- Creating a new insurance marketplace—the Exchange—that allows individuals and small businesses to buy insurance at competitive prices
- Offering a public health insurance option for the uninsured who cannot afford coverage
- Instituting medical malpractice projects to help doctors put their patients first instead of practicing defensive medicine

- Requiring large employers to cover their employees and individuals who can afford it to buy insurance

Those opposed to the Obama administration's reform proposal cite threats to individual liberty and look to the free market for solutions. With criticisms ranging from accusations of socialism to negative assessments of the government-managed healthcare systems of other countries, opponents also argue that the costs associated with government-mandated healthcare reforms are prohibitive. These opponents advocate a free-market approach to health insurance reform, which would include features such as the following:

- Making health insurance personal and portable, controlled by individuals and not by government or employers
- Providing workers with a standard deduction, tax credit, or health savings account for the purchase of health insurance
- Allowing people to purchase health insurance across state lines

It is clear that health insurance reform is a polarizing issue, and it remains to be seen whether or not any of the proposals will receive enough support from citizens and politicians to effect real change or whether this issue will continue to be contested for another 80 years.

See Also: Healthcare Delivery; Health Disparities; Pharmaceutical Industry Reform.

Further Readings

Cato Institute. "Cato Institute Conference on Health Care Reform." http://www.cato.org/events/healthcarereform/index.html (Accessed February 2010).

"The Obama Plan: Stability & Security for All Americans." U.S. Department of Health and Human Services (February 10, 2010). http://www.whitehouse.gov/assets/documents/obama_plan_card.PDF (Accessed February 2010).

Reid, T. R. *The Healing of America: A Global Quest for Better, Cheaper, and Fairer Health Care.* New York: Penguin Press, 2009.

"The Uninsured: A Primer." The Kaiser Commission on Medicaid and the Uninsured (October 2009). http://www.kff.org/uninsured/upload/7451-05.pdf (Accessed February 2010).

Tani Bellestri
Independent Scholar

Highways

Global acceleration of vehicle use is also accelerating highway construction. Expansion and maintenance of highway transportation systems rely upon the manufacture of cement, which, at every stage of the process, takes a toll on the environment. Generally, cement manufacture disrupts biodiversity across ecological spaces by the extraction of fuels and raw materials such as limestone, coal, and petroleum, which further contributes

The global growth of automobile use is accelerating highway construction. The use of cement and asphalt as road construction material has severe implications for the environment.

Source: iStockphoto.com

to greenhouse gas production. Beyond the damage caused by the manufacture of cement, its use as a material in highway construction also has severe implications for the environment. In contrast, emerging green highway concepts and innovations facilitate highway transportation system solutions that are ecologically sustainable. Green highway thinking is evident at the federal, state, and local levels, with highway agencies researching and implementing innovations designed to foster sustainability and to reduce the negative impacts of cement manufacture and highway construction on natural environments. Some of these measures include recycling old pavements, using materials other than cement in highway construction, and protecting watersheds, which suffer serious degradation related to runoff from concrete highways. Further, steps have been taken to involve community members and business leaders in highway planning, construction, and maintenance decisions.

The Federal Highway Administration (FHWA) in 2002 embraced environmental stewardship as a vital component of highway construction, and this decision opened the door to a series of initiatives focused on the ecological sustainability of highway construction and maintenance. The FHWA has implemented a number of programs and methods for creating green highways, including the *Exemplary Ecosystem Initiative*, which recognizes best practices in environmental stewardship as demonstrated at the state level; *Planning and Environment Linkages*, which supports a variety of resources directed at promoting the relationship between planning and the National Environmental Policy Act; and *Recycling* initiatives, which have seen the FHWA encourage an industrial by-products exchange to facilitate the recycling and reuse of materials.

Multidisciplinary partnerships, such as the Green Highways Partnership, have been forged in the effort to create sustainable highways. This initiative represents a voluntary, collaborative effort to create ecologically sustainable, functional transportation systems with minimum negative environmental impacts. Participating agencies include the FHWA, the U.S. Environmental Protection Agency Region 3 (which includes Delaware, the District of Columbia, Maryland, Pennsylvania, Virginia, and West Virginia), various state departments of transportation (DOTs), the National Asphalt Pavement Association, Villanova University, and an assortment of both public- and private-sector representatives. The comprehensive approach fostered by the Green Highways Partnership creates opportunities for uniting existing, complementary agencies and activities in the effort to green U.S. highways.

The best example of green highway logic in action is Maryland's Waldorf Transportation Improvement project, which is poised to become the nation's first truly green highway. Encompassing a section of US 301 that extends from Prince George's County to Charles County, the Waldorf project began applying green highway concepts in the earliest stages of planning. Some of the measures being implemented in relation to the project include evaluation of overall resource conditions and consultation with project partners and public stakeholders, all in an effort to balance and minimize negative consequences on both the natural and built environments. Participants believe that the Waldorf project has the potential to serve as a model for other green highway projects across the United States.

Storm water management is one of the major focuses of green highway construction. An example of the innovative storm water management methods being embraced in green highway initiatives includes the use of biocells, like the one installed in Washington, D.C., in 2004, by the District DOT. A biocell is composed of natural materials such as mulch, soil mix, and various types of vegetation. As opposed to drainage pits, a biocell actually performs like a filtration device and can remove up to 90 percent of the suspended solids from storm water. Using permeable materials to construct highways and their support structures will help keep metals and toxins from leaching into rivers and streams.

Recycling and reuse of materials is another major component of green highway strategies. Successful examples of recycling and reuse include the Pennsylvania DOT's use of shredded tires as lightweight embankment fill in the construction of the Tarrytown Bridge and the West Virginia DOT's use of recycled blast furnace slag for much of its asphalt surface pavements. The blast furnace slag offers a number of benefits, including providing safer driving conditions due to its nonpolishing properties—the roadway does not experience as much spray or misting during rain, providing better visibility and less hydroplaning.

Finally, in efforts to improve conservation and ecosystem management as they relate to highway construction, green highway projects incorporate the concepts of "green infrastructure." Rooted in the way that wetlands, forest preserves, and native plant vegetation naturally manage storm water, this approach not only is practical, holistic, multifunctional, and science based but it usually costs less to install and maintain than do traditional forms of infrastructure. Successes in innovative conservation and ecosystem efforts include Delaware's Blue Ball Properties project. A result of the partnership between the Delaware DOT and the local community, the Blue Ball project used an environmental stewardship approach in its plans to accompany highway repair and construction with the seeding of several large areas as meadows; the restoration of a stream; the creation of a wetland; and the development of a regional storm water management system that includes ecologically sound features.

These efforts to green U.S. highways represent a willingness to embrace new strategies to create highway systems that are efficient and sustainable. By relying on integrated planning, market-based approaches, regulatory flexibility, and environmental streamlining, green highway initiatives represent the next logical step in the evolution of efforts to marry sociocultural needs to ecological sustainability.

See Also: Environmental Protection Agency (U.S.); Government Role in Green Health; Groundwater; Private Industry Role in Green Health.

Further Readings

Green Highways Partnership. http://www.greenhighways.org (Accessed September 2009).
Turley, William. "Pushing Recycling in the Highway Sector." *Construction & Demolition Recycling*, 7/4 (2005).

U.S. Department of Transportation, Federal Highway Administration. "Public Roads." http://www.tfhrc.gov/pubrds/06nov/07.htm (Accessed September 2009).

Elizabeth Nanas
Wayne State University,
Hong Kong University of Science & Technology

Tani Bellestri
Independent Scholar

HOBBY PRODUCTS

Many familiar products used in hobby and art projects, including paints, solvents, and glues and cements, may pose significant health hazards if not used properly. In addition, such materials may damage the environment unless properly disposed of. In some cases, good alternatives are available that should be substituted for the more hazardous materials, particularly if the intended user is a child. In other cases, it is possible to minimize the risk to the user by limiting exposure and/or taking precautions such as working in a well-ventilated studio, wearing a respirator, and wearing skin protection such as gloves.

Even if the solvent in a paint is safe, the pigment in a particular color may be hazardous. Many common pigments in oil paints are highly toxic and/or carcinogenic.

Source: iStockphoto.com

The Labeling of Hazardous Art Materials Act of 1988 (which took effect in November 1990) in the United States requires art materials to be evaluated for their hazard to human health according to standards set by the Consumer Product Safety Commission. Materials that pose a chronic health hazard must carry a warning label and also be labeled as inappropriate for use by children. Not all materials used by artists fall into the jurisdiction of the 1988 act, however. For instance, screen-printing inks developed for industrial applications may also be used by artists and hobbyists but, if labeled for industrial use, are not subject to the terms of the 1988 act unless they are sold to schools or in art supply stores or are advertised as appropriate for artistic use.

Evaluating the Hazard Posed by Specific Substances

The hazard posed by any material depends in part on how it is used, including frequency of use and length of exposure. For instance, a paint thinner used for 15 minutes on a single day to strip a piece of furniture may pose minimal risk, but the same substance might be hazardous if used by an individual for eight hours per day, every day. An additional risk applies to people who use hazardous materials in their home (typical of most hobbyists and artists with a home studio): if the material lingers in the air due to inadequate ventilation, the length of exposure may be much greater than just the time the material was in active use.

There are three main ways toxic substances can enter the body: skin contact, inhalation, and ingestion. The latter may result from causes such as painters pointing their paintbrush with their lips or by the user failing to wash up after working with toxic substances and before eating. Children and adolescents have higher risk of harm from exposure to toxic substances because they are smaller and have faster metabolisms than adults, and fetuses are highly sensitive to many chemicals, even if the exposure is secondhand: for instance, studies have shown that women whose husbands work in the chemical industry have an elevated rate of miscarriages.

Two additional factors must be considered in evaluating the hazard posed by any substance. The first is whether it accumulates in the body over time so that even small exposures over a period of time can lead to a high total body burden (the total amount of the substance present in the body). This is a characteristic of some metals, including lead, mercury, cadmium, manganese, and arsenic, as well as some organic chemicals such as chlorinated hydrocarbons.

The second is whether additive and/or synergistic effects are present. The additive effect means that the consequences of exposure simply add up (e.g., the health consequence of multiple exposures to turpentine is greater than a single exposure), while the synergistic effect means that the combined effect of exposure to two or more hazardous substances is greater than simply adding together the effect of each alone. For instance, alcoholic beverages have a synergistic effect with many solvents so that even a small exposure to the solvent while the users also have alcohol in their body may pose a serious risk to their health. Some prescription drugs also have an additive or synergistic interaction with solvents, so people taking prescription drugs and planning to work with solvents should consult with their physician before proceeding.

Solvents and Pigments

Many solvents (e.g., toluene, turpentine) used in art and hobby applications are highly volatile, meaning that they evaporate quickly and therefore pose a high risk of inhalation. Many solvents are hazardous if ingested or inhaled in quantity, and some are also skin irritants and/or can be absorbed through the skin. Long-term exposure to solvents such as toluene, mineral spirits, and chlorinated hydrocarbons can cause permanent brain damage (this has been observed in people who inhale solvents for recreation). To minimize the risk, gloves, goggles, and adequate ventilation should be used when working with solvents. In addition, because many are highly flammable, they should be kept entirely separate from sources of fire (including smoking) and be stored and disposed of properly.

The safest solvent is water, and this is why water-based paints are preferred for use by children and in schools. For adult artists, a more hazardous solvent may be required (not every pigment will dissolve in water), but there is a hierarchy of risk among them, and it is sometimes possible to substitute less hazardous solvents for more hazardous ones. For instance, heptane is a moderately toxic aliphatic hydrocarbon that is a good substitute for n-hexane, the most toxic of the aliphatic hydrocarbons, and odorless mineral spirits are less hazardous than either turpentine or ordinary mineral spirits and can often be substituted for them. Hazardous solvents should only be used in a well-ventilated workspace, and the solvents (and paints containing them) should be disposed of at a hazardous waste collection facility rather than in the trash or poured down the drain.

Even if the solvent is safe, the pigment (the substance that gives paint its color) may be hazardous. Many common pigments used by artists are highly toxic and/or carcinogenic: examples (and the compounds involved) include antimony white (antimony trioxide), cadmium yellow (cadmium sulfide), chrome orange (lead carbonate), chrome yellow (lead chromate), lead white (lead carbonate), naples yellow (lead antimonite), and vermilion (mercuric sulfide). Examples of pigments that are moderately or slightly toxic include carbon black (carbon), cobalt blue (cobalt stannate), prussian blue (ferric ferrocyanide), and zinc white (zinc oxide). Besides using the least toxic pigments possible, dry pigments should be mixed in a glove box or inside a fume hood to avoid inhaling the powder.

Other Hazardous Products

Some common glues and cements used in art and hobby applications can pose serious health threats. For instance, rubber cement, which is a polymer mixed in a volatile solvent such as hexane or heptanes, poses high risks for inhalation (and inhalant abuse) and is also highly flammable. Adults should use rubber cement in a well-ventilated environment; potential substitutes for children include glue sticks, white glue (e.g., Elmer's), library paste, and mucilage. Contact cement contains toluene, a toxic and flammable solvent, while airplane glue and model cement may contain acetone or toluene, both of which can pose significant hazards. The vapor from instant-bonding glues (e.g., Super Glue) can be intensely irritating to the eyes, and epoxies cause skin irritation and sensitization in some people.

Dust, including that arising from woodworking and wood and stone carving, may also pose a significant inhalation hazard. The hazard is related to the size of the particles involved: particles greater than 5 microns are usually trapped by the defenses of the respiratory system, but those with a diameter less than 5 microns can penetrate into the lungs and cause more serious damage. The specific substance and the amount inhaled are also important factors in the risk. Mineral dusts (including those from silica, asbestosis, and man-made mineral fibers) are particularly dangerous and may lead to pneumoconiosis and cancer; metal dusts (e.g., those raised by grinding or polishing metals) can also be highly hazardous.

Among drawing materials, modern pencils are generally considered safe because they are made using graphite rather than lead. Charcoal and chalk are more hazardous because they can produce a large amount of dust, which may cause irritation if inhaled. Pastel sticks pose a greater hazard because they often use toxic pigments (including compounds of lead and cadmium) as well as producing dust that is easily inhaled. Safer alternatives are crayons and oil pastels that produce almost no dust. It should be noted that children's crayons are safe because they are made from paraffin wax or beeswax and use nontoxic pigments, but industrial crayons may contain toxic pigments. In addition, lead and asbestos have been found in imported crayons.

See Also: Arsenic Pollution; Children's Health; Lead Sources and Health; Lung Diseases; Neurobehavioral Diseases.

Further Readings

Agency for Toxic Substances and Disease Registry, U.S. Department of Health and Human Services. http://www.atsdr.cdc.gov (Accessed July 2010).

Babin, Angela. "Art Painting and Drawing." Center for Safety in the Arts, 1991. http://www.uic.edu/sph/glakes/harts1/HARTS_library/paintdrw.txt (Accessed July 2010).

Goldberg, Jenny. "Art and Hobby Supplies." Washington Toxics Coalition. http://watoxics.org/healthy-living/healthy-homes-gardens-1/factsheets/arthobby (Accessed July 2010).

McCann, Michael. *Artist Beware*, 3rd ed. New York: Lyons Press, 2005.

Rossol, Monona. *The Artist's Complete Health and Safety Guide.* New York: Allworth Press, 2001.

Sarah Boslaugh
Washington University in St. Louis

HOME-GROWN FOOD

When we think about sources of greenhouse gas emissions responsible for global climate change, images of power plants, automobiles, and industry come to mind. Rarely does it occur to us that part of the problem may be on our dinner plates. Yet agriculture is responsible for one-third of greenhouse gas emissions globally. Industrial farming processes used to fill our supermarket shelves—such as mechanization, concentrated animal feeding operations, and agricultural "inputs" (synthetic fertilizers, pesticides, and herbicides) that use fossil fuels as their starting materials—make industrial agriculture an energy-intensive process. But farming activity from planting to harvest takes only one-fifth of the total energy consumed in agriculture. Most of the energy is consumed during transport, storage, and processing. Our food typically travels between 1,500 and 2,500 miles from farm to plate, requiring staggering amounts of fuel. Add in the energy used to process and package those frozen dinners and refrigerate that bag of lettuce, and we begin to see how our modern food production and consumption patterns contribute to climate change.

The industrial food chain has given us unprecedented convenience, year-round availability of produce, and all manner of processed food, but it came at a huge social and ecological price. Industrial farming did not benefit family farms or local farming communities, which, over the last 50 years, could not compete in the marketplace, dried up, and became suburban subdivisions. Industrial (nonorganic) farming methods promote soil nutrient depletion and erosion, generate pollutants that choke waterways, and create "dead zones" where fish and other marine life have disappeared. Growing only varieties of vegetables that have uniformity of size and shape so that they pack and travel well has reduced the nutritional quality and the biodiversity of our food.

Concerns also arise about the safety of the food delivered to our table by the industrial food chain. Food-borne infections can occur regardless where food comes from (even the home garden), but in the industrial food chain, where ingredients in a single product may have come from around the globe, there are more opportunities for food contamination to occur in processing, handling, storage, or transit. The source of contamination is often unexpected, as witnessed in recent food poisonings from bacteria in frozen cookie dough.

In other instances, such as the discovery of melamine in dog food and baby formula ingredients from China, the lack of our capacity to monitor the safety of ingredients entering the food chain is woefully apparent.

The industrial food chain also contributes to the epidemic of obesity. About two-thirds of U.S. adults are overweight or obese and at higher risk, compared with normal-weight adults, for diabetes and heart disease. Indeed, the Centers for Disease Control and Prevention (CDC) estimates that one-third of all American children born in 2000 will develop diabetes as a result of obesity, poor diet (largely a processed, fast food diet), and lack of exercise.

So at a time when Americans are struggling to find solutions to help reduce the costs of healthcare for chronic, preventable diseases, reduce our national dependence on foreign oil, and slow the pace of global climate change, it makes sense to reexamine not just what we eat, but how and where we produce our food. What would it take to transition to a more sustainable and healthy food supply? Fifty years ago, anyone with a sunny backyard put in a vegetable garden. Today, production of home-grown food presents different challenges, since half the population lives in urban areas. Nevertheless, becoming a "locavore"—a person who consumes only food produced locally—is not as difficult as it seems.

The place to begin is the home garden. Whether you start with an acre or two or just an apartment with a small balcony, you can provide some of your own food. You can grow herbs in a windowsill; salad greens in pots; or pounds of tomatoes, peppers, and leafy greens in self-watering containers. Bean sprouts can be grown in a mason jar on the kitchen counter and mushrooms in a kit stashed away on the floor of a closet. A growing number of people with a little land (it may be only a tenth of an acre) are trying backyard homesteading. The most intrepid homesteaders are having success in urban environments raising chickens (hens only) for fresh eggs and bees for honey.

Likewise, city people are reclaiming abandoned lots, underused public green spaces, schoolyards, and golf courses to use as community farms, gardens, and orchards. Originally intended for household food production, community gardens increasingly donate a portion of their crop to local food banks or public schools for healthy school lunch programs. Community gardens have also become training grounds for children to learn about nutrition, ecology, food cultivation, and the skills for a more self-sufficient lifestyle.

The explosive growth in farmers markets and community-supported agriculture (CSAs, farm-to-household subscription schemes) is the clearest indication of public interest and demand for local food. Through these urban street markets, a reinvigorated smallholder farm system has emerged, catering to community demands for fresh, sustainably grown, genetically diverse products. Once just glorified fruit and vegetable stands, farmers markets in many areas are open all year and have become outlets for locally produced fruits, vegetables, dairy, eggs, meat, baked goods, cereal products, flowers, and condiments. Locavores can obtain nearly all of the family "food basket" from such markets, provided they are willing to change their consumption habits to (1) avoid processed foods, (2) feast on what is in season, (3) learn how to preserve (can, freeze, dehydrate, or root cellar) seasonal produce for the winter months, and (4) give up the "exotic" stuff such as jam and mustard from France when there are comparable local products.

In addition to all the environmental and health advantages, local food offers substantial economic benefits. Money invested in a home garden, over time, results in huge savings in the family food bill. Money spent on local food at farmers markets or CSAs stays in the community, creates jobs, promotes resilience and food security, and supports farmers. Beyond the farmer-to-household link, local restaurants, independent grocers, and school cafeterias have established direct market supply arrangements with farmers, thus helping to rebuild local economies.

These trends create optimism that, in time, the advantages of locally produced food may overcome our long-distance industrial food habit. Perhaps one day we will return to being a nation of producers rather than a nation of consumers.

See Also: Obesity; Personal Consumer Role in Green Health; Organic Produce; Urban Green.

Further Readings

Halweil, Brian. *Home Grown: The Case for Local Food in a Global Market*. Washington, DC: Worldwatch Institute, 2002.

Kingsolver, Barbara, Steven L. Hopp, and Camille Kingsolver. *Animal, Vegetable, Miracle: A Year of Food Life*. New York: HarperCollins, 2007.

Madigan, Carleen. *The Backyard Homestead*. North Adams, MA: Sorey Publishing, 2009.

Pollan, Michael. *The Omnivore's Dilemma*. New York: Penguin, 2006.

Pringle, Peter. *Food, Inc.: Mendel to Monsanto—The Promises and Perils of the Biotech Harvest*. New York: Simon & Schuster, 2003.

Weber, Karl, ed. *Food, Inc.: How Industrial Food Is Making Us Sicker, Fatter and Poorer—And What You Can Do About It*. New York: Public Affairs, 2009.

Patricia Stephenson
Independent Scholar

Hormone Therapy

Hormone therapy is used to treat symptoms of menopause in women. Hormone therapy may include forms of estrogen and progesterone, either of which is available in a standard or bioidentical form. The chemicals used in hormone therapy not only include some increased health risks for women, but they can also survive wastewater treatment and affect the health and reproduction of aquatic species. Because of the risk to themselves and to the environment, some women may choose to manage their menopausal symptoms without hormone therapy.

Menopause is the ceasing of fertility and the menstrual cycle within a woman. It occurs naturally for each woman somewhere in her mid-40s to mid-50s. Estrogen and progesterone are produced by the ovaries and regulate the menstrual cycle. During the years before menopause, known as perimenopause, the ovaries begin shrinking, hormone levels fluctuate, and menstrual cycles often become irregular. Once a woman has gone one full year without a period, menopause has likely occurred. Menopause may also be induced surgically through the removal of both ovaries. Common symptoms of menopause include hot flashes, night sweats, insomnia, vaginal dryness, and mood swings. Hormone replacement therapy (HRT) attempts to treat these symptoms by replacing the decreased hormones within the body.

Estrogen is often taken by women using hormone therapy, most commonly in the form of $17\text{-}\beta\text{-estradiol}$. Progesterone, or the synthetic progestin, is often combined with estrogen to thin the uterine lining. Estrogen taken alone can cause the uterine lining to grow excessively, which can cause uterine cancer. Women who have undergone uterine removal may safely receive estrogen alone. Those with a uterus often take progestin alone or combined with estrogen.

Some early studies on the use of hormone therapy seemed to suggest increased protection against osteoporosis, but mixed results were occurring with regard to the effects of hormone therapy overall. In an attempt to determine the effects of hormone therapy on women's health, a study of more than 161,000 healthy postmenopausal women, called the Women's Health Initiative (WHI), began in 1991. The WHI was the largest study of its kind within the United States and included clinical trials and an observational study. The results of hormone trials for estrogen alone and with progestin were as follows:

- Estrogen plus progestin: benefits—37 percent decreased risk for colorectal cancer, 37 percent less hip fractures; risks—26 percent increase in breast cancer, 41 percent increase in stroke, 29 percent increase in heart attack, doubled rates of blood clot
- Estrogen alone: benefits—39 percent less hip fractures; risks—39 percent increase in stroke, 47 percent higher risk for blood clots

The combined hormone trial was stopped early due to the increased risks of cancer and stroke. The estrogen trial was similarly stopped due to increased risks for stroke and blood clot.

Bioidentical hormones have received much public attention as an alternative to traditional hormone therapy. The chemical formula of bioidentical hormones is exactly the same as those produced by the human body. The unique feature of bioidentical hormones is that they are often presented in custom mixes of hormones, tailored for each user after the testing of saliva. These tailored mixes are often advertised as safer than traditional hormone therapies. But the Food and Drug Administration (FDA) has not tested these mixes for purity, safety, potency, or effectiveness; and they have not been approved by the FDA. For this reason, the North American Menopause Society (NAMS) does not recommend using custom-mixed hormones. FDA-approved bioidentical versions of the traditional hormonal therapies used, estrogen (estradiol) and progestin (progesterone), are available in varying doses.

Estrogens are naturally excreted in human waste streams (urine and feces). It should be noted that the total load of estrogens to the environment does not come from HRT, as both premenopausal women and men emit them as well. Those hormones that survive wastewater treatment enter surface water through the effluent of wastewater treatment facilities. Estrogens have been well documented in the effluents of wastewater treatment facilities all over the world. The concentrations are low, in the nanogram per liter range, or parts per billion. But even these low concentrations of estrogens have been shown to have effects on aquatic wildlife. One study by Y. Huang in 2003 examined the occurrence and effects of endocrine-disrupting chemicals (chemicals that alter the normal functions of hormone systems, such as the hormones used in HRT) in the environment. Fish exposed to chemicals that alter the estrogen system, such as those used in HRT, were found to exhibit the following symptoms:

- Elevated levels of vitellogenin (VTG, a precursor for egg yolk normally found in the serum of adult females) in adult male or juvenile fish
- Intersex, such as the growth of both oocytes and testicular tissue within male fish gonads, called ovitestes
- Decreased fecundity, or the rate at which an individual produces viable offspring

Due to the increasing population and trend toward urbanization on the planet, wastewater effluents are increasingly impacting the Earth's surface waters. In fact, in lakes or rivers, it is not uncommon for one municipality's drinking water supply intake to be located downstream of another municipality's wastewater effluent. As a result, the hormones used in HRT, along with many other pharmaceuticals and personal care products, have been found in surface water at concentrations known to affect fish life.

See Also: Antibiotic Resistance; Cancers; Personal Consumer Role in Green Health; Women's Health.

Further Readings

Huang, Y., D. L. Twidwell, and J. C. Elrod. "Occurrence and Effects of Endocrine Disrupting Chemicals in the Environment." *Practice Periodical of Hazardous, Toxic, and Radioactive Waste Management* (October 2003).

Kolpin, D. K., E. T. Furlong, M. T. Meyer, E. M. Thurman, S. D. Zaugg, L. B. Barber, and H. T. Buxton. "Pharmaceuticals, Hormones, and Other Organic Wastewater Contaminants in US Streams, 1999–2000: A National Reconnaissance." *Environmental Science & Technology*, 36 (2002).

U.S. Department of Health and Human Services. "Facts About Menopausal Hormone Therapy." NIH Publication No. 05-5200. Rev. June 2005.

Michelle Edith Jarvie
Independent Scholar

HOSPITALS (CARBON FOOTPRINTS)

The notion of a "carbon footprint"—the measurement, in units of carbon dioxide equivalents, of the greenhouse gases produced by an individual, an organization, or a state—may be a fairly recent phenomenon, but the conservation movement in the United States reaches back to the late 19th century. Today's environmental movement has expanded in numerous ways, and greenhouse gases, green living, and carbon footprints are common topics. While some industries are obviously energy intensive, such as the aluminum or petroleum industries, the fact that healthcare ranks as the second-most-energy-intensive industry in the United States, imprinting the world with a massive carbon footprint, may be surprising.

The American healthcare sector, which includes healthcare facilities, scientific research, and the production and distribution of pharmaceutical drugs, accounts for almost 10 percent of carbon dioxide (CO_2) emissions in the United States. Hospitals account for the largest proportion of these emissions and, according to the U.S. Department of Energy, use 836 trillion BTUs of energy annually and have almost three times the carbon dioxide emissions of commercial office buildings. Part of the reason healthcare's carbon footprint is so large is its reliance on conventional, nonrenewable energy sources, such as oil and coal. Further, in calculating their total carbon footprint, hospitals must include indirect emissions generated by visitor, patient, and staff travel, as well as the carbon dioxide emitted as a result of

In the United States, hospitals account for almost three times the carbon dioxide emissions of commercial office buildings, according to some estimates.

Source: iStockphoto.com

obtaining goods and services. This total footprint is estimated to be as much as three times higher than direct emissions alone, though it is more difficult to calculate.

One of the tools used by researchers and industry analysts to measure carbon footprints is the Environmental Input-Output Life Cycle Assessment model (EIOLCA), developed by the Green Design Institute at Carnegie-Mellon University. The EIOLCA was used in a 2007 study of hospitals that assessed direct and indirect environmental effects of healthcare activities. Most likely due to the high-energy demands of temperature control, ventilation, and lighting, the study found that hospitals were the largest contributor of carbon emissions in the healthcare sector. Following this study, in 2009, Illinois-based Stericycle launched its online Carbon Footprint Estimator as a tool for U.S. hospitals to use in assessing their environmental impact. Stericycle's Carbon Footprint Estimator is designed to help U.S. hospitals determine the amount of plastic and associated cardboard containers they use, as well as how much carbon dioxide they emit. The Carbon Footprint Estimator is free to use, and after inputting certain kinds of information, it helps hospitals determine their effect on the environment and offers suggestions for diminishing this impact. With more than 5,000 hospitals and healthcare facilities in the United States, tools such as the EIOLCA and Stericycle's Carbon Footprint Estimator are useful for measuring negative environmental impact and, in turn, helping hospitals to embrace green initiatives that can help to lessen these negative impacts.

Those who have embraced green initiatives in the healthcare sector say that they are interested in lessening their negative impact on the environment, not only for the environment's sake but also because, in their role as healthcare providers, it makes little sense for hospitals to negatively impact air quality. Some of the measures being taken by various hospitals across the United States include designing more efficient buildings, incorporating renewable energy sources, reducing waste through recycling programs, locally sourcing food and supplies, and using the combined heat and power (CHP) method of generating electricity on-site and then using the heat this produces to heat the hospital. Smaller-scale measures include using motion-sensitive lighting and energy-saving light bulbs and maintaining and insulating boilers and pipes. Other green initiatives in hospitals include encouraging employees to cycle to work with allowance incentives and promoting car-sharing schemes.

Specific examples of these principles in action include the University of Chicago Medical Center's sustainability program, which has implemented a plastic recycling program that diverts more than 500 pounds of waste each day from landfills to recycling plants and ensures that 90 percent of the cleaning supplies used by the hospital have Green Seal certification. Such efforts have actually led to reduced waste costs at the medical center, which is further incentive for other hospitals to embrace similar measures. Texas-based Harris

Methodist Fort Worth Hospital (HMFW) has gone even farther in its efforts to honor the connection between patient care and a healthy environment. Since embracing pollution prevention initiatives in 2005, HMFW has had many successes, including reducing both its overall solid waste stream and its energy consumption. HMFW accomplished these and other goals through the creation of a multidisciplinary Green Team dedicated to public and environmental health; the team spearheaded efforts in areas including the following:

- General recycling
- Hazardous chemical recycling
- Mercury elimination
- Energy efficiencies

Another healthcare facility making giant strides in reducing its carbon footprint is Hartford, Connecticut–based Saint Francis Care (SFC). In recognition of SFC's commitment to improving environmental health, Energy Star, a partnership between government, consumers, manufacturers, retailers, and other industry organizations, first designated SFC an Energy Star business in 2003. SFC was the first hospital in Connecticut to install its own fuel cell, and the hospital recovers the heat produced by the fuel cell to improve the performance and increase the efficiency of its hot water systems, significantly reducing fuel consumption and CO_2 emissions. SFC has also made improvements in its laundry facilities and has spearheaded efforts to make its heating and cooling systems more environmentally friendly and efficient. These improvements have led to a reduced environmental burden as well as significant savings in SFC's operating costs.

The World Health Organization (WHO) has also taken an interest in reducing the carbon footprint of healthcare facilities and encourages a holistic carbon reduction strategy that includes improvements and initiatives in areas including the following:

- Better building design
- Local sourcing of food and supplies
- Encouraging green travel plans that promote healthy modes of transport and help to change travel patterns of patients, staff, and visitors
- Recycling
- Board-level commitment to green efforts
- Investment in large-scale capital improvements that are sustainable
- Increased energy efficiency, including using the CHP model, insulating roof spaces, and incorporating renewable energy sources

In the United States, there exists a broad network of organizations and resources offering support and guidance for healthcare facilities that are interesting in reducing their environmental impact. The U.S. Department of Energy (DOE), in partnership with national healthcare sector leaders, has launched the Hospital Energy Alliance, which aims to increase energy efficiency and the use of renewable technologies in hospital design, construction, operation, and maintenance. Through this alliance, hospitals and healthcare organizations gain access to DOE resources and technical expertise. This level of collaboration connects building owners and operators with research, advanced technologies, and analytical tools emerging from DOE and its national laboratories.

Another collaborative initiative focused on reducing hospitals' carbon footprints, as well as their operating costs, is the Premiere Healthcare Alliance, which is owned by not-for-profit hospitals and includes more than 2,200 U.S. hospitals as well as more than

63,000 healthcare sites. Through Premiere's program known as SPHERE (Securing Proven Healthcare Energy Reduction for the Ecosystem), hospitals in the alliance are focusing on reducing energy costs and greenhouse gas emissions as well as increasing the use of renewable energy. Premiere Healthcare Alliance hospitals have embraced such measures as composting, recycling, local sourcing of food and supplies, and increased energy efficiency. As part of SPHERE, the Premiere Healthcare Alliance is collaborating with Practice Greenhealth, which is the nation's leading organization for institutions in the healthcare community that have made a commitment to sustainable, ecofriendly practices. One of the measures Premiere and Practice Greenhealth have embraced is the Health Care Clean Energy Exchange (HCEE), which has led to the first large-scale reverse auction for energy. HCEE establishes competition among energy suppliers to reduce energy prices. Through these initiatives and collaborative efforts, hospitals are able to share best practices and to set standards with their energy use to positively impact climate change and public health.

There are many organizations and resources, such as those listed above, that are ready to provide guidance to hospitals interested in reducing their carbon footprint. As the examples above show, reduction of negative environmental impacts typically goes hand in hand with cost savings, which suggests that embracing ecofriendly measures is not just good for the environment, but good for the hospitals, too.

See Also: Climate Change; Government Role in Green Health; Healthcare Delivery.

Further Readings

Chen, Joseph S., Philip Sloan, and Willy Legrand. *Sustainability in the Hospital Industry.* Oxford, UK: Butterworth-Heinemann, 2009.

Chung, Jeanette W. and David O. Meltzer. "Estimate of the Carbon Footprint of the U.S. Health Care Sector." *Journal of the American Medical Association*, 302/18 (2009).

U.S. Department of Energy. "Department of Energy Announces the Launch of the Hospital Energy Alliance to Increase Energy Efficiency in the Healthcare Sector." U.S. Department of Energy, April 29, 2009. http://www.energy.gov/news2009/7363.htm (Accessed February 2010).

Tani Bellestri
Independent Scholar

HYDROELECTRICITY

Hydroelectric power offers a cleaner energy alternative to coal- and crude petroleum–powered electricity. Hydroelectricity accounted for 18 percent of global electricity supply in 2006. Currently, 97 percent of global electricity generation from renewable sources is obtained from hydroelectricity. It has been projected that hydropower would generate approximately 1.8 trillion kilowatt-hours (kWh) of electricity by 2030, representing 54 percent of total electricity from renewable energy sources. Hydroelectric power presently occupies significant portions in the electricity mix in both developed and developing countries including (but not limited to) Norway (99 percent hydroelectric),

New Zealand (75 percent), Canada (61 percent), Sweden (44 percent), Brazil (86 percent), Ghana (67 percent), Venezuela (67 percent), China (22 percent), Russia (18 percent), and India (16 percent).

Hydroelectric Facility Classification

Classification of hydroelectric facilities is based on the amount of electricity generated. Hydroelectric facilities could be categorized as large ($\geq$500MW), medium ($\geq$10MW and <500MW), and small (<0.01MW and <10MW). Small hydroelectric facilities comprise mini/micro ($\geq$ 0.01MW and <0.5MW) and pico (< 0.01MW). The challenge in utilizing installed capacity to categorize small hydro facilities could be explained, for example, in the case of a 10MW, low-head small hydro facility. Compared to high head facilities, the 10MW small hydro low head requires greater volumes of water and larger turbines for electricity generation.

Hydroelectric power—like that produced by the Grand Coulee Dam on the Columbia River in Washington state—offers cleaner energy than traditional fossil-fuel electricity. Hydroelectricity accounted for 18 percent of the global electricity supply in 2006.

Source: U.S. Bureau of Reclamation

Hydroelectric Project Components

Construction of a large-scale hydroelectric facility is typically achieved through several years of hydrological studies (usually employing more than 50 years of hydrological data) and a robust environmental impact assessment. Small hydroelectric facilities, on the contrary, require fewer years' data and less stringent environmental assessment.

Techno-economic viability of hydroelectric power projects are very site specific. The economics of an identified hydroelectric project depends on several factors including the power (capacity) to be generated, the existence of feed-in tariffs for generated electricity, and prevailing market price for the generated electricity. Hydroelectric facilities comprise a dam, water passage, and a powerhouse.

The constructed dam helps direct the water into a penstock. The force of the flowing water in the penstock is employed to spin the turbine, resulting in electricity generation. The water exits the powerhouse through the tailrace, back into the river. The water passage consist of the entrance to the penstock; the penstock; the openings and exit of the turbines as well as valves to seal off water flow to the turbine for periodic maintenance work; and the tailrace. Most of mechanical and electrical equipment in a hydroelectric facility are housed in the powerhouse. The mechanical and electrical equipment in the powerhouse include valves for shutting off water to the turbines; river bypass gate to prevent flooding of catchment areas; hydraulic control system for the turbines and valves; electrical switchgear; electrical protection and control system; transformers utilized internally for station service and also for power transmission and distribution systems; water cooling and lubricating system; ventilation system; telecommunication systems; fire alarm systems and backup power supply.

The water flow (Q) as well as the head or drop in elevation (H) determines the power output (P) of the facility, which can be denoted by a simple conservative relationship as $P = 7QH$.

Benefits of Hydroelectric Facilities

Some of the widely acknowledged benefits attributed to hydroelectric facilities are that they are a renewable energy sources that provide the following: very versatile electricity service, addressing both peak load and base load demands; minimal greenhouse gas (GHG) emissions, no particulate emissions; negligible pollution; flood control; and irrigation water. Hydroelectric facilities are not influenced by volatilities in fossil fuel markets. Additionally, small hydro qualifies under the clean development mechanism (CDM) of the Kyoto Protocol as one of the flexible mechanisms to mitigate climate change via north–south partnerships.

Health, Environmental, and Social Impacts

Damming water bodies for large-scale hydroelectric power generation has been reported to be responsible for increase in occurrence in diseases in the tropics and in subtropical regions. Documented diseases in humans associated with large hydroelectric dams and irrigation include malaria, schistosomiasis, encephalitis, gastroenteritis, and onchocerciasis, while river fluke and trypanosomiasis have been reported in livestock. Though the linkages with increased human immunodeficiency virus and acquired immune deficiency syndrome (HIV/AIDS) and dam construction has not been made explicit in the published literature, it has been suggested that the development of major construction projects in prevalent HIV communities (unless aggressive safe sex programs are introduced) could increase the spread of HIV/AIDS among migrant laborers and local sex workers.

A big challenge with large hydroelectric power that leads to public outcry against the facility is the involuntary displacement and resettlement of local communities. Involuntary displacement not only disrupts the prevailing socioeconomic activities of communities but it also results in loss of homes, food resources, land and property, and higher mortality rates in re-settlers due to increased poverty. A study that investigated the mortality rates associated with resettlement found significant increases in mortality rates among Kariba Dam resettlers. In developing countries alone, over 2 million people have been displaced and resettled by large hydroelectric damming, including Three Gorges in China (1,250,000), Upper Krishna II in India (220,000), Sardar Sarovar in India (127,000), Aswan High Dam in Egypt (100,000), Kossou in Côte d'Ivoire (85,000), Akosombo in Ghana (84,000), Longtan in China (73,000), Mahaweli I–IV in Sri-Lanka (60,000), Kariba in Zambia and Zimbabwe (57,000), and Sobradinho in Brazil (55,000).

Damming of rivers for hydroelectric projects not only affects surrounding flora that become submerged but it also impacts fauna negatively. For example, hydroelectric dams have led to significant reductions in salmon populations. This is partly due to difficulties encountered by salmon populations in accessing spawning sites upstream of the dam (even though fish ladders are in place in most dams) and partly as a result of injuries sustained by salmon spawns as they migrate through turbines to the sea.

Compared to large hydro, small hydro is widely acknowledged to provide a better social and ecological and a healthier electricity production option. This is largely due to the fact that small hydro facilities avoid the extensive damming of rivers and waterways associated with large hydroelectric facilities. For example, some small hydroelectric projects are run-of-the-river developments that do not require that water be stored in a reservoir or dam.

Health and Environmental Impact That Hydroelectricity Potentially Offsets

Compared to other competing electricity generation options (excluding nuclear), hydroelectric power is associated with very minimal GHG emissions, unlike coal and crude petroleum. Nuclear electricity, however, results in radioactive waste that poses significant health hazards if not properly treated and stored. Global hydroelectricity has been reported to offset the equivalent combustion of 83.3 billion liters of crude petroleum oil or 122 million tons of coal per annum. The negligible air pollutants from hydroelectric power facilities, compared to coal and crude petroleum, offer significant health hazard reductions from air emissions. Additionally, the reduced GHG emissions from hydroelectricity would possibly have decreased the climatic change effect (and related health problems from climate change) compared to the significant GHG emissions from coal- and petroleum-based electricity globally. Greenhouse gas emissions from large-scale hydropower plants ranged between 35 gCO_2-eq/kWh and 380 gCO_2-eq/kWh, which are lower than the 454 gCO_2-eq/kWh emissions from natural gas and 890 gCO_2-eq/kWh for coal. The minimal GHG emissions associated with hydroelectric facilities are carbon dioxide (CO_2) and methane (CH_4), which emanate from decaying submerged plant biomass in dammed reservoirs. Additionally, there are no sulfur dioxide (SO_2), nitrous oxide (NO_x), and particulate emissions from hydroelectricity, compared with coal electricity. Electricity generation from coal and crude petroleum requires that the energy carrier feedstock be combusted, which results in SO_2 and NO_x and particulate emissions. The utilization of modern sophisticated emissions control systems effectively reduces the emissions associated with competing electricity facilities such as coal-powered energy in developed countries. In developing countries, however, the lack of modern mitigation facilities results in significant emissions of hazardous emissions, which therefore renders hydroelectricity a favorable alternative.

Increased investments in favor of hydroelectricity as opposed to coal electricity eliminate the health hazards that workers are exposed to during coal-mining operations. It also leads to the avoidance of health impacts associated with the combustion of coal for electricity generation. Diseases associated with coal mining and combustion that hydroelectricity potentially offsets include lung cancers in coal miners, coal workers' pneumoconiosis, esophageal cancer, and human selenosis.

Similarly, hydroelectricity eliminates the dangers associated with nuclear accidents and waste management as well as the risk uranium mineworkers are exposed to.

Outlook

Future growth in hydroelectricity generation is mostly expected to occur in developing countries, predominantly Brazil, China, India, and Vietnam. Only marginal increases in hydroelectricity generation could be expected in developed countries, where most of the economically exploitable hydro sources have already been dammed (with the exception of Canada and Turkey).

The public perception of large-scale hydroelectricity could be improved by adequately supporting and improving the socioeconomic status of members of displaced communities from the project revenue. Their increased healthcare requirements also need to be taken care of.

Conclusion

Hydroelectric power is a renewable energy source with less environmental, social, and health burdens compared to the other competing electricity-generation sources. Increased community participation in projects in addition to environmental due diligence will go a long way to improve the public perception of projects.

See Also: Climate Change; Electricity; Gastroenteritis; Malaria; Nuclear Power.

Further Readings

Bakis, R. "The Current Status and Future Opportunities of Hydroelectricity." Energy Sources, Part B. *Economics, Planning and Policy*, 2/3 (2007).

Clark, S., E. Colson, J. Lee, and T. Scudder. "Ten Thousand Tonga: A Longitudinal Anthropological Study From Southern Zambia." *Population Studies*, 49 (1995).

Gish, O. "Economic Dependency, Health Services and Health: The Case of Lesotho." *Journal of Health Politics, Policy and Law*, 6/4 (1982).

Gulliver, J. S. and E. A. Arndt. *Hydropower Engineering Handbook*. New York: McGraw-Hill, 1991.

Lerer, L. B. and T. Scudder. "Health Impacts of Large Dams." *Environmental Impact Assessment Review*, 19 (1997).

Pacca, S. "Impacts From Decommissioning of Hydroelectric Dams: A Life Cycle Perspective. *Climate Change*, 84/3–4 (2007).

RETScreen 2005. http://www.retscreen.net (Accessed May 2010).

Scudder, T. "Social Impacts." In *Water Resources: Environmental Planning, Management and Development,* A. K. Biwas, ed. New York: McGraw-Hill, 1997.

Stanley, N. F. and M. P. Alpers. *Man-Made Lakes and Human Health*. London: Academic Press, 1975.

U.S. Energy Information Administration, Independent Statistics and Analysis. "International Energy Outlook 2009." http://www.eia.doe.gov/oiaf/ieo/electricity.html (Accessed April 2010).

Wilson, E. M. *Assessment Methods for Small-Hydro Projects*. International Energy Agency Implementing Agreement for Hydropower Technologies and Programs, April 2000.

Emmanuel Kofi Ackom
University of British Columbia

Immune System Diseases

The immune system is the means by which the body defends itself from infection and disease. Deficiencies of the immune system (immunodeficiency disorders) mean that the immune system is less active or effective than normal and thus leaves the body at risk for infection from common pathogens (so-called opportunistic infections) that a healthy immune system could fight off. For instance, in the early years of the acquired immune deficiency syndrome (AIDS) epidemic, pneumocystis pneumonia (PCP) was a common cause of death for people with AIDS, while people whose immune system is functioning normally seldom become ill from PCP, even though the pathogen that causes the disease is quite common in the environment.

Immunodeficiency disorders are classified as primary or secondary (the latter is also known as acquired). Primary immunodeficiency diseases are genetic in origin, while secondary immunodeficiency disorders (which are more common) are caused by some external force such as malnutrition, infection with human immunodeficiency virus (HIV), cancers such as leukemia, chemotherapy, use of immunosuppressive drugs, or old age. As with many body systems, the immune system tends to become less effective with age.

Autoimmune diseases are the mirror image of immunodeficiency disorders because they are caused by the body's immune system being too active and attacking healthy tissue within a person's body that it mistakes for a pathogen or foreign invader.

Animal studies have suggested a link between exposure to toxins and immunosuppression, and ongoing research is investigating the possibility that exposure to pollutants, including tobacco smoke, is related to the development of autoimmune disorders in humans. Exposure to radiation is linked with developing some types of cancer, including leukemia (which often causes immunosuppression), and there are ongoing studies investigating whether exposure to pollutants is also associated with the probability of developing cancers such as leukemia. Some researchers have identified "cancer clusters," in which the rate of some type of cancer is unusually high, and then tried to link this observation to environmental exposure to toxins. Such research remains controversial and is prone to methodological difficulties (some call the cluster approach analogous to "shooting the barn and then drawing the target" because the clusters are defined after an unusual rate of disease occurs), although in some cases, such as Love Canal in New York State, the connection has been made sufficiently convincingly to win settlements in court.

Primary Immunodeficiency Diseases

The World Health Organization recognizes over 150 different primary immunodeficiency diseases: some affect a single cell within the immune system, others may affect multiple components in the system, some may produce no symptoms for years, and others can be life threatening at an early age. However, all result from a defect in the immune system. Results of this defect may include increased vulnerability to infection (or difficulty in fighting off infection), inability to regulate the immune response, or inability of the body to distinguish between foreign (nonself) material and itself. If the body cannot tell the difference between self and nonself, the result may be an autoimmune disorder in which the body attacks its own cells or tissues as if they were foreign matter.

Common variable immune deficiency (CVID) is one of the more frequently observed primary immune diseases and is marked by low levels of serum immunoglobulins (antibodies) and correspondingly high susceptibility to infection. Specific deficiencies vary from one patient to the next: some may have decreased levels of all three major types of immunoglobulins (IgG, IgA, and IgM), for instance, while others may be deficient in only IgG and IgA. Some people with CVID may have a normal number of B lymphocytes but which fail to mature into plasma cells capable of creating immunoglobulins and antibodies, while others may have malfunctioning T lymphocytes or excessive cytotoxic T lymphocytes. Clinical signs and symptoms of CVID range from mild to severe, and in most patients the diagnosis (usually made because the patient suffers from numerous and unusual infections) is not made until age 30 or later, although 20 percent of cases are diagnosed in children or adolescents younger than 16. The cause of CVID is unknown, although some studies suggest the involvement of a small group of genes. The most usual symptoms of CVID are recurrent infections in the ears, sinuses, nose, bronchi, or lungs, and patients may also have enlarged lymph nodes and/or an enlarged spleen.

X-linked agammaglobulinemia (also called Bruton's agammaglobulinemia or congenital agammaglobulinemia) is marked by the inability of the patient's body to produce antibodies. It is caused by a gene on the X chromosome, and it therefore occurs almost exclusively in males (females have two copies of the X chromosome, while males have only one; however, women can be carriers of this disease). Patients are prone to infections, particularly at or near the surface of the mucous membranes, including the middle ear, sinuses, and lungs. Infections may invade the bloodstream and spread to other organs including the brain, bones, and joints. Diagnosis is made by evaluating the quantity of serum immunoglobulins and B cells in the patient's blood, as well as by the absence or mutation of BTK protein.

Selective IgA deficiency is marked by an absence of the IgA class of immunoglobulins from the blood, although the patient generally has normal amounts of the other immunoglobulins. It is one of the most common immunodeficiency diseases (occurring in perhaps as many as one in 500 people), and many people with this condition are unaware of it because they remain in normal health. However, some do become sick, and estimates of the number who will eventually develop complications over a 20-year period range from 25 to 50 percent. People with selective IgA deficiency may have increased susceptibility to ear infections, sinusitis, bronchitis, and pneumonia, and a small number will suffer from chronic diarrhea and/or gastrointestinal infections. Allergies and autoimmune diseases such as lupus or rheumatoid arthritis are more common in people with this condition than in the general public.

Hyper IgM syndrome is characterized by normal or elevated levels of IgM and decreased levels of IgG and IgA antibodies. It is associated with a variety of genetic defects, some of which are X-linked while others occur in both males and females. Diagnosis is usually in the first two years of life after recurrent infections and gastrointestinal symptoms. Some patients also have autoimmune disorders and enlarged tonsils, lymph nodes, spleen, and liver.

Severe combined immunodeficiency (SCID) is a rare and serious disease in which the person has no T-lymphocyte or B-lymphocyte function. Currently, 12 different genetic defects are known to result in SCID. The disease is usually diagnosed in infancy because the patient suffers from rare infections such as PCP or because common infections such as thrush may persist much longer than usual. The patient may also suffer from persistent diarrhea, resulting in malnutrition. Diagnosis is made by blood tests and can be performed in utero if a previous child in the family has the disease (because then the specific molecular defect running in the family is known).

Chronic granulomatous disease (CGD) is marked by defects in the phagocytes that leave the patient vulnerable to infection by certain bacteria and fungi. It is usually diagnosed in infancy or childhood after an infection, often with aspergillis (a mold) or *Serratia marcescens* (a type of bacteria), although some patients are not diagnosed until adolescence. Patients with this disease are extremely vulnerable to pneumonia, particularly fungal pneumonias, and may also develop a liver abscess or bone infection. About 85 percent of CGD patients are boys because the most common type of CGD is X-linked recessive.

Wiskott-Aldrich syndrome is an X-linked recessive disease marked by defective T lymphocytes, B lymphocytes, and platelets. Each mutation seems to be unique, meaning that it is different for each family with this disease. Infections are common with this disease and may include respiratory infections, meningitis, and sepsis (blood poisoning). Patients sometimes have bleeding disorders and may have petechiae (small bluish-red spots) or larger spots resembling bruises that indicate bleeding into the skin. They may also suffer from eczema or viral skin infections or from episodes of autoimmune symptoms.

Secondary Immunodeficiency

Secondary immunodeficiency can be caused by many different things. One of the best-known causes of immunodeficiency is infection with HIV, a virus that destroys CD4 cells (also known as T cells) in the bloodstream. CD4 cells are a type of lymphocyte (white blood cell) that helps the body resist infection. A person may be infected with HIV for years and not have symptoms; however, once the disease starts killing large numbers of T cells, the person will become vulnerable to infection. In fact, presence of opportunistic infections such as PCP and candida is one criterion to diagnose AIDS. There is no cure for AIDS, but medications can reduce the viral load to the point where the immune system can function relatively normally.

Some cancers also interfere with normal functioning of the immune system. For instance, in lymphocytic leukemia, the cancer attacks the bone marrow cells that produce lymphocytes—the white blood cells that help the body fight infection. Adult T-cell lymphoma is caused by infection with HTLV (human T-cell lymphotropic virus) and attacks CD4 or T cells, causing them to proliferate abnormally.

Sometimes secondary immunodeficiency is caused by medical treatment. For instance, persons who receive an organ transplant are usually given immunosuppressive drugs to prevent their body from rejecting the new organ. However, because these drugs suppress the body's tendency to reject anything that is foreign, they also diminish its ability to recognize and fight off infectious agents such as bacteria. People are particularly vulnerable to infection in the first days after the transplant because they are receiving high levels of immunosuppressive drugs; normally, these levels are reduced over time, and vulnerability to infection decreases, although the patient will still be more vulnerable compared to a normal person. Chemotherapy may also cause a person to become immunosuppressed, and this must be monitored closely during treatment. Sometimes chemotherapy treatments are postponed to let the patient's immune system recover.

Malnutrition can suppress the normal functioning of the immune system as well. For this reason, disease outbreaks are always a hazard in times of famine or food shortage, and persons who are chronically undernourished may become seriously ill or die from diseases that their bodies could ordinarily have fought off. Individuals can also suffer from malnutrition due to disease (e.g., because they suffer from chronic diarrhea) or from alcohol abuse. In any case, malnutrition is associated with a host of deleterious effects on the immune system.

Autoimmune Diseases

Over 80 diseases have been identified as autoimmune disorders. Overall, they occur more commonly in women and, if considered as a group, it has been conjectured that autoimmune diseases would be among the top 10 killers of women under age 65. Because the mechanism that sets off the autoimmune reaction is often unknown, treatment tends to focus on ameliorating the symptoms.

Rheumatoid arthritis may be the best known of the autoimmune diseases, as it is both a very ancient disease (symptoms similar to rheumatoid arthritis have been identified in Egyptian mummies) and is estimated to affect as many as 1 percent of the world's population. In rheumatoid arthritis, the body attacks the joints, producing inflammation that may be crippling, and may also attack other parts of the body, including the lungs and pericardium (lining of the heart). Treatment includes anti-inflammatory drugs to deal with the symptoms and immunosuppressive drugs to slow disease progression.

Systemic lupus erythematosus (SLE, or lupus) is a chronic autoimmune disorder that may attack many different parts of the body, producing chronic inflammation. Symptoms include a characteristic facial rash (which someone thought resembled a wolf, hence the disease's name), arthritis, fatigue, fever, muscle aches, headaches, seizures, sensitivity to the sun, and abdominal pain. There is no cure, but various treatments are available, including corticosteroids or other drugs to reduce the immune system response and cytotoxic drugs to block cell growth.

Graves' disease is a thyroid condition and the second-most-common disease of the endocrine system (after diabetes). In Graves' disease, the body produces antibodies to the TSH (thyroid stimulating hormone) receptor that stimulates the production of thyroid hormone and causes the thyroid gland to become overactive. Consequences can include goiter (enlarged thyroid gland), bulging eyes, and an accelerated heart rate. Treatment may involve drugs that block the hormones or reduce the size or functioning of the thyroid through surgery or radiation, because the mechanism that produces the autoimmune response itself remains unknown.

The most common endocrine disease, type 1 diabetes (also called juvenile diabetes or insulin-dependent diabetes) is also an autoimmune disease. In type 1 diabetes, the body attacks cells in the islets of Langerhans, the regions of the pancreas that produce insulin. Diabetes develops when the body can no longer produce insulin and thus cannot maintain normal levels of glucose. Insulin must be supplied through injection, and the blood must be regularly tested to see that levels of glucose in the bloodstream are normal.

Celiac disease, a disease that occurs most often in persons of European descent, can damage the lining of the small intestine and can interfere with the body's ability to absorb nutrients. The mechanism is that when people with celiac disease eat foods containing gluten (e.g., wheat, barley, or rye), their immune system attacks the villi, or lining of the small intestine. There is no cure, but if the person avoids gluten-containing foods, the symptoms will cease.

See Also: Children's Health; Men's Health; Radiation Sources; Women's Health.

Further Readings

Centers for Disease Control and Prevention. "HIV/AIDS Fact Sheets." http://www.cdc.gov/hiv/resources/factsheets (Accessed August 2010).

Immune Deficiency Foundation. *Patient & Family Handbook for Primary Immunodeficiency Diseases*, 4th ed. (2007). http://www.primaryimmune.org/publications/book_pats/book_pats.htm (Accessed August 2010).

Krassas, G. E. and W. Wiersinga. "Smoking and Autoimmune Thyroid Disease: The Plot Thickens." *European Journal of Endocrinology*, 54/6 (2006).

National Cancer Institute. "Cancer Clusters." http://www.cancer.gov/cancertopics/factsheet/Risk/clusters (Accessed August 2010).

National Women's Health Information Center, U.S. Department of Health and Human Services. "Autoimmune Diseases: Overview." http://www.womenshealth.gov/faq/autoimmune-diseases.cfm (Accessed August 2010).

Ritz, S. A. "Air Pollution as a Potential Contributor to the 'Epidemic' of Autoimmune Disease." *Medical Hypotheses*, 74/10 (2010).

Sarah Boslaugh
Washington University in St. Louis

INDOOR AIR QUALITY

According to the World Health Organization (WHO), indoor air pollution associated with solid fuel use kills approximately 1.6 million people each year and represented 2.7 percent of the global burden of disease (in disability-adjusted life years, or DALYs). Hence, poor indoor air quality is the second-biggest environmental contributor to morbidity and mortality, behind unsafe water and poor sanitation. In several regions of the world, people spend most of their time indoors. The building design and ventilation in most of the cases influence the indoor air quality concentrations. Due to high population density and land scarcity, and as an effort to conserve energy, building designs are compact, with lower rates of ventilation in some places. While some buildings have natural ventilation, others have mechanical ventilation. Also, the outdoor air enters indoors by infiltration, natural ventilation, and forced ventilation. The natural air exchange through cracks and leaks, such as through door and window frames, chimneys, and exhaust vents, is called infiltration. The exchange of air resulting from opening and closing of windows is termed natural ventilation, while forced ventilation is the air exchange resulting from the use of whole-house fans or blowers.

Gases include the following:

- *Carbon dioxide*: Sources include combustion, tobacco smoke, metabolic activities, garage exhaust, and so forth. Concentrations of carbon dioxide above 15,000 parts per million volume (ppmv) can cause respiratory problems.
- *Carbon monoxide*: The primary sources include boilers, gas or kerosene heaters, gas stoves, wood stoves, fireplaces, tobacco smoke, garage exhaust, outdoor air, and so on. Exposure to carbon monoxide concentrations greater than 700 ppmv causes headache; concentrations greater than 700 ppmv for one hour cause death.
- *Nitrogen dioxide*: Sources include outdoor air, garage exhaust, kerosene and gas space heaters, wood stoves, gas stoves, tobacco smoke, and so on. Concentrations greater than 900 parts per billion volume (ppbv) impact respiratory function.
- *Ozone*: Sources include outdoor air, photocopy machines, electrostatic air cleaners, and so on. At concentrations greater than 0.15 ppmv, ozone causes headaches; concentrations greater than 0.25 ppmv cause chest pains; and concentrations greater than 0.30 ppmv lead to sore throat, cough, and breathing discomfort.
- *Sulfur dioxide*: Important sources include outdoor air, kerosene space heaters, gas stoves, coal appliances, and so on. Concentrations greater than 1.5 ppmv can lead to respiratory infections and impaired lung function.
- *Formaldehyde*: Sources include particle board, insulation, furnishings, paneling, plywood, carpets, tobacco smoke, and so on. Concentrations greater than 0.01 to 2.0 ppm can cause irritation of eyes; greater than 1 to 30 ppm causes respiratory problems; greater than 100 ppmv can cause coma and be fatal.
- *Volatile organic compounds (VOCs)*: Important sources include adhesives, solvents, building materials, combustion appliances, paints, varnishes, tobacco smoke, room deodorizers, cooking, carpets, furniture, draperies, and so on. Some common VOCs include propane, butane, pentane, hexane, n-decane, benzene, toluene, xylene, styrene, and so on. High concentrations can lead to health problems that range from sensory irritation to behavioral, neurotoxic, hepatoxic, and genotoxic effects.
- *Radon*: Diffusion from soil is an important source. A combination of radon and cigarette smoking increases cancer risk.

Aerosol particles include the following:

- *Allergens*: Sources include house dust, domestic animals, insects, pollens, and so on. Allergens cause asthmatic problems.
- *Asbestos*: Sources include fire retardant materials, insulation, and the like. Primary health effects include lung cancer, mesothelioma (cancer of mesothelial membrane lining lungs), and asbestosis (slow degeneration of lungs).
- *Fungal spores, bacteria, and viruses*: Sources include soil, plants, food, human transmission, and so on. Several respiratory diseases and skin allergies are associated with fungal spores, bacteria, and viruses.
- *Environmental tobacco smoke (ETS)*: This is a combination of inhaled and exhaled smoke from cigarettes. ETS consists of a mixture of aerosol particles and gases and is mostly carcinogens.
- *Polycyclic aromatic hydrocarbons (PAHs)*: Sources include fuel combustion, tobacco smoke, mothballs, blacktop, wood preservatives, and so on. Most PAHs are carcinogens.
- *Other*: Sources include combustion processes associated with cooking with wood stoves, fireplaces, outdoor air, and more. Pollutants from these sources cause respiratory ailments.

For evaluating and regulating indoor air quality, pollution standards and guidelines are in use in several countries. Most of them are based on WHO air quality guidelines. In the

United States, the Occupational Safety and Health Administration (OSHA) sets the indoor air pollution standards. Also, several countries follow their own nationally determined minimum guidelines for indoor air pollution control. A multidisciplinary approach involving engineers, material manufacturers, architects, air pollution experts, and other professionals is needed to tackle the indoor air pollution problem.

See Also: Ergonomics; Ozone; Particulate Matter; Radon and Basements; Smog; Smoking.

Further Readings

Godish, Thad. *Air Quality*, 4th ed. Boca Raton, FL: Lewis Publishers, 2003.

Jacobson, Mark Z. *Atmospheric Pollution. History, Science and Regulation.* Cambridge, UK: Cambridge University Press, 2002.

Masters, G. M. *Introduction to Environmental Engineering and Science*, 2nd ed. Englewood Cliffs, NJ: Prentice Hall, 1998.

Nagda, N. L., H. E. Rector, and M. D. Koontz. *Guidelines for Monitoring Indoor Air Quality*, Washington, DC: Hemisphere Publishing, 1987.

Singh, Jagjit. "Impact of Indoor Air Pollution on Health, Comfort and Productivity of the Occupants." *Aerobiologia*, 12 (1996).

Spengler, J. D. and K. Sexton. "Indoor Air Pollution: A Public Health Perspective." *Science*, 221 (1983).

World Health Organization. "Global Burden of Disease Due to Indoor Air Pollution." (2010). http://www.who.int/indoorair/health_impacts/burden_global/en (Accessed September 2010).

Krishna Prasad Vadrevu
The Ohio State University

INDUSTRIAL ECOLOGY

Industrial ecology begins with the examination of the interconnections both within and between industrial and ecological systems. Seeking to transform current industrial processes from linear to cyclical systems, industrial ecology is rooted in the notion that structuring industrial systems to behave as ecosystems do would create sustainable development—where the wastes and by-products created within one system are utilized as energy or raw materials for another system. Just as the death and decay of plant matter provides the nutrients to create rich, fertile soil, the waste and by-products from one industrial process can become the energy and raw materials to fuel another industrial process.

The basic principles of industrial ecology include the following:

- Modeling technological and industrial systems after ecological systems
- Reducing the negative impacts of technology on the environment
- Fostering sustainable technological and societal development
- Using a holistic approach to view environmental problems
- Identifying and tracing flows of energy and materials through various systems

An emergent, interdisciplinary field, industrial ecology draws from subjects as diverse as law, economics, business, public health, natural resources, ecology, and engineering. The roots of industrial ecology reach back to the 1950s and 1960s system dynamics work of Jay Forrester, one of the first researchers to define the world as a collection of interconnected systems. Though Forrester's system dynamics work was first used to address corporate management problems, by the 1970s this framework for viewing human interactions had spawned works that focused on environmental degradation and the lack of sustainability of the industrial system. Building on this early systems analysis work, Robert Ayres in 1989 put forth the concept of *industrial metabolism*—a framework for studying how energy and materials flow through various systems. That same year, a *Scientific American* article written by Robert Frosch and Nicholas Gallopoulos introduced the idea of industrial ecosystems, and soon the term *industrial ecology* entered the lexicon. And so, while Forrester's early work was born of efforts to analyze and address very specific social systems—management and employees—his pioneering of system dynamics lies at the heart of industrial ecology and its ideas regarding the relationship between industrial and natural ecosystems. This correlation between industrial and natural ecosystems is essential to industrial ecology.

The 1990s saw further development in the subject of industrial ecology, with the National Academy of Science holding its Colloquium on Industrial Ecology in 1991. This is considered by many to have marked a turning point for how industrial ecology was to be interpreted and embraced as a field of study, not just by academia, but by industry and government as well. *The Greening of Industrial Ecosystems*, edited by Braden Allenby and Deanna Richards and published by the National Academy of Engineering in 1994, further nurtured the growth of industrial ecology by exploring and defining its core concepts and applications. Beyond examining the implications for industry and the environment, the work also explained how industrial ecology crosses into law, economics, and public policy. By the late 1990s, the *Journal of Industrial Ecology* was in publication, and in 2000, the International Society for Industrial Ecology was formed. The Norwegian University of Science and Technology offered the first degree program for industrial ecology in the mid-1990s, and today, dozens of other colleges and universities offer degrees, graduate certificate programs, and coursework linked to industrial ecology.

Industrial Ecology at Work

An example of industrial ecology in action is Denmark's Kalundborg industrial park, where the participating businesses are integrated to function in a symbiotic manner and where waste and by-products are minimized by finding other uses for them. Here, the excess heat generated by industrial processes is utilized for fish farming, heating of nearby homes, and greenhouse agriculture. Further, as just one example of the way that this industrial park repurposes waste and by-products, the Asnaes Power Station produces gypsum as a by-product of the electricity generation process, and this gypsum then becomes a resource for Gyproc, which produces plasterboards. Thus, just as ecosystems are closed-loop systems that produce no waste, the Kalundborg industrial park has moved from linear-based production systems to circular systems in which waste and by-products become useful materials for other processes.

Some of what the Kalundborg industrial park has achieved includes the following:

- Significant reductions of the consumption of energy and resources such as coal, oil, and water

- Reduction of harmful emissions and reduced volumes of effluent water
- Conversion of traditional waste products into raw materials for production

The success of the Kalundborg industrial park is tied to Kalundborg's embracing of industrial ecology's central tenets, including Ayres's idea of industrial metabolism, which is the process of identifying and tracing flows of energy and materials through various systems. By analyzing material and energy flows and changes, negative impacts on the natural ecosystem can be minimized and resource efficiency can be maximized. Also important to Kalundborg's success, and to any entity practicing industrial ecology, is a focus on the entire life cycle of products, processes, and services. This life cycle assessment (LCA) is key to understanding the environmental aspects of, and potential impacts associated with, a product, process, or service. Predicated on an evaluation of the environmental burdens associated with a product, process, or activity, LCA is achieved by identifying the energy and materials used by a system as well as the resultant wastes released to the environment. In this way, LCA can be instrumental in the creation of strategies to affect not just environmental efficacy but also industrial efficiency.

Closely linked to industrial ecology is the concept of a circular economy, which seeks to balance economic development with environmental and resource protection. Most concretely embraced in China, where, in 2008, the Circular Economy Promotion Law was passed, a circular economy is achieved by interlinking manufacturing and service businesses. In this way, economic performance is enhanced and environmental impacts are minimized through collaborative efforts to manage environmental and resource issues. This level of industrial symbiosis exemplifies how concepts of industrial ecology can affect the domain of politics and economics.

Industrial ecology's impact on environmental and industrial policies is more readily observed elsewhere in the developed world than in the United States. For example, the European Union has proposed an Integrated Product Policy (IPP), that is intended to increase demand for green products while encouraging green design and manufacturing processes. In resource-limited Japan, extensive product recycling and energy efficiency have been embraced to counter decades of environmental degradation and resource exhaustion. Consistent with an eco-industrial approach to achieving sustainable development, Japanese leaders have initiated various types of eco-industrial projects throughout the country.

Driven by a desire both to foster sustainable development and to increase manufacturing efficiency, this international shift in the ways that resources are managed and industry is structured has deep implications for the United States. Today's global economy means that industrial ecology principles are being applied to some of the products, processes, and services created and used by international firms with whom the United States does business. Beyond the way these changes may impact regulatory obligations and trade issues, the United States, whose approach to environmental policy has rested on the evaluation of competing scientific claims and heated, often litigious debate, may approach industrial ecology policies in a different way than has Europe or Japan. Further, it remains to be seen how the principles of industrial ecology may be applied to information technology and to the technological infrastructure that undergirds much of the U.S. economy.

While the United States has not yet displayed the same readiness to accept and apply industrial ecology principles as have some other nations, there do exist examples of industrial ecology in action within U.S. borders. Here, the Environmental Protection Agency (EPA) has been researching and exploring ways to apply industrial ecology concepts, the National Pollution Prevention Center for Higher Education now uses systems analysis to develop pollution prevention educational materials, and the National Science Foundation supports research related to industrial ecology. California's Department of Toxic Substance Control has spearheaded a Green Chemistry initiative, and in the corporate world, AT&T funds an Industrial Ecology Faculty Fellowship program, and Natural Logic, a U.S.-based consulting firm that has worked with dozens of clients including Levi Strauss, Sara Lee, and Rhino Records, provides guidance to firms that are seeking to apply ecological theory to industrial methods.

What More Is Needed?

Still, in order for industrial ecology to truly impact environmental, technological, and economic policy, a rigorous scientific foundation will be essential. Theories, models, research, and experiments are needed to test and advance industrial ecology's assumptions. In order to further develop industrial ecology in the United States and elsewhere, continued research is called for in order to provide a robust framework for understanding how technology and the environment interact with each other. By focusing on environmental, industrial, economic, and human health, industrial ecology is poised to usher in a new age of sustainability. In order to bring this to fruition, industrial ecology needs to further develop as a field, and this can be achieved by strengthening its framework in these ways: clarifying industrial ecology's definition, scope, and goals; examining more intently the relationship between industrial ecosystems and natural ecosystems; fleshing out the definition of "sustainable development"; fostering deeper interdisciplinary collaboration; expanding industrial ecology–focused curriculum development in colleges and universities; refining tools such as life cycle assessment; and implementing government policies that encourage industry to take seriously its role in stewardship of resources and the environment.

See Also: Education and Green Health; Environmental Protection Agency (U.S.); Government Role in Green Health; Green Chemistry; International Policies; Private Industry Role in Green Health.

Further Readings

Allenby, Braden R. and Deanna J. Richards. *The Greening of Industrial Ecosystems*. Washington, DC: National Academy Press, 1994.

Garner, Andy and Gregory Keoleian. "Industrial Ecology: An Introduction." Ann Arbor, MI: National Pollution Prevention Center for Higher Education, November 1995.

Graedel, T. E. and Braden R. Allenby. *Industrial Ecology and Sustainable Engineering*. Upper Saddle River, NJ: Prentice Hall, 2009.

Thomas, Valerie, Thomas Theis, Reid Lifset, Domenico Grasso, Byung Kim, Catherine Koshland, and Robert Pfahl. "Industrial Ecology: Policy Potential and Research Needs." *Environmental Engineering Science*, 20/1 (2003).

Elizabeth Nanas
Wayne State University,
Hong Kong University of Science & Technology

Tani Bellestri
Independent Scholar

INFECTIOUS WASTE

Infectious waste, also known as medical or clinical waste, is produced by healthcare premises such as hospitals, clinics, medical offices, clinical and research laboratories, and nursing homes. Best management practices for infectious waste have been defined by the World Health Organization (WHO) to be adapted according to national and local situations. Many treatment systems are suitable for infectious waste, except encapsulation (which can be used for sharp objects) and inertization to prevent potentially explosive reactions or flammability. Highly infectious waste containing deadly human pathogens should be autoclaved first. As background, healthcare activities produce infectious wastes that lead to adverse health effects and are a risk for patients, healthcare workers, waste handlers, and communities. Used syringes and needles contain residual blood and blood-borne pathogens and may transmit diseases when reused without reprocessing or through accidental needlesticks. Little attention is paid to ensure the budgeting and financing for management of infectious waste.

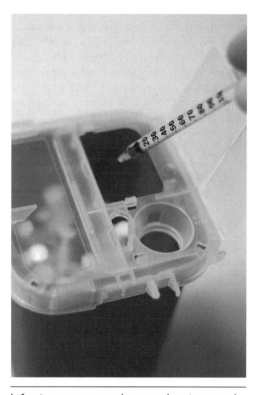

Infectious wastes such as used syringes and needles lead to adverse health effects and are a risk for patients, healthcare workers, waste handlers, and communities.

Source: iStockphoto.com

WHO aims at developing clear guidance, policies, and recommendations for the safety of healthcare workers and the public. The guiding principles of the policy paper on safe healthcare waste management are to prevent the health risk associated with exposure to healthcare workers by promoting sound management policies and to reduce the exposure to toxic pollutants associated with the combustion process through the promotion of appropriate practices. It also acknowledges that until countries in transition and developing countries have access to healthcare waste management options that are safer to the environment and health, incineration may be an acceptable response when used appropriately. Key elements of appropriate operation of incinerators include effective waste reduction and waste segregation, placing incinerators away from populated areas; satisfactorily engineered design; construction following appropriate dimensional plans; proper operation; periodic maintenance; and staff training and management.

To support implementation, a WHO healthcare waste website (http://www.health carewaste.org or http://www.who.int/water_sanitation_health) offers technical options, costing tools, country information, contacts, and 142 reference documents. This includes a practical pocket-size document on managing waste from injection activities, along with two related posters. The posters summarize the strategies proposed for the management of waste from injection activities; a portion of the poster demonstrates on-site and off-site treatments, and another portion shows the use of needle removers, a device not yet approved by WHO because of its potential risk of needlestick injuries for injection providers. Nevertheless, it has added value for preventing needlestick injuries among waste handlers and communities. As of 2010, a study was being conducted in Bangladesh on routine activities, which should soon determine the value of the practices. Meanwhile, WHO has developed prequalification specifications on injection-related equipment, including a needle remover, aimed at providing recommendations for the purchase of equipment, as this equipment has already been introduced by the Ministry of Health in a number of countries (e.g., India). Additionally, recycling or reprocessing plastic syringes into other utensils represents a return value (e.g., in India, where it has been put into practice and is working well).

Partnership is also a key function. In June 2007, at WHO headquarters in Geneva, a three-day meeting on healthcare workers supported by the Gates Foundation brought together 50 participants from 30 nationalities, with country representatives, the United Nations, nongovernmental organizations (NGOs), industry representatives, and WHO country and regional members. An informal healthcare workers alliance was established; it was an opportunity to provide the group with an update on projects and activities and to discuss the content of the new version of the WHO reference document on the safe management of wastes from healthcare activities and the new costing tool, and to develop WHO core principles for achieving safe and sustainable management of healthcare waste.

WHO core principles recognize that safe and sustainable management of healthcare waste is a public health imperative and a responsibility of all. Improper management of healthcare waste poses a significant risk to patients, healthcare workers, the community, and the environment. An investment of resources and commitment will result in a substantive reduction of disease burden and corresponding savings in health expenditures. WHO core principles require that everyone associated with financing and supporting healthcare activities should contribute to the cost of managing infectious waste. Manufacturers also share a responsibility to take waste management into account in the development and sale of their products and services. The establishment and maintenance of effective systems for healthcare waste management depends on the availability of resources.

Activities Taking Place in Various Countries

The objective of the healthcare waste component of the Global Alliance for Vaccine and Immunization (GAVI) was that by the end of 2007, 60 percent of countries receiving GAVI support (36 countries, half of them sub-Saharan African countries) would have adopted national policy and developed plans on infectious waste management. The targeted 2008 objective was 90 percent. Activities supporting the implementation of policies to demonstrate success were proposed for the 2008 GAVI work plan. Meanwhile, under WHO leadership in 2007, the Performance Assessment Tool for Quality Improvement in Hospitals (PATH) was developed to measure the status of injection waste disposal.

The Global Environmental Facility (GEF) Project on Health Care Waste Management aims to demonstrate and promote best techniques and practices for reducing healthcare waste to avoid environmental release of dioxins and mercury. The project involves the United Nations Development Programme (UNDP) as the implementing agency, the United Nations Office for Project Services (UNOPS) as the executing agency, and WHO and Health Care Without Harm (HCWH) as the principal cooperating agencies. The participating countries are Argentina, India, Latvia, Lebanon, the Philippines, Senegal, and Vietnam. Project implementation had not started as of 2010.

Three expanded costing assessment tools have been developed and differentiate between low-, middle-, and high-income countries. They deal with various-size categories of healthcare facilities; allow several treatment options, including centralized or decentralized treatment; and compute potential revenues from the sale of sterilized plastic parts for re-melting. They provide sharps costs per syringe, key indicative value of sharps waste generation rates in kg/bed per day, and more.

In conclusion, there are ongoing and better-structured dynamics taking place that demonstrate that—despite the enormous challenge that safe waste management represents—this is not a hopeless battle to ensure safety.

See Also: Acquired Immune Deficiency Syndrome; Airborne Diseases; Centers for Disease Control and Prevention (U.S.); Seasonal Flu; Tuberculosis.

Further Readings

Cook, Gordon and Alimuddin Zumla. *Manson's Tropical Diseases*. London: Saunders, 2003.
Da Silva, Paulo Sergio Lucas, et al. "The Product of Platelet and Neutrophil Counts (PN Product) at Presentation as a Predictor of Outcome in Children With Meningococcal Disease." *Annals of Tropical Paediatrics*, 27/1 (2007).
Fine, Paul E. "Herd Immunity: History, Theory, Practice." *Epidemiologic Reviews*, 15 (2003).
Folch, Erick, et al. "Infectious Diseases, Non-Zero-Sum Thinking, and the Developing World." *American Journal of Medical Sciences*, 326/2 (2003).
Franco-Paredes, Carlos, et al. "Cardiac Manifestations of Parasitic Infections Part 3: Pericardial and Miscellaneous Cardiopulmonary Manifestations." *Clinical Cardiology*, 30/6 (2007).
Franco-Paredes, Carlos, Ildefonso Tellez, and Carlos del Rio. "Inverse Relationship Between Decreased Infectious Diseases and Increased Inflammatory Disorder Occurrence: The Price to Pay." *Archives of Medical Research*, 35 (2004).
John, T. Jacob and Reuben Samuel. "Herd Immunity and Herd Effect: New Insights and Definitions." *European Journal of Epidemiology*, 16 (2000).
Nuutila, Jari and Esa-Matti Lilius. "Distinction Between Bacterial and Viral Infections." *Current Opinion in Infectious Diseases*, 20 (2007).
Sosa, Anibal. "Resistencia a Antibióticos en América Latina." Boston: Association for the Prudent Use of Antibiotics (APUA), 1998.
World Health Organization (WHO). 51st World Health Assembly. "Emerging and Other Communicable Diseases, Antimicrobial Resistance." Resolution WHA51.17. Geneva, Switzerland: WHO, 1998.
World Health Organization. "Guidelines on Prevention and Control of Hospital Associated Infections." Geneva, Switzerland: WHO, 2009.
World Health Organization. "Health Care Waste Management." Geneva, Switzerland: WHO, 2007.
World Health Organization. "Practical Guidelines for Infection Control in Health Care Facilities." Geneva, Switzerland: WHO, 2000.

Alfonso J. Rodriguez-Morales
Universidad de Los Andes, Universidad Central de Venezuela

INJURIES

In the last few decades, injuries both intentional and unintentional have come to be treated not as random events but as preventable ones, as the World Health Organization (WHO) and other international bodies have become more concerned with the impact of injuries on mortality, disability, and disease. Though the global burden of injuries is not easily estimated, it is certainly significant. For instance, burns, drowning, and falls are among the leading causes of mortality and morbidity among children under 15 and are usually preventable. In industrialized countries like the United States, preventable traffic accidents are the leading cause of death for people aged 15 to 29. In the low- and middle-income countries of the western Pacific, traffic accidents and interpersonal violence (homicide) are the leading causes of death; in the low- and middle-income countries of Europe, suicide and poisoning are. Although mortality is an important indicator of the severity of a problem, morbidity—the need for hospitalization, treatment by emergency personnel, and/or permanent disability—must also be taken into consideration as a significant source of the global health burden and a weight on worldwide health resources.

Among the most significant causes of the burden of disease identified by WHO, seven are preventable injuries: falls, drowning, traffic injuries, fires, war, interpersonal violence, and self-inflicted injury (including but not limited to suicide). The same list, with the addition of poisoning, is found in the leading causes of death. Though suicide among females and homicide among males have decreased as causes of death since 1998, traffic injuries have reached proportions in low- and middle-income countries that international agencies like WHO characterize as epidemic. In poorer countries, even middle-income countries, the economic impact of morbidity is significant: when preventable injuries tax health resources, healthcare becomes more expensive and fewer resources are left available for improving general quality of life. The World Bank and WHO have worked together to try to quantify this impact by creating a measure called "dollars per Disability Adjusted Life Year (DALY)"—calculated as the present value of projected future years of disability-free life that are lost as a result of a disability. The algorithm for determining such a value is sophisticated, calculated according to 107 different causes of death and 483 possible disabling pathological conditions. Studies show that the least impact of this burden is felt in Europe and North America, which combined lose fewer DALYs per 1,000 population than the sub-Saharan African countries do, where the burden is greatest.

Injurious accidents and traffic accidents are among the leading contributors to the global burden of disease; many of the other leading contributors can result from environmental factors, especially anthropogenic factors such as pollution and toxic spills: respiratory infections, chronic respiratory disease, cancer, neuropsychiatric problems, and musculoskeletal conditions.

Irritant gas inhalation injuries are caused by the inhalation of gases that dissolve in the respiratory tract mucus and cause an inflammatory response, typically from sudden acidification or alkalinization. Such injuries can affect any part of the airways, from the mouth to the lungs. Chlorine, sulfur dioxide, phosgene, hydrogen chloride, nitrogen dioxide, ozone, and ammonia are some of the more common irritant gases, and chloramine is a commonly known danger, occurring when ammonia and bleach come into contact in the household. Other gases may also be highly toxic, such as carbon monoxide, or may be harmless in most contexts but cause suffocation when they displace oxygen. The extent of damage caused by a gas is affected by its solubility. Water-soluble gases (such as the aforementioned chloramine, as well as chlorine and ammonia individually) will dissolve in the upper airway, which produces symptoms that are quickly felt, such as mucous membrane irritation that results immediately in a painful cough—provided escape from the area is possible, people exposed to such gases usually do not suffer serious injury because they have an early enough warning impelling them to leave the area and avoid further inhalation. Less soluble gases, though, will not cause upper airway irritations, and if not severely toxic can be inhaled for a long time without noticeable symptoms developing; such symptoms, in fact, will often not develop until hours after the victim has left the area. Workers exposed to irritant gases in the workplace on a Friday may not develop any symptoms until sometime over the weekend. These gases have dissolved in the lower airways, contributing to severe bronchitis, pulmonary edema, pneumonia, and other respiratory problems. Even if the gases are not significantly toxic, prolonged exposure can lead to persistent lung injuries that never fully heal, especially if the exposure is chronic or the victim is a heavy smoker. Chronic exposure to some substances also carries a high risk of cancer.

Risks in the Workplace

Workplace exposures to toxic substances are a serious problem in the United States. Estimates are difficult to substantiate. As early as 1976, a Congressional report on chemical dangers in the workplace estimated almost 400,000 new cases of occupational disease each year and 100,000 deaths—and noted that the estimate was likely to be low because of the difficulty in establishing an occupational cause in many other cases. Furthermore, about two-thirds of cancer cases were believed to be environmentally caused. A later 2001 report found about 133,000 occupational diseases each year in California alone. Many personal injury lawsuits related to such diseases are called toxic torts, in which the plaintiff's argument is that involuntary or unknowing exposure to a chemical resulted in his injury or disease. While workers compensation claims are made against the plaintiff's employer, a toxic tort is usually brought against a third party, such as the owner of the premises where the exposure took place, or the manufacturer or vendor of a product responsible for the exposure. Toxic torts also include cases where the exposure results from consumer products or pharmaceuticals.

Most plaintiffs in toxic tort cases are industrial workers, who are regularly exposed to much higher levels of chemicals than the general population and who work in an environment where safety precautions may not be sufficient or the full extent of the toxicity of a substance may not be known in time to prevent exposure. There continue to be a number of toxic tort cases concerning asbestos, for instance, which was once considered harmless and installed as insulation in many buildings but is now known to be highly dangerous. Workers at factory poultry farms have also found themselves suffering from severe illnesses as a result of the combinations of antibiotics, fertilizers, waste, and other substances they are exposed to in far, far greater concentrations than had ever been found at traditional chicken farms.

Other common toxic tort cases concern benzene (a common industrial chemical, especially in factories working with petroleum products or rubber, from shoe factories to oil refineries; also responsible for soil and water contamination, and in several incidents discovered at unsafe levels in canned and bottled beverages); beryllium (originally used in fluorescent lighting tubes and still in frequent use in nuclear, aerospace, and electronics industries; not only a carcinogen, its inhalation can lead to acute beryllium disease and chronic beryllium disease); polychlorinated biphenyl (PCBs); and manganese (used in many metal alloys and an important nutrient in trace amounts, but toxic in greater amounts, leading to respiratory and organ problems, impaired motor skills, and neurological disorders; a special concern for miners, smelters, and welders, who have been vulnerable to a Parkinson's-like disease called manganism since the early 1800s).

Mercury poisoning is also a significant concern, with a number of possible causes. A heavy metal, mercury is a highly reactive toxin about which little is yet known concerning the mechanism of its toxicity. Once used medicinally to treat headaches and other ailments, it is now known to have severe impact on the central nervous and endocrine systems, to be fatal in large-enough doses or through repeated exposures, and to be responsible for a number of chronic ailments, including neurological syndromes and birth defects. Mercury poisoning can often be caused by problems with amalgam dental fillings, but industrial workplace exposures are also common, and there are perpetual concerns with mercury levels in drinking water and in seafood. Seafood-heavy diets run a real risk of mercury poisoning, but firm guidelines have not been established for how frequently certain seafoods can be eaten without risking crossing a threshold.

There are a number of health conditions that may or may not prove to be influenced by environmental factors. The mechanism of toxicity is so poorly understood in many areas, and there are numerous health conditions that are only vaguely understood. For instance, there has been a significant increase in autism diagnoses in recent years, raising the question of to what extent this is the result of better diagnostics and to what extent there may be some environmental cause for an increase of autism in the human population. Chronic fatigue syndrome, also on the rise, raises the same question—particularly since it seems less likely than autism to have gone unnoticed and undiagnosed in the past. The role of pollution and toxins in the rise of cancer diagnoses is, of course, widely recognized.

See Also: Asthma; Cancers; Fungi and Sick Building Syndrome; Infectious Waste.

Further Readings

Grossman, Elizabeth. *Chasing Molecules: Poisonous Products, Human Health, and the Promise of Green Chemistry.* Washington, DC: Shearwater Press, 2009.
Layzer, Judith A. *The Environmental Case: Translating Values Into Policy.* Washington, DC: CQ Press, 2005.
Levy, Barry S., David H. Wegman, Sherry L. Baron, and Rosemary K. Sokas. *Occupational and Environmental Health: Recognizing and Preventing Disease and Injury.* Boston: Lippincott Williams & Wilkins, 2005.
Malerba, Larry. *Green Medicine.* Berkeley, CA: North Atlantic Press, 2010.

Bill Kte'pi
Independent Scholar

INTERNATIONAL POLICIES

International policies regarding environmental and public health concerns represent the recognition by the global economy for necessary solutions to the many societal problems that cannot be effectively handled by individual nation-states through domestic laws and regulations.

International treaties such as the Basel Convention on the Control of Transboundary Movements of Hazardous Wastes and their Disposal (the Basel Convention) are designed to advance, ameliorate, and augment the ways in which issues facing multiple nations are conducted. These treaties seek to reduce the risk gaps associated with economic differences between developed nations and less developed countries (LDCs).

International policies regarding environmental and health concerns demonstrate that the global economy has made many issues a common concern that cannot be effectively handled by individual nation-states through domestic laws and regulation. International treaties such as the Basel Convention on the Control of Transboundary Movements of Hazardous Wastes and their Disposal (the Basel Convention) are designed to advance, ameliorate, and augment the ways in which issues facing multiple nations are conducted. International policies such as the Basel Convention seek to decrease the movement of hazardous waste between nations. The Basel Convention, for example, specifically seeks to reduce the transfer of hazardous waste from developed nations to less developed countries (LDCs).

International policies are designed to minimize the quantity and toxicity of hazardous wastes produced, to guarantee environmentally friendly supervision of such wastes' disposal, and to ensure that the supervision of waste takes place as close to the source of its generation as possible. International policies also enlist developed nations to assist LDCs in the environmentally friendly supervision of the hazardous wastes the LDCs generate. Decisions related to the development of international policies consider economic, social, and environmental concerns to effect solutions that are equitable, ecologically aware, and efficient.

During the 1970s, legislative bodies of many developed nations passed environmental laws that regulated the disposal of hazardous waste, an example being the U.S. Resource Conservation and Recovery Act (RCRA) of 1976. Environmental regulations at that time sought to protect humans and the environment from hazardous waste, reduce the amount of hazardous waste produced, and ensure that such wastes are disposed of in environmentally friendly ways. As a result of these regulations, disposal costs for hazardous wastes rose dramatically, causing many corporations to seek a less expensive means of dealing with toxic trash. By the early 1980s, developed nations had begun transporting hazardous wastes to LDCs for final disposal. The transportation of hazardous wastes was motivated by the attractive cost differential between disposal of toxic materials in rich nations, where it was very expensive, and in poor nations, where it was not. Increasingly strong communications and transport networks made the movement of hazardous waste to LDCs easy and economical. As nongovernmental organizations (NGOs) and the media began to examine this phenomenon, however, many leaders of developed nations branded the hazardous waste trade immoral and sought international policies to prevent such actions. The Basel Convention was one of the first coordinated international responses to this issue.

The Basel Convention

Convened in 1989, the Basel Convention marked a convergence of industrialized and developing nations. Agreement was swiftly reached that hazardous waste should not be subject to free trade. No consensus existed, however, with regard to how best to proceed. Some parties advocated for an outright ban on the trade of hazardous waste. Others encouraged a system that regulated and controlled such trade, a position that ultimately prevailed. The Basel Convention defined *hazardous waste* as those waste products that are both delineated (e.g., hospital rubbish, certain chemicals) and explosive, flammable, toxic, or corrosive. Stringent conditions regarding the import and export of hazardous wastes are imposed by the treaty, and rigorous requirements govern the notice, consent, and tracking of movement of waste in transit. The Basel Convention governs actions of its 172 parties, although three of these—Afghanistan, Haiti, and the United States—had not ratified the treaty as of August 2009.

Certain limitations of the Basel Convention, such as its refusal to ban the transport of hazardous waste from developed nations to LDCs, have caused other international policies to be designed to deal with these perceived limitations. Examples of such international policies include the Bamako Convention on the Ban on the Import into Africa and the Control of Transboundary Movement and Management of Hazardous Wastes within Africa (the Bamako Convention), which was convened in 1991 and came into force in 1998; the Rotterdam Convention on the Prior Informed Consent Procedure for Certain Hazardous Chemicals and Pesticides in International Trade (the Rotterdam Convention), which was convened in 1998 and came into force in 2004; and the Stockholm Convention on Persistent Organic Pollutants (the Stockholm Convention), which was convened in 2001 and came into force in 2004. The Bamako Convention, for example, is a treaty joined by 23 African nations that prohibits importing any hazardous waste. The Rotterdam Convention is a multilateral treaty that promotes shared responsibility with regard to hazardous chemicals and encourages open exchange of information related to the import of hazardous chemicals. Calling on exporters of hazardous chemicals to use proper labeling, the Rotterdam Convention also demands that exporters include directions on safe handling of hazardous waste and notify purchasers of any known restrictions or bans related to those chemicals. Finally, the Stockholm Convention promulgated an international treaty designed to eliminate or restrict the use of persistent organic pollutants (POPs), which are organic compounds resistant to environmental degradation through biological, chemical, and photolytic processes. Although most POPs were used as pesticides, some are generated as the result of industrial processes or as by-products from the production of solvents or pharmaceuticals. The Stockholm Convention prepared an assessment that identified 12 especially toxic POPs that became known as the "dirty dozen." Signers of the Stockholm Convention ultimately agreed to ban nine POPs that comprised the dirty dozen, to curtail the use of dichlorodiphenyltrichloroethane (DDT) to control malaria, and to limit unnecessary manufacture of dioxins and furans.

International policies related to hazardous waste are transient in their nature. As technology and social mores change, so do the international policies related to hazardous waste. While an agreement may be reached, continual advocacy for certain positions results in continual change and adjustment with regard to the handling of hazardous waste. Many signatories to the Basel Convention, for example, were dissatisfied that it banned hazardous waste exports only to Antarctica, instead requiring only a notification and consent system. Additional concerns arose when many exporters began labeling hazardous waste exports "recycling" in an attempt to circumvent the system. In response to this, certain signatories of the Basel Convention initiated several regional waste bans and the Basel Ban Amendment. The Basel Ban Amendment was especially concerned with the sale of ships for salvage, also known as "ship breaking." Although ship breaking is necessary, and had taken place in European and North American shipyards for centuries, in the late 20th century this work began to move to LDCs in an attempt to avoid environmental regulation. While valuable scrap metal is recovered during ship breaking, many hazardous substances, such as lead, asbestos, and polychlorinated biphenyls (PCBs), also are released, making the endeavor dangerous to workers. Although not yet in force, the Basel Ban Amendment is honored by its signatories and prohibits the export of hazardous wastes from developed nations to LDCs.

International Policies and Health

International policies and coordination of efforts have also proved helpful when dealing with pandemic diseases or general health conditions relating to large population groups. International policies have been generated for many health threats, including influenza, human immunodeficiency virus (HIV), and acquired immune deficiency syndrome (AIDS). International policies with regard to diseases allow global organizations to assist nations with limited resources. For example, the World Health Organization (WHO) has developed a disease staging system for HIV. Disease staging systems represent an approach for use in resource-limited settings such as LDCs and are widely used in Africa and Asia. Disease staging systems have proven to be a useful research tool in studies of progression to symptomatic HIV because they allow resources to be directed where they can provide the most good. For example, while many individuals may have HIV, if they are in an asymptomatic stage, their ailments can be treated as they would be in healthy people, saving scarce medications or other more advanced treatments for those more seriously ill. Coordinated international policies also allow wealthier nations to coordinate assistance to LDCs in a systematic and efficient manner. This cooperation has led to international policies related to such endeavors as needle-exchange and condom-distribution programs as well as low-cost pharmaceutical initiatives that benefit LDCs in controlling certain health threats.

Technological advances have made it possible for developed nations to shift their hazardous wastes to LDCs. International policies have worked to lessen the burden on LDCs and to reduce the potential health hazards to their citizens through the coordination of equitable environmental rules that protect and regulate health concerns.

See Also: Acquired Immune Deficiency Syndrome; Chemical Pesticides; Government Role in Green Health; Health Disparities; United Nations Environment Programme.

Further Readings

Beslier, S. "The Protection and Sustainable Exploitation of Genetic Resources of the High Seas From the European Union's Perspective." *International Journal of Marine & Coastal Law*, 24/2 (2009).

Clapp, J. *Toxic Exports: The Transfer of Hazardous Wastes From Rich to Poor Countries.* Ithaca, NY: Cornell University Press, 2001.

Reed, M. S. "Stakeholder Participation for Environmental Management: A Literature Review." *Biological Conservation*, 141/10 (2009).

Shibata, A. "The Basel Compliance Mechanism." *Review of European Community & International Law*, 12/2 (2003).

Stephen T. Schroth
Jason A. Helfer
Luke L. Karner
Knox College

INTERNATIONAL TRAVEL

The number of people traveling internationally is increasing every year. This increase in traveling is due both to population migration and tourism, including visits to some ecologically inspiring areas (ecotourism). Ecotourism can present a risk for environmental deterioration of sensitive ecosystems as well a risk for diseases acquired by travelers in such areas. This is an important aspect for the global public health that should also be considered in the new perspectives of a more ecological green world.

The movement of individuals, populations, and products across international borders is one of the major factors associated with the emergence and reemergence of infectious diseases. Travelers act as vectors by transporting infectious agents to new areas.

Source: iStockphoto.com

According to data compiled by the World Tourism Organization (WTO), international tourist arrivals in the year 2007 reached 903 million. International tourism receipts rose to $856 billion (625 billion euros) in 2007. International arrivals were expected to reach 1 billion by 2010, and 1.6 billion by 2020. In 2007, just over half of all international tourist arrivals were motivated by leisure, recreation, and holidays (51 percent)—a total of 458 million. Business travel accounted for some 15 percent (138 million), and 27 percent represented travel for other purposes, such as visiting friends and relatives (VFR), religious reasons/pilgrimages, health treatment, and so on (240 million). Slightly less than half of arrivals traveled by air transport (47 percent) in 2007, while the remainder arrived in their destinations by surface transport (53 percent)—whether by road (42 percent), rail (4 percent), or over water (7 percent).

International travel can pose various risks to health, depending on the characteristics of both the traveler and the travel. Travelers may encounter sudden and significant changes in altitude, humidity, microbes, and temperature, which can result in ill health. In addition, serious health risks may arise in areas where accommodations are of poor quality, hygiene and sanitation are inadequate, medical services are not well developed, and clean water is unavailable. Besides this, a significant number of endemic diseases in travel destinations are also of concern for international travel. The World Heath Organization (WHO)—aware of the importance of international health and, specifically, of communicable diseases—provides the International Health Regulations (IHR), which is an international legal instrument. These regulations aim to help the international community prevent and respond to acute public health risks that have the potential to cross borders and threaten people worldwide. Also, WHO provides an important report of information on health risks for travelers.

In this context, travel medicine emerges as an important medical discipline. Travel medicine has emerged in all regions of the world as an important and growing scientific discipline. Travel medicine, aimed at the prevention and treatment of medical complications in travelers and other geographically displaced individuals, embraces specialized areas of tropical medicine, occupational medicine, pharmacology, epidemiology, public health, preventive medicine, and infectious diseases. The primary goal of travel medicine is to keep travelers alive and healthy. This principle applies to tourist travel, students living abroad, corporate travelers, refugees and asylum seekers, and immigrants. One principle for travel medicine should be the same as it is for all medicine: *primun non nocere* (first do no harm). Safety and prevention must be the rule.

Key factors in determining the risks to which travelers may be exposed are mode of transport, destination, duration and season of travel, purpose of travel, standards of accommodations and food hygiene, behavior of the traveler, and underlying health of the traveler.

International movement of individuals, populations, and products is one of the major factors associated with the emergence and reemergence of infectious diseases as the pace of global travel and commerce rapidly increases. Travel can be associated with disease emergence because (1) the disease arises in an area of heavy tourism, (2) tourists may be at heightened risk because of their activities, or (3) because tourists can act as vectors to transport the agent to new areas.

The likelihood of a medical illness developing relates to destination, transportation, duration of travel, level of accommodations, underlying medical condition, immunization history, adherence to indicated chemoprophylactic regimens, and exposure to potential infectious agents during travel. Eliciting a detailed history of the specific locales visited, the timing of travel relative to the onset of symptoms, and specific risk behavior is essential in identifying potential exposure to infectious pathogens and determining the likely incubation period.

Particular groups of travelers are considered at higher risk of developing illness after returning to their place of residence. Adventure travelers and those persons visiting friends and relatives overseas are at greater risk for becoming ill, due to increased exposure to pathogens. Those visiting friends and relatives are also less likely to seek pretravel advice or to take antimalarial prophylaxis or pretravel vaccinations.

Advising international travelers on vaccine-preventable diseases and malaria prevention strategies are the major services provided by travel medicine practitioners. However, they also deal with the post-travel care of patients who become ill during, immediately after, or within defined periods after their return.

Within the topic of international travel and health, migration also should be considered. Today, international immigrants going from developing countries to developed regions carry a significant burden of health issues, particularly those related to endemic diseases from developing countries that can be carried by a migrant population (e.g., schistosomiasis, Chagas disease). Economic hardship and political conflicts constitute main drivers of migration of populations from disease-endemic areas or countries to nonendemic areas of developing countries and to developed countries in North America, Europe, Japan, and Australia. Therefore, infectious diseases threaten to expand rapidly and exponentially, from rural to urban areas and from endemic to nonendemic regions, from poor to rich countries, from the tropics to the rest of the world.

Another important consideration in the setting of international travel and health is the migration of people as refugees. Current estimates suggest that there are around 13 million refugees worldwide. As a reflection of the global trend toward increasing migration of geographically displaced populations, immigration to the United States has progressively increased in recent decades. Notably, among the total number of United States–bound refugees, for example, there has been an increase in the percentage of African refugees, from 8 percent in 1998 to 35 percent in 2005. The so-called Lost Boys of Sudan, named after Peter Pan's band of orphans, were raised in precarious, resource-poor refugee camps in Ethiopia (1987–1991) and Kakuma, Kenya (1990–present). In 2001, the U.S. State Department resettled approximately 3,800 of the Lost Boys in the United States.

Finally, education of travelers about risks and modes of transmission of infectious diseases should be reinforced. Additionally, specific training in travel medicine should be included in the medical studies at national medical schools. In the future, it will also be important to obtain information from airports and regional travel agencies because these are a relevant source of knowledge about significant health risks during travels.

See Also: Acquired Immune Deficiency Syndrome; Airborne Diseases; Antibiotics; Bottled Water; Centers for Disease Control and Prevention (U.S.); Malaria; Seasonal Flu; Severe Acute Respiratory Syndrome; Tuberculosis; Vaccination/Herd Immunity.

Further Readings

Franco-Paredes, Carlos, Roberta Dismukes, Deborah Nicolls, Alicia Hidron, Kimberly Workowski, Alfonso Rodriguez-Morales, Marianna Wilson, Danielle Jones, Peter Manyang, and Phyllis Kozarsky. "Persistent and Untreated Tropical Infectious Diseases Among Sudanese Refugees in the United States." *American Journal of Tropical Medicine and Hygiene*, 77 (2007).

Freedman, David O., Leisa H. Weld, and Phyllis E. Kozarsky. "Spectrum of Disease and Relation to Place of Exposure Among Ill Returned Travelers." *New England Journal of Medicine*, 354 (2006).

Guerrero-Lillo, Lisette, Jorge Medrano-Díaz, Carmen Pérez, Rodrigo Chacón, Juan Silva-Urra, and Alfonso J. Rodriguez-Morales. "Knowledge, Attitudes, and Practices Evaluation About Travel Medicine in International Travelers and Medical Students in Chile." *Journal of Travel Medicine*, 16 (2009).

Lam, M. S. and S. Y. Wong. "Travel Medicine: A Perspective on the Emerging Problem of Travel-related Infections." *Annals of the Academy of Medicine of Singapore*, 26 (1997).

Leggat, Peter A. "Risk Assessment in Travel Medicine." *Travel Medicine and Infectious Diseases*, 473 (2006).

Ostroff, S. M. and Phyllis Kozarsky. "Emerging Infectious Diseases and Travel Medicine." *Infectious Diseases Clinics of North America*, 12 (1998).

Risquez, Alejandro. *Traveler Health Guide, Venezuela 2010*. Caracas, Venezuela: Editorial Arte, S.A, 2009.

Risquez, Alejandro. "World Health and the International Health Regulations." *Internal Medicine (Caracas)*, 25 (2009).

Rodriguez-Morales, Alfonso J., Jesus Benitez, Ildefonso Tellez, and Carlos Franco-Paredes. "Chagas Disease Screening Among Latin American Immigrants in Non-Endemic Settings." *Travel Medicine and Infectious Diseases*, 6 (2008).

Rodriguez-Morales, Alfonso J., Laura Delgado, Nestor Martinez, and Carlos Franco-Paredes. "Impact of Imported Malaria on the Burden of Disease in Northeastern Venezuela." *Journal of Travel Medicine*, 13 (2006).

Rodriguez-Morales, Alfonso J. and Carlos Franco-Paredes. "Travel Medicine." In *Encyclopedia of Global Health*, Y. Zhang, ed. Thousand Oaks, CA: Sage, 2008.

Rodriguez-Morales, Alfonso J. and Hilda Palacios. "Trends in the Publication of Scientific Research in Travel Medicine From Latin America." *Travel Medicine and Infectious Diseases*, 7 (2009).

Rodriguez-Morales, Alfonso J., Julio Silvestre, and Dalmiro Cazorla-Perfetti. "Chagas Disease in Barcelona, Spain." *Acta Tropica*, 111 (2009).

Rossi, M. and H. Furrer. "Reisemedizinische Aspekte bei HIV-infizierten Patienten" [Travel Medicine for HIV/Infected Patients]. *Therapeutische Umschau*, 58 (2002).

Ryan, Edward T., Mary E. Wilson, and Kevin C. Kain. "Illness After International Travel." *New England Journal of Medicine*, 347 (2002).

Venkatesh, S. and Z. A. Memish. "SARS: The New Challenge to International Health and Travel Medicine." *Eastern Mediterranean Health Journal*, 10 (2004).

World Health Organization (WHO). *International Travel and Health*. Geneva, Switzerland: WHO, 2009.

Alfonso J. Rodriguez-Morales
Universidad de Los Andes, Universidad Central de Venezuela

KIDNEY DISEASES

Generally, patients will present with renal problems in two ways—incidentally during a routine exam and/or evidence of hypertension, edema, and hematuria. Evaluation must consider (1) estimation of disease duration, (2) urinalysis, and (3) assessment of the glomerular filtration rate (GFR).

Acute renal failure is a very common entity. It is due to a rapid decrease in renal function, possibly over days or weeks, leading to an increase in nitrogenous products within the blood. This may result from trauma, severe illness, or surgery. A rapid, progressive intrinsic renal disease can lead to acute renal failure very rapidly. It goes into three phases: prerenal due to inadequate renal perfusion, possibly due to depletion by cardiovascular disease; renal due to renal damage or disease—the most common cause being the use of IV iodinated radio contrast agents from renal diagnosis tests, which also causes a decrease in glomerular function; and finally, post-renal, which is commonly due to obstructive nephropathy. Nephropathy consists of various types of obstructions of the voiding and collecting systems.

Cystic kidney disease may be hereditary, congenital, or acquired. Acquired renal cysts are usually simple and are single or multiple within the kidney. Most are clinically insignificant, but they must be distinguished from other renal disorders or renal masses. Multiple cysts are much more common in patients with chronic renal failure, while acquired cysts are only significant within patients who have renal cell carcinoma. Congenital renal cystic dysplasia is a malformation involving the metanephric (embryotic formation of the kidney), malformation, or congenital obstruction uropathies. Nephronophthisis is an autosomal recessive disorder that accounts for 10 to 20 percent of chronic renal failure in children and young adults less than 20 years of age.

There are basically four types of nephronophthisis: juvenile with types 1 and 4, with onset at about age 13; infantile, with the median age of onset being 1 to 3 years old; and adolescent type 3, with the onset age of 19 years old. The causes are seen with gene mutations. Type 1 has extra renal manifestation and disorders.

Polycystic kidney disease is a hereditary disease causing renal cyst formation that will enlarge within both kidneys. It could possibly lead to a progressive renal failure. A familial autosomal-dominant or recessive form causes this illness. The autosomal-dominant form is found mainly in adults, and the recessive form is found in children and is known as infantile polycystic kidney disease (IPKD). Within children, cyst malformations will occur in multiple organ systems. Some children with this type die in the newborn period, but because of improvements in medical management, many progress over an undetermined number of months or years to end-stage renal failure. Autosomal-dominant disease (the adult form) rarely has any clinical significance before the fourth decade. It also may be detected in the newborn period and, depending on the severity, could be fatal.

Idiopathic nephritic syndrome (minimal change disease) is characterized by protein in the urine, low protein production, edema (due to loss of protein), and high lipids in the bloodstream. Affected children are usually under the age of 5, and may show periorbital swelling and low urine formation. This is characterized by glomeruli in the nephron that have a diffuse loss of epithelial foot process.

Acute interstitial nephritis is generally characterized by diffuse or focal inflammation and edema of the renal interstitium and tubules. It is most often related to drugs (antibiotics, mainly methicillin). This inflammation can be severe enough to cause significant deterioration of renal function. Glomerular disease comes in different types. First is post-infection glomerulonephritis. It generally occurs 10 to 14 days after an acute illness, most commonly *Streptococcus*. Urine is tea color, with mild to severe renal insufficiency and edema occurring. There is an excellent outcome, and chronic disease is rare. Membranoproliferative glomerulonephritis shows microscopic blood in the urine. The cause is unknown, but the course of the disease can range from mild to severe, with rapid deterioration in total renal function. The loss of protein in the urine also is severe and leads to edema and low antibody formation.

IgA nephropathy has an asymptomatic presentation of gross blood in the urine, generally during the time after an acute illness, with microscopic blood in the urine between episodes. The cause is unknown, but 90 percent of the initial cases will resolve within one to five years. There can be severe renal insufficiency and also hypertension, and the protein loss in the urine is very severe.

Schonlein-Henoch purpura glomerulonephritis depends on the degree of renal involvement. Generally, microscopic blood in the urine is found. This is all due to the severity of the renal lesion and, in most cases, can progress to rapid severe renal failure. Hypertension varies, and the protein in the urine also can vary from mild to severe terms. Glomerulonephritis of systemic lupus erythematosis (SLE) will show microscopic blood in the urine and protein, too. Renal involvement can be mild to severe. Severe hypertension is present, and the degree of renal insufficiency depends on the level of the disorder. The renal involvement accounts for most of the significant morbidity in SLE. By controlling the hypertension, the renal prognosis can be affected. Hereditary glomerulonephritis (Alport's syndrome) is a transmission of an autosomal-dominant X-linked disorder, with a family history marked by end-stage renal failure, especially in young males. Other associated problems include deafness and eye abnormalities. Females are generally less affected, but they are the carriers of the disorder. Renal hypertension and increasing protein in the urine will occur with advancing renal failure. Males will progress to end-stage renal failure. There is no known treatment.

Focal glomerular sclerosis is due to frequent relapsing nephritic syndrome and also due to corticosteroid-resistant syndrome. These lesions have serious prognostic implications, and as many as 10 to 20 percent of all cases will lead to end-stage renal failure.

Mesangial nephropathy also is due to usage of corticosteroids, and stoppage of the medication will put the disease in remission. Membranous nephropathy is generally due to the exposure to hepatitis B, systemic lupus erythematosus, and congenital and secondary syphilis. Immunologic disorders such as thyroiditis and medications like penicillamine can cause this disorder. The pathogenesis is unknown. It generally occurs in older children.

Hemolytic uremic syndrome is the most common cause of renal failure due to glomerular vascular damage in childhood. This glomerular damage is exacerbated by severe fluid imbalances during a gastrointestinal prodrome. Causes can be due to infections or immunologic or genetic etiologic components. This mainly affects the arteriolar epithelium of the kidney, with the formation of platelet thrombi, leading to a microangiopathic neurolysis. This syndrome is more common under the age of two years, but it is more severe in older children. Usually abdominal symptoms start such as diarrhea, vomiting, oliguria, pallor, and bleeding within the gastrointestinal tract. This is followed by hypertension and seizures and severe renal failure. There is severe anemia due to red cell destruction. Geographic factors may determine the severity of hemolytic uremic syndrome. Most children may recover in two to three weeks, but in severe cases, mortality is of greatest concern during the beginning or acute phase from complications of the central nervous system.

About one-third of adults with nephrotic syndrome have systemic renal disease due to diabetes mellitus, amyloidosis, or systemic lupus erythematosus. The remaining cases are called idiopathic nephritic syndrome and are due to minimal change disease, focal glomerular sclerosis, membranous nephropathy, or membranoproliferative glomerulonephritis. Nephrotic syndrome is urinary excretion of three grams of protein or more per day due to gomerular disease. This disorder is more dangerous in children and is characterized by loss of protein in the urine, hyperproteinemia, edema, and hyperlipidemia. Ninety percent of these children will have some form due to idiopathic nephritic syndrome, minimal change disease in 85 percent, mesangial proliferation in 5 percent, and focal sclerosis in 10 percent. The remaining 10 percent will have some form of glomerulonephritis, membranous and membrane proliferative being the most common cause. This is due to abnormal amounts of proteinuria, which results from an increase in glomerular capillary wall permeability. The edema is from the development of hypoalbuminemia as the result of urinary protein loss. Nephrotic syndrome has also been linked with several extrarenal neoplasms like lymphomas, especially Hodgkin's disease. Nephritic syndrome is hematuria or red cell casts in urinary sediment. This is due to glomerular inflammation that can occur at any age. Generally, postinfectious glomerulonephritis is the most common cause. Other causes include systemic diseases, connective tissue disorders, and paraproteinemias. Hereditary nephritis is genetically heterogeneous and characterized by hematuria, unpaired renal function, sensorineural deafness, and ocular abnormalities. Causes are due to a gene mutation affecting type 4 collagen strands. Although autosomal-recessive varieties exist, the disease is more commonly inherited in X-linked fashion.

In the X-linked transmission, women are usually asymptomatic and have little functional impairment. On the other hand, males will eventually develop renal problems like those in acute nephritic syndrome and will progress to total renal insufficiency between the ages of 20 and 30. Renal tubular acidosis (RTA) is a state of systemic hyperchloremic acidosis leading to the impairment of urinary acidification. There are three types: distal RTA (type I), proximal RTA (type II), and mineralcorticoid deficiency (type IV).

Proximal RTA results from a reduced proximal tubular reabsorption of bicarbonate, leading to a deficiency in carbonic anhydrase production. There is only 60 percent absorption in the proximal tubule, leading to a 15 percent absorption in the distal tubule. There is only 15 percent absorption, a loss of 25 percent of absorption into the urine. With urinary bicarbonate loss, serum bicarbonate levels will fall and cause the urine to become alkaline. This disorder is rare but can be brought on by various drug exposures like tetracycline and sulfonamides. Distal RTA is an impairment of hydrogen ions, leading to acidic urine and to systemic acidosis. This syndrome also is rare but occurs more in women and is due to drugs (lithium), kidney transplants, and chronic renal obstruction. Familial cases are usually present in childhood and are often autosomal dominant. They are associated with hypercalciuria and nephrocalcurosis.

See Also: Children's Health; Men's Health; Physical Activity and Health; Women's Health.

Further Readings

Greenberg, Arthur and Alfred K. Cheung. *Primer on Kidney Diseases*, 5th ed. Philadelphia, PA: Elsevier Health Sciences, 2009.

"Hospitalization Discharge Diagnoses for Kidney Disease—United States, 1980–2005." *Morbidity and Mortality Weekly*, 57/12 (March 28, 2008).

MedlinePlus. National Institutes of Health. "Kidney Diseases, Also Called: Renal Disease." http://www.nlm.nih.gov/medlineplus/kidneydiseases.html (Accessed September 2010).

Moorthy, A. Vishnu, ed. *Pathophysiology of Kidney Disease and Hypertension*. Philadelphia, PA: Saunders/Elsevier, 2009.

Mount, David B. and Martin R. Pollak. *Molecular and Genetic Basis of Renal Disease: A Companion to Brenner and Rector's The Kidney*. Philadelphia, PA: Elsevier Health Sciences, 2007.

National Institute of Diabetes and Digestive and Kidney Diseases, National Institute of Health. "Kidney and Urologic Diseases." http://kidney.niddk.nih.gov/kudiseases/a-z.asp (Accessed August 2010).

National Kidney Foundation, Inc. http://www.kidney.org (Accessed September 2010).

Richard Wills
Independent Scholar

Lead Sources and Health

Plumbum, Pb, or lead, the 82nd element, is usually found in the Earth's crust as lead sulphide or galena and in the atmosphere as lead sulphate or lead carbonate. Lead is malleable, has a low melting point, and therefore has many uses. Lead poisoning may be the oldest industrial disease because humans have used lead for over 8,500 years. There are many sources of lead exposure, and most humans have elevated lead levels or lead stores. Lead poisoning was first described in 250 B.C.E. from a case of lead-induced anemia and colic. Despite this report, it was used to sweeten wine (lead acetate) that was known to cause gout and colic and to make skin ointments and kitchen utensils.

Lead is an almost ubiquitous pollutant. A 2006 New York City study found lead levels in household dust exceeded both the Housing and Urban Development and Environmental Protection Agency indoor levels by three- to 18-fold. Environmental exposure can include lead-based paint, soil, and dust in the vicinity of lead-use industries, ceramic glazes, vehicle exhaust, and plumbing leachate. Industrial sources of lead include metal alloy foundries, glass manufacturers (high-quality crystal glass contains lead), typesetters and printers, roofing manufacturers, mining and smelters, battery manufacturers and recyclers, paint strippers, electronic product manufacturers, furniture refinishers, and demolition companies.

Common activities that may have lead exposure include pottery glazing, stained-glass making, oil painting, target shooting and hunting, lead soldering, painted-furniture stripping, and fishing using lead-weighted lines. Lead is also found in some Ayurvedic, south Asian, and traditional Chinese medicines; cosmetics (e.g., eyeliner); hair dyes; and children's toys. Lead-based paints for the home, children's toys, and household furniture were banned in the United States in 1978, but many countries still manufacture and use lead-based paints. Lead can also be found in foods grown in regions exposed to sources of lead through surface and irrigation water and dust and air pollution from various automotive and industrial sources.

Lead can be absorbed in the gut from water or pica (ingestion of dirt, paint, or particles that contain lead). It binds to red blood cells, preventing the formation of hemoglobin for carrying iron, and becomes distributed throughout the body into soft tissues, bones, teeth, and fat cells. It accumulates over a lifetime, with 95 percent residing in the bones, but is released very slowly. In the nervous system, it can cause paralysis of motor nerves (e.g., painless wrist drop) and lower cognitive function. It can also cause acute abdominal pain, kidney damage, high blood pressure, and reproductive failure. It may also cause irritability; insomnia; metallic taste; lethargy or hyperactivity; headache; and even seizure, coma, and death.

Children under the age of six are especially and permanently susceptible to any level of lead ingested. One hypothesis is that 65 to 90 percent of violent crimes are committed by individuals with elevated preschool lead levels. The average person has 100 parts per billion of lead in his or her blood, a level considered to be much higher than that in ancient times. Virtually everyone is exposed to lead; for example, leaded gasoline was phased out in most but not all countries in 2007, lead solder and brass fittings occur in many water pipes, and dust from diverse industrial sources places lead in the atmosphere.

Exposure to lead at 10 micrograms in a deciliter (one-half cup of human blood) can cause irreversible damage in children, especially to the developing brain. At the level of 25 micrograms per deciliter, the kidneys and nervous system in children and adults can be damaged. Higher levels can cause seizures, unconsciousness, and even death.

During a medical history suspicious for lead exposure in children, your physician will look for anemia, decreased muscle and bone growth, hearing damage, learning disabilities, nervous system and kidney damage, poor muscle coordination and speech, and language and behavior problems. In an adult, the medical history will focus on anemia, cataracts, digestive problems, high blood pressure, memory and concentration problems, muscle and joint pain, nerve disorders, preterm delivery, miscarriage, or stillbirth in women, and infertility in men. A blood test can detect lead exposure.

Elevated blood lead levels are treated with a chelator that binds the lead for excretion in the urine. The most common chelator is calcium disodium EDTA, but DMSA and DMPS are also used. High serum levels of ascorbic acid decrease blood lead levels, especially when combined with these chelators. Calcium supplementation during both pregnancy and lactation not only decreases the mother's intestinal absorption of lead but also increases her lead excretion by 15 to 20 percent.

See Also: Arsenic Pollution; Automobiles (Emissions); Cardiovascular Diseases; Dental Mercury Amalgams; Kidney Diseases; Manganese Sources and Health; Mercury Sources and Health; Reproductive System Diseases.

Further Readings

Agency for Toxic Substances and Disease Registry. "ToxFAQs for Lead" (2007). http://www .atsdr.cdc.gov/tfacts13.html (Accessed May 2010).
Centers for Disease Control and Prevention. "Lead." http://www.cdc.gov/nceh/lead/tips.htm (Accessed May 2010).
Centers for Disease Control and Prevention. "Third National Report on Human Exposure to Environmental Chemicals: Spotlight on Lead." 2005.
U.S. Environmental Protection Agency. "Protect Your Family From Lead in Your Home" (2009). http://www.epa.gov/lead/pubs/leadpdfe.pdf (Accessed May 2010).

Paul Richard Saunders
Canadian College of Naturopathic Medicine

LIGHT BULBS

The use of more energy-efficient light bulbs is a definitive step toward reducing energy consumption, thereby moving to more sustainable resource use. According to the U.S. Energy Information Administration, lighting accounts for 38 percent of end-use electricity consumption in commercial buildings and about 10 percent in residential and manufacturing buildings. A 2006 study by the International Energy Agency estimated that grid-based electric lighting accounted for 19 percent of all global electricity consumption. Consequently, replacing general-purpose incandescent light bulbs can significantly reduce energy consumption; currently, compact fluorescent bulbs are the most economically viable option, and light-emitting diodes (LEDs) are even more energy efficient. In choosing an appropriate light bulb, the application, light intensity, energy efficiency, cost, resource depletion of source materials, and proper end-of-life disposal all need to be taken into account.

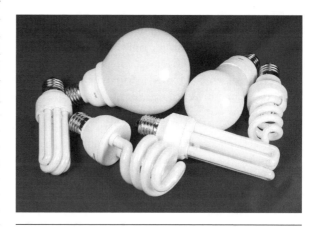

Compact fluorescent light bulbs (CFLs) have an energy efficiency of 20 percent and usually last between 6,000 and 15,000 hours. One drawback is that CFLs contain mercury, which can leach out in landfills and contaminate soil and water resources.

Source: iStockphoto.com

The top light bulb technologies include incandescent, compact fluorescent (CFL), tubular fluorescent, high-intensity discharge (HID), and LEDs. Incandescent light bulbs are widely used in residential, commercial, and portable lighting. Compact fluorescent light bulbs are typically used in residential and lodging applications, replacing incandescent bulbs to reduce energy costs. Tubular fluorescent bulbs are used in commercial applications, particularly office buildings, due to their large area of light coverage. Halogen bulbs are used in display and retail lighting due to their preferred color and spotlight properties. HID bulbs are used for outdoor and streetlight applications due to their high brightness output. LEDs are used for indicator lights, signs, displays, and decorative lighting due to their low light output, color variability, low maintenance, low power consumption, and long lifetime. Recently, advances in high-intensity LED technologies have made them a viable and desirable option for a broader range of lighting applications.

Incandescent light bulbs have not changed substantially in over 100 years and are preferred by consumers due to their familiarity, low cost, and high quality of light. An incandescent bulb uses electricity to heat a filament, typically tungsten, which produces light. An incandescent bulb emits a broad spectrum of radiation, with the light produced rendering colors well. However, incandescent bulbs have low energy efficiency, converting only about 5 to 10 percent of the electricity input into visible light, and the rest emitted as heat. Due to this energy inefficiency, many governments have passed legislation that has either specifically banned certain types of incandescent bulbs, such as the European Union's ban of general-purpose, nondirectional incandescent bulbs, or have passed legislation that requires an increased energy efficiency in general-purpose light bulbs, such as the U.S. Energy Independence and Security Act of 2007, which requires all general-purpose light bulbs to be 25 to 30 percent more energy efficient by 2012 and greater improvements starting in 2020.

CFLs consist of a glass tube filled with an inert gas, usually argon, and small amounts of mercury. Light is created when electricity excites electrons in the mercury vapor, creating UV light that excites phosphors coating the inner surface of the glass tube, which then emits visible light. CFLs have an energy efficiency of 20 percent, compared to the 5 percent of incandescent bulbs. Moreover, CFLs usually have a lifetime between 6,000 and 15,000 hours. However, this lifetime is significantly reduced if they are turned on for only a few minutes at a time. By replacing incandescent bulbs, the increased efficiency and longer lifetime of CFLs can lead to substantial electricity savings, resulting in related reductions in air pollutants and greenhouse gas emissions. While CFLs can benefit householders by reducing electricity costs, as well as reducing overall residential electricity consumption, CFLs have environmental drawbacks because they contain mercury, a toxic heavy metal that can cause brain, kidney, and respiratory system damage. Due to inadequate consumer education, CFLs are typically disposed of with normal household waste and are subsequently broken by the trash compactors in garbage trucks. Consequently, the mercury spills out and contaminates the waste, which customarily is either landfilled or incinerated. In landfills, the mercury can leach out and contaminate soil and water resources, and in incinerators the mercury can enter the atmosphere through smokestack emissions.

LEDs are emerging as an alternative to incandescent bulbs for general purposes due to their low power consumption, long lifetime, and decreasing manufacturing costs. LEDs use semiconductor materials to emit light of a single color. A high-power white LED light can have a 30 percent efficiency, which is more than six times greater than an incandescent bulb, and 50 percent greater than a CFL. An LED bulb with light output similar to a traditional incandescent bulb has a projected lifetime of over 50,000 hours, which is over 50 times greater compared to an incandescent bulb and about five times greater compared to a CFL. Although they are more energy efficient and last longer, LED lights are viewed as economically less favorable for residential and workplace use because of their higher purchase price. LEDs are considered better for the environment because they are more energy efficient and do not contain mercury, as do fluorescent light bulbs. However, with the projected increase in use of LEDs, they have the potential to create significant impacts on the environment because they contain rare and toxic materials such as arsenic, antimony, copper, gallium, gold, indium, iron, lead, nickel, and silver.

See Also: Electricity; Lighting; Personal Consumer Role in Green Health; Solid Waste Management.

Further Readings

ConsumerSearch. "Light Bulbs: Full Report." http://www.consumersearch.com/light-bulbs/led-light-bulbs (Accessed September 2010).

Ferraz, Rob. "A Closer Look at Compact Fluorescents—The Environmental Benefit Versus Health Impact." *Vitality Magazine* (September 2007). http://www.vitalitymagazine.com/ earthwatch_5 (Accessed September 2010).

Humphreys, Colin. "Solid-State Lighting." *MRS Bulletin*, 33 (2008).

Daniel Hsing Po Kang
University of California, Irvine

LIGHTING

As the environment becomes a more pressing concern for society, more and more ways are found for individuals to be environmentally conscious and to assist the environment through more sustainable activities. By examining practices used with the lighting used to illuminate buildings, great energy savings can be obtained through actions that are inexpensive, easy, and reproducible. Over the years, different lighting options have been considered and explored to help people go green. Saving the environment is not the only benefit that comes from sustainable, or green, lighting. Along with helping the planet, refitting lighting fixtures has benefited people's wallets, greatly cutting energy costs for those who pursue green lighting options. While individual action has played a great role in reducing energy usage by replacing incandescent bulbs with more environmentally sustainable alternatives, the U.S. government has also become involved in an effort to speed the process.

Traditional Lighting Methods

For thousands of years, humans have used artificial light sources to illuminate their homes and other spaces, especially at night. Windows, skylights, and other sources of natural light are often used as the main source of light during daytime, chiefly due to their convenience and low cost. In the evenings, artificial lighting is common. Although electric lamps are the most common way to artificially light contemporary homes, gas lighting, oil lamps, and candles have all been used in the past and continue to be used today. Lighting enhances task performance and greatly adds to the aesthetics of many indoor and outdoor situations. Artificial lighting is a major cause of energy consumption and accounts for a considerable part of all energy used worldwide.

During the 19th century, experiments began with the incandescent light bulb. In 1880, Thomas Edison perfected the long-lasting filament that made electrically powered incandescent bulbs a cost-effective alternative to other methods. Incandescent bulbs quickly became the lighting method of choice across the industrialized world. Companies such as General Electric and Sylvania marketed and sold incandescent bulbs that were both reliable and inexpensive. As individuals and government agencies became more concerned about energy consumption, however, incandescent bulbs came under scrutiny. Since approximately 90 percent of the energy used by an incandescent bulb is emitted as heat, rather than used to produce visible light, those concerned with efficient use of natural resources began to seek lighting alternatives.

Alternative Lighting Sources

One of the main alternatives to incandescent lighting are compact fluorescent light (CFL) bulbs. There are many benefits to CFL bulbs, especially when compared to a traditional light bulb. CFL bulbs save energy costs and, as a result, reduce pollution. Light-emitting diodes (LEDs) are also an option. Individual LEDs do not emit much light, so multiple diodes must be used together to supply the amount of illumination needed in most indoor settings. Although LEDs present a promising alternative to incandescent bulbs, LEDs are not very common or easy to use as indoor lighting at this time. However, there is hope for the future, as there have been some light bulbs launched using LED light, and there is great promise and potential for more LED usage in the future. To encourage the switch to nonincandescent bulbs, many nations are passing regulations to improve incandescent bulbs' energy efficiency or phase them out entirely. The United States, for example, in 2007 passed the Energy Independence and Security Act, which requires all light bulbs that produce 310 to 2,600 lumens of light to be 30 percent more efficient beginning in 2012.

Alternative Lighting Strategies

When thinking about lighting and how to be environmentally friendly with lighting, many think only of different types of bulbs. However, bulbs are not the only component when lighting a space and not the only way that people can strive to be environmentally friendly with their lighting. Light fixtures play a big role in making lighting friendly to our environment. Using copper light fixtures, for example, is a way to be environmentally conscious in lighting choices. Copper is recyclable and very long lasting, which greatly reduces landfill waste. Another green choice with regard to lighting is to use products that have already been recycled, such as recycled glass or aluminum, for a lighting fixture. Again, like copper, such choices greatly reduce the energy expended on the creation of lighting fixtures and reduce landfill waste.

A third way to be environmentally friendly when lighting a space is to take the surroundings into consideration and to be conscious of the use of the lighting. When using outdoor lighting, for example, an environmentally friendly choice is to use lights that are controlled by motion sensors. Motion sensors provide light when someone is in the vicinity of the fixture, but automatically turn off when no one is nearby. In this way, outdoor lights are on when needed, but off when they are unnecessary. Inside a house, installing a timer on lights ensures that they will be turned off when not needed, again conserving energy and the environment. Some homes and offices also use motion sensors in interior spaces, although these are sometimes troublesome because there needs to be consistent motion for them to remain illuminated. Other ways to reduce energy consumption inside a home includes using light bulbs of lower wattage, and using curtains at night. Low-wattage bulbs use less energy, yet often provide the necessary level of illumination for those in a given space. If curtains are used, more light is contained in the space where it is needed, instead of needlessly lighting the outdoors.

Conclusion

As the environment becomes an ever more pressing issue on the public agenda, any step that can reduce energy consumption helps. The energy used for lighting represents a major use of energy in most industrialized nations. Environmentally friendly choices in lighting benefit household budgets as well as the environment, so everyone and everything benefit.

See Also: Advertising and Marketing; Electricity; Government Role in Green Health; Light Bulbs.

Further Readings

Brinsky, W. and S. Leitman. *Green Lighting: How Energy-Efficient Lighting Can Save You Energy, Money, and Reduce Your Carbon Footprint.* New York: McGraw-Hill/TAB Electronics, 2010.
Horowitz, M. J. "Economic Indicators of Market Transformation: Energy Efficient Lighting and EPA's Green Lights." *The Energy Journal*, 22/4 (2000).

Stephen T. Schroth
Sarah E. Carlin
Knox College

LIVER DISEASES

The liver lies on the right side of the abdominal cavity beneath the diaphragm. It has many functions, including the metabolism of fats, conversion of glucose to glycogen, production of urea, amino acid synthesis, storage of vitamins and minerals, filtration of harmful substances from the blood, and maintenance of the proper level of glucose in the blood. The liver is also responsible for producing about 80 percent of the cholesterol in the human body. *Liver disease* is a broad term describing any number of diseases affecting the liver. Many are accompanied by jaundice, caused by increased levels of bilirubin in the system. The bilirubin results from the breakup of the hemoglobin of dead red blood cells; the liver normally removes bilirubin from the blood and excretes it through bile.

There are many different kinds of liver diseases that require clinical care by a physician or other healthcare professional. Some of the more common liver diseases are listed below with a brief description of each for general information, bearing in mind that a diagnosis should always be made in person by a physician or healthcare professional:

- Amebic liver abscess is a type of liver abscess caused by amebiasis. The parasite enters the gastrointestinal tract and, once it reaches the bloodstream, it can spread throughout the body, most frequently ending up in the liver, causing liver abscess. Most patients show symptoms of weight loss, earthy complexion, profuse sweating, fever, and pain in right hypochondrium. Patients also show signs of pallor, tenderness and rigidity in right hypochondrium, palpable liver, intercostals tenderness, and basal lung signs. Infections can be prevented by good sanitary practices.
- Autoimmune hepatitis is a cell-mediated immune response to the body's own liver, due to a genetic predisposition or acute liver infection, which has a presentation of human leukocyte antigen class II on the surface of hepatocytes. Diagnosis is best achieved when clinical, laboratory, and histological findings are employed. Treatment is done with glucocorticoids and immunosuppressive drugs.

- Biliary atresia is a rare condition in newborn infants in which the common bile duct between the liver and the small intestine is blocked or absent. Cause of the condition is currently unknown, and the only effective treatments are certain surgeries or liver transplantation. Initially, symptoms are indistinguishable from neonatal jaundice. Symptoms, usually evident between one and six weeks after birth, include clay-colored stools, dark urine, swollen abdominal region, and large hardened liver.
- Cirrhosis is characterized by replacement of liver tissue by fibrous scar tissue as well as regenerative nodules, leading to progressive loss of liver function. Ascites, fluid retention in the abdominal cavity, is the most common complication of cirrhosis, and is associated with a poor quality of life, increased risk of infection, and poor long-term outcome. Cirrhosis is generally irreversible and treatment focuses on preventing progression and complication.
- Hemochromatosis is a hereditary disease that causes the accumulation of iron in the body, eventually leading to liver disease. It is associated to the iron accumulation associated with the HFE gene (hemochromatosis type 1). Hemochromatosis less often refers to the condition of iron overload as a consequence of multiple transfusions. Causes can be primary (generally genetic) and secondary (due to other conditions).
- Hepatitis A is an acute infectious disease of the liver caused by the hepatitis A virus, which is most commonly transmitted by the fecal–oral route via contaminated food or drinking water. Symptoms include fatigue, fever, abdominal pain, nausea, diarrhea, appetite loss, depression, and jaundice. Hepatitis A does not have a chronic stage, it is not progressive, and it does not cause permanent liver damage. Following infection, the immune system makes antibodies against the virus that confer immunity against future infection. The disease can be prevented by vaccination.
- Hepatitis B is a disease caused by the hepatitis B virus, which infects the liver and causes inflammation. The acute illness causes liver inflammation, vomiting, jaundice, and, rarely, death. Symptoms include nausea, loss of appetite, body aches, vomiting, mild fever, dark urine, and jaundice. Hepatitis B is transmitted through body fluids, against which prevention should be taken. Infants may be vaccinated at birth.
- Hepatitis C is an infectious disease affecting the liver caused by the hepatitis C virus. The infection is often asymptomatic, but once established, chronic infection can progress to scarring of the liver, fibrosis, and cirrhosis. Hepatitis C is spread by blood-to-blood contact. Hepatitis C may be suspected based on medical history, unexplained symptoms, and abnormal liver enzymes or liver function test found during routine blood testing. There is currently no vaccine that protects against hepatitis C.
- Liver cancer: multiple types of tumors can develop in the liver because the liver is made of up many various cells. Liver cancer is often characterized by the presence of malignant hepatic tumors. Hemangiomas are the most common type of benign liver tumor that begins in blood vessels. Hepatic adenomas are benign epithelial liver tumors that develop in the liver and are also an uncommon occurrence. Focal nodular hyperplasia is the second-most-common tumor in the liver, and is a result of congenital artiovenous malformation hepatocyte response. There are many cancers that begin in the liver, including hepatocellular carcinoma, hepatoblastoma, cholangiocarcinomas, angiosarcomas, and hemangiosarcomas.
- Primary biliary cirrhosis is an autoimmune disease of the liver, marked by the slow, progressive destruction of the small bile ducts within the liver. When these ducts are damaged, bile builds up in the liver and over time damages the tissue. Patients show signs and symptoms of fatigue, itchy skin, jaundice, xanthoma, and complications of cirrhosis and portal hypertension. Liver function tests, ultrasound, and CT scan aid in diagnosis.
- Reye's syndrome is a potentially fatal disease that causes numerous detrimental effects to the liver as well as many other organs including the brain. The cause is currently unknown. Reye's progresses through five stages. The first-stage signs and symptoms include heavy vomiting, lethargy, nightmares, and mental symptoms (such as confusions). Stage two signs and symptoms include stupor, hyperventilation, fatty liver, and hyperactive reflexes. Stage

three is a continuation of stages one and two, with a possible coma and possible cerebral edema. Stage four is a deepening coma, large pupils with minimal response to light, minimal but still present hepatic dysfunction. Stage five has a very rapid onset following stage four and includes a deep coma, seizures, multiple organ failure, flaccidity, extremely high blood ammonia, and death; thus early diagnosis is vital.

- Wilson's disease, also known as hepatolenticular degeneration, is an autosomal-recessive genetic disorder in which copper accumulates in tissues. This manifests in liver disease as well as neurological and psychiatric symptoms. The liver disease may present as tiredness, increased bleeding tendency, or confusion and portal hypertension. Wilson's disease is caused due to mutations in the Wilson's disease protein gene. A liver biopsy is the most ideal test for a diagnosis of Wilson's disease. A low copper diet is highly recommended for patients, and liver transplantation may be necessary.
- Primary sclerosing cholangitis is a chronic liver disease caused by progressive inflammation and scarring of the bile ducts of the liver. The inflammation impedes the flow of bile to the gut, which can ultimately lead to liver cirrhosis and liver failure. The underlying cause of the inflammation is believed to be autoimmunity. Symptoms include fatigue, severe jaundice, signs of cirrhosis, dark urine, pale stools, malabsorption and steatorrhea, and ascending cholangitis. Imaging of the bile duct makes diagnosis. The definitive treatment is liver transplant.

The symptoms related to liver disease and dysfunction include both physical signs and a variety of symptoms related to digestive problems, blood sugar problems, immune disorders, abnormal absorption of fats, and metabolism problems.

The malabsorption of fats may lead to symptoms that include indigestion, reflux, hemorrhoids, gallstones, intolerance to fatty foods, intolerance to alcohol, nausea and vomiting attacks, abdominal bloating, and constipation.

Nervous system disorders include depression, mood changes, especially anger and irritability, poor concentration, overheating of the body, especially the face and torso, and recurrent headaches associated with nausea.

The blood sugar problems include a craving for sugar, hypoglycemia and unstable blood sugar levels, and the onset of type 2 diabetes. Abnormalities in the level of fats in the bloodstream include elevated LDL cholesterol, reduced HDL cholesterol, elevated triglycerides, clogged arteries leading to high blood pressure, heart attacks, and strokes, buildup of fat in other body organs, lumps of fat in the skin, excessive weight gain, inability to lose weight even while dieting, sluggish metabolism, protuberant abdomen, cellulite, fatty liver, and a roll of fat around the upper abdomen.

A number of liver function tests are available to test the proper function of the liver. These test for the presence of enzymes in blood that are normally most abundant in liver tissue, metabolites, or products. The tests are always utilized in the diagnosis of liver diseases.

See Also: Anti-Cholesterol Drugs; Biomedicine; Cancers; Sexually Transmitted Diseases.

Further Readings

Feldman, Mark, Lawrence S. Friedman, and Marvin Sleisenger. *Sleisenger and Fordtran's Gastrointestinal and Liver Disease.* Oxford, UK: Elsevier, 2003.

Holm, Eggert and Heinrich Kasper. *Metabolism and Nutrition in Liver Disease.* Lancaster, PA: Springer, 1985.

Palmer, Melissa. *Dr. Melissa Palmer's Guide to Hepatitis & Liver Disease*. New York: Avery, 2004.

Worman, Howard J. *The Liver Disorders Sourcebook*. Lincolnwood, IL: McGraw-Hill Professional, 1999.

Alessandra Guimaraes
University of Sint Eustatius School of Medicine

LOW-LEVEL RADIOACTIVE WASTE

Thinking through the issue of nuclear waste as it relates to public health is complicated for a number of reasons. First, the history of nuclear energy is related to nuclear weapons and weapons testing. This legacy is one of devastating health effects and delayed justice for many survivors of nuclear fallout ("downwinders"). Many communities, particularly indigenous peoples and the poor, have also experienced severe health effects as a result of uranium mining. Second, those in positions of authority in nuclear industries and regulatory bodies have not, historically, been transparent with the public regarding the health effects of radioactive materials. As a result, the public has lost trust in official proclamations regarding safety. Third, the debates over nuclear waste and its effects have been deeply polarizing—those in the nuclear industry have pointed to the industry's excellent safety record, while critics of nuclear power and the weapons complex point to the potential for catastrophic risk, industry cover-ups, and problems with security and waste management. Few have sought—or been able to defend—a middle ground.

Perhaps most important, scientists and engineers who study nuclear waste and its health effects must deal with a tremendous amount of uncertainty. Scientists are trained to deal with uncertainty in particular ways: there are always unknowns that must be managed in the performance of science. However, scientists may struggle to effectively engage with the public on this issue, perhaps because they have been educated to communicate and understand risk and uncertainty in ways specific to their disciplines or professions. The public also may have good reasons to mistrust scientific "experts" on the issue of radioactive waste. Anything labeled *nuclear* or *radioactive* can carry a host of negative associations for citizens and communities. Such associations may be rooted in rational responses, lived experience, or communal wisdom and are often misunderstood or dismissed by scientists and engineers. These conflicts can make productive, respectful communication about nuclear waste challenging.

What Counts as Low-Level Radioactive Waste?

The term *low-level radioactive waste* (LLRW) is itself a charged term; various groups disagree about what it means and about the risk LLRW poses. It may clarify things somewhat to begin by defining what LLRW is not, while acknowledging that this dichotomy is not hard or fast. High-level radioactive waste (HLRW) refers to spent fuel from reactor cores and decommissioned weapons as well as any liquid or sludge waste that has been irradiated during the processing of fuel in nuclear reactors. In countries that reprocess their waste, such as France, the solid forms of this liquid or sludge waste are also considered HLRW.

Similarly, many forms of *transuranic waste*, waste that contains radioactive elements heavier than uranium (such as plutonium), are frequently considered HLRW and must be handled as such. The handling of HLRW is subject to a range of safety and regulatory precautions that are much stricter than those placed on the handling of LLRW.

LLRW, by contrast, refers to a wider variety of materials and substances that are by-products of nuclear reactor production, weapons production, medical and research processes and products, and a large number of industrial uses. These materials include parts of nuclear plants that have come into contact with radioactive materials and waste (such as building materials, pipes, and even the clothing worn by reactor operators) as well as materials from medical and science labs that have been irradiated. This can add up to a substantial amount of material, particularly when one considers plant *decommissioning*, or the dismantling and disposing of a plant at the end of its working life. However, proponents of nuclear power argue that the amount of nuclear waste is tiny, constituting only a small fraction of all energy waste created in the United States. Energy from coal, for example, produces much more waste and pollution.

Who Determines What Constitutes LLRW?

In the United States, the handling of nuclear waste is managed by the Nuclear Regulatory Commission (NRC), in conjunction with several other federal agencies such as the Department of Energy (DOE) and the Department of Transportation (DOT). Over time, the NRC has changed its rulings on how to handle certain kinds of waste: for example, some kinds of transuranic waste have been downgraded from HLRW to LLRW. Critics contend this is not because these materials have been proved to pose less risk, but rather that the waste problem itself—where and how to store the waste—has become so politically and socially complex that regulators are simply trying to decrease the amount of HLRW to be managed (debates over the Yucca Mountain waste repository are one example).

Furthermore, from state to state, regulations for dealing with waste may differ. Most LLRW is stored on-site (for reactor waste) or in licensed state-managed or privately managed waste repositories. Depending on the kind of nuclear waste, the NRC generally requires that LLRW be stored until it is no longer radioactive, at which time it can be disposed of like "ordinary trash."

There are a number of reasons this treatment of LLRW is controversial. Medical LLRW, for example, remains radioactive for relatively short time scales—sometimes less than a decade. But LLRW from nuclear reactors can remain radioactive for thousands of years. The NRC policy of "planned leakage" (leakage that cannot be avoided over time) for the latter form of waste is considered particularly troubling by many. Critics contend that these two different kinds of wastes should be managed differently as a result.

What Are the Health Effects of Exposure to LLRW?

Most scientists agree that all humans are exposed to basic levels of what is called *background radiation*—radiation that is produced by nature and that is part of living life on Earth. Exposure to the sun, or even to various kinds of rocks, can lead to higher naturally occurring levels of background radiation. Similarly, workers in nuclear power plants or sailors on nuclear submarines are exposed to low levels of radiation: though some see the findings as controversial, most studies of these nuclear workers show that low-dose exposures such as these have no long-term health effects. Some studies have even shown that health benefits can accrue from long-term low-level radioactive exposure; this phenomenon is called *hormesis*.

However, what is not clear is what kind of cumulative effects the long-term storage (and inevitable leakage) of LLRW might have on the environment and public health generally. Epidemiological studies addressing this question are very expensive and difficult to conduct because it is challenging to isolate specific health and environmental effects caused by specific forms of waste exposure. Many experts advise following the precautionary principle and adopting policies that limit risks, even if those risks are highly uncertain or unproven. Others argue that nuclear waste, in the United States at least, has been successfully and safely managed for many decades now and poses little risk to public health when compared with other forms of energy waste.

See Also: Electricity; Environmental Protection Agency (U.S.); Government Role in Green Health; Radiation Sources.

Further Readings

Cravens, Gwyneth. *Power to Save the World: The Truth About Nuclear Energy.* New York: Alfred A. Knopf, 2007.
League of Woman Voters Education Fund. *The Nuclear Waste Primer.* Washington, DC: League of Woman Voters Education Fund, 1993.
Nuclear Regulatory Commission. "Low-Level Waste." http://www.nrc.gov/waste/low-level-waste.html (Accessed July 2009).
Taylor, Brian C., et al., eds. *Nuclear Legacies: Communication, Controversy, and the U.S. Nuclear Weapons Complex.* Lanham, MD: Lexington Books, 2007.

Jen Schneider
Colorado School of Mines

Lung Diseases

Many factors are associated with the risk of developing a disease of the respiratory system, including genetics and birth factors such as prematurity, but environmental factors also play a major role. Exposure to chemical and particulate pollutants has been shown to be strongly connected to the risk of certain respiratory system diseases that are otherwise rare: examples include tobacco smoke and lung cancer, asbestos and mesothelioma, and flavoring chemicals and bronchiolitis obliterans. Even a common disease like asthma has been shown to be strongly related to exposure to particular matter (e.g., exhaust from diesel-powered vehicles) and inhaled allergens, although the specific allergens vary from one person to another. For these reasons, control of air pollution and of occupational and home exposures is critical if we hope to reduce the incidence of these diseases in the future.

Asthma

About 300 million people worldwide suffer from asthma, a chronic disease marked by bouts of breathlessness and wheezing, and in 2005, about 255,000 people died from asthma. Asthma is the most chronic disease among children, and it occurs in all countries of the world. Over 80 percent of deaths from asthma occur in low- and lower-middle-income countries, in part because of the lack of medical care and pharmaceutical intervention (e.g., inhaled corticosteroids) available to many in those countries.

Exposure to allergens is the greatest risk factor for developing asthma, and inhaled allergens seem to be the most harmful. Some allergens are naturally occurring substances like pet dander, pollen, and molds, while others are created by human activity, including tobacco smoke and other air pollutants. Urbanization is associated with an increase in asthma; although the exact mechanism has not been determined, some theorists think it could be due to greater exposure to air pollution (e.g., from automobiles and other motorized vehicles). Because many cases of asthma go undiagnosed, it is also possible that greater access to healthcare in cities results in previously ignored cases being diagnosed.

In the United States, according to the 2007 National Health Interview Survey, 11.5 percent of Americans suffer from asthma. The prevalence of asthma varies with age, being highest in the late teenage years (16.9 percent for people age 15–19) and lowest in early childhood (8 percent for children under age 5). In adulthood the prevalence is similar for men and women, but in childhood (under age 18) it is more common in boys than girls (14.7 percent versus 11.3 percent). Asthma is somewhat related to race and ethnicity—blacks have higher rates than whites (13.2 percent versus 11.5 percent) and Puerto Ricans have an even higher prevalence (20.3 percent)—and it is higher for people living below the poverty threshold than above it.

Chronic Obstructive Pulmonary Disease

About 210 million people worldwide are believed to suffer from chronic obstructive pulmonary disease (COPD), a lung disease in which airflow is blocked from the lungs, and in 2005, over three million people died of it (about 5 percent of all global deaths). Most COPD deaths (90 percent) occur in low- and middle-income countries, and although historically COPD has been more common in men, as more women worldwide have taken up cigarette smoking, the rates have become more equal between the genders. COPD includes *emphysema* and *chronic bronchitis*, terms no longer used by the World Health Organization (WHO), and is characterized by breathlessness, a chronic cough, and abnormal sputum (matter expelled from the respiratory tract).

In the United States, about 10 million Americans have been diagnosed with COPD, but national surveys have suggested that as many as 24 million people are affected by the disease. In 2006, 120,970 Americans died of COPD, making it the fourth-highest cause of death. COPD rates among women are rising faster than for men, presumably because of the increase in the number of women smoking tobacco, and in 2006, more women died from COPD than men (about 63,00 versus about 58,000).

The primary cause of COPD is exposure (firsthand or secondhand) to tobacco smoke, and as the smoking rate has increased in many countries, particularly among women, COPD deaths are expected to increase substantially unless the rate of smoking and exposure to tobacco smoke is reduced. The American Lung Association estimates that women who smoke are almost 13 times more likely to die from COPD than women who do not smoke, and male smokers face 12 times the risk of death from COPD compared to male nonsmokers. Exposure to indoor air pollution (such as smoke from cooking fires) is also a risk factor (and one that particularly applies to women), as is exposure to outdoor air pollution, occupational exposure to chemicals and dusts, and a history of respiratory infections in childhood. COPD is a slow-developing disease that is most often seen in people over 40 and is diagnosed by a spirometry test that measures how fast air can move in and out of the lungs and how much total air a person can inhale and exhale. There is no cure for COPD, but medicines are available to improve breathing by dilating the air passages.

Lung Cancer

Lung cancer is one of the most common types of cancer in the world: in 2004, there were 1,448,000 cases, with the greatest numbers in the western Pacific and European regions. In the United States, lung cancer is the second most common cancer for both men and women, and more people die from lung cancer than any other type of cancer. In 2006, 106,374 men and 90,080 American women were diagnosed with lung cancer, and 89,243 men and 69,356 women died from it.

There are many risk factors for lung cancer, including genetics, but the strongest relationship is with exposure to tobacco smoke: people who smoke are 10 to 20 times more likely to develop lung cancer than those who do not; about 80 percent of women and 90 percent of men who die from lung cancer are smokers. In addition, there is a clear dose–response relationship between smoking and lung cancer: the risk increases with the amount of smoking (e.g., the daily number of cigarettes smoked) and time of exposure (e.g., the number of years smoked). Other risk factors include exposure to radon gas (the major risk factor among nonsmokers) and various occupational exposures, including asbestos, arsenic, silica, and chromium.

Asbestos-Related Diseases

The term *asbestos* refers to a group of naturally occurring minerals that have been used in building and other industries in the past due to their insulating properties, tensile strength, and resistance to chemicals. Asbestos may take the form of tiny fibers that are easily inhaled deep into the lungs, making exposure a particular risk factor for respiratory diseases. It has been used commercially since the late 19th century, but during World War II, use of asbestos increased substantially. The diseases associated with asbestos exposure may take years to develop (although they have also been seen in people with very brief exposures), and thus the dangers of asbestos were not immediately identified. Current research shows that exposure to asbestosis is associated with many diseases, including mesothelioma (cancer of the mesothelium, or protective lining of the internal organs, including the lungs, larynx, and ovaries), asbestosis (chronic inflammation and fibrosis of the lungs), and pleural plaques (scars in the lungs).

The WHO estimates that 125 million people globally are exposed to asbestos as part of their work and that 107,000 die each year from mesothelioma, asbestosis, and asbestos-related lung cancer due to occupational exposure. Thus asbestos is estimated to be responsible for about one-third of all cancer deaths due to occupational exposure.

Due to increased awareness of the health risks in the United States in the late 1970s, the use of asbestos was phased out of many products, and in 1989, the Environmental Protection Agency banned all new uses of asbestos. In addition, schools were required to remove or encase asbestos within their buildings in order to protect students and staff from exposure. However, many older buildings contain asbestos that may be dislodged during demolition or other destruction. For instance, the World Trade Center in New York City was built in the late 1960s, and asbestos was used in its construction (as was the case for many buildings built during that period); when the buildings were attacked and destroyed in September 2001, many employees, area residents, and responders were exposed to asbestos and many other dangerous pollutants. Most severely affected because of their length of exposure were policemen, firefighters, medics, and people who volunteered to clean up the rubble at Ground Zero. Although it is impossible to separate the effects of the various exposures, one study found that 61 percent of responders developed respiratory symptoms and 28 percent had abnormal lung function tests.

Mesothelioma is a rare form of cancer found primarily in people whose jobs expose them to asbestos: in the United States, 70 to 80 percent of persons with mesothelioma have occupational exposure to asbestos. About 2,000 new cases are diagnosed annually in the United States, more in men than in women (presumably due to differing occupational patterns), and incidence increases with age. Smoking has not been identified as a risk factor in mesothelioma, although the combination of asbestos exposure with smoking has been linked to increased risk of other types of lung cancer.

Early cases linking asbestos and mesothelioma were found among workers in shipyards, the construction industry, the heating industry, and those who produced asbestos products or who worked in asbestos mines and mills. Today, people who work with asbestos are required to wear protective equipment, and the U.S. Occupational Safety and Health Administration has established limits for how much a person may be exposed to asbestos in the workplace. In addition to people directly exposed to occupational asbestos, there is evidence suggesting that their family members are also at higher risk for mesothelioma due to their exposure to asbestos dust on the worker's hair and clothing.

Asbestosis is a lung disease associated with exposure to asbestos (breathing in asbestos particles) that may result in scarring and permanent damage to lung tissue. The scarred tissue cannot perform gas exchange and cannot expand and contract normally; therefore, persons with asbestosis may suffer symptoms such as chest pain, coughing, and shortness of breath. They are also at higher risk for lung cancer, pneumonia, and influenza.

Pneumonoconioses

Exposure to mineral dust is a major cause of work disability, morbidity, and mortality worldwide. One of the most serious health problems related to this exposure is pneumoconiosis, or fibrotic reaction of lung tissue to the dust. There are several different types of pneumonoconiosis: in the United States, the most common are caused by inhalation of asbestos fibers (asbestosis, discussed above), silica dust (silicosis), and coal mine dust (black lung disease, or coal miner's pneumoconiosis). All three are primarily occupational diseases that have only rarely been linked to any other type of exposure.

All three diseases typically develop over a period of time (although some have developed black lung in less than 10 years of mining work) and can be largely or entirely prevented through the use of protective equipment and monitoring of at-risk workers. For example, in 1969, following the passage of the Federal Coal Mine Health and Safety Act, the National Institute for Occupational Safety and Health began close monitoring of U.S. coal miners (e.g., through chest X-rays to detect early signs of disease) and black lung rates, which substantially reduced incidence of the disease.

Infectious Diseases

Tuberculosis (TB) is an infectious disease most often spread through the air from an infected person to other people through coughing, sneezing, spitting, and so on. It was once a disease dreaded worldwide but was largely brought under control in the industrialized world with the use of antibiotics. However, the emergence of drug-resistant strains of TB combined with the emergence of human immunodeficiency virus and acquired immune deficiency syndrome (HIV/AIDS) in the 1980s meant that TB was once again a worldwide risk, although both the incidence of cases and the death rate are greater in poor than in rich countries. Although TB is caused by infection, environmental conditions also play a role because it is more easily spread in crowded indoor areas with poor ventilation and little exposure to sunlight. In addition, smoking increases the risk of TB: the WHO estimates that smoking increases TB risk by two and a half times and that over 20 percent of the incidence of global TB may be attributed to smoking.

The WHO estimates that one-third of the world's population is infected with TB bacilli, although many will not become sick because their immune system can fight off the disease. In fact, only about 5 to 10 percent of infected people will become sick from TB at any point in their lifetime. However, if a person's immune system is weakened, for instance by HIV infection, then the probability of a latent infection becoming active is much greater. The largest number of new cases worldwide occurs in southeast Asia (34 percent of the world total in 2008), but the highest incidence rate is in sub-Saharan Africa (over 350 cases per 100,000 population). In 2008, about 1.3 million people died from TB, with the largest number of deaths in southeast Asia (477,000) and the highest rate in Africa (48 per 100,000). By contrast, in Europe the incidence rate was 48 per 100,000 people, and a mortality rate of 6 per 100,000; in the Americas, the incidence was 31 per 100,000, and the death rate 3 per 100,000.

In the United States, the number of TB cases and rate of infection are both declining: in 2008, there were 12,904 cases of TB, the lowest number since 1953 when national reporting began, and a rate of 4.2 cases per 100,000 population, which represents a 3.8 percent decline from 2007. In 2006, there were 644 deaths from TB, a 46 percent decline from 1996 (1,202 deaths). Interestingly, over half (59 percent) of the TB cases reported in 2008 occurred in people born outside the United States, and the TB rate was 10 times higher for foreign-born as opposed to native-born people (20.3 cases per 100,000 versus 2 cases per 100,000). There were also large differences by ethnicity: 1.1 cases per 100,000 for white Americans, 8.1/100,000 for Hispanics and Latinos, 8.8/100,000 for blacks, and 25.6/100,000 for Asians.

Pneumonia is an acute respiratory infection that causes the alveoli (small air sacs within the lungs) to fill up with pus and fluid. This limits oxygen intake and makes breathing painful. Globally, pneumonia and other lower-respiratory-tract infections are the second most common cause of illness (after diarrheal disease), with over 492 million cases reported in 2004, the greatest number in southeast Asia (134.6 million) and Africa (131.3 million). Worldwide, pneumonia is the largest cause of death among children, killing about 1.8 million children under age 5 annually (20 percent of all deaths in that age group). About 155 million cases of childhood pneumonia are diagnosed each year, with the greatest prevalence in sub-Saharan Africa and southeast Asia. Pneumonia is spread though the air (coughing, sneezing) and through blood (e.g., during or after childbirth), and persons in poor health or with weakened immune systems are most susceptible. In the United States in 2006, about 1.2 million people were hospitalized with pneumonia, and 55,477 died from it.

See Also: Air Filters/Scrubbers; Airborne Diseases; Asthma; Fungi and Sick Building Syndrome; Indoor Air Quality; Occupational Hazards; Radon and Basements; Smoking; Tuberculosis.

Further Readings

American Lung Association. "Chronic Obstructive Pulmonary Disease" (February 2010). http://www.lungusa.org/lung-disease/copd/resources/facts-figures/COPD-Fact-Sheet.html (Accessed August 2010).

National Cancer Institute, U.S. National Institutes of Health. "Mesothelioma: Questions and Answers." http://www.cancer.gov/cancertopics/factsheets/Sites-Types/mesothelioma (Accessed August 2010).

National Institute for Occupational Safety and Health. "Flavorings-Related Lung Disease." http://www.cdc.gov/niosh/topics/flavorings (Accessed August 2010).

Ryan, Frank. *The Forgotten Plague: How the Battle Against Tuberculosis Was Won—And Lost*. Boston: Back Bay Books, 1994.

Wagner, Gregory R. "Screening and Surveillance of Workers Exposed to Mineral Dusts" (1996). http://www.who.int/occupational_health/publications/mineraldust/en/index.html (Accessed August 2010).

World Health Organization. "Asbestos-Related Diseases." http://www.who.int/occupational_ health/topics/asbestos_documents/en/index.html (Accessed August 2010).

World Health Organization. "Respiratory Tract Diseases." http://www.who.int/topics/ respiractory_tract_diseases/en (Accessed August 2010).

World Health Organization. "Tobacco Free Initiative (TFI)." http://www.who.int/tobacco/en/ index.html (Accessed August 2010).

Sarah Boslaugh
Washington University in St. Louis

M

Malaria

Malaria is a life-threatening disease caused by *Plasmodium* parasites that are transmitted to people exclusively through the bites of infected anopheles mosquitoes. These mosquitoes are referred to as *malaria vectors*; *vector* used in this context refers to an organism that transfers a pathogen from reservoir to host. There are four types of human malaria:

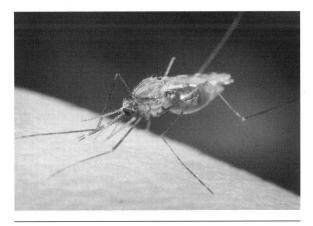

Malaria is a life-threatening disease caused by parasites that are transmitted to people exclusively through the bites of infected anopheles mosquitoes. Approximately half of the world's population is at risk of malaria, with most malaria cases occurring in sub-Saharan Africa.

Source: Centers for Disease Control and Prevention/ James Gathany

- *Plasmodium falciparum*
- *Plasmodium vivax*
- *Plasmodium malariae*
- *Plasmodium ovale*

Plasmodium falciparum and *Plasmodium vivax* are the most common types, and *Plasmodium falciparum* is also the most deadly. Approximately half of the world's population is at risk of malaria. Most malaria cases occur in sub-Saharan Africa; however, Asia, Latin America, and to a lesser extent the Middle East and parts of Europe are also affected. In 2008, malaria was present in 108 countries and territories.

In the human body, the *Plasmodium* parasites multiply in the liver and then infect red blood cells. Symptoms of malaria include fever, headache, and vomiting and usually appear within 10 to 15 days of the mosquito bite. If not treated within 24 hours, malaria can quickly become life threatening by disrupting blood supply to vital organs. In 2008,

269

there were 247 million cases of malaria and nearly 1 million deaths, mostly among young children in sub-Saharan Africa, where the disease claims the life of a child every 45 seconds. Factors related to the parasite, the vector, the human host, and the environment influence the intensity of malaria transmission. All of the main vector species bite at night, and they breed in shallow collections of freshwater. Transmission is more intense in places where the mosquito has a long life span, which allows the parasite time to reach maturity inside the mosquito. Further, vectors that prefer to bite humans rather than other animals also add to the intensity of transmission. Human immunity, developed over years of exposure, never provides complete protection, though it does reduce the risk that malaria infection will cause severe disease. For this reason, most malaria deaths in Africa occur in young children, although in areas with less transmission and low immunity, all age groups are at risk. Transmission also depends on climate conditions that may affect the abundance and survival of mosquitoes, including rainfall patterns, temperature, and humidity.

Growing resistance to antimalarial medicines has spread rapidly, undermining malaria control efforts. The best available malaria treatment is artemisinin-based combination therapy (ACT); when treated with an artemisinin-based monotherapy, some patients may discontinue treatment early, after the malaria symptoms disappear. This results in partial treatment, with persistent parasites remaining in the patient's blood. Without a second drug given as part of a combination (as is done with an ACT), these resistant parasites persist and can be passed on to a mosquito and then to another person. Monotherapies are therefore the primary force behind the spread of artemisinin resistance. If resistance to artemisinins develops and spreads to other large geographical areas, the public health consequences could be devastating because no alternative antimalarial medicines are expected to be available in the near future.

Vector control is the primary public health response for reducing malaria transmission; it is the only intervention that can reduce malaria transmission from very high levels to close to zero. Two forms of vector control are effective in most circumstances, including the following:

- *Insecticide-treated mosquito nets (ITNs)*: the World Health Organization (WHO) recommends universal vector-control coverage, and in most places, the most cost-effective way to achieve this is through provision of long-lasting, insecticide-impregnated nets (LLINs), so that everyone in high-transmission areas sleeps under an LLIN every night.
- *Indoor spraying with residual insecticides*: indoor residual spraying (IRS) with insecticides is the most powerful way to rapidly reduce malaria transmission. Its full potential is realized when at least 80 percent of the houses in targeted areas are sprayed. Indoor spraying is effective for three to six months, depending on the insecticide used and the type of surface on which it is sprayed.

Mosquito control is being strengthened in many areas, but there are significant challenges, including the following:

- Increasing mosquito resistance to insecticides, including DDT and pyrethroids, particularly in Africa
- Lack of alternative, cost-effective, safe insecticides

The health costs of malaria include both personal and public expenditures on prevention and treatment. In some countries, the disease accounts for the following:

- Up to 40 percent of public health expenditures
- Thirty to 50 percent of inpatient hospital admissions
- Up to 60 percent of outpatient health clinic visits

Malaria unduly affects poor people who cannot afford treatment or who have limited access to healthcare, which can make it difficult, if not impossible, for families and communities to escape poverty. Historically, many countries—especially in temperate and subtropical zones—have been successful in eliminating malaria. The global malaria eradication campaign, launched by WHO in 1955, was successful in eliminating the disease in some countries but ultimately failed to achieve its overall goal and was abandoned less than two decades later in favor of the less ambitious goal of malaria control. In recent years, interest in malaria eradication has reemerged. The WHO Global Malaria Program is responsible for evidence-based policy and strategy formulation, technical assistance, malaria surveillance, and coordination of global efforts to fight malaria. WHO is also cofounder and host of the Roll Back Malaria (RBM) partnership, a global framework for implementing coordinated action against malaria. The RBM partnership is composed of more than 500 allies, including malaria endemic countries, the private sector, nongovernmental and community-based organizations, and research and academic institutions.

Regrettably, where some see a dire need for malaria eradication, others see opportunity for exploitation. For example, in 2009, the United Nations Office on Drugs and Crime reported that about 45 million doses of fake antimalarials, valued at $438 million, are trafficked to west Africa from China and India annually by rogue manufacturers. In response to this profoundly disturbing and abusive scheme, mPedigree Network Limited has created a plan to include a scratch-off panel on containers of legitimate antimalarial drugs: the consumer scratches off the panel to reveal a 10-digit code, then uses a cell phone to send the code to a widely advertised phone number and receives a reply confirming or disputing the drug's authenticity. In some African nations, fake antimalarials make up half of the drugs sold for malaria, and worldwide, fake malaria and tuberculosis drugs are responsible for 700,000 deaths annually. mPedigree is working with telephone company Zain, Hewlett-Packard, and various drug manufacturers in order to begin implementing this plan.

Malaria is both preventable and curable, and actions such as education regarding how the disease is contracted, manifests, and spreads, along with the distribution of LLINs and use of indoor residual spraying, offer tools that can be used to eradicate this devastating disease.

See Also: Biomedicine; Cell Phones; Chemical Pesticides; Children's Health; Pest Control.

Further Readings

Centers for Disease Control and Prevention. "Malaria." http://www.cdc.gov/malaria (Accessed May 2010).

Rocco, Fiammetta. *The Miraculous Fever Tree: Malaria and the Quest for a Cure That Changed the World.* New York: HarperCollins, 2003.

World Health Organization. "Malaria." http://www.who.int/topics/malaria/en (Accessed May 2010).

Tani Bellestri
Independent Scholar

MANGANESE SOURCES AND HEALTH

Manganese (Mn) is a grayish-white transition metal with an atomic number of 25 in the periodic table of elements. Manganese is ubiquitous and occurs naturally in rocks and soil, often combined with other substances such as oxygen, sulfur, and chlorine. It can also be combined with carbon to form organic manganese compounds. Manganese is essential for normal physiologic functioning of animals and humans. Manganese is mostly used in metallurgical processes as a deoxidizing and desulfurizing additive and as an alloying constituent. Exposures to manganese can occur through metal fumes and manganese tetroxide, which is generated during the pouring of molten ferromanganese alloys. Health effects of manganese can be acute and chronic. Elimination of manganese exposures, occupationally and environmentally, should be the goal of any green industry.

Ferromanganese, an iron alloy containing manganese metal, is used in steel production. Manganese is also used in other products, including fireworks, dry-cell batteries, pesticides (Maneb, Mancozeb), matches, paints, and cosmetics. Organic manganese compounds such as methylcyclopentadienyl manganese tricarbonyl (MMT) are also used as additives in gasoline to improve the octane rating of the gas. Small amounts of manganese in the form of mangafodipir trisodium (MnDPDP) are used in magnetic resonance imaging of body organs. Confined-space welding can significantly increase exposure to manganese fumes.

Manganese is a cofactor for many enzymes and therefore nutritionally essential for humans. The average daily requirement is around 2 to 5 milligrams per day for adults, but higher requirements have also been suggested. Sources of exposure to manganese are foods such as grains, beans, nuts, tea, green leafy vegetables, or manganese-containing nutritional supplements. Manganese is routinely contained in groundwater, drinking water, and soil at low levels. Releases of manganese into the air occur from industries using or manufacturing products containing manganese, mining activities, and automobile exhaust. Manganese enters the body through ingestion, inhalation, and dermal (skin) contact.

Acute effects include irritation of eyes, skin, and respiratory tract. Skin manifestations can include dermatitis and burns. Acute overexposures to MMT can cause chemical pneumonia and damage the liver and kidneys. Potassium permanganate ($KMnO_4$) is a tissue corrosive. Exacerbations of asthma and bronchitis are seen in manganese-exposed workers.

Chronic occupational exposure to manganese is associated with manganism, a condition reminiscent of Parkinson's disease with tremors, rigidity, and bradykinesia. Earliest manifestations include anorexia, somnolence, fatigue, headaches, and apathy. Episodes of absentmindedness, excitability, hallucinations, garrulousness, and sexual arousal, termed *manganese psychosis*, may develop. Effects on the brain through retrograde transport in nerve cells of the nose have also been proposed. It has been reported that Bangladeshi children exposed to well water with high concentrations of manganese had diminished intellectual function. The researchers noted that the bioavailability of manganese in water was higher than in food. They noted that about 6 percent of U.S. wells have manganese content high enough to potentially put some children at risk for diminished intellectual function. Limited data show that manganese can decrease libido and cause impotence. Chronic manganese exposure can also increase susceptibility to respiratory infections.

With continued exposure, speech can be affected, leading to muteness, clumsiness, "cock walk," and a fixed facial expression.

Studies have found a relationship between manganese pollution and traffic density because of its use in antiknock agents in gasoline. There have been reports of manganese pollution associated with violent crime. Concerns about dispersion of manganese salts via tailpipe emissions from use of MMT have been raised. Some environmental models have suggested that airborne respirable manganese particulates because of MMT in gasoline are low in relation to background levels. However, there is a need for further research to assess health effects in susceptible populations.

Manganese was mined increasingly during the world wars because of its use in machinery production. The Milford mine disaster was associated with manganese mining.

Process enclosures and improved general ventilation are necessary to reduce manganese exposure in industries. Local exhaust ventilation can control exposures to manganese and total fumes, but mechanical ventilation may not. Using HEPA filters in vacuums and personal protective equipment including respirators with proper cartridges are a must. On a global level, manganese is yet another metal, the reduction of which should be the focus of a green health movement, including phasing out from gasoline and stricter regulations on emissions in the industry.

Medical surveillance of workers exposed to MMT and manganese fumes should include examinations of the respiratory system and liver and kidneys. Assessment of ongoing exposures and medical symptoms should be left to specialists in occupational and environmental medicine and other specialists including neurologists and pulmonologists.

See Also: Automobiles (Emissions); Chemical Pesticides; Fertilizers; Green Chemistry.

Further Readings

Agency for Toxic Substance and Disease Registry. "Toxicological Profile for Manganese." http://www.atsdr.cdc.gov/toxprofiles/tp.asp?id=102&tid=23 (Accessed January 2010).

Ewen, C., M. A. Anagnostopoulou, and N. I. Ward. "Monitoring of Heavy Metal Levels in Roadside Dusts of Thessaloniki, Greece in Relation to Motor Vehicle Traffic Density and Flow." *Environmental Monitoring Assessment*, 157/1–4 (2009).

Flynn, M. R. and P. Susi. "Manganese, Iron, and Total Particulate Exposures to Welders." *Journal of Occupational and Environmental Hygiene*, 7/2 (2010).

LaDou, Joseph. *Current Occupational & Environmental Medicine*, 4th ed. New York: McGraw-Hill Medical, 2006.

Occupational Safety and Health Administration. "Manganese." http://www.osha.gov/SLTC/metalsheavy/manganese.html (Accessed January 2010).

Rosenstock L., et al. *Textbook of Clinical Occupational and Environmental Medicine*, 2nd ed. Philadelphia, PA: Elsevier Saunders, 2005.

Szpir, Michael. "New Thinking on Neurodevelopment." *Environmental Health Perspectives*, 114/2 (2006).

World Health Organization, Regional Office for Europe. "Manganese." http://www.euro.who.int/document/aiq/6_8manganese.pdf (Accessed January 2010).

Abhijay P. Karandikar
Pines Health Services

Men's Health

That men and women face different health challenges based both on their physiology and on the kinds of work and play they engage in is probably no surprise, but there is another big difference that profoundly affects men's healthcare experience—how they care for themselves. For example, in the United States, male patients account for only 40 percent of visits to doctors. Further, in a telephone poll conducted by CNN and *Men's Health* magazine, it was found that 76 percent of women in the United States had had a health exam in the previous year, compared to only 60 percent of men. Though these numbers vary with race, class, culture, and ethnicity, in general, women in the United States are more likely both to seek care for health ailments and to ask more questions regarding their care and treatment than are their male counterparts.

Many theorize that men are reluctant to appear helpless and fragile and, thus, either put off seeking medical care or seek it passively, without asking as many questions as do women regarding diagnoses, prognoses, or treatment. Whether it is due to women's accepted function as caregivers or perhaps to the fact that while women have gynecologists—"women's doctors"—there is no comparable "men's doctor," women are often seen as more engaged and savvy consumers of healthcare services than are men. Some in the healthcare field even describe men as being more willing to give up control in a healthcare setting than are women. There is some evidence that suggests that men are less passive when dealing with the healthcare issues of the women in their lives than with their own health crises, and this is often explained as being a function of the fact that men are more comfortable as "protectors" than as "dependents," or, even, as "burdens."

That a man's lifespan is about eight years shorter than a woman's is directly tied to the fact that men take more risks and wait longer to seek treatment for healthcare issues than do women. Whether this is because men see themselves as providers and thus are compelled to ignore or hide their frailties, or for some other reason or combination of reasons, advocates for men's health are focused on empowering men to take their healthcare more seriously and to become the kinds of healthcare consumers who are not just proactive, but who are also dynamically engaged with their healthcare experiences.

Some point to the 1990s as a turning point in the healthcare system and in how people, including men, approach their healthcare. By that time, patients had become "healthcare consumers," a reordering that grew from the actions of various groups, including male homosexuals, women, blacks, Hispanics, and intravenous drug users, who all had begun to push for the right to access to information that would foster making informed decisions regarding their health. These groups, which had traditionally been disregarded by the vast medical establishment, as well as by the culture at large, used the AIDS crisis and concern over breast cancer as a springboard for demanding not just visibility but also quality care.

Galvanized by the actions of these traditionally marginalized groups as well as by examples of strong male figures such as General Norman Schwarzkopf, who used his battle with prostate cancer as an opportunity to urge men to get prostate exams, men today are more engaged with their healthcare needs than in previous decades, though they continue to lag behind women in seeking out preventative care and in being assertive in their demands for information.

Some believe the most effective route to getting men to take their healthcare needs more seriously is to appeal to the women in their lives to badger them into check-ups and follow-ups. Others believe that changing the language of healthcare may be more beneficial—calling check-ups "tune-ups," for example, and working to minimize the use of words like "care" when talking about men's health. Others are uncomfortable with this approach because it relies on notions of gender and does not foster any substantial changes in the healthcare system or in how men may view it or approach it. Still others advocate stressing that men's sexual functioning is directly tied to their health in general and that nurturing a mind–body connection in men (that connects their physical well-being, mental health, and sexual functioning) could compel men to be more proactive in caring for themselves. Again, this approach is criticized by some who find fault in buttressing men's notions of themselves as highly motivated sexual creatures instead of as, simply, human beings who require healthcare for more reasons than simply being able to function sexually.

Some diseases are viewed through a sex lens, such as heart disease, which has received significantly more attention among the male than female population, despite the fact that heart disease is the number one killer of women. A parallel to this phenomenon of focusing on one group over another in regard to certain diseases and health issues can be found in how we diagnose and treat osteoporosis. Long considered a woman's disease, because 80 percent of those affected are women, diagnosis and treatment of osteoporosis in men is woefully inadequate. Until recently, there were no approved medications for male osteoporosis and no research on the disease in men. Today, more than 2 million men in the United States have been diagnosed with osteoporosis, and one in four men will suffer a bone fracture due to osteoporosis, though the disease remains underreported and underdiagnosed among the male population. Again, due to cultural notions about gender, many men have a difficult time accepting being diagnosed with what is considered a "woman's disease." Similarly, men struggle with diagnoses of breast cancer, which claims the lives of about 450 men annually in the United States; about 2,000 new cases of male breast cancer are diagnosed each year. While diagnosing breast cancer in men is not difficult, getting them to go to the doctor after finding a lump or other problem with their breasts is not so simple; either because they are embarrassed at the possibility of having a "woman's disease" or because they do not know that they can actually have breast cancer, too often men ignore the signs and symptoms of this disease.

Heart Disease and Stroke

In the United States, the leading cause of death for all men, except for Asians or Pacific Islanders, is heart disease. For Asians and Pacific Islanders, heart disease is second to cancer as the leading cause of death, while cancer is the number two leading cause of death for other groups of men in the United States (American Indian, white, black, Hispanic). Some researchers have called into question statistics regarding race, ethnicity, and disease, noting that these categories are often misreported, either by patients themselves or by errors in data entry. For all U.S. men as a group, heart disease and cancer account for just over 50 percent of all deaths of men annually. Half of the men who die suddenly of coronary heart disease have no previous symptoms, so absence of symptoms is not an indicator of the heart's well-being. This is significant in light of the fact that men have been shown to be more reluctant than are women to have annual physical exams. A mix of lifestyle choices and medical conditions contribute to heart disease, including the following:

- High cholesterol
- High blood pressure
- Diabetes

- Cigarette smoking
- Overweight and obesity
- Poor diet
- Physical inactivity
- Alcohol use

Strokes are also a leading cause of death among men in the United States, accounting for about 5 percent of all deaths of men annually. Black men and Asians/Pacific Islanders are at higher risk for strokes than are whites, American Indians, or Hispanics. Factors that contribute to stroke prevalence are similar to those for heart disease, with high blood pressure being the leading indicator of the potential to suffer a stroke. Heart disease itself is also a risk factor, as those with heart disease are at twice the risk of having a stroke as are those with healthy hearts.

Since 1998, the U.S. Centers for Disease Control and Prevention (CDC) has spearheaded a national, state-based heart disease and stroke prevention program. Currently, 41 states and the District of Columbia participate in the program. Some of the program's features include the following:

- Facilitating collaboration among the public and private sectors
- Identifying culturally appropriate methods for promoting heart disease and stroke prevention among racial, ethnic, and other priority populations
- Executing and assessing policy, environmental, and educational interventions in healthcare sites, work sites, and communities
- Providing training and technical assistance to healthcare professionals and partners to support primary and secondary heart disease and stroke prevention.

The goals of the program include the following:

- Increasing state capacity by planning, implementing, tracking, and sustaining population-based interventions that address heart disease, stroke, and related risk factors
- Monitoring heart disease and stroke and related risk factors and evaluating policy and environmental support for heart disease and stroke prevention within states
- Identifying favorable strategies for stimulating heart-healthy interventions on the state level
- Promoting cardiovascular health in healthcare, work site, and community settings through education and policy and environmental changes

Examples of the program in action include the following:

- Sponsoring public awareness campaigns about signs and symptoms of heart disease and stroke
- Implementing informational campaigns to educate the public that high blood pressure is a major risk factor for heart disease and stroke—and that high blood pressure can be adjusted
- Raising awareness and providing education about risk factor and lifestyle changes that contribute to the risk of heart disease and stroke

Cancer

Cancer is the second-leading cause of death for men in the United States as a whole, and there are about 700,000 men diagnosed with cancer annually in the United States. Technological advances in detection and treatment have led to an increasing number of cancer survivors, though about 300,000 men die of cancer each year in the United States. Skin cancer is the most prevalent form of cancer among both men and women. The majority of those diagnosed with skin cancer are white men over the age of 50; melanoma is the fifth most common cancer in men and accounts for 5 percent of all cancers in men. Though more women than men are diagnosed with skin cancer, more men die from it. Again, some suggest that there may be some kind of cultural or gender-based proscription at play with these numbers, as it has been found that men are more resistant to using sunscreen than are women, and men take longer to visit the doctor after noticing symptoms of skin cancer—either because they do not recognize the symptoms or are embarrassed to have gotten a disease that is, oddly, perhaps, considered a "women's disease." More than 5,000 men die each year from skin cancer.

Incidence rates of cancer in men, in brief, include the following:

- Lung cancer claims the lives of more men than any other type of cancer—about 100,000 men die of lung cancer each year in the United States. Around 90 percent of the men who die of lung cancer are cigarette smokers.
- Apart from skin cancer, prostate cancer is the most common cancer in men in the United States, and it is the second-leading cause of cancer death in men. Some of the factors that increase the risk of prostate cancer include older age and a family history of prostate cancer; black men are also more likely to develop prostate cancer than are men from other racial or ethnic categories. About 30,000 men die each year in the United States from prostate cancer.
- Colorectal, or, colon, cancer is the third-leading cause of cancer death among men in the United States, and the incidence of colon cancer is rising among young men in their 20s and 30s. Diets high in fat and a family history of colon cancer are risk factors for developing this type of cancer. Though it is considered one of the more preventable and treatable types of cancer, colon cancer still claims the lives of around 30,000 men in the United States each year.

Other Health Issues

In addition to heart disease, stroke, and cancer, men in the United States lose their lives to unintentional injuries, chronic lower-respiratory diseases, diabetes, suicide, homicide, influenza and pneumonia, kidney disease, and Alzheimer's disease. While the events and diseases that claim the lives of men are often induced by factors that cannot be controlled, it is believed that more than 1 million premature deaths of men in the United States occur each year due to lifestyle factors such as smoking, unhealthy diet, and lack of exercise. Men are more likely to smoke than are women, and a study by the CDC found that women of all ethnic backgrounds eat a healthier diet than do their male counterparts, including far more fruits and vegetables in their diets than do men. Men do exercise more often than do women in the United States, though with only about 16 percent of all men and women over age 15 exercising or participating in some kind of sports activity daily, there is clearly a need for both men and women to initiate a daily exercise program into their lives.

Not smoking, eating a healthy diet, and exercising regularly are also crucial to healthy sexual functioning in men. Smoking, high blood pressure, high cholesterol, and diabetes are all related to sexual dysfunction in men. Researchers theorize that because these behaviors and conditions all damage blood vessels, they also affect the minute blood vessels in the penis, causing them to become inelastic and obstructed. It is important for men to remember that all men experience erection problems at some point in their lives, whether due to alcohol, stress, physical or mental exhaustion, or anxiety. Temporary or more long-lasting bouts of impotence can exact a great emotional toll on men, and seeking out information about impotence and trusting and involving your partner in the process of dealing with impotence are vital factors in dealing with it in a positive, effective manner. Further, though sexual functioning may change with age, growing older should not be seen as an end to sexual pleasure—surveys have found that about 30 percent of men in their sixties have sex weekly.

Gay and bisexual men in the United States face particular health challenges, including dramatically higher rates of human immunodeficiency virus and acquired immune deficiency syndrome (HIV/AIDS) than their heterosexual counterparts, as well as increased rates of gonorrhea, anal warts, and hepatitis A and B. While not considered a health condition, per se, gay and bisexual men are also subject to verbal and physical "gay bashings" that range from shouted slurs and threats to violent torture and murders. Many point to this violence as the source of at least some of the mental illness suffered by gay and bisexual men in the United States. HIV/AIDS has received the most attention of all these factors that affect gay and bisexual men's health, and perhaps rightly so—while they comprise a minority of the U.S. population, approximately one half of the 1.1 million persons living with HIV in the United States are gay and bisexual men, and they account for the majority of new HIV infections each year. Among men in the United States diagnosed with AIDS in 2008, 52 percent of black men, 63 percent of Hispanic/Latino men, and 78 percent of white men became infected with HIV through male-to-male sexual contact. Gay and bisexual men of all races are the only group in the United States where the projected number of new HIV infections is rising annually.

Physical health is intimately tied to mental health in both men and women, and men, like women, suffer from a variety of mental illnesses and disturbances. For example, in the United States, there are more than 6 million men each year who suffer from depression. And while more women than men are diagnosed with anxiety disorders, men and women suffer equally from obsessive-compulsive disorder and social phobia. Schizophrenia also affects men and women equally, though in men, it tends to manifest earlier than in women. Post-traumatic stress disorder (PTSD) is an anxiety disorder that can manifest after being a victim of or witnessing a terrifying event or series of events. It is most often associated with rape, sexual abuse, and combat service, though natural disasters, transportation accidents, and violent nonsexual attacks can all trigger PTSD. The rate of PTSD is highest for soldiers, nearly 14 percent of whom develop PTSD. Among the civilian U.S. population, PTSD affects about 4 percent of men and 10 percent of women.

It is obvious that combating many of the health problems that plague men could be greatly buttressed by education that helps to foster lifestyle changes, such as smoking cessation, embracing a healthy diet and exercise regime, and reducing the occurrences of unprotected sex. Further, improving men's health faces a particular challenge in convincing men not only to take symptoms of disease seriously and to seek out healthcare for their illnesses but to be more proactive about preventative healthcare, or "tune-ups."

See Also: Acquired Immune Deficiency Syndrome; Cancers; Cardiovascular Diseases; Centers for Disease Control and Prevention (U.S.); Healthcare Delivery; Physical Activity and Health.

Further Readings

Centers for Disease Control and Prevention. "Men's Health." http://www.cdc.gov/men (Accessed August 2010).

Male Health Center. "Not for Men Only." http://www.malehealthcenter.com (Accessed August 2010).

MayoClinic.com. "Men's Health." http://www.mayoclinic.com/health/mens-health/MY00394 (Accessed August 2010).

New York Times. "Don't Take Your Medicine Like a Man." http://www.nytimes .com/1999/02/17/health/don-t-take-your-medicine-like-man-patients-men-are-impatient-uneasy-both-they.html (Accessed August 2010).

Tani Bellestri
Independent Scholar

MENTAL EXERCISES

Prevention of dementia associated with aging is a public health concern due to its rising prevalence. The general public is also routinely confronted with commercial advertisements for products claiming to be able to prolong cognitive competency, with catchy names such as "neurobics," which are marketed as "good for the brain" and "clinically proven" to reduce the loss of mental function. This entry focuses on the use of mental exercises to improve cognition and slow the onset of dementia, as well as its relevance to health more broadly.

There is an absence of pharmacological treatment with which to counteract dementia, and consequently, lifestyle factors have become an area of attention in this field. Diet, social engagements, physical activity, and cognitive activity are among the factors thought to be involved in the course of brain decline with age. While it has been demonstrated that abilities such as memory and reasoning tend to decline beginning on average in the mid-60s, there is less concrete evidence of what can be done to prevent this. Recent research has shown promising results, indicating that practicing cognitive activities correlates with both a slowing of mental decline and with an improvement in mental function. It has been hypothesized that mental exercises act to maintain brain function with aging by strengthening existing neuronal connections as well as increasing the growth of new neurons in the brain. Study participants were required to use a broad range of mental skills to mimic the cognitive requirements of various activities, including organization, visualization, pattern recognition, reasoning exercises, and processing speed. The largest randomized, controlled trial to date demonstrated improvement in mental function immediately after mental exercise training sessions, as well as diminished rates of cognitive decline five years later compared to what would be expected for the age group.

Studies on the correlation between mental exercise and cognitive function have proved difficult, however, due to factors that are challenging to control and outcome measures that do not adequately gauge the differences between intervention groups. There are very few data to show what form of mental exercise might be most beneficial, the time that must be dedicated to such exercises, or at what age an individual should begin to engage in such activities. Recommendations to physicians based on the available data are able to go only so far as to say that older adults should be encouraged to participate in stimulating activities. Even then, it is unclear if these behaviors will do something positive for mental function in addition to enhancing quality of life.

The studies conducted in this field have limitations that must be taken into consideration should further research be undertaken. First, study methods tend to focus on specific cognitive abilities directly related to the mental exercises performed during the trial, rather than functional outcomes and everyday tasks usually performed by elderly individuals such as housework or shopping. With the currently available data, it is not feasible to say whether cognitive training transfers to the real, commonly performed tasks. Additionally, trial outcomes that did have some connection to real-life activities tended to be measured by self-reporting rather than by performance-based measures. Gathering generalizable data necessitates recruiting individuals with a certain level of brain function; thus, it is difficult to conclude whether similar results would be observed in a population with functional loss already present.

Despite these limitations, there have not been any studies that failed to find an association between cognitive activities and outcomes of mental and/or daily function in at least a specific cognitive domain being measured. Review articles on the topic of mental exercises and cognitive function have expressed optimism toward further research and the possibility that future data will demonstrate a more concrete correlation between brain exercise and better cognitive function in the aging individual.

With no established or straightforward method for evaluating the effectiveness of mental exercises, the benefit of such exercises in older individuals actively engaged in cognitively stimulating activities may be going unnoticed. There is not substantial evidence by which those interested in using mental exercises can gauge what activities or programs will be most advantageous. A simple augmentation of daily cognitive stimulation may or may not be as effective as concrete, systematic mental exercise programs. Additionally, other factors shown to correlate with decreases in cognitive function play a role in the ultimate individual gain from mental exercise. These include physical activity, diet, level of education, occupation, and genetic factors. Because there is not evidence demonstrating that any of these factors alone will protect against dementia, it is sensible to suggest that a concerned or proactive individual should perform some type of cognitive exercise as well.

It is certainly discouraging for the aging population to be told there is nothing specific that can be done to diminish rates of cognitive decline. Those willing to put some amount of time and effort into preventing mental decline surely would benefit from having a better understanding of the effectiveness of mental exercise programs. It is possible this could result in reduced levels of health service utilization, a greater quality of life, and less dependence on caretakers for older adults. Therefore, trials of the effects of cognitive training on the onset and course of dementia should be pursued, and researchers should persevere in finding study designs and outcome measures that can adequately provide information on mental exercises.

See Also: Degenerative Diseases; Healthcare Delivery; Mental Health; Pharmaceutical Industry.

Further Readings

Acevedo, A. and D. Loewenstein. "Nonpharmacological Cognitive Interventions in Aging and Dementia." *Journal of Geriatric Psychiatry & Neurology*, 20 (2007).

Ball, K., D. Berch, K. Helmers, J. Jobe, M. Leveck, M. Marsiske, J. Morris, G. Rebok, D. Smith, S. Tennstedt, F. Unverzagt, and S. Willis. "Effects of Cognitive Training Interventions With Older Adults: A Randomized Controlled Trial." *Journal of the American Medical Association*, 288/1 (2002).

Willis, S., S. Tennstedt, and M. Marsiske. "Long-term Effects of Cognitive Training on Everyday Functional Outcomes in Older Adults." *Journal of the American Medical Association*, 296/2 (2006).

Sylvie R. Boiteau
University of Massachusetts Medical School

MENTAL HEALTH

According to the burden of disease assessments conducted by the World Health Organization (WHO), mental illness, also known as neuropsychiatric conditions, accounts for the largest human burden of noncommunicable diseases at the global level, representing approximately 14 percent of disease burden in all categories. Mental health is a complex concept to define precisely. It generally refers to a person's state of emotional well-being and an absence/presence of neuroses or other psychiatric conditions. A holistic view includes a wider range of factors: multiple interacting biological, environmental, social, and psychological variables. For example, WHO defines good mental health as a state of well-being that signifies the ability to recognize individual abilities, cope with life's normal stresses, work productively, and contribute positively to society.

In addition to established definitions, an emerging body of research shows that an environmentally aware lifestyle, and a healthy interaction with nature, nurtures mental well-being. See, for example, the Biophilia Hypothesis that was proposed and developed by Edward O. Wilson, Stephen R. Kellert, and Oladele A. Ogunseitan's study on the association of topophilia with quality of life. Indeed, nature-based therapies and opportunities to participate in green projects are being recognized as possible complementary approaches to the prevention and management of health problems. The need for effective approaches to understanding and promoting good mental health has never been greater. The number of people worldwide affected by a mental health problem is increasing, and a large number of people are likely to experience an illness at some point in their lifetime. This explores the role that green living can play in mental well-being, beginning with a brief overview of the relationships between mental health and encounters with nature; ecofriendly lifestyles; the use of green care as a complementary therapy; and healthcare systems. This entry also considers the future of green health in terms of health policy and increasing acceptance by health professionals.

Natural Landscapes and Well-Being

Natural landscapes promote a sense of emotional well-being. Contact with the environment, sometimes even minimally, has been associated with positive changes in physical and mental health. Studies have linked broad active and passive exposure to nature to a number of mental health variables, including reduced levels of stress, improved mood, increased social behavior, decreased levels of depression, and improvement in cognitive functioning. Encounters with the natural world may also be important for children's physical, emotional, and intellectual development.

Classic examples of this relationship include greater postoperative recovery in hospital rooms with views of a natural setting; less sickness among prisoners in cells that overlook farmland and trees; greater life satisfaction among workers who have window views of nature; reducing stress by spending time in the wilderness; and relationships between health complaints and exposure to green space in urban areas. Caring for pets has also been shown to have beneficial effects.

Several biologists and psychologists have suggested that the human affiliation with nature may be a product of evolution, refined via experience and culture. This implies an instinctive emotional bond between humans and habitats, activities, and objects in the natural surroundings; and could explain human response and social and cultural connection to the natural world.

The innate relationship between the planet's ecological health and human health has important implications. Current styles of modern living, especially urban living, are often disconnected from nature. Today's widespread environmental destruction might stem from such disconnection, in which people do not acknowledge relationships between their lifestyles and ecological destruction, a situation in dire need of change. Emotional and spiritual values—the foundations of a functioning, happy society—are intricately linked to nature, and could be jeopardized by the current ecological crisis. This calls for urgent action to protect natural resources, promote pro-environmental behavior, and conduct research into how people develop a more positive relationship with the environment.

Green Lifestyles and Mental Well-Being

Healthy lifestyles based on green living are important for mental well-being. Certain foods and nutrients are known to impact on short- and long-term mental health. A lack of certain vitamins and minerals is linked to problems such as Alzheimer's disease, attention deficit hyperactivity disorder (ADHD), depression, and schizophrenia. It has been suggested that dietary inventions can help people to manage their moods and feelings, although further research is needed on this topic.

The relationship between food and mental health hinges on farming methods and food processing. Since World War II, a drive to increase agricultural production has led to increased use of chemical fertilizers and pesticides. This has been linked to the development of cognition in babies and to mental health disorders in older people. For example, reviews of the literature suggest a possible association between exposure to pesticides and Parkinson's disease and other disorders. Modern-day farming methods such as intensive farming have other consequences, particularly, decreased nutritional profile of meat, environmental pollution, and low animal welfare standards. This is in contrast to sustainable agriculture and organic farming, which focus on maintaining long-term ecological health.

Nowadays, attitudes toward food are based more on convenience than on green health. In developed, and increasingly in developing, countries, intake of nutritious food like fresh fruit and vegetables has declined, whereas intake of fast and processed food has increased. Processed foods are high in calories, fat, salt, chemical additives, and are more devoid of nutritional elements. Some researchers have suggested that these changes in nutrition may be a contributing factor to the increasing prevalence of specific mental health problems in society, particularly among children and adolescents.

Ecofriendly lifestyles as well as ethical living in a broader sense are gaining popularity. More people are taking positive action to improve both their health and the health of the environment when making daily decisions. These include responsible shopping and green consumerism; organic food and goods; energy efficiency and recycling; ecological footprints; green home improvements and holidays; as well other social and environmental considerations. There are many guidebooks on the market to help people consider their daily behavior in relation to the environment, health, and ethics.

Green Care Therapy

In the 21st century, the prevailing treatment for a mental health issue is an integrative approach that often involves medication and psychotherapeutic interventions, such as counseling and cognitive behavioral therapy, sometimes alongside complementary methods. An increasing number of healthcare organizations and professionals have suggested that nature-based methods and opportunities to participate in green projects—known as green care therapy—could be used as a complementary way for psychological healing. Preliminary research indicates that green care therapy might promote mental well-being by broadening people's relationships with the natural world, offering a spectrum of cheap, safe, and immediately available alternatives.

An example of green care therapy is green care farming, also known as social and therapeutic horticulture. Care farms provide opportunities for people to access agricultural landscapes, participate in farm work, and interact with animals. They are provided as a service to people with mental health difficulties, disabilities (intellectual or physical), drug/alcohol rehabilitation services, and prisoner and rehabilitation services. In addition to health benefits, social farming benefits farmers by providing rural business and development to maintain the economic viability of their farms. Care farms are new concepts in some countries, although in some European countries (e.g., the Netherlands and Norway) they are more developed and are integrated into the healthcare system.

Animal-assisted therapy (AAT) involves the use of an animal in therapy, either as a treatment by itself or as part of a patient's therapy program and in institutional settings such as schools, hospitals, residential homes, and prisons. Many kinds of animals are used in AAT, including birds, cats, dogs, fish, guinea pigs, pigs, horses, and, perhaps best known of all, dolphins. AAT is used in treatments for anxiety disorders, clinical depression, eating disorders, ADHD, autism, schizophrenia, and others. Studies have associated AAT programs with changes in physical, social, emotional, and cognitive functioning and with improved outcomes in behavioral problems and emotional well-being.

There are, of course, many other ways to experience nature. For example, Mind, a U.K.-based charity, recently launched an ecotherapy campaign to offer opportunities to access green space and participate in green projects. Examples include fishing, horseback riding, canal boating, walking, gardening, and local nature conservation work. Preliminary research indicates that exercising in green environments, such as taking regular walks in the countryside or parks, might improve psychological well-being to a greater extent than exercise alone, or than exercise while viewing images of rural or urban environments.

Mind also argues for a new green agenda to increase awareness of and to promote green mental health. Their recommendations include increased recognition of green therapy as a possible treatment for mental distress; incorporation of referrals to green care projects into health and social care referral systems; funding for complementary therapies; recognition by architects and town planners of the necessity of access to green space; tackling inequality of access to green space; and promoting the benefits of green exercise via public health campaigns.

Mental Healthcare Systems

Like many other business sectors, the healthcare industry has begun to pay more attention to its impact on the environment. Healthcare institutions are in a good position to join the green movement because they have purchasing power to switch to greener practices and a large of number of employees who can participate in corporate and individual social responsibility. Furthermore, the important symbolic significance of hospitals, as powerful symbols of community health, puts healthcare professionals at the forefront of the green movement as suitable advocates to take a lead role in the adoption of green health initiatives.

Healthcare systems adopt six main green approaches: the reduction or elimination of medical waste incineration (which produces dioxin pollutants) by using alternatives; use of safer chemicals, for example, not using mercury products; adoption of green working practices such as resource conservation and pollution; green purchasing; sustainable design and construction of the healthcare industry; and effective environmental management. As mentioned earlier, naturalistic environments in hospitals, including nature images and sounds, have been found to be supportive of patient recovery, and studies of subjective responses to healthcare environments indicate high levels of approval for such designs.

Numerous entities have been established to promote green healthcare. For example, Health Care Without Harm is a global coalition of several hundred organizations around the world, including mental health providers, working to reduce pollution. The coalition aims to transform the worldwide healthcare sector so that it is ecologically sustainable. Their strategies include the creation of markets for safer and alternative healthcare products; use of safer waste treatment practices; design of healthcare facilities to minimize environmental impact; and provision of healthy food and sustainable food production/distribution. These activities will help the healthcare industry to reduce its ecological footprint and to build holistically centered mental healthcare systems.

Limitations

A myriad of challenges hinder the development of green mental health, both theoretically and practically. The important role that nature plays in mental health has yet to be fully understood and embraced. One problem is the lack of rigor of current research. Some claims in the literature, particularly earlier research, are anecdotal or based on preliminary research. Studies are remarkably sparse and raise many questions, suggesting caution in accepting some arguments. Although research to date indicates that green activities might improve mental health and well-being, research is limited and preliminary. More studies are needed to investigate and evaluate whether environmentally based mental health treatments can benefit people with mental distress. In particular, there is a need for more research, such as projects comparing green therapy to traditional and complementary therapies and for well-articulated theoretical formulations.

Despite the huge potential of green mental health therapy, it is currently not widely regarded by healthcare professionals as a serious treatment option. Some people regard it as a money making gimmick. It is rarely acknowledged by mainstream medicine or psychology. How many doctors, for example, would consider a local conservation project or care farming as a complementary treatment option for people with a mental health difficulty? The profession is either unaware of the evidence or skeptical of its quality.

It should also be noted that mental health, and its connection to green living, is complex. The factors discussed above are only some of a large number of factors affecting mental health. Social, cultural, physical, and educational factors are all important. Meanings of health and well-being differ across cultures, disciplines, and theoretical positions, and not everyone agrees that green living is important for well-being. What is provided in this article is general information that is presented as an introduction to the topic and may not apply to individual cases.

Nonetheless, mental health academics and professionals are exhibiting a growing interest in green mental health. Also, new cognate fields are emerging, such as eco-psychology, which adds an ecological dimension to assumptions of Western psychology to help initiate healthy environmental behavior; environmental psychology, which examines the interplay between people, their environment, and behavioral systems; conservation psychology, which focuses on how to encourage conservation of the natural world; and arts for health, a broad movement that utilizes environmental art and design in a number of ways in mental health settings. These growing fields will help to tackle modern society's estrangement from its natural origins.

The Future

The concept of green mental health has far-reaching implications. That green living is a necessary condition for the functioning of individuals and communities shows clearly that people are inseparable from the rest of nature. The examples discussed in this article signify the beginning of a new movement in daily living, academia, and in the health and mental health industries.

The next step is official recognition of the links between green living and mental health, as part of national frameworks and health policies. Access to good quality green space can be an effective strategy to promote health, well-being, and quality of life. Policies that promote green health can complement and share the goals and traditions of other types of health promotion. They can include, for example, promotion methods centered on public information and education; creation of environments to promote healthy lifestyles and mental well-being; and access to green space by making opportunities available, affordable, and attractive.

Taken together, mental health and well-being is a complex combination of variables and factors, including the involvement of green living. The consistent message in the literature is that while the status of green mental health is relatively new, at a preliminary stage, and more research is needed, its potential impact is significant. Tentative findings suggest that direct and indirect exposure to nature does seem to be a critical component for mental well-being and could be a complementary approach to more traditional and conventional approaches.

Given that the world has finally realized that the environmental degradation of our planet poses a serious challenge to humankind, there is a great need to promote psychologically healthy societies that fully integrate nature. More studies are needed to give proper credence to the role that nature plays in the maintenance of good health, prevention of problems, and a functioning society. This also implies that there are health benefits from joining together to mitigate the effects of environmental mismanagement and modern-day lifestyles.

See Also: Anti-Depressant Drugs; Biomedicine; Children's Health; Environmental Illness and Chemical Sensitivity; Environmental Protection Agency (U.S.); Healthcare Delivery; Hospitals (Carbon Footprints).

Further Readings

Buzzell, Linda and Craig Chalquist. *Ecotherapy: Healing With Nature in Mind*. San Francisco: Sierra Club/Counterpoint, 2009.

Clark, Duncan. *The Rough Guide to Ethical Living*. London: Penguin, 2006.

Geary, Amanda. *The Mind Guide to Food and Mood*. London: Mind, 2004.

Hassink, Jan and Majken van Dijk. *Farming for Health: Green-Care Farming Across Europe and the United States of America*. Wageningen UR Frontis Series. Dordrecht, Netherlands: Springer, 2008.

Hillman, Mayer and Tina Fawcett. *How We Can Save the Planet*. London: Penguin, 2004.

Kellert, Stephen and Edward Wilson. *The Biophilia Hypothesis*. Washington, DC: Island Press, 1995.

Ogunseitan, O. A. 2005. "Topophilia and the Quality of Life." *Environmental Health Perspectives*, 113/2 (2005).

World Health Organization. "Mental Health." http://www.who.int/mental_health/en (Accessed September 2010).

Gareth Davey
University of Chester

Mercury Sources and Health

Cycling through both natural and anthropogenic activities, mercury is a persistent substance found globally within the environment. Mercury vapor is the primary form found in the atmosphere. It is released through the weathering of rock that contains mercury ore, the combustion of fossil fuels, and the incineration of waste. Once released, this airborne pollutant can travel long distances to significantly affect the environment and human health. Removed from the atmosphere by precipitation and dry deposit, mercury can be transported to land and water where it undergoes a series of complex chemical transformations. This conversion of inorganic mercury to its methylated, organic form occurs primarily in micro aquatic species, where it can bioaccumulate in the food chain.

Due to its diverse properties, mercury serves an important role in several industrial sectors. Mercury can act as a catalyst for chlor-alkali plants; be used in thermometers, wiring switches, fluorescent lamps, batteries, dental amalgam fillings, and medical devices; and as an antifungal agent. Potential exposure pathways to the general population can occur through inhalation of mercury vapor in ambient air and ingestion of contaminated drinking water and food. Primarily, the dietary intake of methylmercury found in fish and shellfish represents the dominant route of exposure to human health. Large predatory fish species such as tuna, swordfish, and shark pose the most serious of threats due to the bioaccumulation of mercury in the aquatic food web.

Ecological impacts from exposure to mercury include death, impaired reproductive capabilities, and growth abnormalities for fish, birds, and mammals. Aquatic and terrestrial plant life species can also experience developmental damage through decreased chlorophyll content and advanced aging.

The health effects felt by mercury exposure can be most accurately described by the tragic episode that occurred in Minamata, Japan, during the 1960s. A plastics company had dumped mercury-laden wastes into Minamata Bay, causing a net reaction of inorganic mercury that transformed into methylmercury via the bacterium present in the bay. Becoming increasingly concentrated at each aquatic food level, the methylmercury-contaminated fish posed a serious health threat to the community. Since most of the residents depended on fishing for their livelihood and for food, a majority of them suffered devastating health effects from eating the toxic fish. Over 700 deaths occurred, and 9,000 individuals were affected with varying degrees of neurological and brain damage. However, prenatal exposure represented some of the most damaging effects, with some offspring experiencing mental retardation, seizure disorders, cerebral palsy, blindness, and deafness, among other disorders. Overall, vulnerability to the effects of methylmercury depends upon the age of the individual, the dose, and the duration of the exposure, with central nervous damage being the most predominant. In regard to inorganic mercury inhalation exposure, kidney dysfunction can occur in addition to nervous system damage.

The U.S. Environmental Protection Agency (EPA) estimates that fossil-fuel power plants, especially coal-based burners, generate over 40 tons of mercury annually in the United States. Consequently, this type of emission represents the largest source of mercury released into the environment. As a result, regulatory controls are in place to help curb mercury emissions. Under the 1990 Clean Air Act Amendments, mercury is defined and regulated under the hazardous air pollutants (HAPs) listing. Based on the type of industrial source category, mercury emissions are subject to either national emission standards (NESHAPs), maximum achievable control technology (MACT) standards, or general available control technology (GACT) standards. Regulating water point sources, the Clean Water Act defines mercury as a toxic pollutant subject to technology-based effluent limitations. The National Pollutant Discharge Elimination System (NPDES) regulates these effluents by assigning a maximum discharge value for each facility. In regard to occupational standards, the Occupational Safety and Health Administration (OSHA) regulates exposure to mercury in the workplace and enforces permissible exposure limits (PELs) to elementary mercury vapor. Finally, the Resource Conservation and Recovery Act (RCRA) helps prescribe treatment, storage, and disposal requirements for mercury-containing wastes, while the Comprehensive Environmental Response, Compensation, and Liability Act (CERCLA) and Superfund Amendments and Reauthorization Act (SARA) aid in responding to illegal toxic dumpsites containing hazardous substances such as mercury.

Concerning the action level for methylmercury in fish, shellfish, and other aquatic animals, the U.S. Food and Drug Administration (FDA) has set the standard at 1 part per million (ppm). This safety control helps to ensure that the human population and sensitive subpopulations will be adequately protected from the adverse risks associated with methylmercury-contaminated fish. Additionally, ambient water quality criteria regulate acceptable levels of mercury for surface waters. These risk management tools help to provide adequate environmental and product-control safety standards needed to protect the public from mercury exposures.

Both from its natural and anthropogenic emissions, mercury's mobilization and transference throughout the globe has an immeasurable, but obviously adverse, impact on the environment and on human health. Although federal regulations in the form of air and water quality standards are in place to aid in the reduction of mercury discharges, health effects from mercury exposure are still present. Neurological and severe embryological damage can occur through ingestion of mercury-tainted shellfish and fish. Therefore, industrial source control limitations, when actually monitored and enforced at the federal and state level, provide some safeguards against the devastating effects of methylmercury poisonings.

See Also: Dental Mercury Amalgams; Government Role in Green Health; Green Chemistry; Personal Consumer Role in Green Health.

Further Readings

Nadakavukaren, Anne. *Our Global Environment.* Long Grove, IL: Waveland Press, 2006.

National Research Council. *Toxicological Effects of Methylmercury.* Washington, DC: National Academies Press, 2000.

Pirrone, Nicola and Kathryn R. Mahaffey. *Dynamics of Mercury Pollution on Regional and Global Scales.* New York: Springer, 2005.

U.S. Environmental Protection Agency. *Mercury Study Report to Congress, Volume I: Executive Summary.* Washington, DC: U.S. Government Printing Office, 1997.

Hueiwang Anna Cook Jeng
Old Dominion University

Alexandra Meyers
Independent Scholar

METABOLIC SYNDROME DISEASES

In medicine, the term *metabolic syndrome* refers to a cluster of conditions that, particularly if more than one is present, are statistically linked with an increased risk of cardiovascular disease and diabetes. Although several different definitions of metabolic syndrome exist, they generally include obesity and/or central obesity ("apple shape" or belly fat as opposed to "pear shape" or fat on the hips), raised blood pressure, insulin resistance or abnormal blood glucose, and abnormal cholesterol levels. There is no scientific consensus on the causes of metabolic syndrome (and not all scientists and physicians agree that it is a distinct entity), but factors generally considered to play a role include heredity, age, ethnicity, diet, and a sedentary lifestyle. Many scholars have cited the typical conditions of modern life in an industrialized society, including decreased physical activity and increased caloric consumption, as factors in the worldwide increase of metabolic syndrome diseases. Lifestyle changes (such as increased exercise and a healthier diet leading to weight loss) are typically included in a treatment plan for those suffering from metabolic syndrome diseases as well as means to prevent their occurrence.

Prevalence of Metabolic Syndrome and Health Implications

According to a 2009 study by R. Bethene Ervin, based on data collected from 2003 to 2006 in the United States, 34 percent of American adults met the criteria for metabolic syndrome, with the prevalence rates differing by age, ethnicity, and body mass index and with somewhat different patterns for males and females. Ervin used the National Cholesterol Education Program's Adult Treatment Panel III (NCEP/ATP III) definition of metabolic syndrome, which requires presence of three or more of the following factors:

- Abdominal obesity (waist circumference >102 centimeters/40 inches for men; >88 centimeters/35 inches for women)

- Atherogenic dyslipedemia (elevated triglyceride, small LDL particles, low HDL cholesterol)
- Raised blood pressure ($\geq$135/$\geq$85 mm Hg)
- Insulin resistance with or without glucose intolerance (characterized by elevation of C-reactive protein)
- Proinflammatory state (characterized by elevation of C-reactive protein)
- Prothrombotic state (characterized by increased plasma plasminogen activator inhibitor and fibrinogen)

The rate of metabolic syndrome increased with age: about 20 percent of males and 16 percent of females age 20 to 39 had metabolic syndrome as compared to 41 percent of males and 37 percent of females age 40 to 59 years and 52 percent of males and 54 percent of males age 60 and over. Among males, metabolic syndrome was most common among non-Hispanic white males (37.2 percent) and Mexican Americans (33.2 percent) and less common among non-Hispanic blacks; while among females, metabolic syndrome was most common among Mexican Americans (40.6 percent) and non-Hispanic blacks (38.8 percent) and less common among non-Hispanic whites (31.5 percent). Metabolic syndrome was strongly associated with body mass index (BMI), a measure of overweight and obesity calculated by a formula based on an individual's height and weight. For men, only 6.8 percent of underweight and normal-weight individuals were identified as having metabolic syndrome as opposed to 29.8 percent of overweight and 65 percent of obese and extremely obese individuals. Among women a similar pattern was observed, with 9.3 percent of underweight and normal-weight individuals identified as having metabolic syndrome, compared to 33.1 percent of overweight and 56.1 percent of obese and extremely obese individuals.

According to the International Diabetes Foundation (IDF), about a quarter of the world's adults have metabolic syndrome, and the rate is increasing among children and adolescents. The IDF cites metabolic syndrome and diabetes as a greater risk to global health overall than human immunodeficiency virus and acquired immune deficiency syndrome (HIV/AIDS) due to the increased risk of cardiovascular disease: persons with metabolic syndrome are three times as likely to have a heart attack or stroke and twice as likely to die from heart attack or stroke as people without the syndrome. Additionally, people with metabolic syndrome are also five times as likely to develop type 2 diabetes as those without the syndrome, and diabetes it also a significant risk factor for cardiovascular disease.

Diseases Related to Metabolic Syndrome

Metabolic syndrome is strongly associated with the risk of diabetes and cardiovascular disease. Over 14 million Americans have been diagnosed with diabetes, a medical condition in which either the pancreas fails to produce any insulin, it produces insufficient insulin, or the body becomes insensitive to insulin. Insulin is necessary to regulate blood sugar in the body and the health consequences of diabetes can be severe, including kidney failure, blindness, cardiovascular disease, nerve damage, and death. An additional 6 million Americans are estimated to have undiagnosed diabetes. Given current rates of increase, it is projected that 29 million Americans will have diabetes by the year 2050. About 1.4 million Americans have type 1 (insulin dependent) diabetes (the remainder have type 2 or noninsulin resistant diabetes), as do an estimated 10 to 20 million people globally. The incidence of type 1 diabetes is increasing globally by 3 to 5 percent annually and the incidence of type 2 diabetes is expected to double in the next 25 years.

Cardiovascular disease comprises many conditions including stroke and coronary heart disease (CHD): the latter includes myocardial infarction (heart attack), angina pectoris (chest pain due to lack of oxygen supply to the heart muscle), and sudden coronary death. Cardiovascular disease is the leading cause of morbidity (illness) and mortality in most industrialized countries and is rapidly becoming a leading cause of morbidity and mortality in developing countries as well. The prevalence of cardiovascular disease and death from cardiovascular disease varies widely across countries and U.S. states and also differs by gender and age. For instance, in the United States the death rate for males from cardiovascular disease is 307 per 100,000 versus 170 per 100,000 in Japan (the lowest reported) and 1,167 per 100,000 in the Russian Federation (the highest reported). For women, death rates for cardiovascular disease were 158 per 100,000 in the United States, while the lowest rate was in France (59 per 100,000) and the highest in Russia (540 per 100,000).

In the United States, Mississippi has the highest cardiovascular disease mortality (421 per 100,000) and Puerto Rico the lowest (234 per 100,000). Cardiovascular disease rates in the United States increase with age and are higher among men than women in every age group (e.g., 11.6 percent for men versus 3.6 percent for women in the 55 to 64 age group and 16.8 percent versus 10.3 percent for people age 75 and older). Rates of cardiovascular disease in the United States also differ by race/ethnicity. Among men, non-Hispanic white men have the highest incidence of cardiovascular disease at 8.8 percent, followed by African Americans at 7.4 percent and Mexican Americans at 5.6 percent. Among women, blacks had the highest rate at 7.5 percent, followed by non-Hispanic white females (5.4 percent) and Mexican Americans (4.3 percent).

Diabetes is a major risk factor for cardiovascular disease: over 75 percent of people with diabetes will die of cardiovascular disease, including stroke and myocardial infarction. The combination of metabolic syndrome and diabetes further increases the risk: for instance, a study of U.S. adults by S. Malik and colleagues in 2004 found that mortality from cardiovascular disease among U.S. adults was 5.3 per 1,000 person-years for a person with neither metabolic syndrome nor diabetes, 7.8 per 1,000 for a person with metabolic syndrome only, and 8.6 per 1,000 for a person with both diabetes and metabolic syndrome.

Lifestyle Interventions in Preventing and Treating Metabolic Syndrome

Both in the United States and globally, prevention and primary intervention for individuals with metabolic syndrome are based on lifestyle modifications aimed at reducing obesity and/or abdominal obesity through increased physical activity and changes in diet including caloric restriction. A review by Peter Janiszewski and colleagues found that such interventions improved all aspects of metabolic syndrome and successfully reduced abdominal obesity, dyslipedemia, increased blood pressure, and dysfunctional glucose metabolism. In addition, common recommendations to prevent or treat metabolic syndrome include quitting smoking and increasing the consumption of fiber-rich foods such as fruits, vegetables, beans, and whole grains, which can help control insulin levels as well as promote weight loss.

Many specific diet plans are available that have shown to be effective in reducing metabolic syndrome including the DASH (Dietary Approaches to Stop Hypertension) diet and the Mediterranean diet. Both emphasize high consumptions of fruits, vegetables, and grains and reduced consumption of sugar and animal fat. The DASH diet is structured around foods common in the typical American diet and includes several servings of nonfat dairy and meat, poultry, or fish per week, while the Mediterranean diet is based on the typical diet of countries such as Spain and Greece and emphasizes olive oil as the principal source of fat and legumes, nuts, fish, poultry, and dairy products as the principal protein source with red meat consumed rarely.

Recommendations for the prevention of metabolic syndrome are similar to those for treating it: a diet that does not result in weight gain and that is high in fruits, vegetables, and whole grains and low in sugar and animal fats along with sufficient physical activity. Various recommendations for physical activity have been issued by different organizations at different periods but most commonly the recommendation for adults is in the range of 30 to 60 minutes of moderate to vigorous activity daily or almost daily.

Metabolic syndrome and related diseases provide a good example of health conditions that respond well to changes in lifestyle and also demonstrate how trans-individual factors such as urban design may play a role in promoting either healthy or unhealthy lifestyles. Many scientists have noted that the modern world seems in many ways to have been designed to promote inactivity and obesity and thus to increase rates of metabolic syndrome and the diseases associated with it. For instance, physical activity was once a normal part of most people's lives, but changes in work and transportation patterns have meant that many people get no significant physical activity unless they choose to do so by participating in a sport or exercise program. Cities and suburbs designed for automobiles rather than for walking and bicycling contribute to this trend as does an absence of inviting green space such as parks in many areas. The ready availability of cheap processed foods high in fat, sugar, and salt and low in fiber also contributes to metabolic syndrome. Some inner-city neighborhoods have been termed *food deserts* because a lack of grocery stores and greenmarkets combined with the presence of fast food chains means that it may be nearly impossible for residents to eat a healthy diet rich in fruit and vegetables and whole grains and low in sugar, salt, and saturated fat. In addition, even if available, healthy foods may cost more than unhealthy foods, making them less accessible to people with limited resources.

See Also: Calorie Labeling for Restaurants; Fast Food; Obesity; Physical Activity and Health; Urban Green.

Further Readings

Ervin, R. Bethene. "Prevalence of Metabolic Syndrome Among Adults 20 Years of Age and Over, by Sex, Age, Race and Ethnicity, and Body Mass Index: United States, 2003–2006." *National Health Statistics Reports*, 13 (2009). http://www.cdc.gov/nchs/data/nhsr/nhsr013 .pdf (Accessed July 2010).

Grundy, Scott M., Bryan Brewer, Jr., James I. Cleeman, Sidney C. Smith, Claude Lenfant, et al. "Definition of Metabolic Syndrome: Report of the National Heart, Lung, and Blood Institute/American Heart Association Conference on Scientific Issues Related to Definition." *Circulation*, 109 (2004). http://circ.ahajournals.org/cgi/reprint/109/3/433.pdf (Accessed July 2010).

International Diabetes Foundation. "IDF Worldwide Definition of the Metabolic Syndrome." http://www.idf.org/metabolic_syndrome (Accessed July 2010).

Janiszewski, Peter M., Travis J. Saunders, and Robert Ross. "Themed Review: Lifestyle Treatment of the Metabolic Syndrome." *American Journal of Lifestyle Medicine*, 2/2 (2008).

Malik, S., N. D. Wong, and S. S. Franklin. "Impact of the Metabolic Syndrome on Mortality From Coronary Heart Disease, Cardiovascular Disease, and All Causes in United States Adults." *Circulation*, 110 (2004). http://circ.ahajournals.org/cgi/content/short/110/10/1245 (Accessed July 2010).

Mayo Clinic. "Metabolic Syndrome." http://www.mayoclinic.com/health/metabolic%20 syndrome/DS00522 (Accessed July 2010).

Steinberger, Julia, Stephen R. Daniels, Robert H. Eckel, Lauran Hayman, Robert H. Lustig, Brian McCrindle, and Michele L. Mietus-Snyder. "Progress and Challenges in Metabolic Syndrome in Children and Adolescents: A Scientific Statement From the American Heart Association: Atherosclerosis, Hypertension, and Obesity in the Young." Committee of the Council on Cardiovascular Disease in the Young; Council on Cardiovascular Nursing; and Council on Nutrition, Physical Activity, and Metabolism. *Circulation*, 119/4 (2009). http://circ.ahajournals.org/cig/reprint/CIRCULATIONAHA.108.191394 (Accessed July 2010).

Sarah Boslaugh
Washington University in St. Louis

METHANE/BIOGAS

Biogas is a renewable natural gas containing approximately 70 percent methane (CH_4) and roughly 30 percent carbon dioxide and trace amounts of other gases. Potential agricultural feedstocks for biogas production include manure (hog, dairy, beef, and poultry), food processing (by-products of meat processing, potato, dairy, cheese whey, sugar beet, and vegetables), and energy crops cut as silage (wheat, barley, triticale, clover, alfalfa, ryegrass, turnips, and corn). Commercial products from biogas production include methane, electricity, heat, steam, fertilizer, chemical recovery, odor reduction, water recycling, carbon dioxide, and, potentially, carbon credits and greenhouse gas credits.

Animal manure, human sewage, or food waste can produce methane during anaerobic digestion. Here, an animal scientist measures a cow for respiration of oxygen and carbon dioxide and production of heat and methane.

Source: U.S. Department of Agriculture, Agricultural Research Service/Keith Weller

Anaerobic digestion of wastes provides biogas. Biogas contains about 70 percent methane that can be used to generate electricity or as fuel for vehicles or for heating. Any animal manure, human sewage, or food waste can produce methane during anaerobic digestion. Biogas can also be "cleaned" to yield purified methane that can be used in natural gas pipelines. Methane from biogas is an excellent alternative energy source. Using methane for energy helps the environment by replacing the use of nonrenewable fossil fuels with renewable energy and by taking methane out of the atmosphere. Methane is a greenhouse gas (GHG) with 21 times the heating effect of carbon dioxide. Biogas methane is renewable, unlike natural gas that is mined from underground wells and is a nonrenewable fossil fuel. Methane yields from agricultural feedstocks are in the range of 50 to 70 percent. Manure has the lowest yield, while energy crops and food processing have the highest yields. Blending feedstocks can achieve desirable methane yields while solving environmental issues.

Incentivizing the production of renewable natural gas, or biomethane, from sources that include animal manure, landfills, renewable biomass, and agricultural wastes will support expanding the role of renewable energy sources in the existing energy sector, where little opportunity exists today. It will also create new business investment prospects for renewable project developers and the potential to expand rural economies while supporting existing industrial jobs and dramatically reducing carbon emissions. The benefits of using biomethane are as follows:

- Renewable biomethane is a versatile form of bio-energy. It can be used directly at the site of production or placed in the pipeline to support a variety of residential, commercial, or industrial applications.
- Renewable biomethane produced from renewable sources including animal manure, landfills, renewable biomass, and agricultural wastes can be produced at high efficiencies ranging from 60 to 70 percent. Additionally, the technology components to produce renewable gas from this variety of sources exist today.
- Renewable biomethane can be delivered to potential customers via any existing pipeline infrastructure.
- Renewable biomethane can provide a renewable option for many heavy industries, which could save existing industrial jobs in a carbon-constrained economy while creating new rural green jobs to produce renewable biomethane.
- Renewable biomethane production in digesters provides the agricultural sector additional environmental benefits by improving waste management and nutrient control.

Biogas is produced from organic wastes by concerted action of various groups of anaerobic bacteria. Anaerobic biodigesters can be either wet fermentation or dry fermentation. Wet biodigesters, with a life expectancy of approximately 10 years, need to be cleaned out every one to three years. Dry biodigesters, however, do not require cleaning as frequently as wet biodigesters and have a life expectancy of approximately 20 years. Capital cost for a biogas facility that produces one megawatt of power is in the range of $4 to $6 million, depending on the level of infrastructure at the site. Both solid and liquid systems need to develop the infrastructure to increase manure collection and handling to feed the digesters. The frequencies of manure collection need to increase to at least four times a year to feed digesters with relatively fresh material. However, the availability of turnkey technology, expected to develop in the future, will help to reduce the costs involved.

Biogas technology provides an excellent alternate source of energy and is hailed as an archetypal appropriate technology that meets the basic need for cooking fuel in rural areas. Using local resources like cattle waste and other organic wastes, energy and manure are derived. Microbial conversion of organic matter to methane has become attractive as a method of waste treatment and resource recovery. Realization of this potential and the fact that most agriculturally dependent countries support large cattle wealth can lead to the promotion of biogas in a major way as an answer to the growing fuel crisis.

Although biogas production technology has established itself as a technology with great potential that could exercise major influence in the energy scene in rural areas, it has not made any real impact on the total energy scenario in the global context. One of the serious limitations is the availability of feedstock, followed by defects in construction and microbiological failure. This indicates that the technology transfer is not complete, and that it requires coordinated efforts of scientists and engineers to overcome such limitations in order to translate this high-potential technology into high-performing technology.

See Also: Alternative Energy Resources (Solar); Biodiesel; Firewood and Charcoal; Hydroelectricity; Nuclear Power; Organic Produce; Rural Areas; Solid Waste Management.

Further Readings

Chawla, O. P. *Advances in Biogas Technology*. New Delhi: Indian Council of Agricultural Research, 1986.

Doraisamy, P., V. Udayasurian, K. Ramasamy, and G. Oblisami. *Biological Nitrogen Fixation and Biogas Technology*, S. Kannaiyan, K. Ramasamy, K. Ilamurugu, and K. Kumar, eds. Coimbatore, India: Tamil Nadu Agricultural University, 1992.

Jagadeesh, K. S. *Proceedings of the International Conference on Biogas Energy Systems*. Conference held at the Energy and Resource Institute, New Delhi, India, January 22–23, 1996.

Khendelwal, K. C. and S. S. Mahdi. *Biogas Technology: A Practical Technology*. New Delhi, India: Tata McGraw-Hill, 1986.

McInerney, M. J. and M. P. Bryant. *Fuel Gas Production From Biomass*, D. L. Wise, ed. West Palm Beach, FL: Chemical Rubber Co. Press Inc., 1981.

Mohua Guha
International Institute for Population Sciences

METHICILLIN-RESISTANT *STAPHYLOCOCCUS AUREUS*

Staphylococcus aureus (*S. aureus*) are gram-positive cocci-shaped bacteria that are commonly found throughout the environment, including on the skin of healthy individuals. *S. aureus* is essentially benign on the skin, although if allowed to enter the bloodstream or internal tissues, *S. aureus* can cause a range of potentially serious infections. A person who has had some sort of skin trauma (including a minor cut) may be susceptible to a variety of different skin and soft-tissue infections caused by *S. aureus*, but the infection can usually be successfully treated with penicillin-derived antibiotics. MRSA, or methicillin-resistant *S. aureus*, are bacteria that have become highly resistant to many commonly used antibiotics, such as methicillin. MRSA infections are more difficult to treat because traditional antibiotics are ineffective, and patients must therefore be treated with alternative antibiotics. Until the 1990s, identified cases of MRSA were confined primarily to hospitals and healthcare facilities, although MRSA can now be found worldwide in community settings. MRSA infections can range in severity from minor to very serious and even fatal. In 2005, it was estimated that about 20 percent of those MRSA infections categorized as "serious" resulted in the death of the patient. As a result, MRSA have emerged as very serious nosocomial (hospital-acquired) and community-acquired pathogens.

Methicillin is a semi-synthetic beta-lactamase-resistant penicillin antibiotic that was introduced in 1959 to aid in the fight against bacterial infection. Remarkably, *S. aureus* strains resistant to methicillin were reported soon after the introduction of this antibiotic. In the early 1960s, outbreaks of MRSA infections were reported in Europe. Since then, scientists around the globe have identified different strains of MRSA and, using genetic testing, three of those can be traced back to the original MRSA strains first isolated in Europe in the 1960s.

All known *S. aureus* strains that are resistant to methicillin (MRSA strains) carry a gene known as *mec* on the bacterial chromosome. Strains of *S. aureus* that do not have the *mec* gene are sensitive to methicillin and are therefore termed methicillin-sensitive *S. aureus* (MSSA). The *mec* gene is a component of the larger Staphylococcal chromosomal cassette *mec* (SCC*mec*). To date, at least six different SCC*mec* types have been identified (types I–VI). The *mec* gene encodes a penicillin-binding protein 2a (PBP-2a), which ultimately gives *mec*-carrying microorganisms resistance to methicillin and many other beta-lactam antibiotics such as nafcillin, oxacillin, and cephalosporins. Penicillin-binding proteins (PBPs) are essential protein enzymes in the bacterial cell wall that catalyze the production of an important structural barrier of the cell wall known as peptidoglycan. As their name suggests, some PBPs also bind to penicillin-derived antibiotics (beta-lactams), which results in inactivation of the PBP enzyme activity and ultimately death of the bacterial cell. The protein encoded by the *mec* gene, PBP-2a, has a low affinity for beta-lactams and therefore, even in the presence of most antibiotics, PBP-2a will continue to catalyze the synthesis of the bacterial cell wall. This continued function of PBP-2a in the presence of antibiotics such as methicillin results in growth and survival of MRSA strains even in the presence of prolonged antibiotic treatment.

MRSA infections can be broadly divided into two main groups, healthcare-associated MRSA (HA-MRSA) and community-acquired MRSA (CA-MRSA). Before the emergence of CA-MRSA in the 1990s, all MRSA was considered to be HA-MRSA. HA-MRSA infections, as the name suggests, are primarily in individuals who are hospitalized, individuals who are in long-term healthcare facilities such as nursing homes, immune-compromised patients, or individuals who are predisposed to infection (including the very young and the very old). CA-MRSA infections, on the other hand, are commonly seen in younger populations such as athletes and school-aged children. All *S. aureus* (including MRSA) can be transmitted by a variety of different methods including direct contact (i.e., touching the skin of someone who is colonized with the bacteria), aerosols (i.e., tiny airborne droplets), or fomite transmission (touching a bacteria-laden object or surface). As a result, CA-MRSA infections sometimes occur as outbreaks among individuals in close contact such as members of sports teams, military personnel, prisoners, children in day care, or schoolchildren. The consequences of acquiring a CA-MRSA infection include an increase in skin and soft tissue infections and necrotizing pneumonia. Those at risk include healthy and physically fit individuals who are more prone to skin trauma such as "turf burns," cuts, and sores, such as football players.

In general, CA-MRSA strains tend to be more virulent (i.e., cause more serious infection-related symptoms) than HA-MRSA strains. This is illustrated by the fact that CA-MRSA strains are readily able to infect otherwise healthy individuals while HA-MRSA strains are not. Instead, HA-MRSA strains are typically only capable of infecting individuals who have risk factors that make them more susceptible to bacterial infection in general (e.g., individuals who are hospitalized and have compromised immune systems). Intriguingly, CA-MRSA strains tend to be more susceptible to other antimicrobials, while HA-MRSA strains are typically more resistant to multiple antibacterial agents, making CA-MRSA infections generally easier to treat than those caused by HA-MRSA strains. Another difference between CA-MRSA and HA-MRSA strains is that most HA-MRSA clones are associated with SCC*mec* types I, II, and III, while CA-MRSA clones are associated with SCC*mec* type IV and sometimes type V.

CA-MRSA is the number one cause of community-acquired skin infections. It has been observed that only a few *S. aureus* strains are responsible for most CA-MRSA infections in the United States, with strain USA300 being the most common. A study of 11 university emergency departments in the United States in August 2004 revealed that 59 percent of *S. aureus* infections were caused by CA-MRSA and that strain USA300 made up 97 percent of those. Furthermore, a study of MRSA infections that were causing skin abscesses of football players for the St. Louis Rams professional football team in 2003 showed that eight MRSA infections were identified on five of 58 players (9 percent) caused by turf abrasion. All of these MRSA isolates were identified as being SCC*mec* type IV and CA-MRSA clone USA300.

Infections due to MRSA are becoming more prevalent worldwide and are of great concern due to the inherent resistance of these organisms to most routinely used antibiotics. Of further concern are the CA-MRSA strains that are able to cause infection in otherwise healthy individuals. Because of the increasing resistance of *S. aureus* against multiple antibiotics and the limited treatment options available, MRSA have emerged as serious hospital-acquired and community-acquired pathogens.

See Also: Antibiotic Resistance; Antibiotics; Nosocomial Infections.

Further Readings

Benner, E. J. and F. H. Kayser. "Growing Clinical Significance of Methicillin-Resistant *Staphylococcus aureus*." *The Lancet*, 2/741 (1968).

Boyce, J. M., et al. "Patient Information: Methicillin-Resistant *Staphylococcus aureus* (MRSA)." http://www.uptodate.com (Accessed August 2009).

Enright, M. C., et al. "The Evolutionary History of Methicillin-Resistant *Staphylococcus aureus* (MRSA)." *Proceedings, National Academy of Science USA*, 99/7687 (2002).

Kazakova, S. V., et al. "A Clone of Methicillin-Resistant *Staphylococcus aureus* Among Professional Football Players." *New England Journal of Medicine*, 352/468 (2005).

Lowy, F. D., et al. "Microbiology and Pathogenesis of Methicillin-Resistant *Staphylococcus aureus*." http://www.uptodate.com (Accessed August 2009).

Moran, G. J., et al. "Methicillin-Resistant *S. aureus* Infections Among Patients in the Emergency Department." *New England Journal of Medicine*, 355/666 (2006).

Tracey A. H. Taylor
Kansas City University of Medicine and Biosciences

METRICS OF GREEN HEALTH

Green health is the goal of individual and societal investment in environmentally and socially sustainable products and processes that influence human health. The achievement of green health may be represented by both procedural standards and perceived destinations in various aspects of human endeavors. In situations where variation is expected in the level of performance or compensation, there is a tendency to count discrete incidences or to develop measures for comparative assessments and for fine-tuning selectivity and decision making. The need to maintain such measurements across temporal, geographic, and political boundaries has engendered the quest for standardization, or the invention of indicators and milestones toward green health.

International organizations are best positioned to develop standardized metrics in healthcare systems, in part because of the high level of variation in socioeconomic and sociocultural factors that affect public health. In this regard, the collaborations between the World Health Organization (WHO) and other international agencies, such as the World Bank, the World Meteorological Organization (WMO), and the United Nations Environment Programme (UNEP), have been very influential. The collaborations have been guided by the comprehensive definition of health adopted by WHO since its inception, that "health is a state of complete physical, mental and social well-being and not merely the absence of disease or infirmity." The extensive scope of this definition of health complicates the development of universal metrics of health status in populations. Several investigators have attempted to reconcile metrics developed from different perspectives.

There are three major categories of metrics that are applicable to green health. The first category, titled Status Metrics, attempts to capture the level of deviation of a population from perfect health in terms of epidemiological assessments of mortality and morbidity. An example of metrics in this category is the widely cited disability-adjusted life years (DALYs), a composite measure of disease burden developed through extensive research by the WHO, the World Bank, and several academic investigators. The second category, Process Metrics, deals with the preventive processes that are designed to protect against morbidity and premature mortality. An example of metrics in this category is the series of performance standards developed for business, governments, and society by the International Standards Organization (ISO). The third major category, titled Product Metrics, compares the impacts of materials on social and environmental systems through the use of quantitative tools of life cycle assessments (LCA) methods.

Ideally, all three categories in concert are necessary to understand and characterize risks to human health, to manage the risks through the implementation of targeted procedures and programs, and to make informed decisions on consumer products that prevent the development of new risks. In the remaining sections of this article, these three categories and examples are described in more detail.

Green Health Status Metrics

The burden of disease in a population manifests in several dimensions. These include numerical estimates of cases of different diseases; geographical (spatial) distribution of diseases and mortality across a regional landscape; and the distribution of socioeconomic factors that correlate with specific disease conditions and their presentation in diverse communities. In the 1990s, the World Bank commissioned a series of studies with the aim of producing quantitative measures of health in populations. These studies built upon earlier work of Richard Zeckhauser and Donald Shepard, who explored legal and economic perspectives of comparative health status of populations in their 1976 publication titled "Where Now for Saving Lives?"

Since these seminal works, several composite measures of burden of disease have been developed. These measures include quality-adjusted life years (QALYs), which reflect individual preferences for time spent in different states of health; disability-adjusted life expectancy (DALE), which estimates the years of life expected to be lived without any disease; and healthy life years (HeaLYs), originally developed to explore the association between investments in health infrastructure in developing countries and the trajectory of vital statistics and which measures the number of life years lost prematurely combined with the amount of healthy life years lost due to morbidity. Among the most prominent of these measures is the DALY, which combines estimates of morbidity in terms of years of life lived with disability (YLD) with estimates of premature mortality or years of life lost (YLL). Hence, the disability-adjusted life years lost due to any particular disease (i) is represented arithmetically as:

$$\text{DALY}i = \text{YLL}i + \text{YLD}i$$

YLL and YLD are estimated from several parameters according to the following formula:

$$\text{YLL} = (KCera)/(r+\beta)2\ [e-(r+\beta)(L+a)\{-(r+\beta)(L+a)-1] \\ -e-(r+\beta)a[-(r+\beta)a-1]\} + (1-K)/r(1-e-rL)$$

where:

a = age at death

r = discount rate, usually set as 0.03 (but this is sometimes disputed)

β = age-weighting function parameter (contentious)

K = modulating constant of age weighting

C = constant of age-weights adjustments and equals 0.17

L = life expectancy from a model life table at the age of death (a)

The formula for estimating YLD includes the multiplication of the formula for YLL by the weight given to disability from the specific disease, and standard life expectancy is replaced by the number of years lived with the disability, as follows:

$$\text{YLD} = DW\ \{(KCer)a/(r+\beta)2\ [e-(r+\beta)(L+a)[-(r+\beta)(L+a)-1] \\ -e-(r+\beta)a[-(r+\beta)a-1]] + (1-K)\}/r(1-e-rL)$$

where:

DW = weight given to disability from disease i

a = age at the onset of disability

r = discount rate, usually used as 0.03

β, K, and C are constants as indicated in the formula for YLL

L = number of years lived with the disability (based on the maximum life expectancy, typically set at 82.5 years and 80 years for Japanese women and men, respectively, representing the highest among populations, globally)

The initial results of the global burden of disease assessments conducted by WHO revealed, with a few surprises, the leading causes of mortality and morbidity in the world. Coronary heart disease and cerebrovascular disease were the two leading causes of deaths, representing 12.2 percent and 9.7 percent, respectively, of total annual deaths. However, neuropsychiatric conditions are the leading cause of disease burden in terms of total DALYs, highlighting the gap between realistic health status of populations and societal investment in research and development to find cures and support disease prevention.

The burden of disease assessments were also conducted within seven epidemiological regions (Africa, eastern Mediterranean, Europe, the Americas, southeast Asia, and western Pacific; these regions are further delineated into 14 total subregions). The regional assessments revealed great disparity in health status among nations broadly categorized as industrialized countries and developing countries. The environmental burden of disease, representing the portion of total disease burden that can be reduced through the modification of environmental risk factors, is greatest in relatively poor countries. Diarrheal diseases, for example, continue to exert high burden in countries where the population has limited access to piped water supply and water purification systems. Diarrheal diseases kill an estimated 2.2 million people every year, most under the age of 5 years, mostly in Africa and Asia. The proportion of total DALYs attributable to diarrhea is among the highest at 4.8 percent globally. This level is lower than the 6.2 percent associated with lower-respiratory infections, which are also overrepresented in populations of developing countries.

Through the assessments of the global burden of disease and rank ordering of disease categories, it is possible to develop strategies through which investments in green technologies can reduce disease burden in the areas of greatest need. For example, investments in green energy resources that are associated with low levels of toxicity and particulate matter production should support preventive strategies to reduce the burden of respiratory diseases. Green approaches to urban planning and rural development can ensure better water supplies as a strategy to reduce the burden of diarrheal diseases. Investments in green park systems may support the reduction of the burden of neuropsychiatric conditions while encouraging physical exercise as a means to reduce the burden of heart disease.

In their essay titled "Environmental Metrics for Community Health Improvement," Benjamin Jakubowski and Howard Frumkin discuss various ways in which measurements of various environmental parameters in environmental media can inform policy formulation in preventive healthcare through rewarding tangible and sustainable improvements in community health. The most useful metrics in this respect are designed to be "simple, sensitive, robust, credible, impartial, actionable, and reflective of community values." These values are, to different extents, accommodated by metrics such as the DALYs, which are data intensive. Other metric systems represent the bridge between status indicators and process indicators. These include "quality of life indicators," "sustainability indicators," and "environmental public health indicators" that emphasize air and water quality.

Green Health Process Metrics

The commercial and domestic processes that support human standards of living contribute in large part to the risks to population health. The risks include toxic exposures, pathogen infections, and occupational hazards leading to physical injuries and disability. Newer inventions such as computing and activities supporting information and communication technologies (ICT) are not immune from contributions to health risks. The recognition of the cumulative effects of society engagement with these processes has led to a movement toward greenness in how we make and process things. To assess progress on this front, it is essential to develop standards of measurement and procedures. The International Standards Organization (ISO), a network consisting of the national standards institutes in 163 countries, has one of the most widely cited frameworks for establishing standards for green processes and for producing guidelines for implementation across different commercial sectors.

According to the ISO, the cooperative development of standards for products and processes or services across national boundaries provides assurance that process characteristics such as quality, greenness (environmental friendliness), safety, reliability, efficiency, and interchangeability can be achieved affordably. Moreover, the centralization of process standards supports rapid dissemination of innovation, especially with respect to environmental sustainability criteria. So far, the ISO has developed more than 18,000 international standards for various processes; of these, approximately 570 are closely related to environmental or green issues ranging from agriculture and building construction to transportation, medical devices, ICT, and management procedures. One of the potential limitations of ISO standards in driving innovation toward sustainability and green health targets is that the organization works only on standards for which there is a market requirement, and the standards are then established by employees of sectors that have requested the standards. The worst-case scenario is that this practice is biased toward the status quo, and innovations that are in favor of green health but for which economic profitability remains uncertain are retarded in the process.

Environmental procedures that are relevant to green health are covered in one of the widely known ISO standards (the ISO 14000 series). These standards are designed to minimize environmental impacts and to maximize environmental performance in various sectors. For example, ISO 14001:2004 and ISO 14004:2004 provide organizations with the requirements and guidelines, respectively, for implementing environmental management systems. In concert, these standards support a "framework for a holistic, strategic approach" toward environmental policies and plan of actions. This allows organizations to measure the environmental impacts of their activities and subsequently improve environmental performance to meet preset targets. The added benefit is that organizations adopting these standards may simultaneously comply with existing jurisdictional statutes and environmental regulations and provide consumers and the general public with assurances of being proactive toward sustainability and green health.

The construction of houses and other buildings and their occupation generally consume nearly two-thirds of global supply of electricity and 40 percent of raw materials and energy, respectively. Therefore, sustainability initiatives have long recognized the need to develop standards for green buildings, and activities in this direction have accelerated since the United Nations Declaration on Environment and Development in 1992. Such green buildings are not only able to conserve resources, but many studies have demonstrated that they are beneficial for the health of the inhabitants. The ISO has established a set of international standards that support sustainability principles in building construction (ISO 15392:2008, Sustainability in Building Construction—General Principles).

Additional ISO standards that are directly relevant to green health include bioenergy (under development as ISO/PC 248, Sustainability Criteria for Bioenergy; ISO standard 13065) and aquaculture (ISO/TC 234, Fisheries and Aquaculture). The scope of TC 234 includes the maintenance of specified biotic and abiotic conditions appropriate for raising healthy fish cultures, criteria for environmental monitoring and waste disposal. Given that fish supply approximately 6 percent (approximately 30 percent for about 1 billion people in Asia) of protein in the human diet worldwide, it is very important to establish sustainable processes for fish recovery and fish farming. This will also ensure that toxic contaminants such as mercury, organic chemical pollutants, hormones, and enteric viruses are kept from affecting public health.

Green Health Product Metrics

Educating consumers of commercial products to select products that support environmental sustainability without compromising human health and well-being is essential for the perpetuation of green health culture in society. Traditionally, functional suitability and common availability have long been the determining considerations for selecting materials to be used for designing and manufacturing consumer products. The bottom line for manufacturers has been economic profitability. The costs to public health and environmental impacts of raw materials recovery, refining and finishing, and ultimate utilization of chemicals and alloys in products have typically been externalized. The sustainability movement has encouraged progress toward resource conservation, but little or no progress has been made to include material toxicity and health impacts as major determining factors in product design and manufacturing. Recently, the introduction of sophisticated models of material life cycle impacts has allowed researchers to develop tools for quantitative assessment of impacts on various resources, including not only raw materials and energy but also impacts on occupational health during manufacturing and on population health during use and disposal of products at the end of useful life.

The U.S. Environmental Protection Agency (EPA) hosts one of the most comprehensive databases on materials toxicity and the use of the information in life cycle assessments (http://www.epa.gov/nrmrl/lcaccess).

Life cycle assessment (LCA) is a crucial tool in green health metrics because it produces quantitative estimates of how products affect the environment and health at every stage of the life cycle, namely from "cradle to grave," or even better, "from cradle to cradle," to emphasize the importance of recycling and reuse. The outcome of LCA assessments of different products with similar functions can assist consumers in making purchases that encourage greenness and sustainability. LCAs conducted at the product development stage can also help manufacturers design greener products through frameworks such as Design-for-Environment (DfE) (http://www.epa.gov/dfe). According to the EPA, "Design for the Environment allows manufacturers to put the DfE label on household and commercial products, such as cleaners and detergents that meet stringent criteria for human and environmental health. Using these products can protect your family's health and the environment."

Part of the DfE program involves alternative assessments, which is a comparative analysis of alternate products' impacts on the environment and human health. EPA provides support for alternative assessments for manufacturers and industries seeking to identify safer chemicals than what they are currently considering (http://www.epa.gov/dfe/alternative_assessments.html). In situations where there is a lack of input data for comprehensive LCAs, the EPA recommends the "Best Practices Program," which provides support for occupational and community health through strategies that minimize toxic exposures (http://www.epa.gov/dfe/best_practices.html).

LCA is data intensive, and several software programs have been developed to facilitate the process. Examples of prominent software include *SimaPro*, produced by PRé Consultants in the Netherlands (http://www.pre.nl/simapro.html); the BEES (Building for Environmental and Economic Sustainability) hosted by the National Institute for Standards and Technology (NIST) (http://www.bfrl.nist.gov/oae/software/bees), specializing in environmentally preferable building products; Carnegie Mellon University's Economic Input-Output Life Cycle Assessment (EIO-LCA; http://www.eiolca.net); and Ecoinvent, hosted by the Swiss Center for Life Cycle Inventories (http://www.ecoinvent.ch).

In general, LCAs proceed through four stages, beginning with a scoping exercise whereby the product, process, or activity being subjected to LCA is described and the limits of the assessment process are defined in terms of environmental or health end points to include or exclude, especially where data may not be available. The second stage includes inventory analysis, in which potentially impacted resources and health effects are identified and quantified. For example, these may include water and energy, raw materials extraction, and release of wastes into the environment. Following this phase, the impact assessment stage commences with an assessment and rating of the effects on human health and the ecosystem. The fourth and final stage is the interpretation of the results and articulation of the uncertainties.

The use of LCAs as a metric tool in green health has been undergoing development for several decades, but despite several successful applications and endorsements, some controversy persists in its application in various sectors. For example, in the United States, some state attorneys general took a position in 1991 against the use of LCA results to advertise products for which major uncertainties remain in the methodology and results. This resulted from the practice by certain product manufacturers to tout their products as environmentally friendly or green based on unverified LCA outcomes. Subsequently, LCA standards were developed by the ISO and included in the 14000 series of standards.

Among the most exciting developments in the use of LCAs as a green metric is the collaboration fostered by the United Nations Environment Programme (UNEP), titled "International Life Cycle Partnerships for a Sustainable World" (http://lcinitiative.unep.fr). The key partners are UNEP and the Society for Environmental Toxicology and Chemistry (SETAC), with the mission to "bring science-based Life Cycle approaches into practice worldwide." According to the initiative, the specific advantages of the collaboration include the following:

- The ability to access and mobilize an established and growing global network of over 2000 interested members who have been and continue to be interested in understanding and advancing life cycle approaches worldwide. These experts represent industry, government, academics, and the service sectors and are the leaders in developing and applying life cycle assessment and life cycle management worldwide.
- The ability to gather and manage examples of best practices and life cycle achievements across the world.
- The status of being considered as a one-stop shop for life cycle approaches.
- The opportunity to connect science and decision making in policy and business with the supply and demand side of life cycle approaches. Therefore, an opportunity exists to become the global authority for consensus building and peer review on methodological questions and environmental assessments of natural resources, materials, and products in the field of science.

The first phase of the collaboration (2002–2006, focused on four goals, namely, Life Cycle Management, Life Cycle Inventory, Life Cycle Impact Assessment, and Life Cycle Cross-cutting Activities. The second phase of the collaboration (2007–2012) is organized according to five working groups focusing on Life Cycle Approaches for Methodologies and Data; Resources and Impacts; Consumption Clusters (e.g., housing, transportation, food, and consumer products); Capability Development (including institutional empowerment, training, curricular development, etc.); and Life Cycle Management in Businesses and Industries.

Prospects for Green Health Metrics

Growing awareness of the dependence of human health and quality of life on ecological integrity has spurred the development of products and processes that make claims of sustainability and greenness. The extreme diversity of products and the need to assess the veracity of the claims has also engendered the need to develop indicators or metric systems for sustainable development that minimizes impacts on public health. It is also important for consumers to participate in the process and to be informed about the choices that they make through their purchases. Improvements to public health can only be noticed if we have good internally consistent measures of the status of health in populations. The DALY approach developed by the WHO is a very good metric system, but it is data intensive and has several contested assumptions. We should expect continuing research and improvement in the DALY metric and other composite measures of disease burden. The International Standards Organization's set of standards in environmental sustainability is widely subscribed and will likely continue to be popular among businesses and governments wanting to adopt new processes toward greenness, while also wanting to protect their investments in research and development toward innovation. As a methodology, LCAs have provided quantitative approaches for alternatives assessments for consumer products. This is the cornerstone of the green health metric system that is closest to consumer decisions and policy formulation. There remain uncertainties about the scope of impacts to be considered in LCAs and how trade-offs should be accomplished, for example, among product toxicity, energy consumption, and functionality. These are topics for further research that will support the sustainable future of green health metrics.

See Also: Education and Green Health; Environmental Protection Agency (U.S.); Government Role in Green Health; Industrial Ecology; Personal Consumer Role in Green Health; World Health Organization's Environmental Burden of Disease.

Further Readings

Hendrickson, Chris, T. Lester, B. Lave, and H. Scott Matthews. "Environmental Life Cycle Assessment of Goods and Services: An Input-Output Approach." Washington, DC: Resources for the Future Press, 2006.

Horne, Ralph, Tim Grant, and Karli Verghese. "Life Cycle Assessment: Principles, Practice and Prospects." Collingwood, Australia: CSIRO Publishing, 2009.

International Standards Organization. http://www.iso.org/iso/home.html (Accessed July 2010).

Jakubowski, Benjamin and Howard Frumkin. "Environmental Metrics for Community Health Improvement." *Preventing Chronic Disease: Public Health Research, Practice, and Policy*, 7/4 (2010).

Prüss-Üstün, A. and C. Corvalán. "Preventing Disease Through Healthy Environments: Towards an Estimate of the Environmental Burden of Disease." Geneva: World Health Organization, 2006.

Tomlinson, Bill. "Greening Through IT: Information Technology for Environmental Sustainability." Cambridge: MIT Press, 2010.

U.S. Environmental Protection Agency. "Life Cycle Assessment (LCA)." http://www.epa.gov/nrmrl/lcaccess (Accessed July 2010).

World Bank. "World Development Report 1993: Investing in Health." http://files.dcp2.org/pdf/WorldDevelopmentReport1993.pdf (Accessed July 2010).

World Health Organization. "Global Health Risks: Mortality and Burden of Disease Attributable to Selected Major Risks." http://www.who.int/healthinfo/global_burden_disease/GlobalHealthRisks_report_full.pdf (Accessed July 2010).

World Health Organization. "Preamble to the Constitution of the World Health Organization," as adopted by the International Health Conference, New York, June 19–22, 1946; signed on July 22, 1946, by the representatives of 61 States (Official Records of the World Health Organization, no. 2, p. 100) and entered into force on April 7, 1948.

Zeckhauser, R. and D. Shepard. "Where Now for Saving Lives?" *Law and Contemporary Problems*, 40/4 (1976).

Oladele Ogunseitan
University of California, Irvine

MICROWAVE OVENS

Microwaves are a form of electromagnetic radiation; that is, they are waves of electrical and magnetic energy moving together through space. While microwaves are used in numerous applications, including detecting speeding cars, sending telephone and television communications, and drying plywood and curing rubber, the most common consumer use of microwave energy is in microwave ovens. Microwaves have the three following characteristics that allow them to be used in cooking:

- They are reflected by metal.
- They pass through glass, paper, plastic, and similar materials.
- They are absorbed by foods.

Microwaves are produced inside the oven by an electron tube called a magnetron. The microwaves are reflected within the metal interior of the oven and absorbed by food, causing the water molecules inside the food to vibrate, which produces the heat that cooks the food. Although heat is produced directly in the food, microwave ovens do not cook food from the inside out. When thick foods are cooked, the outer layers are heated and cooked primarily by microwaves, while the inside is cooked mainly by the conduction of heat from the outer layers. In the decades since microwave ovens started becoming a common feature of modern kitchens, there have been many claims made about the dangers of using microwaves for heating and cooking food. Much of the research cited to support these claims is based on work done by Russian, Japanese, and Swiss scientists, though some American researchers, including those at Stanford University and Penn State University, have suggested that the "microwave effect" does, indeed, exist. In addition to the "microwave effect"—the notion that certain dangerous transformations, which cannot be replicated with conventional heating methods, occur in both microwaved food and in the humans who eat it—there are other claims about microwaves and microwaved foods, including the following:

The Food and Drug Administration admits that less is known about the dangers of exposure to low levels of microwave radiation than is known about high levels. The U.S. federal government limits the allowable amount of microwave radiation that can leak from an oven throughout its lifetime.

Source: iStockphoto.com

- Microwaves leak radiation.
- Microwaved food causes cancer.
- Microwaved food loses some of its nutritional value.
- Microwaved food causes "microwave sickness," which is marked by symptoms including insomnia, night sweats, headaches, dizziness, impaired cognition, depression, and nausea.

While some of the claims of the dangers of microwave ovens are based on a misunderstanding of the differences between ionizing radiation (the nuclear, dangerous kind, such as that used in X-rays and gamma rays) and nonionizing radiation (the kind that is used in microwave ovens), many believe that the debate over the safety of microwave ovens is one that has yet to be settled. Still, according to the World Health Organization (WHO) and the U.S. Food and Drug Administration (FDA), which has regulated the manufacture of microwave ovens since 1971, microwave ovens that meet the FDA standard and are used according to the manufacturer's instructions are safe for human use. The FDA requires that all microwave ovens have a label stating that they meet the FDA safety standard. To ensure the standard is met, the FDA tests microwave ovens in its own laboratory and also evaluates manufacturers' radiation testing and quality control programs at their factories.

As for the claims made by microwave oven foes, both the WHO and the FDA contend that microwave energy is changed to heat as it is absorbed by food, and that this does not make food radioactive or contaminated. Further, they contend that microwave cooking not only does not reduce the nutritional value of foods any more than does conventional cooking, but that foods cooked in a microwave oven may keep more of their vitamins and minerals, because microwave ovens can cook more quickly and without adding water.

Despite those precautions and assurances, even the FDA admits that less is known about the dangers of exposure to low levels of microwave radiation than is known about high levels. The U.S. federal government limits the amount of microwave radiation that can leak from an oven throughout its lifetime to 5 milliwatts of radiation per square centimeter, at approximately two inches from the oven's surface. Microwave energy decreases significantly as you move away from the source of radiation, so that a measurement made 20 inches from an oven would be approximately one one-hundredth of the value measured at two inches. Painful skin burns, cataracts, damage to testes, and altered sperm count are all associated with high levels of exposure to microwave radiation—much higher than the 5 milliwatts limit for microwave oven leakage—but in regard to the low levels of microwave radiation associated with microwave ovens, no controlled, long-term studies involving large numbers of people have been conducted.

The FDA continues to reassess the safety standard's adequacy as new information becomes available, and the fact that many scientific questions about exposure to low-levels of microwave radiation remain unanswered requires the FDA to continue enforcement of radiation protection requirements. Microwave oven foes sum up this strategy by conceding that although the use of microwave ovens may not make you glow in the dark, the cumulative effect on the body of years of microwave oven use remains uncertain.

See Also: Government Role in Green Health; Personal Consumer Role in Green Health; Radiation Sources.

Further Readings

Mercola, Joseph. "Why Did the Russians Ban an Appliance Found in 90% of American Homes?" http://articles.mercola.com/sites/articles/archive/2010/05/18/microwave-hazards .aspx (Accessed May 2010).

U.S. Food and Drug Administration. "Radiation-Emitting Products." http://www.fda.gov/ radiation-emittingproducts/resourcesforyouradiationemittingproducts/consumers/ucm142 616.htm (Accessed May 2010).

World Health Organization. "Electromagnetic Fields & Public Health: Microwave Ovens." http://www.who.int/peh-emf/publications/facts/info_microwaves/en (Accessed May 2010).

Tani Bellestri
Independent Scholar

Musculoskeletal Diseases

The musculoskeletal system consists of bones, muscles, ligaments, tendons, joints, cartilage, and other connective tissues. These components all work together to provide form, stability, and movement to the human body. Diseases of the musculoskeletal system may result in the inability to walk, sit, or even breathe, and are generally accompanied by pain, as well as limitations in physical functioning and fatigue. The functional and social limitations associated with musculoskeletal diseases often have emotional consequences as well.

There are many musculoskeletal diseases that require clinical care by a physician or other healthcare professional. Some of the more common musculoskeletal diseases are listed below with a brief description of each for general information, bearing in mind that a diagnosis should always be made in person by a physician or healthcare professional.

This entry provides a brief overview of many musculoskeletal diseases, from those that have high impact on functional health to those with low impact.

Repetitive strain injury refers to a range of conditions caused by repetitious movements and force. This can result in functional limitations in the hand, elbow, and shoulder, as well as the health state of chronic back pain. Typically, patients present with weakness and lack of endurance as well as worsening pain with activity.

Osteoarthritis is characterized by the breakdown of cartilage in a joint. Symptoms may include joint pain, tenderness, stiffness, inflammation, creaking, and locking of joints. The patient presents with pain upon weight bearing, including standing and walking, and pain worsens throughout the day as the joints are used. Due to decreased movement because of the pain, regional muscles may atrophy and ligaments may become more lax. In osteoarthritis, a variety of factors may initiate the process leading to loss of cartilage, such as hereditary, developmental, metabolic, and mechanical.

Rheumatoid arthritis is a chronic systemic inflammatory disorder that may affect tissues and organs, but principally attacks the joints, producing an inflammatory synovial membrane, which lines joints and tendon sheaths. Joints become swollen, tender, and warm, and stiffness limits their movement. Rheumatoid arthritis typically manifests with signs of inflammation, and stiffness in the early morning upon waking or following prolonged inactivity. Gentle movements may relieve symptoms in early stages of the disease. As the disease progresses, the inflammatory activity leads to tendon tethering and erosion and destruction of joint surface. This impairs range of movement and leads to deformity.

Systemic lupus erythematosus is a chronic autoimmune connective tissue disease that can affect any part of the body, resulting in inflammation and tissue damage. While systemic lupus erythematosus affects the heart, kidneys, lungs, and other aspects of the human body, the most common reason for seeking medical attention is joint pain. Usually the small joints of the hand and wrist are affected, although all joints are at risk, and lupus arthritis is less disabling and usually does not cause severe destruction of the joints, as seen in other forms of arthritis. Being that systemic lupus erythematosus is a chronic disease with no known cure, the treatment is usually symptomatic, usually preventing flare-ups and reducing their severity.

Fibromyalgia is a chronic condition of the soft tissues characterized by widespread pain. Other core symptoms are debilitating fatigue, sleep disturbance, and joint stiffness. There is no recognized cure for fibromyalgia, but some prescription treatments have been shown to be effective in reducing symptoms. However, fibromyalgia is a controversial diagnosis. Many members of the medical community consider it a nondisease because of a lack of abnormalities on physical examination and the absence of objective diagnostic tests.

Osteoporosis is a disease of bone that leads to an increased risk of fracture. In osteoporosis, the bone mineral density is reduced, and the amount and variety of noncollagenous proteins in bone are altered. The underlying mechanism in all cases of osteoporosis is an imbalance between bone reabsorption and bone formation. Osteoporosis itself has no specific symptoms; its main consequence is the increased risk of bone fractures. Osteoportic fractures are those that occur in situations where healthy people would not normally break a bone. These types of fragility fractures occur in the vertebral column, rib, hip, and wrist. Treatment is usually prescribed with medication, exercise, calcium, and vitamin D.

Bursitis is the inflammation of one of more bursae of synovial fluid in the body. The bursae rest at points where internal functionaries, such as muscles and tendons, slide across bone. Healthy bursae create smooth, painless movement. Movement with bursitis is difficult and painful. Repetitive movement and excessive pressure on the joints typically cause it. Symptoms of bursitis may vary from local joint pain and stiffness to burning pain that surrounds the joint around the inflamed bursa. In this condition, pain is usually worse during and following activity. Bursitis that is not infected is treated with rest, ice compresses, anti-inflammatory drugs, and pain medication.

Cherubism is a rare genetic disorder that causes prominence in the lower portion of the face. Patients with this symptom present with a loss of bone in the mandible, which the body replaces with excessive amounts of fibrous tissue. In most cases, the condition fades as the child grows. This also causes premature loss of primary teeth. The genetic mutation associated with cherubism is said to be autosomal dominant.

Osteogenesis imperfecta is an autosomal-dominant genetic bone disorder. People with osteogenesis imperfecta are born with defective connective tissue or without the ability to make it. The bone is altered in this disease, causing brittleness and loose joints. There are types I to VIII, ranging from mild to severe. All share bones that fracture easily. At the time of this writing, there is no cure for osteogenesis imperfecta.

Gout is a disease hallmarked by elevated levels of uric acid in the bloodstream. In this condition, crystals of uric acid are deposited on the articular cartilage of joints, tendons, and surrounding tissues. Excruciating, sudden, unexpected burning pain, as well as redness, swelling, warmth, and stiffness in the affected joint characterize gout. Hyperuricemia is a common feature of gout; however, gout can occur without hyperuricemia. Ultrasound imaging is helpful in diagnosis of gout. Treatment for gout has three main objectives: to lower serum uric acid, to prevent acute attacks, and to manage the symptoms of acute attacks when they occur.

Chondrocalcinosis is a rheumatologic disorder with varied clinical presentations due to the precipitation of calcium pyrophosphate dihydrate crystals in the connective tissues. Patients usually present with inflammation of one or more joints; this often results in pain in the affected joints. Radiography plays a large role in the diagnosis of chondrocalcinosis.

Septic arthritis is the invasion of a joint by an infectious agent, which produces arthritis. Septic arthritis should be considered whenever one is assessing a patient with joint pain. Usually only one joint is affected. The diagnosis can be difficult because there is no test available to completely rule out the possibility. Diagnosis is done by aspiration of fluid from the joint. Treatment is usually done with intravenous antibiotics and aspiration of the joint to dryness.

Juvenile idiopathic arthritis (JIA) is the most common form of persistent arthritis in children. Juvenile idiopathic arthritis differs significantly from arthritis commonly seen in adults and other types of arthritis that can present in childhood, which are chronic conditions. Symptoms of juvenile idiopathic arthritis are nonspecific initially and include lethargy, reduced physical activity, and poor appetite. The first manifestation may be limping. The main symptom is persistent swelling of the affected joint, which commonly includes the knee, ankle, wrist, and small joints of the hands and feet. Pain is also an important symptom; however, young children may not be able to communicate the pain directly. Other effects are joint contracture and joint damage. There are three kinds of juvenile idiopathic arthritis: oligoarticular, polyarticular, and systemic. Oligoarticular JIA affects four or fewer joints. Polyarticular JIA affects five or more joints. Arthritis, fever, and a pink rash characterize systemic JIA.

Polymyositis is a type of chronic inflammatory myopathy. Polymyositis is usually evident in adulthood, presenting with bilateral muscle weakness. It is mostly noted in the upper legs due to early fatigue while walking, being unable to rise from a seated position without help, and inability to raise arms above head. There is a loss of muscle mass, particularly in the shoulder and pelvic girdle. Thickening of the skin on the fingers and hands is a frequent feature. The cause of polymyositis is unknown, but appears to be related to autoimmune factors, genetics, and in rare cases is known to be infectious. Typically, patients are treated with high-dose steroids; unresponsive patients may be placed on immunosuppressive medication.

Systemic scleroderma is a systemic connective tissue disease. The most visible symptoms are in the skin. Scleroderma causes hardening and scarring, and the skin may appear red and scaly. In regard to musculoskeletal, the first joint symptoms that patients with systemic scleroderma have are nonspecific joint pain. This can lead to arthritis or cause discomfort in tendons or muscles. Calcinosis or skin thickening may restrict mobility of the joints, especially of the small joints in the hand. There is no clear cause for systemic sclerosis. Genetic predisposition appears to be limited. There is no cure for systemic scleroderma; however, there is treatment for the some of the symptoms.

Osteomalacia is a disease in which there is a defect in the bone building process of the bones. The disease is commonly caused by a deficiency in vitamin D, which is obtained through diet and sunlight exposure. Patients typically have a waddling gait as well as pain and aches beginning in the lumbar region and spreading symmetrically. Body pains, muscle weakness, fatigue, and fragility of the bones, also referred to as softening of the bones, are also signs of this disease. Osteomalacia in children is referred to as rickets.

Osteomyelitis is the inflammation of the bone and bone marrow, caused by an infection. In children, long bones are usually affected. In adults, pelvis and vertebrae are most commonly affected. Osteomyelitis often requires prolonged antibiotic treatment. It may also require surgical debridement and severe cases may lead to the loss of a limb. Diagnosis is based on radiological results and a culture of material taken from the bone biopsy to identify the specific pathogen.

Tietze's syndrome is the benign inflammation of one or more of the costal cartilages. Patients present with acute pain in the chest, along with tenderness and some swelling of the cartilages affected. Pain is often exacerbated with respiration. The condition is usually benign and resolves in 12 weeks; however, it can be a chronic condition. It is common for some patients to mistake the pain for that of a myocardial infarction. Tietze's syndrome does not progress to cause harm to any organs.

Avascular necrosis is a disease resulting from the temporary or permanent loss of the blood supply to an area of bone. Without blood, the bone tissue dies and the bone collapses. If avascular necrosis involves the bones of a joint, it often leads to destruction of the joint articular surfaces. It primarily affects bones of the hip, knee, and shoulder. MRI and bone scintigraphy are the best choices for diagnosis because in early stages, X-rays appear normal. Total joint replacements are common, especially in the hip. Other treatment methods include core decompression and free vascular fibular graft.

Conclusion

Because many other body systems including nervous, vascular, and integumentary systems are interrelated, disorders of one of these systems may also affect the musculoskeletal system and complicate the diagnosis of the disorder's origin. Diseases of the musculoskeletal system mostly encompass functional disorders. The severity of the problem determines the level of impairment. The most common musculoskeletal diseases are articular, pertaining to joints. Disorders of muscles from another body system can display irregularities such as impairment of ocular motion and control, respiratory dysfunction, and bladder malfunction.

Musculoskeletal diseases have their own origins, however they sometimes are signs and symptoms of other disease: due to that fact, it is always important to check with a physician or other healthcare professional about any disease and treatment options.

See Also: Antibiotics; Cardiovascular Diseases; Degenerative Diseases; Immune System Diseases; Supplements.

Further Readings

Dequeker, Jan. *Medical Management of Rheumatic Musculoskeletal and Connective Tissue Diseases.* New York: Informa Health Care, 1997.

Reilly, Thomas. *Musculoskeletal Disorders in Health-Related Occupations.* Amsterdam: IOS Press, 2002.

Von Schulthess, Gustav Konrad, and Christoph L. Zollikofer. *Musculoskeletal Diseases: Diagnostic Imaging and Interventional Techniques.* New York: Springer, 2005.

Alessandra Guimaraes
University of Sint Eustatius School of Medicine

Neurobehavioral Diseases

The term *neurobehavioral disease* (ND) refers to a cluster of disorders, primarily with behavioral (and cognitive) dysfunctions that develop as a result of underlying neurobiological pathology. Classically included in the Axis II type of disorders in the *Diagnostic and Statistical Manual of Mental Disorders* (DSM-IV, 1994), these diseases form a substantive part of pediatric psychiatry. Although the term *ND* is relatively novel, initial references to such disorders were made as early as the beginning of the 20th century, with terms like *minimal brain dysfunction* and *hyperkinetic disorder of childhood*, to name a few. In educational settings, the term *emotional and behavioral disorders* is also frequently used. Etiology and pathogenesis of ND is still a matter of much debate, and a wide range of factors ranging from genetic anomalies to environmental and psychosocial factors have been considered to play a role. The majority of ND can be broadly included in three groups: (1) disorders of excessive motor activity and thought, attention deficit hyperactivity disorder being perhaps the most common disorder encountered in daily practice, with its predominance reaching almost 50 percent in the child psychiatric clinic populations; (2) pervasive developmental disorders, for example, autism spectrum disorders; and (3) specific learning disorders. Some of the more common disorders will be discussed in further detail.

For purposes of disambiguation, the term *ND* is also used on several occasions to refer to organic brain syndromes, a concept used in neurological practice that will be elucidated further.

Disorders of Excessive Motor Activity and Thought: Attention Deficit Hyperactivity Disorder

Attention deficit hyperactivity disorder (ADHD) is a debilitating psychiatric disorder manifested by paucity in the attention span and/or inappropriate hyperactivity and impulsivity. The DSM-IV distinguishes three forms of this disorder: exclusive disruption of attention span (ADHD, predominantly inattentive type); exclusive involvement of hyperactivity and impulsivity span (ADHD, predominantly hyperactive-impulsive type); or both (ADHD, combined type), with all symptoms being present in at least two situations (e.g., school, play, or home) and for longer than six months. Hallmarks of inattention include difficulty to attend to a task or to sustain attention for a longer period of time, forgetful behavior, and easy distraction. Hyperactivity is manifested by fidgetiness, excessive talking, and restlessness, while symptoms of impulsivity might range from tendency to interrupt conversations and difficulty in awaiting turn to acting without thinking about subsequent consequences.

Even though several theories have been contemplated for a potential explanation of this disorder, much still remains unknown. The possible genetic component of ADHD can be explained from twin studies and studies with siblings, which show a greater concordance of ADHD in monozygotic twins than in dizygotic twins and a higher risk in siblings of ADHD children to develop the disorder as compared to the general population. Apart from genes, the faulty development of the different brain areas during the pre- and perinatal period under the influence of various detrimental factors, such as hypoxia, toxic, and metabolic damage to the developing brain, have been also discussed. Recently, findings about functional differences in brain area activation patterns with the help of modern brain imaging techniques show a decreased activation of the frontal lobes (a structure responsible for inhibitory control of lower, more voluntarily acting systems) of the brain in children with ADHD, thus suggesting a reason for possible behavioral disinhibition.

With the advent of newer imaging techniques, new concepts like default mode network (DMN) have been introduced. DMN is described as a network of brain areas that are connected and are active when the brain is in a resting state (a state defined when the subject is awake and alert without performing any goal-directed tasks, such as calculating). Such neuronal networks are mostly involved in a "broad based continuous sampling of external and internal environments," and in patients with ADHD, altered connectivity between the DSN possibly explain attention deficits.

The central nervous system (CNS) stimulants methylphenidate and dextroamphetamine are the two commonly prescribed drugs in ADHD. The mechanism of action of these drugs can be explained by the alterations in neurotransmitter (NT) levels dopamine and norepinephrine in brain areas involved in impulsivity (nucleus accumbens) and inhibitory control (frontal cortex). Essentially, these drugs act by three mechanisms: (1) inhibition of presynaptic reuptake of NT, (2) strengthening in release of NT, and (3) inhibition of monoamine oxidase (an enzyme that catabolizes NT at the synaptic cleft and thus increases NT levels). Contrary to earlier belief that the response to CNS stimulants in children with ADHD was paradoxical to that in children without the disorder, modern data show that at similar doses, these drugs act in a similar fashion in both adults and children with or without ADHD, reducing levels of activity and strengthening attention.

Pervasive Developmental Disorders: Autism Spectrum Disorders

More predominant in the male population, *autism spectrum disorders* (ASD) is an umbrella term for at least four types of disorders: autism, Asperger syndrome, Rett syndrome, and childhood disintegrative disorder, the following points being common for all of them: (1) impairments in reciprocal social interactions: lacking of social smile in infants, abnormal eye contact, inability to recognize and differentiate familiar and unfamiliar people, difficulty in making friends; (2) impairments in communication: deficits in spoken language development, pronominal reversals ("you" instead of "I"); and (3) restricted repetitive and stereotyped patterns of behavior: similar ritualistic play behavior, stereotypes, and abnormal movements.

The most frequent disorder in this group is autism, which shows all the classical features together with mental retardation in up to 70 percent of cases. Various factors, from genes and immunogenic factors to altered and incomplete brain development—especially regions responsible for cognition (prefrontal cortex) and emotion (limbic system)—have been speculated to be responsible for autism. The concept of a genetic marker responsible for autism is currently under investigation. Copy number variation in genes (until it was recently deemed that human beings only have two copies of a single gene, but this concept has been extended and different numbers of a single gene have been shown) and single nucleotide polymorphisms (variation in genetic sequence relatively common in the population) are two vital candidates. The possibility of these genetic factors to cause aberrant synaptic connections during the development of the brain is a subject of study.

Rett syndrome is seen only in females and is an X-linked dominant disorder with behavioral symptoms similar to autism in addition to a variety of physical anomalies: microencephaly (small circumference of head), small hands and feet, scoliosis (abnormal lateral curvature of the spine), and a high incidence of concurrent seizures. Complete concordance in monozygotic twins confirms the genetic component of the disorder.

Asperger syndrome is often considered as a milder variant of autism. It also shares most of the features common to all ASDs, together with minimal language and cognitive malfunctioning, thus distinguishing it from autism.

Childhood disintegrative disorder is a comparatively rare disorder. It has overlapping features with autism and differs by a later onset and loss of previously acquired development.

The underlying goal of treatment of ASDs is mostly symptomatic, centering primarily on individual and group psycho-, behavior, speech, and physical therapy. Even though no specific drug has been used in ASD, the use of haloperidol remains familiar in the clinical setting. Care is also taken to educate parents of autistic children.

Specific Learning Disabilities

Specific learning disabilities (LDs) involve specific deficits in the acquisition and execution of reading, writing, spelling, reasoning, or carrying out mathematical calculations, and together with normal intelligence are classified together as LD. Reading disorder (dyslexia or alexia), one of the commonest disabilities of this group, is characterized by notably poor reading skills despite normal intelligence. Other LDs frequently encountered in pediatric psychology include mathematics disorder, dysgraphia, and dyspraxia.

The ability to recognize phonemes (the smallest contrastive unit in a spoken language), to understand grammatical constructions, and to recall words is exclusively hampered in reading disorder. Theories about underactive right hemisphere, malfunctioning of the magnocellular pathway (neuronal pathway carrying information from the eyes to higher cortical structures), and cerebeller dysfunctioning have been put forward to elucidate the causes of reading disorder. Functional magnetic resonance imaging scans during performance of various reading tasks in dyslexic and normal controls show the lower activation of the right hemisphere in dyslexic subjects as compared to healthy controls. Contrary to reading disorder, disorder in written expression (or dysgraphia) involves the inability to understand and construct grammatically correct sentences and phrases relevant to one's level of intelligence- and education-appropriate norms. Analogously, mathematics disorder is characterized by the inability to manipulate and conceptualize numbers, numeric sequences, or to perform mathematical calculations. In many cases, LD goes unnoticed until the child is at a higher level of education, when it becomes increasingly difficult to keep up with other students in the same level, culminating in feelings of guilt and self-pity, a hotbed for future mood disorders. Furthermore, difficulty in keeping up with one's peers further leads to decline in the wish and finally neglect of reading, writing, or mathematics.

Treatment mainly focuses on remedial education, a common term in many Western countries, in which more attention is paid to underprepared children to uplift their level of skills in which they lack competence.

To sum up thus far, despite the striking differences in the clinical presentation of the above-mentioned disorders, a few common qualities glue them together: the deviation from "normal" behavior, the predominant early onset in life, and relatively higher risk for development of concurrent psychiatric disorders in these patients.

Neurobehavioral Disorders: Organic Brain Disorders

In neurological practice, delineating it from psychiatry, the term *ND* often refers to organic brain disorders (OBDs). Robustly it implies diseases with an underlying neurobiological pathology with symptoms in the behavioral and cognitive domain. All OBDs can be divided into the following three broad groups:

- *Acute confusional state or delirium*: a frequent neuropsychiatric condition characterized by impairment of consciousness and cognition together with a rapid onset of clinical symptoms. Usually a disorder secondary to numerous pathologies, delirium can have both intracranial (seizures and meningitis) and extracranial (poisoning of various origin, endocrinological dysfunctions, liver and kidney pathologies) causes. Symptomology of delirium includes psychiatric symptoms like dissociation in time and space and incoherent speech coupled with physical manifestations like increased heart rate, respiratory rate, flushing, and sweating. Treatment of the underlying condition still remains the primary goal.
- *The dementias*: a global term used to describe an extremely wide range of neurobehavioral disorders with the core symptom of dementia being unique to all of them. Dementia is a progressive disorder manifested by gradual decline of cognitive functions like memory, attention, language, and decision-making and problem-solving skills, while the consciousness domain remains unaffected. Alzheimer's disease, vascular dementia, fronto-temporal dementia, and dementia in Parkinson's disease and Huntington's disease are a few well-known examples of dementia.
- *OBD secondary to head trauma and focal brain damage*: behavioral disruptions can also be the principal manifesting signs of brain trauma and damage of brain tissue as a result of tumors, stroke, or vascular malformations. Depending on the area of the brain involved in the damage, the clinical signs are varied, manifesting in the forms of aphasias (disorder in speaking or interpreting language), amnesias (complete or partial loss of memory), or other neurobehavioral impairments. Treatment of the underlying pathology remains the fundamental objective.

See Also: Children's Health; Men's Health; Mental Exercises; Mental Health; Women's Health.

Further Readings

Broyd, S. J., et al. "Default-Mode Brain Dysfunction in Mental Disorders: A Systematic Review." *Neuroscience and Behavioural Reviews*, 33/3 (2009).

Kaplan, I. H., B. J. Saddok, and J. A. Grebb. *Synopsis of Psychiatry*, 7th ed. Baltimore, MD: Williams and Wilkins, 1994.

Maisog, J. M., E. R. Einbinder, D. L. Flowers, P. E. Turkeltaub, and G. F. Eden. "A Meta-analysis of Functional Neuroimaging Studies of Dyslexia." *Annals of the New York Academy of Science*, 1145 (2008).

Marsh, R., A. J. Gerber, and B. S. Peterson. "Neuroimaging Studies of Normal Brain Development and Their Relevance for Understanding Childhood Neuropsychiatric Disorders." *Child and Adolescent Psychiatry*, 47/11 (2008).

Solanto, M. V. "Neuropsychopharmacological Mechanisms of Stimulant Drug Action in Attention-Deficit Hyperactivity Disorder: A Review and Integration." *Behavioral Brain Research*, 94/1 (1998).

Tharper, A., J. Holmes, K. Poulton, and R. Harrington. "Genetic Basis of Attention Deficit and Hyper Activity." *British Journal of Psychiatry*, 174 (1999).

Weisberg, L. A., C. Garcis, and R. Strub. *Essentials of Clinical Neurology*, 3rd ed. St. Louis, MO: Mosby Year Book, 1996.

Yudofsky, S. C. and R. E. Hales. *Textbook of Neuropsychiatry*, 3rd ed. Washington, DC: American Psychiatric Press, 1997.

Rahul Pandit
Utrecht University

NOSOCOMIAL INFECTIONS

Hospital-acquired (nosocomial) infections are a significant concern to the health of patients requiring a hospital stay. These infections affect millions of Americans, lead to tens of thousands of deaths yearly, and cost billions of dollars. They may arise from devices, such as intravenous tubes and catheters, and often affect older and critically ill patients. There are a variety of possible causative agents. Some of these pathogens have developed resistance to common treatment therapies, complicating treatment options. Many steps to minimize nosocomial infection rates have been taken that have proved marginally successful. Proper prevention and treatment are necessary to reduce morbidity and mortality, as well as to decrease the costs of treating these preventable infections.

Causes of Nosocomial Infections

Nosocomial infections are defined as infections not present or incubating at the time the patient was admitted to the hospital, and that develop 48 hours or more after admission. Common infection sites include the following:

- *Bloodstream (most common site)*: often from indwelling catheters, surgical wounds, and abscesses
- *Urinary tract (second most common)*: usually associated with a Foley catheter
- *Respiratory system*: pneumonia arising in intubated patients or those with a decreased level of consciousness
- *Gastrointestinal tract*: *Clostridium difficile* colitis (inflammation of the colon) secondary to antibiotic use, which eliminates the normal gut flora

Infection can develop after contact with an infected or colonized (uninfected person carrying a pathogen) individual, which may include healthcare workers, visitors, or even other patients.

Fomites (inanimate objects that may harbor pathogens) can lead to nosocomial infections. Studies have demonstrated that fomites such as physicians' neckties and handbags often have bacteria that can cause nosocomial infections. Researchers were able to grow bacteria after swabbing them. Although it is unusual to receive a nosocomial infection this way, it underscores the importance of maintaining strict hygienic procedures in hospitals.

Infections can be transmitted via respiratory droplets. Coughing, sneezing, or even talking closely with an infected person could lead to disease.

Many infections are directly related to invasive devices used in the hospital to monitor the patient or to provide therapy. Catheters (intravenous or bladder) and surgical drains or tubing can lead to infection. Proper sanitary technique and prompt changing/removal can reduce nosocomial infections.

Poor hand hygiene by hospital staff can lead to nosocomial infections. There has been a trend in hospitals and clinics over the past several years to install hand sanitizers outside each patient's room to make it easier to remember to wash hands after and before seeing patients.

A number of risk factors have been identified for development of hospital-acquired infections. These include but are not limited to the following:

- Mechanical ventilation
- Advanced age
- Immunosuppression (by either medications or disease)
- Chronic disease
- Recent surgery
- ICU admission
- H_2 or antacid therapy (decreases the level of acid in the gut, which normally can fight off some bacteria)
- Decreased level of consciousness
- Paralytic agents
- Intubation
- Indwelling device placement

Primary Involved Pathogens

Common causative pathogens are the following:

- *Staphylococcus aureus*
- *Staphylococcus epidermidis*
- *Escherichia coli*
- *Pseudomonas aeruginosa*
- *Acinetobacter baumannii*
- *Clostridium difficile*
- *Klebsiella pneumonia*
- *Mycobacterium tuberculosis*
- *Enterococcus faecium*
- *Legionella pneumonia*

After a period of time, and facilitated by antibiotic misuse/overuse, many pathogens are able to adapt and become resistant to previously effective therapies. Some become resistant to multiple drugs, leading to difficulty treating these infections. Some prevalent multi-drug-resistant pathogens include the following:

- Methicillin-resistant *Staphylococcus aureus* (MRSA)
- Vancomycin-resistant *Enterococcus* (VRE)
- *Pseudomonas aeruginosa*
- *Clostridium difficile*
- Multidrug-resistant *Acinetobacter baumannii* (MRAB)
- *Legionella pneumonia*

Treatment

When a patient presents to the hospital with signs and symptoms of an infection (non-nosocomial), it is often appropriate to initiate broad-spectrum, multidrug therapy as an initial treatment. After test results (cultures) return in a few days, the pathogen and susceptibility can be identified. Then therapy is narrowed to cover the infecting pathogen with as little antibiotic usage as possible beyond what is necessary to treat.

Treatment strategies vary, depending on risk factors for multidrug-resistant (MDR) bacteria. In a patient with no known risk factors, treatment may consist of the following:

- Ceftriaxone
- Ampicillin-sulbactam or piperacillin-tazobactam
- Levofloxacin or moxifloxacin
- Ertapenem
- Trimethoprim/sulfamethoxazole
- Amoxicillin
- Nitrofurantoin
- Ciprofloxacin
- Metronidazole

Problems With Resistance

In patients with risk factors for MDR bacteria, combination therapy is necessary to ensure proper antibiotic coverage. Possible combinations include the following:

- Cephalosporin (cefepime or ceftazidime) or carbapenem (imipenem or meropenem), with antipseudomonal coverage PLUS
- Antipseudomonal fluoroquinolone, that is, ciprofloxacin or levofloxacin PLUS
- Aminoglycoside, that is, gentamicin or tobramycin or amikacin PLUS
- Linezolid or vancomycin, if MRSA is suspected

It is also important to consider that positive cultures do not necessarily mean that a patient has an infection. A positive culture without clinical or radiologic evidence of infection may simply mean the person is chronically colonized with the pathogen and should not be treated at the present time.

Prevention

The single most effective and easiest method to prevent nosocomial infections is proper hand washing. Traditional soap and water are effective, but quick drying, alcohol-based antiseptic foams and gels have gained favor due to ease of use and reduced application time, as well as improved effectiveness of decontamination.

Timely removal of indwelling devices can help control nosocomial infections as well. Peripheral intravenous lines should be changed every three days, and arterial lines changed every four days. Central venous lines may be left in place as long as they are functional and show no signs of infection. Many catheters are impregnated with antibiotics to reduce infection rates, albeit at a higher cost.

Competent nursing care goes a long way toward prevention of infections. Maintaining clean and dry dressings, moving patients to minimize decubitus ulcers, and taking aspiration precautions help reduce infections. Proper isolation of infectious patients is also important.

Numerous vaccines are available to aid in the prevention of nosocomial infections. Hepatitis A and B, varicella, pneumococcus, and influenza vaccines are all effective in decreasing the incidence of hospital-associated infections. While these do not guarantee complete protection, they are another step in prevention.

Proper education of staff, patients, and their families can be an important step in curtailing hospital infections. A little knowledge about the importance of good hygiene is an easy step that can make a significant difference.

Conclusion

Nosocomial infections are a significant problem in hospitals, affecting approximately 5 percent of patients. These infections lead to increased hospital stays, sizable costs, increased morbidity, and up to a 5 percent mortality rate. The sources of infection are often well known and include numerous gram-positive and gram-negative organisms. Some of these have developed resistance to therapies that have been previously effective, leading to difficulty in treatment. Therapy varies depending on the identified pathogen and its resistance to standard medication. Numerous steps are available to prevent nosocomial infections, with the easiest being hand washing. Appropriate prevention practices and timely identification and treatment of infections can improve hospital outcomes, reduce morbidity and mortality, and decrease extraneous costs in the hospital.

See Also: Antibiotic Resistance; Healthcare Delivery; Infectious Waste; Personal Hygiene Products.

Further Readings

Dotan, I., M. Somin, and A. Basevitz, et al. "Pathogenic Bacteria on Personal Handbags of Hospital Staff." *Journal of Hospital Infection*, 72/1 (2009).

Fauci, A., E. Braunwald, and D. Kasper, et al. "Health Care–Associated Infections." *Harrison's Principles of Internal Medicine*, 17th ed. New York: McGraw-Hill Professional, 2008.

File, T., Jr. "Epidemiology, Pathogenesis, and Microbiology of Hospital-Acquired, Ventilator-Associated, and Healthcare-Associated Pneumonia in Adults." http://www.uptodate.com (Accessed July 2009).

McPhee, S. and M. Papadakis, et al. "Hospital Associated Infections." *CURRENT Medical Diagnosis and Treatment*, 48th ed. New York: McGraw-Hill, 2009.

Gautam J. Desai
Stephen M. Derrington
Kansas City University of Medicine and Biosciences,
College of Osteopathic Medicine

NUCLEAR POWER

There are a number of deeply politicized approaches to the issue of nuclear power, which provides nearly 20 percent of electricity in the United States and 15 percent worldwide. Nuclear scientists and engineers frequently argue that nuclear power poses no discernible impact on public health; antinuclear activists point to the potential risk and pollution produced at all stages of the nuclear fuel cycle; advocates for a transition to a less carbon-intensive energy economy often argue in favor of "carbon-free" nuclear; and public opinion remains fairly divided on the issue. For example, there is extensive debate over whether nuclear power plants are "safe." To understand these debates, it is important to know that the history of nuclear power is embedded in the history of nuclear weapons. Nuclear power emerged, at least in part, as a response to the dropping of American atomic bombs on Hiroshima and Nagasaki, Japan, in 1945. Following the devastation of both cities, many U.S. scientists and leaders believed they needed to make sense of and justify continued research into nuclear technologies. In 1953, President Dwight D. Eisenhower publicly promoted the Atoms for Peace campaign; following that, he and others argued that the atom should be used not only to make weapons but also to produce vast amounts of affordable electric power, promote international agreement, and develop medical breakthroughs.

During the 1986 nuclear plant disaster in Chernobyl, Ukraine, there was a significant release of radioactive materials, and thousands of people are believed to have died or contracted serious cancers as a result of radiation exposure. Here, a view of Chernobyl in 2005 taken from the roof of a building in the still-deserted city of Pripyat.

Source: Wikipedia/Jason Minshull

Despite these efforts, many Americans mentally associate the risks of nuclear power with the risks posed by nuclear weapons. This sense of risk has been heightened by certain historical events, such as the Three Mile Island incident of 1979, in which a large reactor in Pennsylvania suffered significant technical problems and threatened to release substantial amounts of nuclear radiation. Nuclear scientists and engineers did not immediately understand the problem and poorly communicated the risks; widespread media coverage of the event was frequently inaccurate or confusing, causing unnecessary panic in some cases. The majority of nuclear scientists and engineers today agree that the Three Mile Island incident was not an accident, and that a crisis of public and environmental health was averted. Yet public perceptions of the event as an accident continue.

Even more troubling to antinuclear activists was the 1986 Chernobyl disaster, which occurred in what is now the Ukraine. Unlike Three Mile Island, there was a significant and severe release of radioactive materials during the Chernobyl accident, and the effects on public health and the environment in that part of the world (and in fallout zones) was disastrous. Thousands of people are believed to have died or contracted serious cancers as a result of radiation exposure. Furthermore, the land surrounding the Chernobyl plant is still highly radioactive, causing serious ecological, environmental, and economic devastation. It can be argued that accidents such as Chernobyl provide evidence that nuclear power production poses risks that far outweigh its benefits.

And more recent was the crisis at the Fukushima Daiichi plant in Japan as a result of the 9.0 earthquake on March 11, 2011. There is significant debate about whether the radiation leaks out of Fukushima Daiichi will cause long-term health impacts in Japan and, if so, what those impacts will be. Further, the extent of the environmental impacts of the radiation releases is not yet known. The scientific community does seem to agree that the accident will have negligible-to-minor health and environmental impacts internationally: Most agree that radiation that reaches other continents will be widely dispersed and have little impact on health or environment. Whether this holds true in the long term remains to be seen.

Another argument against nuclear power has to do with the problems associated with the nuclear fuel cycle, which includes every step of nuclear power production, from the mining of uranium ore to the decommissioning (taking apart and disposing) of nuclear power plants at the end of their use-life. Critics of nuclear power are particularly concerned with the two end points of the cycle: uranium mining and nuclear waste disposal. The United States does not currently mine significant amounts of uranium; instead, it imports uranium. But the history of uranium mining in the United States—and in other countries, even at present—is mostly an ugly one. There are examples of uranium tailings leaching into groundwater, uranium miners (often indigenous peoples) being exposed to radiation and suffering sickness and death as a result, and notable detrimental environmental effects when mines are not properly operated or decommissioned. Critics warn that a nuclear renaissance will give rise to further problems as uranium becomes a more sought-after and potentially dwindling resource.

Nuclear waste poses an even greater concern than uranium mining for many in the United States. Unlike in France, where most nuclear waste is reprocessed (recycled) and reused again as fuel, in the United States, there is a ban on reprocessing. This means that all high-level nuclear waste is stored near nuclear reactors in temporary storage sites. The United States has been slow to develop national, long-term repositories for this waste. Many activists, scientists, politicians, and others are worried about the safety and security of the temporary storage sites, fearing that they could be breached, or that the holding receptacles may eventually begin to degrade and leak into water supplies. They also worry about the impact of nuclear waste on future generations.

It is important to note that there are convincing arguments in favor of the relative safety of nuclear power production as well. There have been no widespread, serious nuclear accidents in the United States since the beginning of nuclear power production in the 1960s (there have been some incidents at research reactors, but no major commercial power plants have had significant accidents). Temporary storage sites have not leaked or been breached. Furthermore, many nuclear scientists and engineers would argue strenuously that living with nuclear power poses no major risks: residents living near reactors are exposed to less radiation from a power plant than they receive from natural forms of radiation, such as being exposed to solar radiation, living at higher altitudes, or living near geological formations that produce radiation naturally. Some even argue that some radiation exposure is healthy—in a process called hormesis, exposure to low levels of radiation can lead to positive health effects.

Another persuasive argument is a comparative one. Nuclear power may pose risks, but so do other forms of electricity production. The United States gets approximately 50 percent of its electricity from coal-fired power plants. These plants pose significant public health and environmental risks, including mine accidents and disasters, coal-ash spills and flooding, and mountaintop removal and stream burial. Finally, the burning of coal leads to the release of carbon dioxide into the atmosphere and is a significant contributor to global warming. There are other forms of pollution created by coal burning as well that can lead to public health problems such as asthma.

Compared with the destructive effects of coal, nuclear advocates argue, the relatively clean safety record of nuclear power in the United States suggests that it is a useful and safe source of electricity generation. It can also provide fairly large amounts of baseload power, which technologies such as wind and solar cannot. In France, for example, 90 percent of electricity generation comes from nuclear. Therefore, some see nuclear as a potentially appealing replacement for coal and can imagine it as an important option—along with wind, solar, and other renewable sources of energy—for "decarbonizing" our economy.

See Also: Government Role in Green Health; Low-Level Radioactive Waste; Radiation Sources.

Further Readings

Caldicott, Helen. *Nuclear Power Is Not the Answer*. New York: New Press, 2006.
Cravens, Gwyneth. *Power to Save the World: The Truth About Nuclear Energy*. New York: Alfred A. Knopf, 2007.
Elliott, David, ed. *Nuclear or Not? Does Nuclear Power Have a Place in a Sustainable Energy Future?* New York: Palgrave Macmillan, 2007.
The Future of Nuclear Power: An Interdisciplinary MIT Study. Cambridge: Massachusetts Institute of Technology, 2003 (updated 2009). http://web.mit.edu/nuclearpower (Accessed January 2010).

Jen Schneider
Colorado School of Mines

Nursing, Lack of

Nursing has long recognized the importance of the environment on health, hence its current focus on green health. The profession of nursing approaches health and illness from an integrated and holistic perspective and considers the health determinants to be critical factors that interact with one another to affect the health of individuals and populations. The environment—the physical, psychosocial, and biological environment—is a primary

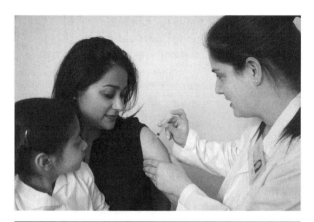

As the largest group of healthcare providers globally, nurses are in a unique position to promote and protect health and to reduce environmental hazards in homes, hospital settings, and communities. The nursing profession actively works to address green health through new initiatives in education, research, and practice.

Source: Centers for Disease Control and Prevention/Judy Schmid

determinant of health and one that nurses are particularly well suited to address. As the largest and most trusted group of healthcare providers globally, nurses are in a unique position to promote and protect health and to reduce environmental hazards in homes, hospital settings, and the community. The nursing profession actively works to address green health through new initiatives in education, research, and practice.

Historical Perspective

Florence Nightingale, the founder of nursing, believed in a holistic approach to health promotion and disease prevention, and that adequate light, clean air and water, and sanitary living conditions were essential to health. Lillian Wald, who founded the Visiting Nurse Service in the late 1800s, recognized the importance of proper hygiene and sanitation and taught new immigrants—who lived in cramped, squalid conditions in New York tenement buildings—the relationship between cleanliness and health. However, nursing's emphasis on living conditions and the immediate environment gradually changed throughout the 1900s.

Environmental health was particularly relevant to nursing at the turn of the century, when nurses worked primarily in the community and living conditions were not to today's standards. However, as public health improved, most notably after World War II, nursing became more hospital based and patient focused. In the 1960s, authors such as Paul Ehrlich, who wrote *The Population Bomb,* and Rachel Carson, who wrote *Silent Spring,* precipitated a shift in attitudes by bringing public attention to issues of overpopulation and environmental degradation. Slowly, the general public became aware of the importance of taking personal responsibility for the impact of its behavior on human health and the environment, and the environmental movement was born. As the environmental movement became more mainstream, the healthcare industry became aware of its role in contributing to environmental pollution.

Within the last 25 years, nursing has renewed its interest in the environment and acknowledged the significant role the profession has to play in improving health in this area. Since the 1980s, when the terms *green health* and *sustainable development* first came into common usage, the profession of nursing has gradually evolved to define its role more broadly in addressing environmental health through nursing practice, research, education, and advocacy at the individual and population levels.

Nursing's Renewed Focus on Environmental Health

In 1986, the International Council of Nurses issued a position statement, "The Nurse's Role in Safeguarding the Human Environment." The statement defined the nurse's role as helping to detect the ill effects of the environment on human health and vice versa. However, it was not until the Institute of Medicine (IOM) released *Nursing, Health, & the Environment: Strengthening the Relationship to Improve the Public's Health,* in 1995 that guidelines were established that began to define the role of nursing in addressing chemical, physical, biological, and psychosocial environmental health hazards. The report called for the inclusion of environmental content in nursing education at all levels, and that general competencies should be achieved by all nurses related to (1) basic knowledge and concepts of environmental health, (2) assessment and referral for conditions with suspected environmental causes, (3) knowledge of the role of advocacy, ethics, and risk communication in patient care and community intervention, and (4) an understanding of policy and legislation and regulation related to environmental health.

Nursing's efforts to address environmental health in the United States expanded considerably after the IOM report was issued. A series of working groups and workshops were convened between 1996 and 2002 to identify knowledge gaps in the area of environmental health and how they could best be addressed. In 2002, a roundtable organized by the National Institute of Environmental Health Sciences (NIEHS), the Agency for Toxic Substances and Disease Registry (ATSDR), and the National Institute of Nursing Research (NINR) assessed the then-current status of environmental health and nursing and made recommendations for research, education, and translation to practice. With respect to research, it was recognized that the number of nurse researchers in environmental and occupational health was insufficient; a research agenda and priorities were needed along with resources to build research capacity; and a journal on environmental health nursing should be established. Best practices for environmental health nursing were needed to incorporate the topic into nursing education at all levels. Reference materials and a central clearinghouse were recommended for translation into practice. Resources/sources of funding were needed to support the development of all three areas.

2002 to the Present

Since the 2002 roundtable, nursing has made serious efforts to define its role in environmental health. Green healthcare as it relates to reducing toxic medical waste, using safe medical supplies and materials, conserving resources and recycling, designing green environmentally friendly buildings, and limiting negative effects on the environment is now clearly within the purview of nurses' responsibilities. The American Nurses Association (ANA) became a partner with both Healthcare Without Harm (HCWH), an international coalition with a mission to transform the healthcare industry so it does not harm the environment, and the Hospitals for a Healthy Environment (H2E) project, now part of Practice Greenhealth, a networking organization committed to sustainable, environmentally friendly healthcare practices. The mission of the H2E project is to work in partnership with the American Hospital Association and its affiliated hospitals to prevention pollution and reduce mercury and other waste generated by hospitals.

A number of position statements were issued in the years following the 2002 roundtable to guide nursing practice with respect to green health and healthcare. In 2004, the ANA issued a resolution on "Environmental Health Principles in Nursing Practice," acknowledging the nurse's responsibility to be aware of broader health concerns, such as environmental pollution, and endorsing government adoption of the precautionary principle, which states that precautions should be taken against any activity that threatens to harm human health or the environment, even if there is no scientific proof that the threat may cause harm. In that same year, the International Council of Nurses (ICN) released a position statement on

"Medical Waste: Role of Nurses and Nursing." The position statement highlighted the role of nurses in reducing/eliminating medical waste and stressed nurses' responsibility to be influential in policy decisions related to medical waste. Specific initiatives nurses were encouraged to engage in included purchasing recyclable materials, segregating waste, minimizing incineration and use of toxic materials, and patient education on environmental pollution. In keeping with the ANA and ICN resolutions, in 2006, the Association of periOperative Registered Nurses (AORN) released a statement to guide perioperative nurses on environmentally responsible practices related to resource conservation and environmental protection. In 2005, the Public Health Nursing section of the American Public Health Association (APHA) presented 12 Environmental Health Principles for Public Health Nursing at the Annual APHA Meeting, which included recommendations for public health nursing practice, education, research, and advocacy. Included among the resolutions was an endorsement of the precautionary principle and the need for safe and sustainable environments for public health.

Research, Education, and Practice

Below are examples of successful efforts that have been made to achieve the goals of the 2002 roundtable with respect to research, education, and translation to practice for environmental health and nursing.

Johns Hopkins University is the first doctoral program in the world to offer nurses doctoral preparation in occupational and environmental health. This program addresses the need for nurse educators in the field as well as scholars prepared to conduct research and design interventions for workplace and environmental health–related issues.

The first book for nurses on how the environment affects nursing practice, *Environmental Health and Nursing* by Barbara Sattler and Jane Lipscomb, was published in 2002. Nursing programs have developed content to integrate environmental health into nursing curricula at all levels through classroom lectures, specific courses, summer conferences, field trips, community projects, and full degree programs with support from the NIEHS, ATSDR, and NINR. The American Association of Occupational Health Nurses has broadened its scope to include occupational and environmental health nursing. Occupational/environmental health master's degree programs in nursing are now offered by several universities in the United States.

Environmental Health Nursing Research Archive, an online database to archive environmental health articles and topics of interest for nurses, was established by the University of Washington, School of Nursing, funded by a grant from Health Care Without Harm. The archive is intended to support nurses conducting research on topics related to environmental health. The University of Maryland has created an online resource center, EnviRN (http://www.envirn.umaryland.edu), also funded with a grant from Health Care Without Harm, to serve as a vehicle for nurses to share information about environmental health issues.

While these are important outcomes of the nursing profession's efforts to address the need for an increased focus on environmental health, unfortunately, many in the field would state that success in expanding the role of nurses in environmental health since the 1995 IOM report was issued has been limited. Additional work has yet to be done to define environmental health concepts for inclusion in nursing curricula and to develop faculty expertise so they are better prepared to offer such content. Additional sources of funding are needed to support environmental health research, the role nursing can play in green health, and to facilitate the role of nurses in educating for a healthy, safe, healthcare environment.

See Also: Education and Green Health; Healthcare Delivery; Hospitals (Carbon Footprints); Mercury Sources and Health; Solid Waste Management.

Further Readings

American Nurses Association. "Environmental Health Principles in Nursing Practice." http://www.noharm.org/lib/downloads/nurses/ANA_Res_Env_Health.pdf (Accessed August 2009).

Green Guide for Health Care. http://www.gghc.org (Accessed August 2009).

Health Care Without Harm. http://www.noharm.org (Accessed August 2009).

International Council of Nurses. "Medical Waste: Role of Nurses and Nursing." http://www.noharm.org/lib/downloads/waste/Medical_Waste_Role_of_Nurses.pdf (Accessed August 2009).

International Council of Nurses. "The Nurse's Role in Safeguarding the Human Environment," Geneva, Switzerland, 1986.

Pope, A. M., M. A. Snyder, and L. H. Mood, eds. *Nursing, Health, & the Environment: Strengthening the Relationship to Improve the Public's Health*. Institute of Medicine, 1995.

Practice Greenhealth. http://www.practicegreenhealth.org (Accessed August 2009).

University of Washington. Pub Hub: Environmental Health Nursing Research Archive. http://healthlinks.washington.edu/howto/env/index.html (Accessed August 2009).

Connie Currier
Michigan State University

OBESITY

Obesity is a complex disease with multiple interacting and causal factors in which an imbalance between caloric intake and caloric requirement leads to excess body fat and adverse impacts on health. The condition leads to physical disability; cardiovascular disorders such as increased risk for coronary artery disease, high blood pressure, myocardial infarction, and stroke; and metabolic diseases such as type 2 diabetes. Even certain forms of cancer are associated with obesity. Depending on the severity of obesity, different bone and joint disorders, with mechanical- and inflammatory-based origin may manifest. Last but not least, multiple psychological consequences may result from obesity.

The origins of obesity are mainly attributed to increased consumption of energy-dense foods high in saturated fats and carbohydrates combined with reduced physical activity. From a public health perspective, origins of such a complex metabolic problem are also currently related to

Obesity is a complex disease in which excess body fat has accumulated to the point where it may have adverse effects on health. The origin of such metabolic imbalance is mainly due to an increased consumption of energy-dense foods high in saturated fats and carbohydrates, along with reduced physical activity.

Source: iStockphoto.com

epidemiological transitions in many countries. Economic growth, modernization, urbanization, and globalization of food markets (even in the context of so-called Westernization) have had profound changes in society and in behavioral nutritional patterns of communities over recent decades. Although genetic predisposition is important, energy imbalance related to those societal changes and worldwide nutrition transition are of utmost relevance when evaluating the obesity epidemic.

In many countries, malnutrition problems represent a double burden of disease. Although in the past, nutritional deficits were the most important of them, today over-weight and obesity have become highly relevant, not only in adults but also in children, and not just in developed countries but also in many developing countries, where, it has been stated, the increase is often faster than in the developed nations. This worldwide problem has increased in frequency, particularly in the last three decades. Globally, there are more than 1 billion overweight adults, with at least 300 million of them being obese. This complex condition implies serious social and psychological dimensions, affecting virtually all ages and socioeconomic groups. Recently, it has been considered that, given the current prevalence of overweight and obese children, in the future, the extent of such pandemic will be even higher than today.

Although often not calculated, the economical burden of obesity will impose large pri-mary healthcare costs and investments derived from related morbidity medical care. Different studies have found that the increase in obesity prevalence has led to even a three-fold rise in hospital costs. In 2001, Canada's economic burden in terms of overweight and obesity was estimated at $4.3 billion ($1.6 billion in direct costs and $2.7 billion in indirect costs). In the United States, medical expenses accounted for 9.1 percent of total U.S. medical expenditures in 1998 and may have reached as high as $92.6 billion.

In children, the prevalence of obesity has a relatively high variability around the world. This problem of overweight and obese childhood has been increasing during the last 40 years. As has been stated, such increasing prevalence of childhood obesity could be strongly related to adult obesity, which should be identified early and treated in order to reduce its related morbidity and mortality during childhood and adulthood. Particularly in people with certain risk factors, such as those with a family history of obesity, low birth weight, low physical activity, type 1 diabetes, and fetal macrosomy, excessive weight and obesity should be screened and followed up, including not just anthropometrical variables but also biochemical and hormonal variables. Disease prevalence in children has doubled or more than doubled in many countries, including in North America in Canada and the United States; in South America in Brazil and Chile; in the western Pacific region in Australia and Japan; and in Europe in the United Kingdom, Finland, Germany, Greece, and Spain. It is estimated that in the decade 2010–2020, around 45 percent of children in the Americas will be categorized as overweight.

Diagnosis of obesity is mainly based on the values of the weight according to height of the individuals. This can be easily measured and calculated as the body mass index (BMI), an equation that compares weight and height, dividing weight (in kilograms) by height (in meters). According to the World Health Organization (WHO), people are defined as over-weight (pre-obese) when their BMI is between 25 kg/m^2 and 30 kg/m^2 and obese when their BMI is greater than 30 kg/m^2. Additionally, obesity has been classified in three levels or classes: class I obesity includes those with a BMI between 30.0 and 34.9 kg/m^2; class II obesity, 35.0 to 39.9 kg/m^2; and class III obesity, $\geq$ 40.0 kg/m^2.

In children, although still controversial and very discussed, three approaches to the categorical analysis of BMI have been stated: the traditional approach, whereby an "abnor-mal" group is taken to be more than two standard deviations (>2SD) from the mean; the new International Obesity Task Force (IOTF) approach, which relates BMI categories in childhood to the accepted classification in adults; and a new Centers for Disease Control and Prevention (CDC) set of standards, whereby obesity is specified as >95th BMI percen-tile of carefully selected representative data from the United States. National percentile and Z-score distributions for BMI and waist circumference have been proposed in order to address in a more specific and adjusted method the proper cut-off values for each ethnic and country reference for diagnosis of overweight and obese children.

Management of obesity is oriented primarily to dieting and physical exercise. To supplement this, or in case of failure, anti-obesity drugs may be taken to reduce appetite or inhibit fat absorption. In severe cases, bariatric surgery is performed, or an intragastric balloon is placed to reduce stomach volume and/or bowel length, leading to earlier satiation and reduced ability to absorb nutrients from food. In addition to these strategies, lifestyle changes should be of utmost importance in reducing obesity and its consequences. Recently, the theory that the behavioral environment strongly influences changes in lifestyle has gained importance; this should be considered when adjusting the management of obesity in different population groups. In different age groups, interventions should be tailored to that group because, naturally, diet and physical activity usually differ. In most cases, such interventions are effective when they are oriented to the restriction of caloric intake and the increase of physical activity with exercise, as a combination in the goal of reduction of overweight and obesity.

In order to apply such interventions, once properly diagnosed as obesity with its potential metabolic consequences, a multidisciplinary approach should be initiated. Healthcare professionals from internal medicine, pediatrics, surgery, endocrinology, psychiatry, psychology, nutrition, physical rehabilitation, social work, and many other fields should be involved.

For different reasons, particularly esthetic ones, today many medical and nonmedical companies and organizations have been developing diets and products that intend to reduce weight, particularly in a fast manner. However, many of them lack scientific validity. These should be avoided; a balanced diet should be correctly designed and adjusted to each individual by certified healthcare professionals. Finally, different strategies and interventions should be balanced and integrated in order to reduce weight and to stay healthy over time.

An emerging topic, particularly in developing countries in many regions, especially in the Americas, is the appearance of overweight and obesity in aborigines. This has been observed in Brazil, Chile, Peru, Argentina, and Venezuela. This should be kept under surveillance in order to prevent its consequences.

Current epidemiology of this problem in children and adults should be seriously considered by national and international health authorities and policy makers in order to design strategies for collective prevention of excessive weight and obesity. Nutritional programs should consider the current nature of the problem and the representative data related to the consequences of being overweight and obese in different population groups. Early interventions in this setting should be a priority, particularly in those countries where the disease is rapidly increasing.

Finally, it is important to highlight that, currently, obesity is a public health problem that is even considered a pandemic disease. Obesity is a leading preventable cause of death worldwide, with increasing prevalence in adults and children. Even more, beyond consideration as an esthetic and/or psychological problem, excessive weight and obesity can lead to serious metabolic complications and life-threatening conditions during childhood and adulthood. For these reasons, increased efforts should be made to decrease its burden globally.

See Also: Cardiovascular Diseases; Children's Health; Liver Diseases; Men's Health; Metabolic Syndrome Diseases; Physical Activity and Health; Women's Health.

Further Readings

Amigo, H. "Obesity in Latin American Children: Situation, Diagnostic Criteria and Challenges." *Cadernos de Saude Publica*, 19 (2003).

Kelishadi R. "Childhood Overweight, Obesity, and the Metabolic Syndrome in Developing Countries." *Epidemiologic Reviews*, 29 (2007).

Lindgärde, F., M. B. Ercilla, L. R. Correa, and B. Ahrén. "Body Adiposity, Insulin, and Leptin in Subgroups of Peruvian Amerindians." *High Altitude Medicine & Biology*, 5 (Spring 2004).

Mattevi, V. S., C. E. Coimbra, Jr., R. V. Santos, F. M. Salzano, and H. M. Hutz. "Association of the Low-density Lipoprotein Receptor Gene With Obesity in Native American Populations." *Human Genetics*, 106 (May 2000).

Orden, A. B. and E. E. Oyhenart. "Prevalence of Overweight and Obesity Among Guaraní-Mbyá From Misiones, Argentina." *American Journal of Human Biology*, 18 (September 2006).

Pérez, F., E. Carrasco, J. L. Santos, M. Calvillán, and C. Albala. "Prevalence of Obesity, Hypertension and Dyslipidemia in Rural Aboriginal Groups in Chile." *Revista Medica de Chile*, 127 (October 1999).

Pinzo Serrano, Estefania. "Obesity in Pediatrics." *Continuous Education Program in Pediatrics*, 7 (2008).

Risquez, Alejandro, Alejandro Echezuria, and Alfonso J. Rodriguez-Morales. "Epidemiological Transition in Venezuela: Relationships Between Infectious Diarrheas, Ischemic Heart Diseases and Motor Vehicle Accident Mortalities and the Human Development Index (HDI) in Venezuela, 2005–2007." *Journal of Infection and Public Health*, 3 (2010).

Wang, Y. and T. Lobstein. "Worldwide Trends in Childhood Overweight and Obesity." *International Journal of Pediatric Obesity*, 1 (January 2006).

World Health Organization (WHO). *Global Strategy on Diet, Physical Activity and Health: Obesity and Overweight*. Geneva, Switzerland: WHO, 2003.

Alfonso J. Rodriguez-Morales
Universidad de Los Andes, Universidad Central de Venezuela

Yadira Vera
Universidad Simón Bolivar

OCCUPATIONAL HAZARDS

An occupational hazard is something that can cause harm in the work environment if not controlled. Occupational hazards include physical hazards such as exposure to ionizing radiation or noise; biological hazards such as exposure to bacteria or infectious diseases; chemical hazards such as exposure to asbestos or highly reactive fumes; mechanical hazards such as falls or collisions; and psychosocial hazards such as stress, overwork, or the threat of violence. Occupational illnesses and injuries are defined by the Occupational Health and Safety Administration (OSHA) as any abnormal condition or disorder caused by exposure to factors associated with employment. An occupational disease is typically identified when it is shown that it is more prevalent in a given body of workers than in the general population or in other worker populations. Subsequently, the goal of occupational safety and health is to promote a safe work environment by protecting the safety, health, and welfare of people engaged in work or employment by identifying opportunities for mitigating the effects of and preventing exposure to occupational hazards. The protection of workers is an important aspect of public health because occupational diseases and injuries are estimated to account for a large portion of the worldwide degradation, temporary or permanent, of the physical, mental, or social well-being of workers.

In the United States, the workforce is made up of 110 million people, with a gender distribution of 46 percent women and 54 percent men. Occupational hazards contribute to morbidity and mortality by causing approximately 900,000 illnesses and 60,000 deaths each year in the United States. The 10 most frequent work-related diseases and injuries in the United States are the following:

- Lung disease
- Musculoskeletal injuries
- Cancers
- Severe trauma
- Cardiovascular disorders
- Disorders or reproduction
- Neurotoxic disorders
- Noise-related hearing loss
- Dermatologic conditions
- Psychological strain and boredom

Worldwide, it is estimated that each year there are 160 million new cases of work-related illnesses and 1.7 million work-related deaths (3 percent of all deaths). Various occupational hazards are estimated to be responsible worldwide for 37 percent of back pain, 16 percent of hearing loss, 13 percent of chronic obstructive pulmonary diseases, 11 percent of asthma, 8 percent of injuries, 9 percent of lung cancer, and 2 percent of leukemia. Workplace needlestick injuries account for about 40 percent of hepatitis B and C infections and 4.4 percent of HIV infections in healthcare workers. In addition, studies in industrialized countries demonstrate that psychosocial hazards and work-related stress affect one-fifth of the working population. However, it should be noted that deaths and disease incidence from occupational illness for most diseases are hard to enumerate. For many diseases, such as those from chemical exposure, various occupational cancers, and other problems, individual deaths and diseases are difficult to recognize and document. Therefore, it is possible that the national and global estimates of the burden due to occupational hazards may be conservative.

Despite statistics that describe improvements in worker safety over time, in most countries, hazardous exposures that have adverse effects on the health of workers are still found in high numbers of workplaces. And in addition to the major traditional occupational health needs that still prevail among the global workforce, changes in economic structures and technologies also present new occupational health exposures and needs at a rapid pace. A subset of these new technologies and associated jobs are green, or part of an industry that creates products and processes to reduce environmental impact and improve use of resources such as energy efficiency, water conservation, and use of environmentally preferred materials. However, green jobs, which have been defined broadly as jobs that help improve the environment and enhance sustainability, offer opportunities as well as challenges for workers and for occupational health.

Examples of green jobs include installation and maintenance of solar panels and generators; construction and maintenance of wind energy turbines; jobs related to recycling; jobs related to the manufacture of green products; and jobs where green products are used in traditional fields such as agriculture, healthcare, and the service sector. In some instances, the hazards to workers may be similar to those in established industries. However, some green and sustainable practices may pose new health concerns for workers, such as the introduction of green substitutes for cleaning solvents. As policies promoting the protection of workers rely on an assessment of such occupational hazards, it is likely that there will be a lag period in which new green jobs will present unknown and uncontrolled risks for workers.

Green jobs as of 2010 constituted approximately 700,000 positions across the United States, or approximately 0.5 percent of total employment, but the number is expected to grow dramatically. There are two government agencies in the United States whose responsibility it is to protect persons working in green industries. The National Institute for Occupational Safety and Health (NIOSH) of the U.S. Department of Health and Human Services has the responsibility for research and prevention activities with regard to workers' health. NIOSH advises as to allowable levels of exposure, based upon scientific review. NIOSH and its partners launched the Going Green: Safe and Healthy Jobs initiative to make sure that green jobs are good for workers by integrating worker safety and health into green jobs and environmental sustainability. Within the Department of Labor is the Occupational Safety and Health Administration, which has the responsibility of actually setting and enforcing workplace regulations. One of the difficulties that OSHA faces is an insufficient number of inspectors to evaluate what actually goes on at most workplaces; given the number of inspectors in the United States at this writing, it would take more than two decades for all workplaces to be inspected even once. As such, OSHA's resources are focused on fatalities rather than epidemics of occupational illnesses.

At the global level, the World Health Organization (WHO) has recommended that national policy and programs for the further development of occupational health should be reviewed and prepared in collaboration with government and social partners. WHO recommends that to accomplish this, national programs for developing occupational health, in light of changing green industries, should include updating of legislation and standards. It notes that the approach to accomplish this should include strengthening the role of authorities; emphasis on the primary responsibility of the employer for health and safety at work; collaboration between government, employers, and trade unions for implementation of national occupational health programs; education and training; development of occupational health services; analytical and advisory services; research; development of registries to document occupational accidents, diseases, and exposures; and action to ensure collaboration between employers and workers at workplace and enterprise levels. Comprehensive control of occupational hazards in the workplace is a multifaceted and complicated goal that requires the coordination and expertise of many different groups and individuals.

In short, there is a wide range of occupational hazards and diseases and, subsequently, opportunities for prevention interventions. As green industries expand, new opportunities for exposure to hazards are increasing despite the improvements in sustainable and environmentally responsible products and processes. In the United States, the number of investment dollars in the development of green technologies is increasing. As such, it is important that the impact of existing and new occupational exposures affecting worker health be evaluated and controlled appropriately to reduce worker-related morbidity and mortality. Due to the magnitude of the global workforce and the prevalence of workplace hazards, identifying opportunities for prevention should be a public health priority.

See Also: Computers and Printers (Ink); Environmental Illness and Chemical Sensitivity; Ergonomics; Fungi and Sick Building Syndrome; Green Chemistry; Indoor Air Quality; Injuries; Lighting; Particulate Matter.

Further Readings

Cherrie, John, Robin Howie, and Sean Semple. *Monitoring for Health Hazards at Work.* Hoboken, NJ: Wiley-Blackwell, 2010.
Collins, Larry R. and Thomas D. Schneid. *Physical Hazards of the Workplace (Occupational Safety & Health Guide Series).* Boca Raton, FL: CRC Press, 2001.
U.S. Centers for Disease Control and Prevention. "Workplace Safety & Health." http://www.cdc.gov/workplace (Accessed September 2010).
U.S. Department of Labor, Occupational Safety and Health Administration. http://www.osha.gov (Accessed May 2010).

Briseis A. Kilfoy
Yale University School of Medicine

ORAL DISEASES

Healthy teeth, mouth, and gums are very important to a person's overall health. The term *oral* refers to the whole mouth, including the teeth, gums, hard and soft palate, tongue, lips, salivary glands, and chewing muscles. Good oral health means being free of tooth decay and gum disease, as well as being free of chronic oral pain conditions, oral cancer, and birth defects such as cleft lip and palate. Tooth decay and gum disease are largely caused by plaque, a sticky material that develops on the exposed portions of the teeth, consisting of bacteria, mucus, and food debris. Plaque accumulates on teeth after eating, and if it is not removed thoroughly each day, it will mineralize and harden into a hard deposit called tartar, which becomes trapped at the base of the tooth. A number of problems are associated with plaque and tartar, including the following:

- Cavities
- Gingivitis
- Periodontitis
- Oral cancer
- Trench mouth
- Bad breath (halitosis)

A cavity is a small hole that forms on the surface of a tooth. Cavities are caused when sugars in the food we eat and bacteria in our mouths mix together, producing a mild acid that eats away at the outer layer of our teeth, called enamel. Small cavities can easily be filled but, if left untreated, can become larger and cause pain. The tooth may even have to be removed and replaced with a false (or artificial) tooth.

Gingivitis is the first stage of gum disease. Signs of gingivitis are puffy, nonpainful gums; traces of blood on your toothbrush; or a change in the color of your gums from pink to slightly red. Gingivitis is a form of periodontal disease involving inflammation of the gums (gingiva) and can be caused by the long-term effects of plaque deposits. Plaque and tartar irritate and inflame the gingiva.

The acid from the mixture of bacteria in the mouth and the food you eat can attack the surfaces of the teeth and gums, leading to infected, swollen, and tender gums. Trauma to the gums, commonly seen in excessive brushing or flossing of the teeth, can also cause gingivitis. The risks for developing gingivitis include poor dental hygiene, uncontrolled diabetes, pregnancy, tobacco use, and certain medications. Gingivitis can be prevented, and if started, it can be reversed.

Untreated or delayed treatment of gingivitis can lead to periodontitis. Inflammation causes a pocket to develop between the gums and the teeth, which fills with plaque and tartar. Soft tissue swelling traps the plaque in the pocket. Continued inflammation spreads from the gums and destroys the soft tissue and bone that support the teeth. Loss of support causes the teeth to become loose and eventually fall out. This disorder is not common in children but increases during adolescence. Among adults, periodontitis is the primary cause of bleeding, pain, and tooth loss. Damage caused by periodontal disease is not reversible, but it is largely preventable. Daily brushing and flossing and regular professional dental cleanings can greatly reduce the chance of developing periodontitis.

Oral cancer is a disease resulting from abnormal cell growth in the mouth, most commonly involving the tissue of the lips or the tongue. Most oral cancers are classified as squamous cell carcinomas, a very malignant cancer that tends to spread quickly. The actual cause of oral cancer is not known, but smoking and other tobacco use are associated with 70 to 80 percent of oral cancer cases. Smoke and heat from cigarettes irritate the mucous membranes of the mouth. Use of chewing tobacco causes irritation from direct contact with the mucous membranes. Heavy alcohol use and prolonged, repeated exposure of the lips to the sun are also high-risk activities associated with oral cancer. It is especially dangerous to combine smoking and alcohol. People, especially men, over the age of 45 are most at risk. Oral cancer can be treated successfully if caught early enough.

Trench mouth is a painful bacterial infection and ulceration of the gums. The term *trench mouth* comes from World War I, when the disorder was common among soldiers. The incidence of trench mouth is now mostly restricted to those with an impaired immune system. Normally, there is a balance of different microorganisms in the mouth. Trench mouth occurs when there is an overabundance of the normal mouth bacteria, resulting in infection of the gums. The infection leads to the development of painful ulcers inside the mouth (mouth ulcers are sores or open lesions within the mouth caused by various disorders, including canker sores and herpes simplex virus). A very distinctive symptom of trench mouth is that the breath smells extremely foul. The tips of the gums between the teeth erode and become covered with a gray layer of dead tissue. The gums bleed easily, and eating and swallowing cause pain. Often, the lymph nodes under the jaw swell, and a mild fever develops. Viruses may be implicated in allowing the bacteria to overgrow. Poor oral hygiene usually contributes to the development of trench mouth, as do physical or emotional stress, poor diet, and lack of sleep.

The term *halitosis,* or bad breath, describes noticeably unpleasant odors exhaled while breathing. Most adults suffer from bad breath occasionally, although it may affect up to a quarter of adults on a regular basis. Bad breath is usually brought on by the breakdown of proteins by bacteria somewhere in the mouth, which means that there are just about as many causes of bad breath as there are sources of bacteria in the mouth. Halitosis may be caused by the following:

- Certain foods (such as garlic and onions)
- Poor oral care (food that is allowed to collect on the teeth, gums, and tongue may cause an unpleasant odor and taste in the mouth)
- Improper cleaning of dentures
- Peridontal disease
- Xerostomia/dry mouth (decreased saliva makes it hard for the mouth to cleanse itself and remove particles left behind by food. Xerostomia may be caused by certain medications, a salivary gland disorder, or by continuously breathing through the mouth instead of the nose.)
- Tobacco products

Bad breath may be an indicator or symptom of any of the named conditions. However, it might also be indicative of a more serious condition, such as a respiratory infection, diabetes, a gastrointestinal disorder, or a liver or kidney disorder.

The mouth is an integral part of human anatomy and plays a major role in our overall physiology. Thus, oral health is intimately related to the health of the rest of the body. Current research suggests that infections in the mouth such as periodontal (gum) diseases may increase the risk of heart disease, may put pregnant women at greater risk of premature delivery, and may complicate control of blood sugar for people living with diabetes. It is important to understand oral diseases—what they are, how to recognize them, and how to prevent them.

See Also: Cancers; Personal Consumer Role in Green Health; Personal Hygiene Products; Smoking.

Further Readings

American Academy of Periodontology. "Parameter on Plaque-Induced Gingivitis." *Journal of Periodontology*, 71/5 (2000).
Centers for Disease Control and Prevention. "Chronic Disease Prevention and Health Promotion." http://www.healthcentral.com/channel/408/1122.html (Accessed January 2010).
Ferri, F. F. *Ferri's Clinical Advisor: Instant Diagnosis and Treatment.* St. Louis, MO: Mosby, 2005.
Marx, J. *Rosen's Emergency Medicine: Concepts and Clinical Practice*, 5th ed. St. Louis, MO: Mosby, 2002.

Monique Mostert
Independent Scholar

ORGANIC PRODUCE

As a relatively new industry, the organic food movement is a green alternative to more conventional food production. Organic food is believed by its advocates to be healthier, ecologically balanced, and free of pesticides. While quantity and mass production are the

Organic farming has many benefits for the environment; namely, the practice does not release synthetic pesticides that can harm soil, water, and wildlife. Here, organic corn has more weeds late in the season, since no herbicides were used.

Source: U.S. Department of Agriculture, Agricultural Research Service/Bob Nichols

major concerns of nonorganic food producers, the major concern of organic food producers is quality. As the market for organic produce increased exponentially over the last few decades of the 20th century, confusion arose as to what precisely constitutes "organic" produce. To meet the needs of consumers, trade groups and government regulatory agencies have worked to define what organic produce is and provide special certification to protect consumers, and have conducted studies to ensure that organic farms are less damaging to the environment and provide the greater nutritional value and taste expected by consumers.

To receive the classification of "organic," a product must abide by certain rules and regulations. Produce cannot have been genetically modified, nor can nonorganic herbicides, insecticides, or pesticides have been used during the growth process. Nonorganic fertilizers are sometimes used in raising organic produce, however. In most countries, including the United States, organic produce cannot be genetically modified, meaning that produce whose genetic material has been altered using recombinant DNA technology cannot be considered organic. Certain artificial methods, such as chemical ripening and food irradiation, also preclude produce from being labeled organic. Organic farms focus on improving soil output and controlling pests through composting, using green manure, rotating crops, controlling pests biologically, and cultivating crops mechanically. Although organic produce was historically grown on smaller, family-owned farms, as demand for the product has grown, many corporations have entered the organic field. Larger producers have sometimes engaged in practices that were controversial compared with traditional practices of organic production, such as limited use of pesticides. As organic produce sales have increased by over 20 percent annually since the early 1990s, government certification of precisely what constitutes organic produce has become more widespread.

Many nations, including Canada, China, the European Union, Japan, and the United States, have fully implemented programs for the certification of organic produce. Certification standards vary from country to country and control standards for planting, growing, harvesting, storing, processing, packaging, and shipping organic produce. In the United States, certification originally began as the result of a private cooperative effort. Beginning in 1973, the California Certified Organic Farmers (CCOF) was founded as a membership organization to promote and attest to its members' organic produce. As organic produce became more popular and widespread, more consistent definitions of what determined the types of organic produce that were sought by both producers and consumers. Ultimately, in 2002, the U.S. Department of Agriculture (USDA) promulgated a regulatory framework governing organic produce and other foods. This framework is known as the National Organic Program, and it sets out all aspects of production, processing, delivery, and sale to consumers of food that wishes to be labeled "organic." All producers with sales greater than $5,000 per year must register with the USDA to have their crops labeled "USDA Organic." The National Organic Program accredits over 100 domestic and international bodies, including cooperatives such as CCOF, that are able to offer organic certification.

Organic produce, and indeed organic farming as a whole, has many benefits for the environment. Conventional pesticides and fertilizers have many negative effects on the environment, farm workers, and consumers. These negative impacts are the primary reasons for minimizing their usage. Specifically, organic farming practices do not release synthetic pesticides into the environment, substances that can harm soil, water, and wildlife. Organic farming also uses less energy per unit of production and produces less waste, such as chemicals and their packaging materials. Critics of organic farming, however, assert that organic farming requires a greater amount of land to be cultivated to produce the same yield as conventional practices. Proponents of organic produce and farming practices counter this assertion by insisting that reduced energy and pesticide usage more than make up for this discrepancy. Indeed, practices used to cultivate organic produce greatly reduce the effects and side effects that pesticides inflict on farm workers' health.

Organic produce is promoted as having better nutritional value and taste than that produced by conventional farming practices. Organic produce has been shown to have a higher degree of nutritionally valuable vitamins and antioxidants. Produce cultivated using organic farming practices also has demonstrated lower levels of such harmful substances as pesticide residues, heavy metals, and mycotoxins or other harmful fungi. Blind taste tests have shown certain organic fruits to taste sweeter. Organic produce also presents certain drawbacks. The lack of food preservatives tends to cause faster spoilage in organic produce, and some consumers prefer the more uniform look of conventionally cultivated produce. Organic produce generally costs from 10 to 40 percent more than its conventionally produced rivals, which makes it too expensive for some. Costs of shipping organic produce are generally higher as well, in part because organic products are produced in smaller amounts in a more geographically diverse area than conventionally produced options.

As consumer demand for organic produce and other healthier food options has grown, the number of organic options has exploded. Government regulation of what can be labeled organic produce has been undertaken to increase consumer confidence and safety. As consumers become increasingly interested in minimizing the negative environmental impact of their purchasing decisions, the popularity of organic produce seems likely to increase. The perception that organic produce, and organic farming in general, assists in the maintenance of diverse ecosystems also makes it accepted. And certain consumers remain convinced of organic produce's taste and health benefits. Efficient and consistent regulation of organic produce must be maintained to preserve consumer confidence.

See Also: Advertising and Marketing; Bottled Water; Fast Food; Food Allergies; Home-Grown Food; Urban Green.

Further Readings

Hinrichs, C. C. and T. A. Lyson. *Remaking the North American Food System: Strategies for Sustainability.* Lincoln: University of Nebraska Press, 2007.

Nestle, M. *What to Eat*. New York: North Point Press, 2007.
Planck, N. *Real Food: What to Eat and Why*. New York: Bloomsbury Group, 2007.

Stephen T. Schroth
Jason A. Helfer
Bridget G. Dooley
Knox College

OZONATION BY-PRODUCTS

Ozone is a highly reactive gas that can be used for both the destruction of taste- and odor-causing compounds and for the disinfection of drinking water. Ozone forms a variety of organic and inorganic by-products. Many of the by-products caused by ozone are less harmful than the disinfection by-products formed by chlorine. Careful application of ozone can make it a powerful tool for reducing the total amount of harmful by-products produced when treating drinking water.

Water may be safe to drink but still contain natural organic matter (NOM) that may cause undesirable taste, odor, or color. In the industrialized world, consumers may object to water with discernable tastes and odors, or even mistake it as unsafe for consumption. As NOM is primarily found in surface water sources, utilities that obtain their water from these sources often use some means to destroy taste and odor compounds. Ozone readily reacts with these chemicals in water to oxidize them, causing the chemicals to break down.

This same reactivity also causes ozone to act as a disinfectant by attacking and inactivating pathogens within water. Although ozone can cause complete disinfection, it evaporates quickly from water, leaving behind no residual disinfection ability. Within the United States, water distribution systems are required to have a disinfection residual, referred to as a secondary disinfectant. Residual disinfectant will inactivate any pathogens that water may encounter on its way from a treatment facility to the tap. Some form of chlorine is generally used to provide this residual. But regulation of the disinfection by-products formed by chlorine has caused increased use of ozone as a primary disinfectant for drinking water, used earlier in the treatment process.

The disinfection by-products formed by ozone occur in two main forms: nonhalogenated and halogenated. Of these two by-products, the halogenated by-products form the greatest health risk. Halogens refer to the nonmetal elements of fluorine, chlorine, bromine, iodine, astatine, and ununseptium. Because they are highly reactive, halogens exist in the environment only as compounds or within ions.

When ozone reacts with natural organic matter, it does not form halogenated by-products. In this case, it typically forms aldehydes and acids. Nonhalogenated by-products include formaldehyde, acetaldehyde, glyoxal, methyl glyoxal, oxalic acid, succinic acid, formic acid, acetic acid, pyruvic acid, and hydrogen peroxide. None of these are regulated contaminants for which maximum contaminant levels exist for drinking water within the United States.

Although the nonhalogenated by-products form some health concern, of greater concern are halogenated by-products, which form when ozone is used to treat water that has the bromide ion present. The halogenated by-products formed include bromate, bromoform, brominated acetic acids, bromopicrin, and brominated acetonitriles. Because of the formation of these more hazardous substances, ozone may not be suitable to treat water that contains bromide. Many of the halogenated disinfection by-products are regulated within drinking water within the United States due to known cancer risks.

Within the United States, the Environmental Protection Agency's Stage I Disinfection and Disinfection Byproducts Rule (Stage I DBPR) was promulgated to regulate the allowable concentrations, known as maximum contaminant levels (MCLs), of various disinfection by-products within drinking water, including bromate, halogenated acetic acids, and trihalomethanes. The MCL for bromate is 0.010 mg/L due to increased cancer risks.

The Maximum Contaminant Level (MCL) for the halogenated acetic acids is 0.060 mg/L due to increased cancer risks. For the purposes of drinking water regulation, the haloacetic acids are often referred to as HAA5 because they include the following five acids: monochloroacetic acid, dichloroacetic acid, trichloroacetic acid, bromoacetic acid, and dibromoacetic acid. The last two acids on this list are brominated acetic acids, which may form when ozone is used in waters containing bromide. The first three acids in the list are chloroacetic acids, which may form when chlorine is used for disinfection. The MCL is applied to the sum of the concentrations of all HAA5 within treated drinking water.

HAA5 are not the only halogenated by-product of ozone regulated in drinking water. The MCL for total trihalomethanes (TTHM) is 0.080 mg/L due to increased cancer risks and liver, kidney, or central nervous system problems. The TTHMs regulated in drinking water include chloroform, bromodichloromethane, dibromochloromethane, and bromoform. The chlorine-containing by-products are a concern when chlorine is used for drinking water treatment. The MCL for TTHM applies to the sum of the concentrations of all these disinfection by-products.

Ozone is often used as a primary disinfectant as a means of reducing the total amount of chlorine used for disinfection and minimizing the harmful by-products formed when chlorine is used in drinking water treatment. However, when bromide is present in the water, the use of ozone in water treatment can result in the formation of various cancer-causing chemicals as well. For this reason, ozone use is best restricted to bromide-free waters.

See Also: Bottled Water; Chlorination By-Products; Tap Water/Fluoride; Waterborne Diseases.

Further Readings

Richardson, S. D., A. D. Thruston, Jr., T. V. Caughran, P. H. Chen, T. W. Collette, K. M. Schenck, B. W. Lykins, Jr., C. Rav-Acha, and V. Gleze. "Identification of New Drinking Water Disinfection By-Products From Ozone, Chlorine Dioxide, Chloramine, and Chlorine." *Water, Air, & Soil Pollution*, 123 (October 2000).

U.S. Environmental Protection Agency. *EPA Guidance Manual: Alternative Disinfectants and Oxidants.* April 1999.

U.S. Environmental Protection Agency. *Stage I Disinfectants and Disinfection Byproducts Rule: A Quick Reference Guide.* EPA 816-F-01-010. May 2001.

Michelle Edith Jarvie
Independent Scholar

OZONE

In 1840, ozone was proposed as a distinct chemical compound by Christian Friedrich Schönbein, who named it after the Greek verb *ozein* (to smell) from the peculiar odor in lightning storms. Twenty-five years later, in 1865, Jacques-Louis Soret suggested the formula for ozone, O_3, which was confirmed by Schönbein in 1867.

Cars, industrial plants, and power plants are significant sources for one class of air pollutant—nitrogen oxides—and are largely responsible for tropospheric, or ground-level, ozone.

Source: iStockphoto.com

Ozone, or O_3, is a triatomic molecule and an allotrope of oxygen that is less stable compared to the diatomic O_2. Ozone is an extremely reactive gases which is constructed of three oxygen atoms. It is a powerful oxidizing agent, far better than dioxygen. It is very unstable at high concentrations and decays into regular diatomic oxygen. The half-life of ozone in atmospheric conditions is about 30 minutes. This reaction proceeds more rapidly with increasing temperature and decreasing pressure. Deflagration of ozone can be triggered by a spark and can occur in ozone concentrations of 10 percent or higher. Ozone might have a different effect depending on its location in the atmosphere. This reactive gas has good and bad effects on life on Earth.

This reactive gas can be formed either at the stratospheric level or ground level. The stratospheric level of ozone is formed naturally through the interaction of solar ultraviolet (UV) radiation with molecules of oxygen (O_2). The atmospheric ozone layer can extend up to 30 miles above the Earth's surface and reduces the amount of harmful UV radiation reaching the Earth's surface. The stratosphere is the region of the atmosphere with the highest levels of ozone. This region is also known as the ozone layer, which is located about 6 to 30 miles above the Earth's surface.

Photons with shorter wavelengths (less than 320 nm) of ultraviolet light (UV) rays (270 to 400 nm) are harmful to most forms of life in large doses. These harmful photons are filtered out by the stratosphere layer. Ozone in the stratosphere region is mainly produced from UV rays reacting with oxygen. Therefore, this layer of ozone acts as a shielding layer against the UV rays and absorbs these harmful rays.

Solar UV rays make up part of the electromagnetic or photonic spectrum of light and radiant energy. A part of this spectrum can be broken down into different wavelengths. The electromagnetic spectrum within the wavelength region ranges from the vacuum ultraviolet to the far infrared. The UVs cannot be seen by human eyes since they have a shorter wavelength compared to visible light. There are three different types of UVs as explained in the following:

- UVA wavelengths (320–400 nm) are only slightly affected by ozone levels. Most UVA radiation is able to reach the Earth's surface and can contribute to tanning, skin aging, eye damage, and immune suppression.
- UVB wavelengths (280–320 nm) are strongly affected by ozone levels. Decreases in stratospheric ozone mean that more UVB radiation can reach the Earth's surface, causing sunburns, snow blindness, immune suppression, and a variety of skin problems including skin cancer and premature aging.
- UVC wavelengths (100–280 nm) are very strongly affected by ozone levels, so that the levels of UVC radiation reaching the Earth's surface are relatively small.

The effects of UV radiation on the Earth's ecosystems are not completely understood. Even isolating the effects of UVA versus UVB is somewhat arbitrary. All UV radiations can be damaging. This knowledge has prompted many manufacturers of sunscreens and sunglasses to offer products that protect against both UVA and UVB wavelengths.

Two major classes of air pollutants—volatile organic compounds (VOCs) and nitrogen oxides (NO_x)—are responsible for tropospheric, or ground-level, ozone. These reactions are expedited in the presence of heat and sunlight. Therefore, more ozone forms in the summer months. Cars, industrial plants, and power plants are the most important NO_x resources. Gasoline pumps, chemical plants, oil-based paints, auto body shops, print shops, consumer products, and some trees are significant sources of VOC emissions.

Some hydrocarbons such as aldehydes can directly interact with ozone and get removed from the air; however, the products themselves are key components of smog. Ozone can be photolyzed by UV light and produce the hydroxyl radical OH groups that play a major role in reactivity of the product. This reaction leads to the removal of hydrocarbons from the air, but it is also the first step in the creation of components of smog such as peroxyacyl nitrates.

Balance between sunlight that creates ozone and chemical reactions that destroy it specifies the natural level of ozone in the stratosphere. Ozone is created when the kind of oxygen (O_2) we breathe is split apart by sunlight into single oxygen atoms. Single oxygen atoms can rejoin to make O_2, or they can join with O_2 molecules to make ozone (O_3). Ozone is destroyed when it reacts with molecules containing nitrogen, hydrogen, chlorine, or bromine. Some of the molecules that destroy ozone occur naturally, but people have created others.

The total mass of ozone in the atmosphere is about 3 billion metric tons. That may seem like a lot, but it is only 0.00006 percent of the atmosphere. The peak concentration of ozone occurs at an altitude of roughly 20 miles above the surface of the Earth. At that altitude, ozone concentration can be as high as 15 parts per million (0.0015 percent) or 15 PPM.

The atmospheric lifetime of tropospheric ozone is about 22 days and its main removal mechanisms are being deposited to the ground, the above mentioned reaction giving OH, and by reactions with OH and the peroxy radical HO_2.

Ozone is strongly oxidizing and is a primary irritant to human eyes, affecting especially the eyes and respiratory systems, and can be hazardous at even low concentrations. Therefore its biological effects should be taken very seriously, and it demands having a safety rule for the workers and people who are exposed to doses of ozone.

While humans can choose various courses of protection and shielding, for instance, avoiding mid-day sun, plants and animals are not so fortunate. Studies have shown that increased UV radiation can cause significant damage, particularly to small animals and plants. Phytoplankton, fish eggs, and young plants with developing leaves are particularly susceptible to damage from overexposure to UV.

Therefore, in the air we breathe, closer to Earth, the ozone would be a harmful pollut-ant that causes biological damage to lung tissue and plants. Near the Earth's surface, where the ozone has a chance to come into direct contact with live animals, plants, and humans, it primarily shows a destructive behavior. Therefore, motor vehicle exhaust and industrial emissions, gasoline vapors, and chemical solvents are some of the major sources of NO_x and VOCs, also known as ozone precursors.

Devices generating high levels of ozone, some of which use ionization, are used to sanitize and deodorize uninhabited buildings, rooms, ductwork, woodsheds, and boats and other vehicles.

Work environments where ozone is used or where it is likely to be produced should have adequate ventilation, and it is prudent to have a monitor for ozone that will alarm if the concentration exceeds the allowed dose.

The unit for total ozone measurement is the Dobson unit (DU). There is another physical definition for DU: If a column of air stretching from the surface of the Earth to the space in the atmosphere filled with ozone, and if this column is brought to equilibrium under the stan-dard condition, which is 0 degrees Celsius in temperature and has a pressure of 1013.25 mil-libars, or 1 atmosphere, or atm, the thickness of this column would be about 0.3 centimeters.

Thus, the total ozone would be 0.3 atm-cm. To make the units easier to work with, the Dobson unit is defined to be 0.001 atm-cm. Therefore, a 0.3 atm-cm would be the equiv-alent to 300 DU.

Actually, the chemical reactions involving chlorine and bromine cause ozone. Studies in the past decades have shown that in the southern polar region these chemical reactions are destroyed rapidly and severely each year during the southern hemisphere spring. Therefore, a depletion layer or region is formed, which is known as the *ozone hole*. A map of total column ozone is a factor for determination of the area of the ozone hole. This map is cal-culated from the area on the Earth that is enclosed and surrounded by a line with a con-stant value of 220 Dobson units. The value of 220 Dobson units is chosen since ozone values of less than 220 Dobson units were not measured or observed in the historic obser-vations over Antarctica prior to 1979. Also, from direct measurements over Antarctica, a column ozone level of less than 220 Dobson units is a result of the ozone loss from chlorine and bromine compounds. Therefore, this simple definition helps to specify the area of the ozone hole using a practical measurement. Annual records since 1979 show that the sever-ity of the ozone hole varies somewhat from year to year. These changes and fluctuations are superimposed on a trend extending over the last 30 years.

See Also: Automobiles (Emissions); Smog; Suburbs; Ultraviolet Radiation.

Further Readings

NASA. "Ozone Hole Watch." http://ozonewatch.gsfc.nasa.gov (Accessed September 2010).

Rubin, Mordecai B. "The History of Ozone. The Schönbein Period, 1839–1868." *Bulletin for the History of Chemistry*, 26/1 (2001).

Soret, Jacques-Louis. "Recherches sur la Densité de l'Ozone." *Comptes Rendus de l'Académie des Sciences*, 61 (1865).

Stevenson, D. S., et al. "Multimodel Ensemble Simulations of Present-Day and Near-future Tropospheric Ozone." *Journal of Geophysical Research*, 111 (2006).

Hassan Bagher-Ebadian
Independent Scholar

PAIN MEDICATION

Pain medications can be divided into several different groups. These groups range from herbal and over-the-counter (OTC) pain medications, which are available without a healthcare provider's prescription, up to highly controlled opioid medications. All pain medications should be taken exactly as recommended by the manufacturer or prescribed by a healthcare provider, as all these medications have the potential for side effects or adverse reactions. It is important when prescribing these medications that healthcare providers ascertain all medicines the patient is taking, including other over-the-counter medications and herbal supplements, to avoid possible drug interactions.

A systematic review of 10 randomized controlled trials of patients with low back pain has found that herbal preparations of devil's claw (*Harpagophytum procumbens*), white willow bark (*Salix alba*), and cayenne plasters (*Capsicum frutescens*) may be as effective as pain medication for short-term pain control. However, the medical community still holds major concern around herbal remedies, as the manufacturing process of these substances is not regulated and there seems to be a lack of consistency of the preparations. A second concern is the potential for drug interactions with traditional medicine, which has yet to be fully determined.

OTC pain medications are typically found in three basic groups: topical preparations, nonsteroidal anti-inflammatory drugs (NSAIDS), and acetaminophen. Topical corticosteroids are mainly used for local inflammation in the skin, such as acute flairs of eczema, and are rarely used for control of pain other than superficial pain. Capsaicin, a substance found naturally in chili peppers, is used in some topical analgesic creams. It is best used with superficial pain, as it can change pain signals at the level of the skin without blocking other sensations. Acetaminophen (called "paracetamol" outside North America and goes by the brand name Tylenol) was first used in medicine in 1894 but did not become popular as a pain medication until the late 1940s. Acetaminophen is available in multiple dosages and preparations, such as tablets and elixirs, so as to be tailored to the individual by weight and age. Users of acetaminophen should follow the directions exactly as they are stated because of the risk of liver damage and potential for liver failure. Acetaminophen is three times as likely to cause liver failure as all other drugs combined. It is the most common cause of acute liver failure in the United States. Most often, liver failure is associated with overdosing, but even at recommended doses, if it is taken with alcohol, it can cause irreversible liver failure.

NSAIDS include medications such ibuprofen (Advil and Motrin), naproxen sodium (Aleve), and ketoprofen (Orudis KT). Many of these medications are also available in a stronger prescription-strength dosage too. These medications tend to reduce inflammation, and the pain associated with inflammation, by preventing the body from manufacturing prostaglandins. Prostaglandins are substances that act as mediators of pain and inflammation. However, these same prostaglandins also protect the lining of the stomach. Use of NSAIDS should be monitored closely for stomach upset and gastrointestinal bleeding, the risk of which goes up with long-term use of these medications. The risk of liver failure is much less with NSAIDS than with acetaminophen. Aspirin is classified as an NSAID but is unique in some of its properties. In addition to reducing pain and inflammation, it acts as a blood thinner and can help prevent blood clots. Aspirin, too, has some risk with use, and it should not be taken by children under 16 who have chicken pox or flu symptoms due to the risk of Reye's syndrome, characterized by acute encephalopathy and liver failure.

Opioid pain medications have been used for centuries to relieve pain—early civilizations used the seedpods of the poppy plant (*Papaver somniferum*). Morphine and codeine are derivatives of opium, whereas other opioids such as meperidine are synthetics. Opioids work to relieve pain in two separate manners. They attach to opioid receptors in the brain, spinal cord, and gastrointestinal tract and interfere with the transmission of pain signals to the brain. They also alter the sensation of pain. It should be understood that opioids do not take away pain; they only alter a patient's perception of the pain.

Chronic use of opioids may result in tolerance to the drug, meaning higher doses are needed to obtain the same pain-relieving effects. It is this tolerance that can cause abusive behavior and eventually lead to addiction to the drug. For this reason, the prescribing of opioid pain medications is highly controlled by regulatory organizations such as the Drug Enforcement Administration in the United States. Side effects common to opioid medications are drowsiness, nausea, vomiting, constipation, dilated pupils, and euphoria. Euphoria is the leading factor in the abuse of opioids. Opioids can also cause respiratory system depression, which can be additive with other sedative medications and potentially lead to respiratory arrest and death.

Opioids are available in a variety of dosing options. Most commonly, opioids are dosed as tablets; however, they also come in liquid preparations for injection, transdermal patches, and oral absorption forms such as suckers and sponge sticks. With whichever dosage form is used, opioids should be started at a low dose and titrated up to effect. When removing someone from use of opioid medications, particularly when on higher doses of opioids, the healthcare provider should taper the doses slowly so as to avoid withdrawal effects of the medication. Common adverse events with abrupt withdrawal can include anxiety, restlessness, tachycardia, hypertension, nausea, vomiting, abdominal cramps, diarrhea, muscle aches, and pains.

Because of problems with diversion of opioid medications for illicit use, much attention has been given to other prescription medications that can be used in an effort to control pain. Anti-depressant medications have recently been used for chronic pain, including headaches and menstrual pain. Anti-depressant medications work by adjusting the levels of available neurotransmitters in the brain. This can increase the body's signals for well-being and relaxation, allowing for improved pain control when patients do not respond adequately to usual treatments. Examples of anti-depressants that have been used for pain are the selective serotonin reuptake inhibitors (SSRIs) such as citalopram, fluoxetine, and paroxetine; tricyclic anti-depressants such as amitriptyline, nortiptyline, and imipramine;

and selective serotonin and norepinephrine reuptake inhibitors (SSNRIs) such as venlafaxine and duloxetine. Tricyclic anti-depressants have an additional benefit of blocking sodium channels within nerve tissues and this blockage can create a local anesthetic effect. Duloxetine has received an approval for use in peripheral neuropathy, a type of pain caused by injury to nerve tissue, by the U.S. Food and Drug Administration.

One disadvantage of the use of anti-depressant medications in the treatment of pain is that these medications require regular doses in order to build up in the body over a period of time to achieve an effect. An additional problem is that the use of anti-depressant medications is also associated with unwanted side effects. Potential side effects include sedation, lowered blood pressure, constipation, diminished sexual function, and dry mouth. When patients suffer from chronic pain, it is important to discuss any symptoms of depression that they may also be feeling, as this treatment approach may be most appropriate when an individual suffers from concomitant depression.

One final group of medications used in the treatment of pain are anticonvulsant drugs, which are typically used to treat seizure disorder. The mechanism of action in pain control from these medications is not well understood but is thought to be related to minimizing the effects of nerves on the sensation of pain and possibly reducing ectopic discharge of nervous tissue. Carbamazepine was the first anticonvulsant medication studied in clinical trials and has been shown effective in treating such chronic pain conditions as trigeminal neuralgia, diabetic neuropathy, and postherpetic neuralgia. Gabapentin has been shown to have the most clearly demonstrated analgesic effect in the treatment of neuropathic pain. Due to a favorable side effect profile and positive research results, gabapentin is quickly becoming a first-line therapy in the treatment of neuropathic pain.

As the understanding of the underlying mechanisms of pain evolves, so too will the treatment of pain. Newer drug compounds are currently being explored for use in the treatment of pain. The source of pain and any co-morbid conditions should be taken into consideration when selecting a medication, and any potential risks or side effects should be weighed against the benefit of treating with a specific medication.

See Also: Anti-Depressant Drugs; Pharmaceutical Industry; Pharmaceutical Industry Reform; Prescription Drug Addiction; Supplements.

Further Readings

"Depression. Understand It, Treat It, Beat It." http://www.clinical-depression.co.uk (Accessed May 2010).

Dick, I., R. Brochu, Y. Purohit, G. Kaczorowski, W. Martin, and B. Priest. "Sodium Channel Blockade May Contribute to the Analgesic Efficacy of Antidepressants." *Journal of Pain: Official Journal of the American Pain Society*, 8/4 (2007).

Gagnier, J. J., M. Vantulder, and B. Berman, et al. "Herbal Medicine for Low Back Pain." *Cochrane Database Systematic Review*, 19/CD004504 (2006).

MedicineNet.com. "Drug Induced Liver Disease." http://www.medicinenet.com/drug_induced_liver_disease (Accessed May 2010).

Meldrum, M. *Opioids and Pain Relief: A Historical Perspective.* Progress in Pain Research and Management, vol. 25. Seattle: International Association for the Study of Pain Press, 2003.

National Reye's Syndrome Foundation. http://www.reyessyndrome.org (Accessed May 2010).

Schiodt, F., F. Rochling, D. Casey, and W. Lee. "Acetaminophen Toxicity in an Urban County Hospital." *New England Journal of Medicine*, 337 (1997).

Tremont-Lukats, I. W., C. Megeff, and N. M. Backonja. "Anticonvulsants for Neuropathic Pain Syndromes: Mechanisms of Action and Place in Therapy." *Drugs*, 60/5 (2000).

U.S. Drug Enforcement Administration. http://www.dea.gov (Accessed May 2010).

WebMD. "Pain Management Guide." http://www.webmd.com/pain-management/guide/ (Accessed May 2010).

Rance McClain
Independent Scholar

PAINT

Paint is a decorative and protective finish that can be applied to a variety of surfaces to improve both their appearance and longevity. The pigment is what gives paint its color; pigments come in the form of a ground solid powder that imparts a color to the paint. Pigments can be natural, as in the form of mined minerals or clays, or synthetically produced. Fillers are a form of pigment that serves to "thicken" the paint and add body and depth. They may be made of cheaper materials than the pigments. A binder is what produces a "film" of paint; it has adhesive properties that hold the pigments together and bond them with the surface. The binder is also largely responsible for characteristics of the surface finish such as the gloss potential and durability and flexibility. Binders can be made from a variety of natural and synthetic products and are classified by the way in which they dry or cure (where drying refers to solidification by solvent evaporation and curing refers to polymerization).

Paints that cure by catalyzed polymerization tend to be two-part paints of resin and hardener that when mixed together polymerize into a hard plastic finish. This category contains two-pack, epoxy, and polyurethane finishes. Paints that coalesce are water-based latex or vinyl emulsion paints. They have an aqueous dispersion of tiny polymer particles, which are produced through a process of emulsion polymerization. They contain a large quantity of water and a coalescing solvent in trace quantities. First, the water evaporates, by far the larger part of the volume of the paint.

Low-impact paints attempt to match the performance and application methods of traditional paints but with a product that has less deleterious effects on the environment. Most major paint retailers offer no—or low—volatile organic compound (VOC) paint alternatives, including Home Depot's Fresh Aire line.

Source: HomeDepot.com

Then, the coalescing solvent brings together the tiny polymer particles, which join and form a matrix that is impervious to the water/solvent that formed the original carrier. Paints that cure by oxidative cross-linking polymerize when exposed to oxygen in the atmosphere. This category contains enamels and alkyd paints. Paints that cure by solvent evaporation are classed as lacquers; they consist of a solid binder dissolved in a solvent—the binder then hardens as the solvent evaporates. The type of solvent chosen can have a significant effect on the environmental impact of the paint.

In addition to the three main components of paint, paints may have other additives to modify the properties of the paint and impact special characteristics. These may include an adhesion promoter, biocides, catalysts, deglossing agents, an emulsifier, liquid crystals, photochromic additives, photochromic chemicals, a stabilizer, a texturizer, thickeners, and a UV stabilizer. Traditional paints have impacts at the beginning of their life cycle due to the chemical products consumed during their manufacture; there is a high product-to-waste ratio with paint manufacture (the production of 1 liter of paint can result in up to 30 liters of waste). Furthermore, paints can have an environmental impact during application due to the volatile chemicals that evaporate as the paints dry. This off-gassing can last for years after the paint has been applied. The paint can also have an impact during disposal—either the disposal of dried paint on painted surfaces or the disposal of unused paint in its liquid state. Traditional paints made from petrochemicals contain synthetic chemicals not found in nature; as a result, at the end of their life cycle, they are not easily reassimilated into the natural environment. In the past, toxic substances such as lead have been used in paint manufacture, and items painted with such products can be hazardous to human health.

During the application of traditional paints, insufficient ventilation can cause the user to experience nausea, headaches, and dizziness. These effects are caused by the solvents used in the paint's manufacture, which aid drying. The World Health Organization (WHO) stated as far back as 1989 that professional decorators were 40 percent more likely to suffer from lung cancer. There are a number of different chemicals used in traditional paint manufacture that can off-gas once the paint has been applied, including benzene, formaldehyde, ammonia, and toluene. Eco-paints seek to replace synthetic petrochemicals with green alternatives that are found in nature. In low-energy building design, it is often a goal to improve the airtightness of buildings and to improve the heat retention of the building fabric. The product of traditional paints' off-gassing is less easily dispersed in a building with less frequent air changes. There are a number of alternatives to traditional paints, including the following:

- Green paints using natural solvents for synthetic chemicals, which may be plant derived instead of petrochemical derived
- Ecological paints using natural fillers such as chalk
- Sustainable paints without biocides or fungicides in their products, although natural finishes that are alkaline may act as a biocide
- Green finishes using water as the solvent rather than volatile compounds
- Green paints that do not require extensive processing to produce a finished product with minimal embodied energy content

However, there can be some disadvantages with eco-paints, including the following:

- Eco-paints may not offer the color range that synthetic paints can produce.
- Natural pigments may not produce the brilliant color traditional paints offer.
- Eco-paints that use orange and lemon oils as solvents also have the potential to cause skin irritation and headaches.

Low-Impact Paints

Low-impact paints are an improvement on traditional paint products in that they attempt to match the performance and application methods of traditional paints but with a product that has less deleterious effects on the environment. They are water-based and do not share the problems associated with high-volatile organic compounds (VOC) paint; however, they still use synthetic chemicals in their manufacture, so they still have some environmental impact.

Natural Paint

Minerals cannot be considered a renewable resource because they are extracted from the ground. However, they represent a more sustainable solution than synthetic paints. Typically, brightly colored compounds that occur as the result of certain geological formations are extracted and ground to produce a pigment, which can then be mixed with lime or plaster and applied to walls, or used with other natural compounds to form a suspension in paint. Mineral paints may use sodium or potassium silicate, sometimes known as *waterglass*, to produce a compound that will bond to silicate materials in the surface to which the finish is being applied.

A traditional method of paint manufacture that is undergoing a resurgence of interest as a result of environmental awareness is milk-based paints. Milk-based paints use the proteins found in milk, casein, combined with lime to produce a calcium-caseinate, which allows the paint to adhere to a range of surfaces.

Plant-based paints rely on natural compounds that can be sourced from plants to provide dyes and colors. It is harder to find the same palette of colors that commercial synthetic paints can produce, so the choice is limited to a range of "natural" hues.

Typically, paint is sold in containers premixed with the required solvent, so it can be applied directly from the can. Some manufacturers of eco-paints prefer to sell their formulation as a powdered mixture, which is then mixed by the user with either water or natural oil to act as a solvent, depending on the formulation. This reduces the bulk of the paint during shipping and transportation, as well as the carbon emitted in moving the paints to market.

Advanced Finishes

There are some advanced finishes and surface coatings under development that, while largely made from synthetic chemicals using complex processes—and not falling under the umbrella of simple eco-paints—have the potential to offer an environmental improvement by way of a painted finish. Technologists have examined, for example, how solar power generation could be accomplished through a painted finish or how a painted finish used on the exterior of buildings in cities could be used to capture nitrogen dioxides. However, many of these approaches are still "in the lab" and far from commercialization.

See Also: Indoor Air Quality; Lead Sources and Health; Occupational Hazards; Phaseout of Toxic Chemicals.

Further Readings

Hall, Keith Dennis. *The Green Building Bible: Essential Information to Help You Make Your Home & Buildings Less Harmful to the Environment, the Community & Your Family.* San Rafael, CA: Green Building Press, 2005.

Nayar, Jean. *Green Living by Design: The Practical Guide for Eco-Friendly Remodeling and Decorating.* New York: Hachette Filipacchi, 2009.

Gavin D. J. Harper
Cardiff University

PAPER PRODUCTS

The majority of the raw fiber to generate paper products is harvested from a few key areas in the world, which creates additional pressures on forests in those areas and results in local resource concerns. The paper bleaching process can also cause the formation of dioxins, which can cause specific health concerns and may particularly affect the users of feminine hygiene products. Although paper products are used throughout the world, companies in the United States, Sweden, Finland, and Japan dominate the world paper products supply. The United States is the largest consumer of paper in the world, although the raw materials for these products come from the world's forests.

The United States is the largest consumer of paper in the world, and the raw materials for these products come from forests around the world. Here, a stack of lumber sits outside a paper mill.

Source: iStockphoto.com

The Food and Agriculture Organization (FAO) of the United Nations tracks global forest supply and demand and issues a report every other year detailing the state of the world's forests. According to the FAO, pulp and paper products utilize about 12 percent of the world's harvested wood. Availability of harvested wood is a concern due to diminishing supplies and the increasing global population.

The FAO reported that the world's forested area was reduced by 3 percent from 1990 to 2005, but the rate of deforestation has decreased in recent years. The world's forests lost 0.22 percent annually during 1990 to 2000, and the rate of annual loss slowed to 0.18 percent in 2000–2005. However, loss of resources is not uniform across the globe:

- Africa experienced the world's largest annual forest losses of 0.64 percent in the period 1990–2000 and 0.62 percent in 2000–2005. Africa's main wood product output is industrial round wood, not pulp and paper.
- Latin America and the Caribbean experienced annual forest losses of 0.46 percent in 1990–2000 and 0.51 percent in 2000–2005. Although this region did experience less loss than Africa, it showed a trend of increasing deforestation. Additionally, this region consumes more paper products than it produces.

- Asia experienced lower annual forest losses of 0.17 percent in 1990–2000 and 0.09 percent in 2000–2005. Almost three-fourths of the wood harvested in Asia is burned as fuel, and Asia is a net importer of paper products.
- Europe experienced even lower annual forest losses of 0.09 percent in 1990–2000 and 0.07 percent in 2000–2005, although these rates vary across Europe and are much higher in western Europe. As one of the greater paper-producing areas, Europe is a net exporter of paper and paper products.
- North America was reported by FAO as experiencing the lowest annual forest losses of 0 percent in 1990–2000 and 0.01 percent from 2000–2005. However, data from Canada, where most of North America's forest resources exist, was not complete, and it was assumed that no change occurred in forest stock from 1990 to 2005. This assumption most surely skewed the reported results. But the United States had no loss of forests, and instead experienced annual forest gains of 0.12 percent in 1990–2000 and 0.05 percent in 2000–2005. Mexico reported comparatively high forest losses of 0.52 percent annually in 1990–2000 and 0.40 percent annually in 2000–2005. Overall, North America produces more paper products than it consumes.

Due to the varying rates of forest depletion, the impact of wood harvesting has differing regional effects. In addition to the environmental effects of deforestation, the paper-making process has certain effects, and its production of dioxin has been of particular concern.

Pulp is created by breaking down wood, either mechanically or chemically. Colors naturally present, from lignin and other chemicals in wood, must be removed by bleaching. Mechanical pulp retains most of the wood lignin, while chemical pulp has very little lignin. Mechanical pulp does not cause much environmental concern because most of the organics are retained in the pulp and it is typically whitened with hydrogen peroxide and sodium dithionite, which do not have harmful end products. The bleaching of chemical pulps, on the other hand, has significant environmental impacts.

Chemical pulps can release large amounts of organics. When bleached with chlorine, these form chlorinated organics, such as dioxins, which are known to be persistent environmental pollutants. Dioxins are carcinogens and can cause reproductive, developmental, immune, and hormonal problems in humans. Dioxins can be released into the environment through the water effluent of paper plants that use free-chlorine bleaching and bioaccumulate in the environment, concentrating in the fatty tissues of animals. The primary route of exposure of humans to dioxin is through consuming animal products.

However, other possible routes of exposure to dioxin also exist. Much debate exists regarding possible exposure of women through the use of tampons that are bleached. In the United Sates, tampons are regulated by the Food and Drug Administration (FDA). A statement released by the FDA in 2009 explained that the exposure to dioxin of women using tampons is approximately 5.4 pg/month or 0.108 pg/kg body weight in a month. The following month, a different FDA statement concluded that the risks of adverse health effects from this exposure were negligible. But a 2003 review of U.S. Environmental Protection Agency methods for calculating the dose–response relationship between dioxin and cancer found that there is no safe dose (threshold) for exposure to dioxin. Even within the scientific community, debate exists as to whether there is a safe dose for dioxin exposure below which humans will suffer no ill effects. Since there are bleaching methods available that do not use chlorine, women concerned about possible dioxin exposure can select tampons labeled as chlorine free.

See Also: Biodiesel; Climate Change; Firewood and Charcoal; Personal Consumer Role in Green Health.

Further Readings

Environment Canada. Health Canada. "Effluents From Pulp Mills Using Bleaching—PSL1." Ottawa, ON: Health Canada and Environment Canada, 1991

Mackie, David, Liu Junfeng, Loh Yeong-Shang, and Valarie Thomas. "No Evidence of Dioxin Cancer Threshold." *Environmental Health Perspectives*, 111/9 (2003).

PricewaterhouseCoopers, LLP. *Global Forest, Paper and Packaging Industry Results*, 2009 ed. Results of 2008 Survey, Downloadable. http://www.pwc.com/gx/en/forest-paper-packaging/2009-fpp-survey (Accessed September 2010).

UN Food and Agriculture Organization. *State of the World's Forests 2009*. United Nations: Rome, 2009.

U.S. Food and Drug Administration. "Dioxin in Tampons" (2009). http://www.fda.gov/ScienceResearch/SpecialTopics/WomensHealthResearch/ucm134825.htm (Accessed June 2010).

U.S. Food and Drug Administration. "Tampons and Asbestos, Dioxin, & Toxic Shock Syndrome" (2009). http://www.fda.gov/MedicalDevices/Safety/AlertsandNotices/PatientAlerts/ucm070003.htm (Accessed June 2010).

World Health Organization (WHO). "Dioxins and Their Effects on Human Health." Fact Sheet No. 225. Geneva, Switzerland: WHO, 2010.

Michelle Edith Jarvie
Independent Scholar

PARTICULATE MATTER

Airborne particulate matters (PMs) consist of a mixture of solids or liquids suspended in the air. PMs are conveniently classified based on their aerodynamic diameter sizes, ranging from nanometers to 100µm. Thoracic particles (PM10) are those with aerodynamic diameters less than 10µm, with similarities applicable to coarse particles (PM2.5–10), fine particles (PM2.5), and ultrafine particles (PM0.1). Although the ultrafine particles are smaller and weigh less, their numbers are much larger than those of the rest. Generally speaking, coarse particles constitute about 90 percent of suspended particles, while fine and ultrafine make up 1 to 8 percent of the total mass. Large particles come from such things as stirring up road dust, agricultural processes, mining operations, sea spray along coastlines, plant pollen, and insect parts. Ultrafine particles are formed by condensation of nuclei, which are the substances formed by high-temperature vaporization or chemical reactions such as fossil fuel combustion. Since large variations in the sources, masses, and compositions are found in particulate matter due to differences in the surrounding environment, health effects of particulate matter vary among different regions, climates, and seasons.

Particulate matter comes from organic carbon (OC), elemental carbon (EC), transition metals, ions, reactive gases, biological materials, and minerals, among other items. OC and EC make up the largest percentage of compositions, since a significant fraction of ambient particles is derived from combustion processes. Certain groups of OC components have been shown to be involved in PM-induced oxidative stress and its related carcinogenic characteristics. Quinone and polycyclic aromatic hydrocarbons (PAHs) are well known to be toxicologically active on many biological targets. Similar to the quinines, transition metals can be involved in a Fenton-type reaction and potentially produce reactive oxygen species. Methods for determining inorganic compounds in ambient PMs have been established by the U.S. Environmental Protection Agency (EPA) since 1999. Furthermore, metals in ambient PMs can be determined using inductively coupled plasma/mass spectrometry (ICP/MS).

The characteristics of air pollution mixtures can be altered by changes in the sources of air pollution and meteorology. A study of seasonal analyses of air pollution and mortality was applied to the database of the National Morbidity and Mortality Air Pollution Study in 100 U.S. cities for the period 1987 to 2000. At a national level, mortality has increased as a result of a 10 μg/m³ increase in PM10 at a one-day lag in all seasons. A strong seasonal pattern in the Northeast has been observed with a peak in summer, while little seasonal variation has been found in the southern region of the country.

Epidemiologic studies have demonstrated a significant association between exposure to ambient PMs and the acute as well as chronic adverse impact on public health. Some studies have shown that the toxicity and carcinogenicity of PMs are related to their size, which represents their capability to penetrate the gas-exchange region of the lungs. As such, smaller particles are associated more with adversely affecting heart function than larger ones. Evidence on short-term exposure to PMs shows that fine particles are much more likely to affect public health than coarse particles. Based on the study of a national database comparing daily time series data from 1999 through 2002, a short-term increase in hospital admission rates was associated with fine particles for all health outcomes except injuries. Heart failures tended to show up as the highest risk factor, with a 1.28 percent growth in occurrence per 10 μg/m³ increase during same-day PM2.5 exposure. Cardiovascular effects have been found at higher levels in areas located in the eastern region of the United States, including the Northeast, the Southeast, and the Midwest. A study of six U.S. cities by Harvard University scientists showed that long-term PM2.5 exposure shows a high correlation with lung cancer and cardiopulmonary mortality. The researchers also found that reducing the amount of PM2.5 was associated with people living longer after an eight-year follow-up. Though the mechanisms of how PMs induce diseases are not fully known, several studies have demonstrated that PMs can cause oxidative stress, which leads to several health concerns. The aim of studying PMs is to determine the toxicity of specific classes of airborne particles.

To protect public health and welfare, the U.S. Congress passed the Clean Air and Air Quality Acts in 1963 and 1967. The EPA issued National Ambient Air Quality Standards (NAAQS) to set standards for six criteria pollutants in 1971. In 1997, the U.S. National Ambient Air Quality Standards (NAAQS) for airborne PMs were revised to include annual and 24-hour standards for PM2.5 mass in ambient air, which were 15 μg/m³ and 65 μg/m³, respectively. In 2006, the standards of PM2.5 were further revised to limit the 24-hour standard to 35 μg/m³ and to retain the existing annual standard at 15 μg/m³. Based on the 2003–2005 monitoring data, 143 counties exceeded revised PM2.5 standards across the United States, in which 56 counties exceeded both annual and 24-hour standards, 70 counties exceeded the 24-hour standard only, 17 counties exceeded the annual standard only. The EPA has revoked the annual PM10 standard due to a lack of evidence that long-term exposure to coarse particles may cause health problems, but the agency has decided to retain the 24-hour PM10 standard of 150 μg/m³. While the EPA is working on public hearings and comments collection, further revision will be considered in the next review of PM standards.

See Also: Environmental Illness and Chemical Sensitivity; Environmental Protection Agency (U.S.); Government Role in Green Health; Lung Diseases.

Further Readings

Dominici, F., et al. "Fine Particulate Air Pollution and Hospital Admission for Cardiovascular and Respiratory Diseases." *Journal of the American Medical Association,* 295 (2006).

Laden, F., et al. "Reduction in Fine Particulate Air Pollution and Mortality. Extended Follow-up of the Harvard Six Cities Study." *American Journal of Respiratory and Critical Care Medicine,* 173 (2006).

"Particulate Matters: Properties Related to Health Effects, COST Action 633." http://cost633 .dmu.dk (Accessed January 2010).

Peng, R. D. "Seasonal Analyses of Air Pollution and Mortality in 100 U.S. Cities." *American Journal of Epidemiology,* 161 (2005).

Schlesinger, R. B. "The Health Relevance of Ambient Particulate Matter Characteristics: Coherence of Toxicological and Epidemiological Inferences." *Inhalation Toxicology,* 18 (2006).

Hueiwang Anna Cook Jeng
Liang Yu
Old Dominion University

PERSONAL CONSUMER ROLE IN GREEN HEALTH

The green movement assumes both individual responsibility for lifestyle and health choices and collective understanding regarding outcomes that may affect the collective well-being of groups throughout the world. The personal consumer's role in green health has helped empower individuals to take seriously their health and healthcare choices. Consumers affect green health through their assessment of potential benefits and limitations of the Western medical model, their choices regarding foods and other substances they ingest, and how these choices support notions of sustainability. Individuals interested in living in a healthy as well as a sustainable manner must traverse a complex terrain in which healthcare and lifestyle choices are negotiated through information provided by the media, insurance companies, employers, schools, and the Internet, to name just a few. The integration of personal habits and services that can assist the individual to live a full and healthy life is a central concern for the ecologically sensitive consumer.

The pharmaceutical industry's research and development programs have been responsible for many of the current advances in Western medicine. Environmental groups and concerned consumers have challenged pharmaceutical companies, however, with regard to production practices that have despoiled the environment. For example, the pharmaceutical industry has been charged with burying over 500 million pounds of compounds used in research, development, and manufacturing of products in landfills and releasing over 250 million pounds of hazardous waste into waterways, many of which supply drinking water. Some of the wastes released include compounds found in anticonvulsants, mood stabilizers, and sex hormones. It is known that certain hazardous wastes affect lower life forms and it is assumed that the potential of pervasive ingestion of these compounds might have an effect on humans as well. Such practices have made the pharmaceutical industry the target of much criticism. Proponents of the pharmaceutical industry, however, maintain that such occurrences are isolated and ignore tremendous contributions to public health that the industry as a whole has made. Advances in pharmacology have allowed, for instance, individuals with the acquired immune deficiency syndrome (AIDS) to manage their condition as a chronic rather than as a terminal illness. So, too, have powerful antibiotics assisted many in overcoming serious infections.

Modern medicines sometimes also cause problems that have little to do with their manufacture. The proliferation of antibiotics results in the drugs breaking down as they pass through humans or animals. When excreted, such waste may result in the development of microbial drug resistance. Researchers use the enzyme-linked immunosorbent assay (ELISA) or liquid chromatography-mass spectrometry (LC-MS) to measure the level to which food and meat contain drug residues and if human ingestion of such residues is leading to the development of microbial resistance. Concern for the development of microbial resistance and the by-products of pharmacological manufacturing are not only a human concern, but also a cause for alarm that antibiotics may lead to potential genetic mutations in animals and plant life.

Personal consumers must weigh the costs and benefits associated with Western healing practices and the effects the received treatments have on the environment. Further, green personal consumers need to consider multiple information sources in order to make an informed choice. Much of this information is culled from traditional scientific paradigms such as double-blind studies, but also advocacy from public relations outlets of professional organizations or trade and advocacy groups. Personal consumers considering engaging in antibiotic therapy to treat an ear infection, for instance, must be aware of the most reasonable scientifically validated ideas on antibiotic use. In 2004, the American Academy of Pediatrics (AAP) and the American Academy of Family Physicians (AAFP) released a report on the proper protocol for diagnosis and treatment of ear infections. The report emphasized the need to properly distinguish between types of ear infections (acute otitis media and otitis media with effusion) and the proper use of antibiotics. A central purpose of the work of AAP and AAFP was to better ensure that antibiotics are used only when necessary. So, too, personal consumers must consider the place, if any, for alternative treatments sometimes advocated by groups outside mainstream Western medicine. Alternative treatments, however, necessitate that personal consumers be aware of the limitations of the suggested treatments and their efficacy in treating the particular disease. Acupuncture, for example, has not only become increasingly popular in the past 25 years, it also has a body of research that buttresses oft-repeated claims for its ability to control, cure, or reverse the progress of a variety of diseases.

Habits, Health, and the Environment

Green personal consumers must also be aware of how their habits of consumption affect not only their health but also that of the environment. There are a variety of organizations and movements that support personal consumers in making lifestyle choices related specifically to the foods they eat. The slow food movement (SFM), for example, emphasizes the need for individuals to be aware of where food comes from, how it is grown, the environmental cost of immediate access, and the potential eradication of certain foods due to demand in particular locales. Thus, SFM demands consumers forgo easy access to foods that are not locally grown for foods that are available seasonally. SFM has benefited from the Internet insofar as interested consumers are provided a way to converse about SFM ideas across time zones and geographic locations. Cookbooks, such as those by Alice Walker, assist the novice and seasoned SFM participant in cooking thoughtfully, healthfully, and aesthetically while embodying the central tenets of the SFM.

Farmers markets are a related outgrowth of SFM thinking, and have proliferated nationally. Farmers markets have also been a boon for the ecologically concerned personal consumer insofar as they enable interested consumers to purchase locally grown produce, meats, and baked goods. Many of these products are organic. Even those that are not, however, still have a lower carbon cost than imported produce, meats, and other goods because they are produced locally. Moreover, farmers markets support local growers and provide concerned personal consumers a space in which to meet and share ideas about best practices for growing produce and other topics germane to living a sustainable lifestyle.

Changes that support an ecologically sound lifestyle are not limited to individuals living in rural areas. Many urban areas have support groups for container gardening. Personal consumers who engage in container gardening use planters, pots, and wooden boxes to cultivate produce. This practice has many obvious advantages for the urban dweller such as flexibility, mobility, and ease of pest control. Food cooperatives are yet another way for personal consumers to better control which foods they ingest and support companies interested in sustainable practices. The typical food cooperative is locally owned and run by members. The food cooperative may operate as a store where all can purchase items or as a buying group where groups of individuals join forces to buy in bulk to save money. There is often, if not always, an educational arm to a food cooperative that is charged with providing workshops and other informational sessions for members and nonmembers alike. Often, the topics covered in the information sessions focus on how to live sustainably and how the sustainable lifestyle choice affects personal consumers' health.

Although often overlooked, the development of ecologically sound thinking among personal consumers is of great importance to the green movement. To be sure, much of this development can occur during childhood and in the home. Children who are provided opportunities to visit farms and farmers markets, participate in food cooperatives, and understand the cost and benefit of certain lifestyle choices become more savvy consumers and are often more committed to an ecologically friendly lifestyle. Because many children live outside areas where it is feasible for them to participate in such activities, public schools often have taken the lead in introducing children to sustainable practices. Many schools provide organic gardens, composting, recycling receptacles, and programs that expose children to notions central to living an ecologically sound life. Earth Day is a ubiquitous celebration in public schools each April. Earth Day celebrations often include contests to see which grade level or class can produce the least amount of trash or eat the most healthful foods during a certain period. Intimately related to the aforementioned practices is the emphasis on taking control of one's personal health through exercise. Public and private K–12 schools have also taken seriously the current obesity epidemic. Programming inside and outside physical educational classes attempts to provide both a space where children can relax and recharge as well as instilling a lifelong fitness routine. So, too, do K–12 public schools often provide basic dental and medical care to children who are otherwise lacking access to such services. Much of such basic care focuses on prevention. Such programs meet children's needs while concurrently providing instruction in preventative measures of personal and dental hygiene.

The Role of Higher Education

Higher education is also taking on an increasing role in providing personal consumers with the tools to be greener and healthier. Many colleges and universities have sustainability initiatives that focus on reducing consumption, often seeking to minimize campus energy usage, water consumption, and trimming the use of other resources across campus. Colleges and universities also frequently provide interested participants with opportunity to plant and cultivate organic produce. These opportunities provide excellent resources at little cost to the institution, but are also excellent ways of enabling students to see the benefits of sustainable farming practices and offering access to high-quality, fresh produce. The role of educational programming in developing the informed consumer is essential for the advancement of the green movement. Public and private schools, colleges, and universities are very much at the center of developing the practices that will allow young people to develop into thoughtful consumers.

Green personal consumers are also educated through local, state, and federal legislation, regulations, and programming. For example, federal requirements exist for the certification of a farm or other food industry to be considered "organic." This recent federal involvement into sustainable lifestyle choices has had many effects in terms of practices. More important, the cost of attaining certification, and the relaxation of standards over time, has led many organic farmers and other interested individuals to resist and even reject becoming federally certified. For example, the federal government's process focuses only on those retailers who package goods for sale. Companies that sell less than $5,000 worth of goods per year are exempt from certification. This is helpful insofar as many farmers are thus exempt from federal intrusion. However, the low dollar amount required for an exemption makes it difficult or impossible for small and middle-sized independent producers, who do indeed live and work in a sustainable environment, to receive federal certification. The National Organic Program (NOP) is alleged to privilege large agribusinesses and to pressure some smaller farms and companies to opt out of being certified organic.

Even more troubling than the privileging of large businesses, however, are numerous reports suggesting that the stringent requirements of the federal government have been eased in recent years so that many "certified organic" products actually contain additives. Multinational corporations have, for all intents and purposes, bought up the once independent farms. When enforcement of regulations is lax, even personal consumers with the best intentions are challenged in their attempts to live an ecologically friendly existence. The organic market is the fastest growing segment of the food industry. When regulations of the industry are suspect or confusing, personal consumers who believe in ecologically friendly practices can become unable or unwilling to practice or support local growers, food cooperatives, and the like because of uncertainty regarding which vendors to support. When this occurs, both personal consumers and truly green producers are placed in a vulnerable position. Here, too, it becomes imperative that advocacy groups provide educational programming, community outreach, and opportunities that allow people to control, as best as they are able, what they purchase and consume.

Advocacy for sustainable living must include personal consumers and their choices as well as consideration of how those choices affect the larger world. Perhaps most important, advocacy must focus on what it means to live within one's environment as a sustainable consumer. Such advocacy might include emphasizing the eating of seasonal foods or supporting local growers through participation in a food cooperative. Indeed, many local organizations advocate for these actions and behaviors. With the influx of media messages, however, the various perspectives provided as to what constitutes "green living" present distinct challenges to personal consumers with the best intentions. Coupled with the limited time that many personal consumers have to consider their eating habits and what they can reasonably expect from their limited choices for medical care, many who wish to make green choices are precluded from doing so.

Personal consumers interested in empowering themselves as discriminating customers in control of their health choices have made great strides in advocating for alternative possibilities in healthcare. Many insurance companies, for instance, now will pay for acupuncture and chiropractic care. State and federal regulations have forced many companies to be more forthright with regard to packaging information, thus allowing the personal consumer to make more informed choices. Even so, it is up to the individual to sculpt his or her vision for what it means to live healthfully and sustainably.

The grassroots movements that once anonymously peppered the landscape of America have become flourishing framers markets, SFMs, and food cooperatives. A greater number of individuals are demanding more from their local leaders in terms of support and recognition for individuals who take seriously their role as a consumer. The demands of personal consumers have led to many positive changes in business practices related to the environment. Indeed, many businesses now specifically cater to the needs of the ecologically aware consumer. Dry cleaners have begun using ecologically friendly materials, fast food restaurants have altered packaging to produce less waste, and food manufacturers have started providing more healthful choices to consumers. As these options become more prevalent, personal consumers are able to make choices that allow them to contribute to the green movement in positive ways. Personal consumers have a great deal of power to ensure their health and the environment's well-being. As a result, the green movement empowers personal consumers to take actions for healthy choices in an ecologically sound manner.

See Also: Acquired Immune Deficiency Syndrome; Antibiotic Resistance; Dry Cleaning; Home-Grown Food; Obesity; Oral Disease; Physical Activity and Health.

Further Readings

Ehlers, M. M., C. Veldsman, E. P. Makgotlho, M. G. Dove, A. A. Hoosen, and M. M. Kock. "Detection of Antibiotic Resistance Genes in Randomly Selected Bacterial Pathogens From the Steve Biko Academic Hospital." *FEMS Immunology & Medical Microbiology*, 56/3 (2009).

Tiemann, T. K. "Grower-Only Farmers' Markets: Public Spaces and Third Places." *Journal of Popular Culture*, 41/3 (2008).

Verkooijen, K. T., G. A. Nielsen, and S. P. J. Kremers. "Leisure Time Physical Activity Motives and Smoking in Adolescence." *Psychology of Sport & Exercise*, 10/5 (2009).

Stephen T. Schroth
Jason A. Helfer
Daniel O. Gonshorek
Knox College

PERSONAL HYGIENE PRODUCTS

Most people have a false sense of security when it comes to the vast array of personal hygiene products available on the market today. For the most part, consumer concerns center around truth in advertising claims that products will actually whiten teeth, diminish wrinkles, or leave hair soft and manageable. Yet anyone who has ever studied the label of a shampoo, liquid soap, deodorant, or toothpaste realizes that commercially available personal products contain a chemical soup of compounds including perfumes, preservatives to reduce bacterial growth, detergents, surfactants, and color additives. This entry will explore the safety of these chemicals and what other issues environmentally conscious buyers may consider when choosing personal hygiene products.

A number of common ingredients in personal hygiene products and cosmetics are suspected of having adverse health or environmental impacts. Dangerous substances often found in these products include mercury, lead, phthalates, petroleum products, and formaldehyde.

Source: iStockphoto.com

Regulatory Context

The agency charged with oversight of cosmetic and personal hygiene products, the U.S. Food and Drug Administration (FDA), has no authority to require premarket safety assessment as it does with drugs, and does not subject such products to premarket approval. With the exception of color additives, the FDA does not review, nor does it have the authority to regulate, what goes into these products before they are marketed for consumer use. Although manufacturers are not required to register their cosmetic establishments with the FDA, the FDA can inspect cosmetic manufacturing plants and conduct import inspections to monitor whether products are adulterated or misbranded under the Federal Food, Drug and Cosmetic Act (FD&C) or the Fair Packaging and Labeling Act (FPLA). FDA also follows up on reports of adverse reactions.

To discourage congressional legislation, the cosmetics industry trade group (now called the Personal Care Products Council) created a system of voluntary self-regulation in 1976 through the Cosmetic Ingredient Review (CIR). Therefore, the manufacturers themselves are responsible for substantiating the safety of their products and ingredients before marketing. According to the Environmental Working Group Campaign for Safe Cosmetics, a national organization of nonprofit health and environmental organizations, the CIR has reviewed the safety of only a small fraction of ingredients used to formulate personal care products. This is hardly surprising given the lack of regulatory requirements for rigorous product scientific testing and post-marketing surveillance. Consequently, the burden of determining which products contain ingredients that may have adverse health effects largely falls on the consumer.

Safety Concerns

The average adult uses numerous personal care products every day, each with hundreds of ingredients including common industrial chemicals. Unlike in water and food where exposures are measured in parts per million, cosmetics and other personal care products may contain chemicals as base ingredients or significant contaminants of base ingredients. Many of these chemicals can be absorbed through the skin. Over the long term, exposures can be cumulative. Women are exposed more often than men because they use more personal care products and cosmetics.

A staggering number of commonly used ingredients in personal hygiene products and cosmetics have known or suspected adverse health or environmental impacts. A few examples of ingredients that can have serious health impacts are the following:

- Extracts of human and cow placenta, found in hair and skin products as softeners and conditioners, can contain high levels of hormones.
- Mercury, found in mascara as the preservative thimerosol, is able to damage nerve and brain function at low levels.
- Lead, found in some brands of hair dye for men, is a potent neurotoxin.
- Synthetic fragrance and perfumes contain unknown and unregulated chemical compounds. The synthetic fragrances used in cosmetics can have as many as 200 ingredients. There is no way to know what the chemicals are, since on the label it will simply read "fragrance."
- Hydroquinone, found in some hair color and acne products, can cause a disease called ochronosis with disfiguring blue-black lesions that may become permanent.
- Phthalates, used in plastics to increase durability, flexibility, transparency, and longevity, may be found in nail polish and are associated with birth defects, asthma, early puberty, and low sperm counts in men.
- Petroleum products, found in some bar soaps, lotions and moisturizers, facial cleansers, vapor rubs, eczema treatment, lip gloss, hair relaxers, dyes and bleaches, sunless tanning products, and a host of other personal hygiene products, may contain a variety of impurities such as 1,4 dioxane classified as a probable human carcinogen by the Environmental Protection Agency.
- Formaldehyde, found in shampoo, liquid hand wash, body washes, makeup, lotions and facial cleansers, sunscreens, and many other personal products, is a carcinogenic impurity released by a number of cosmetic preservatives, including diazolidinyl urea, imidazolidinyl urea, DMDM hydantoin, quaternium-15, 2-bromo-2-nitropropane-1,3-diol, and sodium hydroxylmethylglycinate.

Other products are linked with skin and eye irritation, rashes, dryness of hair and skin, and allergic reactions, and include the following:

- *Methyl, propyl, butyl, and ethyl paraben*: Often used as preservatives because of their ability to inhibit growth of microorganisms and extend the shelf life of products.
- *Diethanolamine (DEA), triethanolamine (TEA)*: Often used in cosmetics as emulsifiers and/ or foaming agents, DEA and TEA are ammonia compounds that can form cancer-causing nitrosamines when they come in contact with nitrates. They are toxic if absorbed into the body over a long period of time.

- *Sodium lauryl/laureth sulfate*: A detergent used in shampoos for its cleansing and foam-building properties. Often derived from petroleum, it is frequently disguised in pseudonatural cosmetics with the phrase "derived from coconuts."
- *Stearalkonium chloride*: Developed by the fabric industry as a fabric softener, SKC is a quaternary ammonium compound used in hair conditioners and creams to add shine and improve manageability. It is a lot cheaper and easier to use in hair conditioning formulas than proteins or herbs, which are beneficial to the hair.
- *Synthetic colors*: Used to make cosmetics attractive to buyers, synthetic colors along with synthetic hair dyes will be labeled as FD&C or D&C, followed by a color and a number. Example: FD&C Red No. 6/D&C Green No. 6.

One must take care in drawing conclusions that using cosmetics and personal care products will make you sick. Risks in humans associated with low-level exposure to toxic chemicals are notoriously difficult to assess, especially serious risks such as cancer and birth defects that may occur with long-term exposure. It takes considerable investment in research to determine whether exposure to a chemical causes serious and measurable harm. Even rigorous scientific research can be challenged, since it is usually impossible to prove cause and effect. Not all people exposed to a product become ill. Moreover, people exposed to a product and who become ill could have also been exposed to many other offending substances or have a genetic predisposition to the condition. This casts doubt on the true risks of exposure to the chemical in question and limits the certainty with which we draw conclusions. Nevertheless, informed consumers may be able to live with this uncertainty and take a conservative approach, opting to reduce their exposures to potential toxic chemicals.

Simple Solutions

Without a Ph.D. in chemistry, or misplaced faith in the beneficence of industry, the informed consumer can try to determine which products are safe to use and protect him or herself from excessive exposure to potentially harmful chemicals by doing the following:

- *Do more than read the label*. The cosmetic industry has no clear definition of the words *natural*, *pure*, or *safe*, nor are these words defined in law. Thus, manufacturers have considerable latitude in marketing and may use such terms as marketing ploys. Even products that claim to be organic can be suspect if they do not have a U.S. Department of Agriculture (USDA) seal on the label.
- *Consider reducing the number of products you use each day*. For example, one can avoid risks of respiratory irritation by eliminating use of body powder and aerosols such as hair spray and deodorant. Consider eliminating the use of nail polish and switch to makeup that does not contain harsh chemicals. Consider eliminating use of hair dyes or switch to a brand with fewer harsh chemical ingredients.
- *Make your own shampoos and body wash*. Products can be made from liquid, biodegradable castile soap. Organic essential oils or a few drops of almond or olive oil can be added to enhance emollient qualities for dry hair and skin.
- *Switch to a plain bar soap*. One can use plain soap (such as castile or Ivory) to avoid dyes, synthetic fragrances, antibacterial additives, deodorants, and other ingredients that you may not want to be exposed to on a daily basis.
- *Research*. The environmental working group Skin Deep has a searchable database containing 46,861 products and ratings on their ingredients. It helps consumers navigate the impenetrable amount of information on product safety and choose the best products for their needs.

See Also: Animal Products; Antibiotic Resistance; Personal Consumer Role in Green Health; Petrochemicals; Skin Disorders.

Further Readings

Campaign for Safe Cosmetics. http://www.safecosmetics.org (Accessed May 2010).
Cosmetic Safety Database by Environmental Working Group. http://www.cosmeticsdatabase .com (Accessed May 2010).
Organic Natural Health. http://www.health-report.co.uk/ingredients-directory.htm (Accessed May 2010).
U.S. Food and Drug Administration. "FDA Authority Over Cosmetics." http://www.fda.gov/ Cosmetics/GuidanceComplianceRegulatoryInformation/ucm074162.htm (Accessed May 2010).

Patricia Stephenson
Independent Scholar

PEST CONTROL

Pest control includes attempts to manage the population or location of a nuisance species. The species is considered a pest due to its impacts on the economy (including agriculture), human or animal health, or the local ecosystem. Pests may include a nuisance plant, such as a weed threatening to take over an agricultural field. But most often, the term *pests* refers to insects and animals. For nuisance animals, human behavioral changes can often act as a form of pest control. Other types of pest control include biological means, elimination of breeding grounds, physical destruction by fire, hunting/trapping, and chemical repellants and poisons. Of these methods, the development and use of DDT has had the greatest impact on human health. However, in recent years, less persistent pesticides have become more common.

When human activities cause pests to proliferate, the simplest solution is to change the human behavior. For example, when household or other organic waste is not properly disposed of, rats and other scavengers can thrive on this food stream. Careful management of household refuse controls localized rodent populations naturally by eliminating a food source. In fact, even if all the pests in an area are destroyed by chemical means, the same species will likely recolonize unless the local environment is modified to make food and shelter less readily available. Thus, even when other means of pest control are utilized, alteration of human behavior may still be necessary.

Humans can also sustain pests in the form of nuisance animals through intentional feeding. By feeding animals such as bears and raccoons, humans may cause the wildlife to become acclimated to interaction with humans. These animals will then be more likely to approach both humans and their possessions. This may result in a nuisance animal and may ultimately result in the animal's destruction. Education can play an important part in preventing humans from creating these nuisance animals, as many people are unaware that feeding animals can lead to nuisance creatures that may eventually be put down.

Biological control is a means of controlling pests by using natural predators. Insects are naturally controlled by predators, parasitoids, and pathogens. Weeds are naturally controlled by herbivores and plant pathogens, while plant diseases are naturally controlled by antagonists. Predators, such as the lady beetle, utilize their prey as a food source, reducing its population. Parasitoids, such as the wasp, lay their eggs inside a host organism, which is consumed by the developing parasitoid from the inside, most often resulting in death of the host. Pathogens are disease-causing microorganisms, such as bacteria. Biological control is usually used to fight an invasive species. However, it must be used carefully, as introduced biological control agents may also target native species and affect local biodiversity. Biological control agents that feed on or infect a wide range of species are not a good choice, and careful research must be done to select a control agent that will target only the pest that needs to be controlled.

Pests may also be controlled by eliminating their breeding grounds. Mosquitoes lay their eggs in standing water, and elimination of these breeding sources will reduce the number of mosquitoes in a localized environment. Small containers such as bottles and buckets can provide enough water to breed hundreds to thousands of mosquitoes. Eliminating the breeding grounds near homes can reduce the incidence of diseases transmitted by mosquitoes, such as West Nile virus or malaria. In developing countries in warm environments, efforts to prevent standing water, such as the building of soak pits for wash water away from wells, can significantly improve local health.

One of the oldest methods of pest control used by humans is physical destruction by fire. If a crop has been overcome by pests, a controlled burn of the field may prevent the pests from spreading through transported harvested crops and destroy any pests that may remain in the field post harvest. Field burning is used in the western United States to dispose of leftover materials after harvesting, maintain grass seed purity, and control the spread of insects and disease. Although it is effective for controlling pests, field burning may have fatal effects on humans and other wildlife. Smoke from field burning has drifted over roadways, causing traffic collisions. Smoke exposure effects in humans range from eye irritation, headaches, runny nose, and scratchy throat to bronchitis or death. It is very important that careful attention be paid to meteorological conditions and that the controlled burning be implemented when the conditions are optimal for smoke dispersal. Controlled burning may be used for both insects and crop pathogens but is not usually used for controlling animals that may flee a burning field on foot.

Animal pests may be controlled by trapping or hunting. The two basic types of traps are live traps and killing traps. Killing traps usually have a spring-loaded mechanism that activates an arm that crushes the animal's spine or head. Live traps come in various forms, some of which only capture an animal's foot or paw, while others enclose the whole animal in a cage. For either type of trap, food is often used as bait. Trapping mice and rats in homes is a common practice to deal with the pests in a chemical-free way. Killing traps are often used for this purpose. However, traps must be placed carefully to avoid being triggered by pets or children. Live traps are also commonly used to catch nuisance animals for relocation in areas where pets may be likely to eat poisoned bait and the human population is too dense to safely use firearms.

Hunting in the United States generally includes the killing of an animal by firearm. However, in other parts of the world, dogs, or even farm tools may be used to chase down animals feeding on crops. To offer incentive for the destruction of specific species by hunters and trappers, bounties may be offered by local governments. Bounties have been blamed for the reduction in specific species in the past, such as wild cats and wolves in the United States. However, bounties are not a thing of the past and today are most often offered when a specific species maintains a persistent risk to local agriculture or biodiversity. In the early part of the 21st century, a controversial fox bounty was debated in Victoria, Australia.

Chemical repellants and poisons are means of controlling pests that may cause the most long-term damage to human health and the environment. DDT, or dichlorodiphenyltri-chloroethane, is one of the most infamous pesticides in production today. DDT use as a pesticide began in World War II to control malaria and typhus. After the war, it was implemented as a highly effective insecticide for agriculture and became widely used.

In 1962, Rachel Carson wrote the book *Silent Spring*, in which she argued that DDT effects were not limited to pests, and that it also killed beneficial insects, birds, and was likely a human carcinogen. This book has often been credited with the birth of the modern environmental movement in the United States. Public outcry led to the banning of DDT for agricultural use in 2004 under the Stockholm Convention on Persistent Organic Pollutants (POPs are chemicals that do not breakdown easily in the environment but bioaccumulate within plants and animals) but still allowed its use for the control of disease-carrying insects. However, DDT may still be used in countries that have not signed the Stockholm Convention, such as North Korea. Additionally, it is an important tool for malaria control in Africa.

Concern for these persistent organic pollutants has led to the development of chemical pesticides that are easily biodegradable, called biopesticides. Some of the most well-known biopesticides come from the insecticidal proteins produced by the bacterium *Bacillus thuringiensis* (Bt). Various strains exist, each producing a unique insecticidal protein or delta-endotoxin. Some of the Bt-derived biopesticides are as effective as organophosphate pesticides. But organophosphates are not selective in their effects, while the Bt biopesticides' effects are specific to certain harmful insects. In Europe, government limits exist on the amount of pesticide residue in food, but due to their comparative safety, most biopesticides have no residue limit.

See Also: Biological Control of Pests; Chemical Pesticides; Environmental Illness and Chemical Sensitivity; Genetically Engineered Crops; Home-Grown Food; Malaria.

Further Readings

American Mosquito Control Association. "Mosquito Prevention and Protection." http://www.mosquito.org/pdfs/Mosquito-Prevention-Facts.pdf (Accessed July 2010).
Biopesticide Industry Alliance. "Residue Management." http://www.biopesticide industryalliance.org/benefitsresidue.php (Accessed July 2010).
Carson, Rachel. *Silent Spring.* Boston: Houghton Mifflin, 1962.
Neppl, Camilla C. "Management of Resistance to *Bacillus Thuringiensis*." Environmental Studies Program, University of Chicago, May 26, 2000.
State of Oregon Department of Environmental Quality. "Fact Sheet: Open Field Burning in the Willamette Valley (2007)." 07-AQ-019a.
Stockholm Convention on Persistent Organic Pollutants. http://chm.pops.int/default.aspx (Accessed July 2010).
U.S. Department of Agriculture and Forest Service North Central Forest Experiment Station. "How to Live With Black Bears." NC-HT-66.
Weeden, Catherine, Anthony Shelton, and Michael Hoffmann. "Biological Control: A Guide to Natural Enemies in North America." Cornell University. http://www.nysaes.cornell.edu/ent/biocontrol (Accessed July 2010).

Michelle Edith Jarvie
Independent Scholar

PETROCHEMICALS

Petrochemicals are derived from a variety of sources, including coal and natural gas, though the majority of the petrochemicals we use today are made from petroleum. Understanding how petroleum—also known as *crude oil*—is transformed into petrochemicals begins with a basic understanding of petroleum itself. Petroleum is found deep in the Earth's crust, beneath land and ocean alike, and it is formed from decaying plant and animal matter. While not considered a chemical compound in and of itself, petroleum is instead composed of scores of individual chemical compounds, all of which contain carbon and hydrogen atoms, which are referred to as *hydrocarbons*. These hydrocarbons can either be simple or quite complex, ranging from a few atoms to nearly 100 atoms. Once petroleum is extracted from the Earth, it is sent, via pipeline or ship, to refineries. Through various physical and chemical processes, such as fractional distillation—which is the heating of petroleum to different temperatures in order to "boil off" specific hydrocarbons—refineries alter the petroleum so that the more complex hydrocarbons can be separated from the less complex ones. The resulting compounds are known as *fractions*, and it is from these fractions that petrochemicals are derived.

Petrochemicals are derived from coal, natural gas, and petroleum and converted into useful compounds in refineries like this one. A mere handful of petrochemicals contains the building blocks for creating more than 4,000 other compounds and products.

Source: iStockphoto.com

These initial fractions, often referred to as *feedstocks*, include gasoline, kerosene, diesel oil, lubricating oils, and heavy gas oils. In the fractional distillation tower, the lighter, shorter-chained hydrocarbons, like gasoline and kerosene, make their way to the top, while the heavier, longer-chained hydrocarbons settle near the bottom. These heavier hydrocarbons—the lubricating and heavy gas oils—are converted into other chemical compounds through a process known as *cracking*. Cracking uses heat and steam, and sometimes a catalyst, to alter the molecular structure of these heavier fractions into an assortment of simpler fractions, or *petrochemicals*; these petrochemicals now become the feedstocks used for creating an astounding array of other chemical compounds and products. Petrochemical feedstocks fall into one of three following classes:

- *Olefins*: propylene, ethylene, and butadiene
- *Aromatics*: benzene, toluene, and xylene
- *Synthesis gas*: used to make ammonia and methanol

This handful of petrochemicals, in combination with each other or with other chemicals, contains the building blocks for creating more than 4,000 other compounds and products.

In the nearly 100 years since petrochemical manufacture began in earnest, these petroleum-derived compounds have profoundly impacted almost everything we interact with in our daily lives, including clothes, food, cosmetics, plywood, water hoses, food packaging, artificial limbs, paint, disposable diapers, crayons, water and soda bottles, toothpaste, flooring, blankets, building insulation, heart valves, candles, fertilizers, surgical gloves, kitchenware, toys—the list goes on and on. It is no exaggeration to say that petrochemicals are found in nearly every conventionally made product found inside and outside our homes.

History

Recovering petrochemicals from crude oil and manipulating the molecular structure of hydrocarbons was under way by the 1860s, by which time crude oil had been discovered in the United States, giving rise to refineries that used fractional distillation to recover kerosene for heating. Though initially ignored by oil refineries as useless, by about 1900, as electric lighting replaced kerosene and as automobiles became more and more prevalent, gasoline soon gained demand, and by 1913, refineries were utilizing thermal cracking to increase the yield of gasoline recovered in fractional distillation. Not wanting to waste the chemical by-products created in the cracking process, oil refineries began to produce petrochemicals.

World War I helped to nurture the growth of the petrochemical industry, particularly with the ever-increasing demand for gasoline to power ships and planes; further, it was during this era that the British began extracting benzene and toluene from crude oil. By the time of World War II, the petrochemical industry in the United States was rapidly expanding in order to meet rising demand for petrochemicals, and by the 1950s and 1960s, European scientists were synthesizing chemicals that could replace natural products. By the 1980s, petrochemical costs were rising along with the prices of oil and gas; it was also during that same time period that Middle Eastern countries began to develop their own petrochemical plants. Some of the petrochemicals that we most often come into contact include the following:

- *Ammonia*: An irritant that affects the skin, eyes, and respiratory passages, adds nitrogen to the environment, which can disrupt the ecosystem, and is included as a toxic chemical on the Environmental Protection Agency (EPA) Community Right-to-Know list. It is found in conventional window cleaners.
- *Benzalkonium chloride*: A synthetic disinfectant and bactericide that is biologically active. Widespread indiscriminate use of bactericides is linked to the emergence of new strains of bacteria that are resistant to them. It is found in conventional spray disinfectants, disinfecting cleaners, hand soaps, and lotions.
- *Benzene*: Benzene is classified by the International Agency for Research on Cancer as a carcinogen, is listed in the 1990 Clean Air Act as a hazardous air pollutant, and is on the EPA's Community Right-to-Know list. It is found in conventional oven cleaners, detergents, furniture polish, and spot removers.
- *Chlorine*: Chlorine was used as a powerful poison in World War I; is the household chemical most frequently involved in household poisonings in the United States; ranks first in causing industrial injuries and deaths resulting from large industrial accidents; and is found in conventional laundry bleach, dishwasher detergent, scouring powders, and basin, tub, and tile cleaners.
- *Diethanolomine*: The chemical is linked with kidney, liver, and other organ damage according to several government-funded research studies; has been proved to cause cancer in rats when applied to the skin; builds up in fatty tissues of the liver, brain, kidneys, and

spleen; and is found in over 600 home and personal care products, including soaps, detergents, and surfactants.

- *Ethylene*: Ethylene is an irritant with links to breast cancer and bone cancer; is known to cause damage to cardiopulmonary system; and is found in wires, cables, packaging containers, and various plastic items.
- *Perchloroethylene*: This chlorinated solvent is used most commonly in the dry cleaning process; is implicated in 90 percent of all groundwater contamination; and is found in conventional degreasers, spot removers, and dry cleaning fluids.
- *Propylene glycol*: Propylene glycol is implicated in kidney damage and liver abnormalities; can cause nausea, headache, and vomiting; and is found in antifreeze solutions, brake and hydraulic fluids, solvents, pet foods, processed foods, cosmetics, toothpastes, shampoos, deodorants, and lotions.
- *Sodium lauryl sulfate (SLS) or sodium laureth sulfate (SLES)*: These compounds are capable of altering the genetic material found in cells, corroding hair follicles, and damaging the immune system, and are found in concrete floor cleaners, engine degreasers, car wash detergents, and nearly every soap and shampoo sold.

Plastics

And then there's plastic—the most widely used synthetic in the world. The most well-known plastic products are Teflon, Tupperware, nylon, synthetic rubber, and PVC. The manufacture of plastics unintentionally forms by-products such as dioxins, which are linked to cancer and endocrine disruption. The National Institutes of Health has determined that endocrine disruptors can leach out of plastics and into our bodies, accumulating in our body fat. Plastics also leach into the foods they contain, such as meats and cheeses, so that we also "consume" plastic without knowing it. Phthalates are also found in plastics, such as toothbrushes, toys, and food packaging, as well as in aspirins and cosmetics, and they also act as endocrine disruptors and are potentially carcinogenic. The Centers for Disease Control and Prevention (CDC) has found alarming rates of phthalates in urine and blood samples.

Children are particularly at risk of petrochemical exposure, and not just because toys and other products contain petrochemicals—human breast milk is loaded with more than 100 industrial chemicals, including dioxins and pesticides. Some argue that it would not even be approved for sale by the FDA because of its alarmingly contaminated nature. Unfortunately, many of the alternatives to breast milk are even more toxic.

With petrochemicals showing up in everything from food to water to building materials to cleaning products to cosmetics to toys to breast milk, knowing the impact of these chemicals on human and ecological health is incredibly important. Petrochemicals can be absorbed through the skin and scalp, and, once absorbed, they can affect human organs and tissues, leading to nerve, liver, and brain damage as well as asthma, birth defects, and cancer. Further, petrochemicals are released into the ground, water, and air, which can profoundly affect both human health and the health of the environment. Citizens, watchdog groups, and even petrochemical manufacturers themselves acknowledge the potentially harmful nature of petrochemicals, though, as of yet, the U.S. government has not tested and regulated petrochemicals as diligently as many would like. While the EPA, the Food and Drug Administration (FDA), and the Occupational Safety and Health Administration (OSHA) all play a part in regulating, to a degree, petrochemicals and the products that contain them, there is a growing movement of those who believe that these agencies do not go far enough in protecting people and the environment from the potentially harmful effects of petrochemicals.

For example, most of the conventional household cleaning products used today are toxic. They leave behind chemical trails and traces that stay in our homes for hours, and, sometimes, days. Further complicating matters is the fact that we often unintentionally mix cleaners—the glass cleaner we spray on the bathroom mirror falls onto the sink where it reacts with the cleaner we use to scrub that area, and these unintended combinations actually react to create another toxic compound. Researchers theorize that the average house contains gallons and gallons of toxic materials—and the fact that there is no law that requires manufacturers to list their ingredients or to test their products for safety means that protecting our health is up to each individual consumer. Though many cleaning products list warnings of being poisonous or toxic, most of us assume that those warnings only apply if one is not using the product properly or if it is accidently ingested; as it turns out, most cleaning products are toxic even when used "safely."

Petrochemicals in Foods

Petrochemicals are also found in many of the foods we eat, such as frozen and canned products. Further, fruits and vegetables grown conventionally are loaded with petrochemical-based fertilizers and with chemicals designed to make fruit and vegetables more appealing, that is, more "shiny." Many of the most commonly used pesticides are classified by the EPA as possible carcinogens, and many have been shown to cause damage to the nervous, reproductive, and immune systems in laboratory animals. EPA regulations of pesticides do not consider the potential effects that cumulative, low-level exposure can cause. Further, the EPA ignores the combined effects of simultaneous exposure to multiple chemicals. Tap water is treated with petrochemicals including chlorine and chlorine dioxide, both of which can form chloroform, which has been linked to liver and nervous system damage and kidney failure. Many who do not trust the safety of their tap water buy bottled water, though not only are regulations lax concerning such water, the bottles that contain it are made of petrochemicals.

The U.S. government does not require makers of bath and hygiene products to disclose their ingredients, yet toothpaste, mouthwash, soap, shampoo, and other personal care products are loaded with petrochemicals. Hair dyes, both permanent and temporary, contain petrochemicals that not only damage the hair and have the potential to cause allergic reactions, but that are harmful to the body in other ways. And, as with cleaning products, no one knows the long-term effects of mixing these products—for example, shampoo followed by styling gel. Even more frightening than these products that we put on our bodies are the products that we interact with more intimately, such as tampons. Tampons are made from rayon, which is chlorine-bleached wood pulp, or from cotton that has been treated with DDT. Using pads that are not subjected to chlorine bleaching is considered much safer than using tampons. Also as with cleaning products, choosing personal care products that are made from natural, nontoxic ingredients, and that list their ingredients on their label, is the safest way to avoid the potential consequences of petrochemicals.

Another petrochemical, polyvinylchloride (PVC), is also considered toxic and is used in many applications, including children's toys. Phthalates are also used on toys, and the fumes that they emit are toxic. The Consumer Products Safety Commission has asked that toy makers stop using PVC, but, thus far, the market is still loaded with PVC toys. Children today are exposed to a dizzying array of petrochemicals and their by-products, and the cumulative effects of these compounds have yet to be unraveled.

Petrochemicals have also contributed to our quest for cleaner, brighter clothes. To that end, detergent manufacturers have embraced the use of "optical brighteners" in their laundry soaps. In truth, though, these brighteners only create the illusion that clothes are brighter. Optical brighteners work by absorbing ultraviolet light and emitting it back as blue light, which masks yellowing, making clothes appear brighter and cleaner, when, in fact, their dinginess is simply masked. These brighteners are known to create skin reactions and are not only nonbiodegradable, but are toxic to the environment. Still, in our seemingly never-ending pursuit of efficiency and cleanliness, even the illusion of brighter clothes is considered by some to be a sign of progress.

Another mark of progress has been the advent of more energy-efficient homes that save money on utilities; unfortunately, these homes are so tightly constructed that fresh air flow is impeded—studies have found that air inside the average home is two to five times more polluted than is outside air. Cleaning products, personal care products, paint fumes, and furniture—all of which contain petrochemicals—fill our homes' air with toxic fumes. Carpet is an especially dangerous emitter of toxic petrochemicals; the carpet backing, its pad, and the glue used to secure it are all made from petrochemicals. Furniture made from materials treated with formaldehyde emits toxic fumes for as many as five years after purchasing it.

Progress, then, is a relative idea, and our notions of it are informed by the beliefs that petrochemical-based products are helping to keep us and our homes cleaner, that petrochemical-based products are not only useful in creating our built environment and material possessions but that they make our lives easier, and that petrochemical-based products are the evidence of our dominion over science and chemistry. In order to accept petrochemical-based products as a part of our daily lives, we must also accept the idea that the daily cocktail of chemicals that we come into contact with does not affect our bodies. Converse to this idea is the fact that petrochemicals are linked to increases in cancers, endocrine disruption, asthma, and environmental illness. Further complicating matters is the fact that most people believe that the government is protecting us from harm. This notion is debatable, particularly in light of the fact that the majority of the more than 2,000 chemicals that come onto the market every year are not subjected to even the simplest tests to determine toxicity, nor are these chemicals tested in order to determine how they react with each other and with our bodies.

What Can Be Done?

Those who are alarmed by our reliance on petrochemicals and on their prevalence in almost everything with which we interact offer the following guidelines for diminishing the negative effects of petrochemicals:

- Allow as much fresh air into your home as possible.
- Use paints that are low in volatile organic compounds (VOCs).
- Buy furniture with wood frames and cotton or wool cushions.
- Try to avoid dry cleaning (if you must dry clean your clothes, allow them to air out for three to four days before wearing them).
- Try to find toys that are made from polyethylene or polypropylene—not PVC.
- Burn beeswax or vegetable wax candles as opposed to paraffin, which is petroleum-based.
- Use an air purifier and a vacuum cleaner fitted with a HEPA (High Efficiency Particulate Air) filter.
- Use natural cleaning products.
- Buy vegetables and fruits that are organic.

While petrochemicals are clearly almost impossible to avoid, most people are not exposed to doses large enough to kill immediately; unfortunately, the effects of the type of chronic exposure to which we are all subjected remain, for the most part, unknown. Many argue that the U.S. government needs to mandate full disclosure of products that contain petrochemicals and that their known and suspected effects also need to be disclosed. Without pressure from the public, many doubt that the government will make any moves to further regulate the petrochemical industry. There are more than 75,000 chemicals registered with the Environmental Protection Agency, yet only a handful have actually been thoroughly tested in regard to their effect on human health and the environment. Much like guilt in the American justice system, petrochemicals are assumed to be safe until proved dangerous; the problem is that this "proof" often arrives in the form of workplace injury or wildlife poisoning or environmental damage—rigorous testing before allowing the use of petrochemicals makes much more sense than the blind faith that seems to guide regulatory system guidelines in relation to petrochemicals.

In order to meet the challenge of protecting human and environmental health, many cite the Precautionary Principle as a relevant approach to safety. Using this principle, it would be up to petrochemical manufacturers and end-use manufacturers to bear the burden of proof in demonstrating that products are safe. This process must be open and democratic, involving all parties who stand to be affected—including, of course, consumers and the American public. As we have embraced more and more petrochemical-based products, many that are known carcinogens, it makes deadly sense that cancer rates have more than quadrupled over the past 100 years. It remains to be seen what effects the next 100 years of reliance on petrochemical-based products will bring.

See Also: Cancers; Centers for Disease Control and Prevention (U.S.); Chemical Pesticides; Children's Health; Environmental Illness and Chemical Sensitivity; Environmental Protection Agency (U.S.); Fertilizers; Plastics in Daily Use.

Further Readings

Centers for Disease Control and Prevention. "National Report on Human Exposure to Environmental Chemicals." http://www.cdc.gov/exposurereport/chemical_information .html (Accessed August 2010).

Ecology Center. "The True Costs of Petroleum: Body Map." http://www.ecologycenter.org/ erc/petroleum/body.html (Accessed August 2010).

Thomko Petro Chemical Blog. "What Are Petrochemicals?" http://thomko.squarespace.com/ what-are-petrochemicals (Accessed August 2010).

Tani Bellestri
Independent Scholar

Pharmaceutical Industry

The pharmaceutical industry creates products that ameliorate and cure many diseases and conditions that affect the health of humans. The process of creating pharmaceuticals and bringing them to market includes several steps in research and development, clinical tests, manufacturing, packaging, marketing, and distribution. As part of this process, the pharmaceutical industry uses many chemicals and processes that create wastes, chemical syntheses, and by-products that can adversely impact the environment. The pharmaceutical industry is widely considered to be one of the most highly regulated existing commercial markets, with governmental oversight extending into aspects of research, product development, manufacturing, marketing, distribution, and sales. Until recently, however, much of this regulation has focused on the safety and reliability of the pharmaceuticals manufactured and not on the environmental impact of the process. The green movement has impacted the pharmaceutical industry in that it seeks to reduce the amount of negative ecological impact resulting from the synthesis, manufacture, packaging, and distribution of drugs that are approved as medications. As environmental concerns have attained increasing attention, governments, private industry, and consumers have all sought ways to ensure that the pharmaceutical industry engages in sustainable practices when producing items for consumption.

While many cutting-edge developments within Western medicine have greatly improved the general quality of life for a significant portion of the global populace, pharmaceutical manufacturers have been accused by environmental activists and sympathizers of burying over 500 million pounds of research-grade chemical compounds, and releasing over 250 million pounds of toxic waste into public waterways, thereby contaminating drinking water supplies. Also contributing to this problem is the over-prescription of certain medications and use of drug therapy treatments in place of, rather than as a supplement to, behavioral therapy practices. Though the effects of these discarded drugs on humans are not known, such contaminations can adversely affect plant and animal species in the natural environment. Once powerful antibiotics and other medications are excreted from the human body, they may still have the potential to be damaging to the drinking water supply and natural biosphere. These practices, along with the perceived influence of pharmaceutical industry lobbyists advocating stronger patent laws and tax benefits within political spheres, have helped foster some criticism and skepticism directed toward Western medicinal practices among the general public. Nevertheless, many individuals and institutions continue to place faith in the pharmacological development of new drug treatments, as advances within Western medicine now render manageable many diseases once considered fatal.

The pharmaceutical industry comprises a collection of international corporations that develop, manufacture, patent, license, and market drugs that possess positive medicinal attributes. Unlike many other industries, pharmaceutical manufacturers are regulated by government agencies to ensure consumer health and safety. To this end, government regulators such as the U.S. Food and Drug Administration (FDA) promulgate rules that are intended to protect and promote their citizens' health and well-being. Government regulators require the pharmaceutical industry to demonstrate that their products are both safe and effective before their sale to consumers is permitted. As a result, companies within the pharmaceutical industry must carefully work with regulators and comply with their directives to ensure consumer safety as well as ethical businesses practices. Failure to comply with the regulatory system may result in fines, some of which are for hundreds of millions of dollars, and open the manufacturer to potential litigation.

Considerations of both consumer safety and ethical issues pertain to social and environmental considerations. Regulation is of paramount importance, and as a result, such resolutions are staffed by teams to prevent decisions being made by individuals, with consensus on determinations being a goal. Decisions to approve or disapprove a drug can have huge repercussions with regard to a manufacturer's viability. Strict adherence to the regulator's processes is required on the part of manufacturers who wish to have a drug approved. Due to the burdens associated with this adherence, however, businesses within the pharmaceutical industry often feel they are subject to the whim of sometimes-shifting definitional standards. Regulations that seek to make the process of pharmaceutical manufacture greener are often viewed by pharmaceutical corporations and lobbyists as largely intrusive and unnecessarily complex. As a result of this, the promotion of green practices within the pharmaceutical industry must reckon with these objections and the subsequent obstacles placed along the path toward the adoption of sustainable practices by interests within the pharmacological field hoping to avoid any extra expenditures in the processes of their commerce.

Challenges of Green Practices

Green practices often are challenging for the pharmaceutical industry to adopt because many such practices conflict with other considerations designed to protect the health and safety of consumers. For example, regulators and the public have the expectation that medication will be able to withstand degradation resulting from exposure to time and the elements. Attempts to make pharmaceutical packaging more environmentally friendly often seek to reduce the amount of packaging used. Green packaging practices often encourage the use of benign, organic compounds and solvents that will easily degrade without causing damage to the ecosystem. However, this type of packaging, some have speculated, may be incompatible with the need to protect medications from degradation, and this is one of the many difficulties associated with encouraging the adoption of green practices within the pharmaceutical industry. This is but one example of the difficult and sometimes contradictory expectations placed upon the pharmaceutical industry. Environmental considerations must be balanced with those of the consumer. This fundamental problem underlies many decisions relating to whether to pursue green chemistry and engineering within the field. Pharmaceutical manufacturers are forced to negotiate these complicated issues and to try to make happy multiple constituencies who are motivated by goals and objectives that are sometimes very different. Companies also have difficulty justifying to consumers their decision to use anything but the least expensive possible resources in producing life-saving medications, especially in light of the constant criticism regarding the cost of their products. When this occurs, however, these same companies are berated for their contribution to the destruction of the natural biosphere.

Two terms commonly associated with the push for green practices within the pharmaceutical industry are *green chemistry* and *green engineering*. These seemingly self-explanatory terms, as they pertain to the pharmaceutical industry, describe the use of environmentally friendly procedures and resources in the design and production of both medicinal drugs and their packaging. Green chemistry refers to an approach that reduces and prevents pollution at its source. Pharmaceutical companies that adopt green chemistry principles therefore strive to prevent waste, design safer chemicals and products, devise less hazardous chemical syntheses, use renewable materials, employ catalysts rather than stoichiometric reagents, avoid chemical derivatives, maximize atom economy, apply safer solvents and reaction conditions, increase energy efficiency, fabricate chemicals and products to degrade after use, analyze in real time to prevent pollution, and minimize the potential for accidents. Using green chemistry allows the pharmaceutical industry to adopt practices that are as environmentally friendly as possible while also ensuring consumer safety.

Green engineering focuses on steps designers need to take to ensure minimal environmental harm from products they help to manufacture. Steps that can be taken to help support green engineering include the following:

- Using as many nonhazardous materials as possible
- Preventing rather than treating waste
- Designing separation and purification operations to minimize energy use
- Maximizing mass, energy, space, and time efficiency
- Ensuring that products, processes, and systems are output pulled rather than input pushed through use of energy and materials
- Conserving complexity through embedded entropy and other choices
- Making durability rather than immortality a design goal
- Meeting needs while minimizing excess capacity or capability
- Minimizing material diversity to promote disassembly and value retention
- Integrating material and energy flows in all design
- Choosing products, processes, and systems for commercial afterlife
- Selecting material and energy inputs that are renewable rather than depleting

While such a focus would require the pharmaceutical industry to alter some of its engineering processes, doing so at the beginning of the process would allow a minimization of environmental harm.

Another challenge associated with the adoption of environmentally considerate practices within the pharmaceutical industry is the predominant belief that the importance of developing adequate medications is so great that any environmental degradation caused by the industry should be tolerated by society as just a small price to pay. In order to overcome this issue, companies must negotiate complex issues of priority. As the pharmaceutical industry has made assessments of competing interests throughout its history, considering environmental issues is merely another instance of making decisions that balance multiple stakeholders' claims. The history of pharmacology and the pharmaceutical industry is inextricably linked with the natural environment. Indeed, nearly all major breakthroughs in the field come from the isolation of active compounds found within naturally occurring plants and animals. One development that has caused considerable public controversy and uproar among environmentalist groups involved a treatment for ovarian cancer.

In 1964, extracts from the bark of the Pacific yew, a tree native to the old-growth forests in the northwestern coastal regions of the United States and in British Columbia, were tested on cancer cells and found to show promising signs for use in treatment. In 1969, whereupon a large-enough amount of extracts were prepared, the isolation of the active compound taxol was possible. After decades of study and clinical trials, positive effects were reported. The marketing of taxol as a treatment for ovarian cancer began in 1992. Originally, the Pacific yew bark was the only known source of this chemical compound. Environmental groups claimed that the harvesting of the yew's bark would endanger its survival, since the treatment for just one single patient necessitated the collection of bark from between three and 10 century-old plants. It was also claimed that this extraction might cause serious ecological degradation to the region and potentially destroy much of the habitat of the endangered spotted owl. Eventually, scientists discovered a way to convert an organic precursor found in a more renewable source into taxol.

Animal Testing

While environmentalist groups feared, in this instance, that the harvesting of bark might have secondary ramifications for the environment and endangered species, there are many instances of revolutionary advancement within the pharmaceutical industry that have been manifested only alongside the direct expenditure of animal life. Animal testing has historically been viewed by science as a necessary stage in the preliminary testing of newfound pharmaceuticals. Every year, as many as 100 million vertebrate animals are used worldwide in scientific experiments, after which most are euthanized. Nongovernmental animal rights activist groups such as People for the Ethical Treatment of Animals (PETA) and the British Union for the Abolition of Vivisection (BUAV) criticize the use of nonhuman animals in scientific testing, claiming the practice is too costly, inadequately regulated, of dubious scientific merit, and a violation of the fundamental rights of any living creature. Proponents of animal testing argue that nearly every significant medicinal breakthrough in the 20th century relied upon animal testing.

For example, as the identification, definition, and synthesis of biological hormones began to occur in the second half of the 19th century when, in 1889, German scientists Oskar Minkowski and Joseph von Mering discovered in experiments that by cutting the pancreas from a dog, significant elevated blood glucose levels and metabolic changes occurred, which are symptoms akin to those of human diabetes mellitus. This discovery facilitated the identification of insulin, because due to this instance of animal testing, the function of islet cells was learned. This discovery has assisted in the production of extracted pancreatic insulin medicaments used in the treatment of diabetes, which began in 1922. For the next 50 years, most of the insulin used in the treatment of diabetes patients was extracted from the pancreases of pigs and cows. While the difference in chemical structure between the species' insulin was not significant, due to advances in recombinant deoxyribonucleic acid (DNA) technology, most insulin sold by pharmaceutical companies on the market today is synthetically produced and is identical to human insulin.

For much of the 20th century, the majority of pharmacological screenings hinged solely upon the use of whole animals such as rats and mice. Nobel laureate Paul Ehrlich conducted chemical screening on mice, which led to the discovery of arsphenamine, the first effective drug used in the treatment of syphilis. Animal rights activism and protests became a contributing source of increasingly strenuous regulatory pressures upon designers and manufacturers of medicinal drugs toward the end of the 20th century. The concerns of these animal rights activists are balanced by the needs of those suffering from chronic and other debilitating diseases who seek drug treatment for their conditions.

The average patron of the pharmaceutical industry who is interested in the promotion of environmentally considerate practices within the field is often more concerned with a general reduction in carbon footprint and contribution to global warming. However, these and similar instances of animal testing as well as instances of environmental extraction serve to underscore a prevailing attitude within the pharmaceutical industry. The foundation and history of the pharmaceutical industry clearly demonstrates a primary interest in the preservation of human life. This interest is typically given higher priority by the pharmaceutical industry than the interests of the natural biosphere or animal rights.

Historically, the environmental movement readily has demonstrated that most frequently the trigger for the successful augmentation of and advocation for ecologically sustainable practices may be directly attributed to concerns regarding the safety of humans. While there certainly exists no shortage in the amount of rhetoric appealing to the promotion of animal rights and environmental restoration efforts, these movements have gained the most momentum following the exponential increase in the belief that if sustainable practices are not adopted, human survival will become threatened. Hence, some of the greatest catalysts for the adoption of sustainable practices within the pharmaceutical industry are instances of fomenting public fears, such as those related to reports of trace levels of pharmaceuticals found in drinking water.

Conclusion

While competition between private pharmaceutical companies has largely driven the discovery, development, and production of technological advancements within the field of pharmacology, cooperation between private, public, and governmental interests is necessary to facilitate the advent of and adherence to environmentally sustainable practices. At the time of this writing, there are over 10,000 different drugs being sold in the pharmaceuticals market. The vast majority of these drugs are not made by methods that could be considered green. Corporations seeking to adhere to the principles of environmentally conscious manufactured pharmaceuticals must prevent waste, design safer substances with little or no toxicity, and use renewable matter in the syntheses of medicinal drugs. Chemicals should be designed to break down to benign states in order to prevent environmental degradation. Packaging must be designed in a manner that provides for the least amount of excess waste as possible. To that end, although competition has driven many significant advancements within the field, cooperation between private and public interests is necessary in order to facilitate the distribution of adequate informational resources to enable the expedited identification of hazardous compounds.

Other general barriers that are causing delay to the widespread adoption of green practices within the pharmaceutical industry include the belief that the practice of green chemistry in the design of medicaments is too costly and arduous in implementation, the belief that simply meeting the established regulatory obligations ought to prevent a pharmaceutical corporation from having to maintain an environmentally conscious perspective in its commercial endeavors, as well as the widespread practice of outsourcing stages in the production, marketing, and distribution of pharmaceuticals. The unlimited outsourcing of various stages of design, manufacture, marketing, and distribution within the pharmaceutical industry produces unnecessary pollution and expenditure of natural and human resources. Multinational corporations are often criticized for their lack of obligation toward any one country; rather, investments are directed toward the cheapest labor demographic, and work is outsourced from one country to the next until all exploitable resources are consumed. While temporary fiscal earnings might be attributable to similar management tactics, the outsourcing of labor is inevitably no replacement for scientific innovation and ethical practices in commerce, which are informed by genuine concern for one's impact upon and contribution to the degradation of the Earth's biosphere.

As consumers and government regulators begin to demand a greener approach from the pharmaceutical industry, changes to make the process more environmentally friendly will occur. Adopting principles of green chemistry and green engineering early in the development process of new drugs would greatly promote green concerns, especially with regard to minimizing hazardous wastes and promoting recycling of facilities, equipment, and packaging. Certain tensions will always exist, however, as concerns for the environment are constantly balanced with the need and desire to develop effective cures for the many conditions and diseases causing human suffering. Similarly, with regard to animal testing, few decisions made will satisfy all constituencies. Outreach to and training for hospitals, long-term care facilities, and other end users of pharmaceuticals also is necessary to minimize the risks of hazardous wastes caused by improper disposal of drugs and other materials.

See Also: Advertising and Marketing; Environmental Illness and Chemical Sensitivity; Government Role in Green Health; Green Chemistry; Health Insurance Industry; International Policies; Pain Medication; Prescription Drug Addiction.

Further Readings

Alfonsi, K., J. Colberg, P. J. Dunn, T. Fevig, S. Jennings, T. A. Johnson, H. P. Kleine, C. Knight, M. A. Nagy, D. A. Perry, and M. Stefaniak. "Green Chemistry Tools to Influence a Medicinal Chemistry and Research Chemistry Based Organisation." *Green Chemistry*, 10/1 (2008).
Khetan, S. K. and T. J. Collins. "Human Pharmaceuticals in the Aquatic Environment: A Challenge to Green Chemistry." *Chemical Reviews*, 107/6 (2007).
Kümmerer, K. "The Presence of Pharmaceuticals in the Environment Due to Human Use—Present Knowledge and Future Challenges." *Journal of Environmental Management*, 90/8 (2009).
Slater, C. Stewart and M. Salveski. "A Method to Characterize the Greenness of Solvents Used in Pharmaceutical Manufacture." *Journal of Environmental Science & Health*, 42/11 (2007).

Stephen T. Schroth
Jason A. Helfer
Daniel O. Gonshorek
Knox College

PHARMACEUTICAL INDUSTRY REFORM

The prospect of healthcare reform in the United States was a hot topic of debate during the 2008 presidential election, and as of 2010, American citizens were still grappling with this complex and polarizing issue. Inherent in all discussions of how best to provide quality healthcare is the subject of pharmaceutical industry reform. Americans spend $200 billion on prescription drugs annually, and that rate is growing at more than 10 percent per year. Those most interested in reforming the pharmaceutical industry argue that drug companies have too much influence over regulatory laws, and these hopeful reformers take issue with the rising cost of prescription medicines, the unfair practices embraced by drug companies eager to rake in massive profits, and the aggressive marketing campaigns spearheaded by mammoth, profit-rich companies commonly referred to as Big Pharma.

After two years of fiery debate, healthcare reform legislation passed on March 21, 2010, and, for the most part, the pharmaceutical industry stands to benefit more than it will lose. Big Pharma has agreed to contribute $85 billion toward the bill's cost, which mostly will be distributed as industry fees and lower prices paid to pharmaceutical companies under the new program. In addition, Big Pharma will see tens of billions of dollars in increased revenue as more prescriptions are written for the huge influx of newly insured citizens. Further, because healthcare reform does not include drug price control measures or increased regulation, Big Pharma has largely embraced the legislation, spending about $100 million in marketing geared toward promoting reform.

Generic drug makers were not quite as lucky as name-brand drug makers, since healthcare reform gives name-brand manufacturers marketing exclusivity for 12 years. This, of course, is bad news for consumers, many of whom can only afford generic versions of name-brand drugs. Another win for name-brand manufacturers—and loss for consumers and generic drug makers: Congress omitted a provision from the legislation that would have placed new restrictions on patent settlement agreements between generic and name-brand drug manufacturers. Big Pharma defends these settlement agreements as helping to foster innovation.

The pharmaceutical industry claims that its efforts to create innovative drugs require massive spending on research and development, and they defend their business practices by citing free-market values and invoking notions of American free enterprise. In contrast, Marcia Angell's 2004 book, *The Truth About the Drug Companies*, describes a pharmaceutical industry rife with corruption and greed, bolstered by a complicit U.S. government that grants drug companies patents and exclusive marketing rights while creating a system that allows those firms to rely on taxpayer-funded research carried out at universities and the National Institutes of Health. In answer to Big Pharma's argument that research and development requires massive spending, Angell noted that research and development expenditures typically consume 10 to 15 percent of Big Pharma budgets, while marketing and administration costs devour close to 40 percent of the budgets. Many note that the aggressive marketing carried out by Big Pharma is necessary precisely because the market is flooded with so many similar drugs, with very little actual innovation.

Angell also noted that pharmaceutical industry profits are astoundingly high—in the United States, Big Pharma's profits hover around 20 to 25 percent of sales; in comparison, median profits for other Fortune 500 companies are less than 4 percent of sales. Angell pointed out that in 2002, the combined profits for the 10 drug companies in the Fortune 500 ($35.9 billion) were more than the profits of the other 490 businesses combined ($33.7 billion). The astounding financial success of the U.S. pharmaceutical industry began in the 1980s, fueled by the pro-business shift ushered in by the Reagan administration. Legislation passed under this administration allowed Big Pharma to begin relying heavily on taxpayer-funded research and made it easier to extend their patents and exclusive marketing rights. Beginning around 2000, things began to shift for Big Pharma, as employers, private insurers, and state governments began taking measures to cut prescription drug costs. The public also began to protest, especially when it became known that they pay more for prescription drugs than do Europeans or Canadians.

Ten years later, many Americans are still struggling to pay for their medications, and though healthcare reform may yet prove to be a boon to them, it remains to be seen how this legislation will truly affect them. Further, reformers are still hoping to see a reduction in prescription drug prices as well as at least some of the following reforms, which were not included in the healthcare reform legislation that passed:

- A focus on creating truly innovative drugs—currently, the FDA approves a drug if it is better than only a placebo, and testing of new drugs against existing drugs is not required for FDA approval
- More transparency in the pharmaceutical industry's research and business practices
- A loosening of Big Pharma's control over evaluation of their own products

See Also: Government Role in Green Health; Health Insurance Reform; Pharmaceutical Industry.

Further Readings

Angell, Marcia. *The Truth About the Drug Companies: How They Deceive Us and What To Do About It.* New York: Random House, 2004.

Carroll, Jamuna, ed. *The Pharmaceutical Industry.* Opposing Viewpoints Series. Detroit, MI: Greenhaven, 2008.

Pharmaceutical Research and Manufacturers of America (PhRMA). http://www.phrma.org (Accessed February 2010).

Tani Bellestri
Independent Scholar

Phaseout of Toxic Chemicals

In the 21st century, global efforts have emerged toward phasing out the use of toxic and environmentally persistent chemicals. The United Nations has led this effort, focusing on chemicals known to linger in the environment and those that have adverse effects on human and ecosystem health. Most of these chemicals were developed as insecticides but have been found to affect more than their target species.

The largest global action to phase out the use of toxic chemicals is the Stockholm Convention on Persistent Organic Pollutants. This action began in 1995 when the United Nations Environment Programme (UNEP) called for global action to address chemicals that linger in the environment, bioaccumulate, and may adversely affect animals and humans. As a response, a list of 12 chemicals, called the "dirty dozen," was developed. The original convention was negotiated in 2001 but was not ratified until 2004. There were 151 signatory nations that agreed to the elimination of the following chemicals:

- *Aldrin*: An organochloride insecticide used to treat seed and soil, aldrin has a similar structure to dieldrin. Aldrin quickly breaks down to Dieldrin in the environment and in bodies. It has been banned from use in the United States since 1978.
- *Chlordane*: An organochloride pesticide used on lawns, gardens, and for crops that was limited to termite control within the United States by the Environmental Protection Agency in 1983. It has been associated with breast, prostate, brain, and blood cell cancers. It is suspected that more people may suffer the noncancerous health effects of chlordane, which include migraines, respiratory infections, diabetes, anxiety, depression, and activated immune system.
- *Dieldrin*: A chlorinated hydrocarbon developed as a pesticide alternative to DDT, dieldrin has been associated with Parkinson's, breast cancer, and damage to the immune, reproductive, and nervous systems. It has been banned from use in the United States since 1978.
- *Endrin*: An organochlroide used to control insects and rodents, endrin has been associated with respiratory disease. Acute poisoning affects the human nervous system, and it has caused mass poisonings worldwide, with children especially susceptible. Endrin has not been produced in the United States since 1986.
- *Heptachlor*: Heptachlor is an organochloride insecticide used in homes, buildings, and crops. Since 1988, its use has been limited in the United States to the control of fire ants in

underground transformers. It is a possible human carcinogen and has been associated with liver tumors. When exposed in vitro or early infancy, heptachlor may affect the immune and nervous systems, result in lower weight, or even death.

- *Hexachlorobenzene*: A cholorocarbon fungicide used until 1965 to protect seeds or control wheat bunt (a fungal infection), hexachlorobenzene has other uses, including the manufacture of ammunition, synthetic rubber, and fireworks. In Turkey, in the 1950s, high death rates occurred among young children and nursing infants of mothers who ate grain exposed to hexacholorbenzene. It has been associated with damage to the liver, bones, kidneys, blood, and the immune, endocrine, and nervous systems when ingested over time.
- *Mirex*: Mirex is a chlorinated hydrocarbon used in the past to control fire ants and also as flame retardant (under the name Dechlorane) for electrical goods, paint, paper, rubber, and plastic. In 1978, the use and production of mirex was banned in the United States. Mirex has been associated with harmful effects to the liver, kidneys, stomach, intestines, eyes, miscarriage, and thyroid, nervous, and reproductive systems. Acute exposure may cause irritability, trembling, blurry vision, and headaches.
- *Toxaphene*: All uses of Toxaphene, an insecticide that contains more than 670 chemicals, formerly used to control insects on cotton, livestock pests, and undesired lake fish, were banned in the United States in 1990. It can damage the lungs, nervous system, adrenal glands, immune system, liver, and kidneys and can cause death. It is a probable human carcinogen and may affect fetal development.
- *PCBs (polychlorinated biphenyls)*: PCBs are a class of organic compounds including up to 209 possible congeners. PCBs were widely used as dielectric fluids in transformers, capacitors, and coolants. Their manufacture has been banned in the United States since 1977. PCBs are probably carcinogenic to humans and can cause cancer in animals. They have also been associated with adverse developmental, reproductive, dermatologic, hepatic, immunologic, and endocrine effects.

The nations also agreed to only utilize DDT (dichlorodiphenyltrichloroethane) for malaria control, and eliminate its use as a pesticide. DDT was used to control malaria and typhus during World War II. After the war, it was used widely as a pesticide. Rachel Carson's 1962 book *Silent Spring* proposed that DDT not only killed pests but also killed beneficial insects and birds and could cause cancer in humans. The resultant public outcry led to a 1972 ban of DDT for use in the United States. Due to its effectiveness at controlling mosquitoes, the Stockholm Convention agreed to limit its use to malaria control. Several developing countries had almost eliminated malaria in the 1960s through the use of DDT, but when use of DDT ceased, the disease returned. In 2006, the World Health Organization endorsed the use of DDT to control malaria. Limited application of DDT for malaria reduces environmental exposure and includes only spraying the inside walls of houses, and is supported by several environmental groups, such as the Sierra Club. Since 2006, the use of DDT in Africa is thought to be on the rise, and some scientists have expressed concerns of the possible health effects to humans exposed to DDT in their homes.

Another goal of the convention was to curtail the production of dioxins, PCBs, and hexachlorobenzene through unintentional means. These chemicals may be produced through the breakdown or interactions of other chemicals in the environment and the destruction of waste through incineration.

In addition to the 2001 agreements, the Stockholm Convention held a Fourth Conference of Parties in 2009, where it was agreed that another nine chemicals would be added to the original "dirty dozen." These nine include several insecticides and fire retardants, as well as the stain repellant formerly used in Scotchgard and voluntarily phased out by 3M.

Global consensus is important to prevent any one country from producing and disseminating a chemical. The Stockholm Convention provides a platform where the global community can reach consensus on the necessary actions regarding specific persistent environmental chemicals.

See Also: Biological Control of Pests; California's Green Chemistry Initiative; Cancers; Chemical Pesticides; Environmental Illness and Chemical Sensitivity; Fabrics; Green Chemistry; Pest Control.

Further Readings

Agency for Toxic Substances and Disease Registry. "Heptachlor/Heptachlor Epoxide: Toxicological Profile" (August 2007). http://www.atsdr.cdc.gov/ToxProfiles/TP.asp?id=746&tid=135 (Accessed July 2010).

Agency for Toxic Substances and Disease Registry. "Toxicological Profile for Chlordane" (May 1994). http://www.atsdr.cdc.gov/ToxProfiles/TP.asp?id=355&tid=62 (Accessed July 2010).

Stockholm Convention on Persistent Organic Pollutants. http://chm.pops.int/default.aspx (Accessed July 2010).

World Health Organization. "Malaria Programme." http://www.who.int/malaria/en (Accessed July 2010).

Michelle Edith Jarvie
Independent Scholar

PHYSICAL ACTIVITY AND HEALTH

The evolutionary history of humans suggests the influence of the strong adaptive pressures involved in the physical activities of hunting and gathering. Beginning around 10,000 B.C.E., agriculture made it possible for hunting-gathering tribes to obtain larger amounts of food yet remain in a smaller area. This symbolized the beginning of a more sedentary lifestyle.

After the emergence of cultural differentiation from 3,000 to 200 B.C.E., it became apparent that different cultures viewed physical activity differently. The ancient Chinese recognized the association of inactivity with certain diseases. They believed that regular physical activity could help prevent the disease process. Chinese citizens participated in Kung Fu, gymnastics, dancing, fencing, and wrestling. In India, the pursuit of fitness was secondary only to the pursuit of spiritual enlightenment. The Indian Hindu philosophy incorporated Yoga, which aimed to bring together and develop the body, mind, and spirit simultaneously. Other Asian cultures, as well as ancient Greek and Roman cultures, viewed physical activity and fitness as paramount to society's ability to use its military to defend and expand its culture.

Perhaps no other civilization has held fitness as highly as the ancient Greek culture. The Greeks believed the development of the body was of equal importance to the development of the mind. The eventual fall of the Roman civilization in the late 400s C.E. paralleled a decline in physical activity and fitness in the culture, where materialistic conquest became a higher priority than physical conditioning. From the fall of the Roman civilization to modern day society, there have been many instances of waxing and waning of physical activity within all societies. However, all societies have known that without physical activity the health of the individual and also of society falters. As science and technology have improved from generation to generation, so too has our knowledge of the importance of physical activity to an individual's health.

Contemporary Perspectives

A daily program of 30 to 45 minutes of moderate-intensity exercise or even less time of vigorous-intensity exercise is sufficient to lower an individual's risk of developing heart disease, high blood pressure, diabetes, and many other disease processes. One area of growing interest is how the interrelationships of various characteristics of exercise affect one's fitness level and health. Research has shown promising results in cardiorespiratory fitness gains when subjects undertook several shorter bouts of exercise daily in place of a single longer bout of exercise. It is important to note that the multiple shorter bouts of exercise had the same total time and intensity as the single longer bout of exercise. It appears as though the gains in cardiorespiratory fitness are nearly identical with both types of physical activity. This is promising news for the individual who desires improved fitness but is not able to schedule a single prolonged period of exercise due to a busy daily schedule.

There have been many clinical trials demonstrating the lowering of an individual's risk of coronary artery disease and high blood pressure, especially when combined with a low-fat diet. Physical activity also lowers one's risk of developing diabetes mellitus and colon cancer, most likely because of a lowered risk of becoming overweight or obese. Obesity, a major concern in many areas of the world, can be prevented by daily exercise and the increased caloric expenditure associated with physical activity. Physical activity early in life can lead to the development and maintenance of peak bone mass in young adults, which may translate into a lower incidence of osteoporosis later in life. There are promising findings that falls are less likely in elderly people who perform regular physical activities, such as strength training. Contrary to popular belief, physical activity has not been associated with a higher incidence of osteoarthritis later in life. However, competitive athletics may have some association to the development of osteoarthritis, though this is thought to be due to sports-related injuries and not to physical activity itself.

Physical activity may reduce symptoms of depression and anxiety and improve one's mood. Some studies have even shown promising evidence that physical activity and exercise can lower the risk of developing depression altogether. Physical activity appears to possibly improve health-related quality of life measurements by enhancing psychological well-being and improving physical functioning.

In 2004, the World Health Organization (WHO) adopted the Global Strategy on Diet, Physical Activity and Health in an effort to reduce death and the burden of disease worldwide. The Global Strategy has the following four main objectives:

- To reduce risk factors for chronic diseases that stem from unhealthy diets and physical inactivity through public health actions
- To increase awareness and understanding of the influences of diet and physical activity on health and the positive impact of preventive interventions
- To develop, strengthen, and implement global, regional, national policies and action plans to improve diets and increase physical activity that are sustainable, comprehensive, and actively engage all sectors
- To monitor science and promote research on diet and physical activity

The WHO realizes that this is an effort that will take many years, if not several decades, to adopt worldwide. It has established regional offices around the world and developed implementation toolboxes for use by global partners in their quest to improve the health of the world.

See Also: Cardiovascular Diseases; Men's Health; Musculoskeletal Diseases; Obesity; Women's Health.

Further Readings

American College of Sports Medicine. "Position Stand. The Recommended Quantity and Quality of Exercise for Developing and Maintaining Cardiorespiratory and Muscular Fitness and Flexibility in Healthy Adults." *Medicine & Science in Sports & Exercise*, 30 (1998).

Anderson, J. K. *Hunting in the Ancient World*. Berkeley: University of California Press, 1985.

Felson, D. T., J. Niu, M. Clancy, B. Sack, P. Aliabadi, and Y. Zhang. "Effect of Recreational Physical Activities on the Development of Knee Osteoarthritis in Older Adults of Different Weights: The Framingham Study." *Arthritis & Rheumatism*, 57 (2007).

Garnsey, P. *Food and Society in Classical Antiquity*. New York: Cambridge University Press, 1999.

Grant, M. *A Short History of Classical Civilization*. London: Weidenfeld and Nicolson, 1991.

Harris, H. A. *Sport in Greece and Rome*. Ithaca, NY: Cornell University Press, 1972.

Schuler, G., R. Hambrecht, G. Schlierf, J. Niebauer, K. Hauer, J. Neumann, E. Hoberg, A. Drinkmann, F. Bacher, and M. Grunze. "Regular Physical Exercise and Low-Fat Diet: Effects on Progression of Coronary Artery Disease." *Circulation*, 86 (1992).

Strawbridge, W. J., S. Deleger, R. E. Roberts, and G. A. Kaplan. "Physical Activity Reduces the Risk of Subsequent Depression for Older Adults." *American Journal of Epidemiology*, 156 (2002).

World Health Organization. "Diet and Physical Activity." http://www.who.int/dietphysical activity/en (Accessed February 2010).

Wuest, D. A. and C. A. Bucher. *Foundations of Physical Education and Sport*. St. Louis, MO: Mosby, 1995.

Rance McClain
Independent Scholar

PLASTICS IN DAILY USE

Plastics play a tremendous role in daily modern life. Plastic polymers are found in a countless number of manufactured products used on a daily basis throughout the world. According to the Environmental Protection Agency (EPA), global plastic production has increased from less than 1 million tons per year in 1960 to nearly 25 million tons per year in 2000. In 2008, the United States alone generated approximately 13 million tons of plastics in the form of containers and packaging, approximately 7 million tons as nondurable goods, and 11 million tons in the form of durable goods. The largest category of

The increased use of plastics over the past several decades has resulted in numerous concerns about the consumption of natural resources used in their production; the toxicity associated with plastic production; and the enormous environmental impact resulting from the amount of disposed plastic products that end up in landfills or as litter.

Source: National Biological Information Infrastructure/ Annette L. Olson

plastic is the type found in disposable containers and packaging (e.g., drink bottles and food containers); however, a large amount of plastic also is produced annually in the form of both durable (e.g., appliances) and nondurable goods (e.g., diapers, grocery bags). The increased production and use of plastics over the past several decades has resulted in numerous concerns about the consumption of natural resources used in their production, the toxicity associated with plastic production and use, and the enormous environmental impact resulting from the amount of disposed plastic products that ultimately end up in landfills.

Plastics can generally be divided into two major categories, thermosets and thermoplastics. Thermosets are extremely strong and are most often used in automobile parts and construction applications because of their durability. Alternatively, weaker bonds that result in soft, pliable plastics like milk and food containers hold thermoplastic polymers together. The manufacturing of plastic materials requires the use of fossil fuels, up to approximately 80 million tons of petroleum annually in the United States alone. The manufacture of plastic materials also can be dangerous, at times requiring exposure by workers to toxic and hazardous chemicals. For example, worker exposure to extremely toxic vinyl chloride vapor during the production of polyvinyl chloride (PVC) has resulted in a number of reforms cleaning up the way that plastics are made.

Improvements have been made in developing a number of greener processes that eliminate the use and production of many of the harmful chemicals and by-products.

There has been some progress in replacing conventional plastics with plant-derived alternatives that are cleaner to produce and biodegradable; however, this technology remains relatively new. The main approaches being examined in this realm involve the conversion of plant sugars into plastic polymers, "growing" plastic inside microbes, as well as producing plastic polymers directly within corn and other crops.

The widespread use of plastic materials is unprecedented, and therefore, requires ongoing management of the resultant waste. Plastics are recycled for a number of economic and environmental reasons. It is estimated that recycling 1 ton of plastic can save more than 7 cubic yards of space in landfills. The increase in plastic recycling also decreases the use of valuable natural resources, like domestic natural gas, used in the production of new plastic materials. An estimated 832,394,000 pounds of thermoplastic waste were recycled in 2008, representing a 28 percent increase in recycling since 2005. It is estimated that approximately 80 percent of American households have access to a plastic recycling program, and awareness of these programs continues to rise. To assist in the recycling of plastic waste, the Plastic Bottle Institute of the Society of Plastics Industry devised this method of marking plastic products using the following classifications:

- #1 = PET (PETE), *polyethylene terephthalate*: soft drink bottles, disposable water bottles
- #2 = HDPE, *high-density polyethylene*: detergent bottles, milk jugs
- #3 = PVC, *polyvinyl chloride*: pipes, floor tiles, industrial packaging
- #4 = LDPE, *low-density polyehtylyene*: dry-cleaning bags, produce bags
- #5 = PP, *polypropylene*: bottle caps, drinking straws, yogurt containers
- #6 = PS, *polystyrene*: packing materials, disposable cups and tableware, take-out "clam-shell" food containers
- #7 = OTHER, *other*: any plastic other than the named #1–#6, food containers and reusable drink bottles

Most plastic bottles sold in the United States are made from polyethylene terephthalate (PET, #1). Polyethylene terephthalate is a plastic resin formed by combining modified ethylene glycol and purified terephthalic acid. In general, plastics labeled #1, #2, #4, and #5 are considered to be harmless to human health when used as directed, while those labeled #3, #6, and #7 should generally be avoided. Of most concern to human health and the environment are the following types of plastic, coded for recycling as #3, #6, and #7:

- #3 (*Polyvinyl Chloride*): Companies have begun to phase out the use of polyvinyl chloride (PVC), a type of plastic widely used in construction and consumer goods, because it poses serious health threats at every stage of its life cycle. The production of PVC requires toxic and carcinogenic chemicals. Additionally, the incineration of PVC results in the emission of carcinogenic dioxins, and PVC thrown away in landfills results in toxic chemical leaching into groundwater.
- #6 (*Polystyrene*): Polystyrene is a plastic derivative used in foam food trays, egg cartons, and a large number of take-out food containers. Chemical styrene can leach from the plastic container into food and beverages, resulting in widespread disorders of the nervous system, long-term liver and nerve damage, and even different types of cancers.
- #7 (*Polycarbonates*): The #7 designation is for plastics that do not fit into the other coded categories. A large number of these products are polycarbonates, which are commonly used in plastic baby bottles and the plastic lining of metal food cans. The chemical bisphenol-A (BPA) is an endocrine disruptor that is widely used in making polycarbonates. It has been found that this chemical can leach from the material, resulting in a number of alterations to human immune and reproductive systems.

In addition to recycling, programs aimed at reducing the daily use of unnecessary plastic materials, such as plastic grocery bags, are gaining popularity and seem to be successful in significantly reducing local plastic waste. For example, in 2008, the city of San Francisco banned the use of plastic grocery bags entirely within the city limits, requiring shoppers to instead incorporate the use of reusable cloth bags. This past January 2010, Washington, D.C., implemented a five-cent tax on plastic grocery bags, thereby reducing the monthly average use of plastic bags from almost 23 million bags in 2009, to fewer than 3 million in January 2010. The five-cent tax in Washington, D.C., has additionally generated an estimated $150,000 in revenue, which is being directly used to clean up and restore the polluted Anacostia River.

The ever-increasing role of plastics in daily use throughout the world requires technology to keep pace with its proper handling and disposal. Increased awareness about reducing the use of plastic waste and avid recycling of plastic materials will need to continue to combat the tremendous amount of waste that ultimately ends up in landfills, polluting the environment, and negatively affecting human health. Recent research, however, has raised doubts about the utility of these approaches. The future of green plastics will ultimately require scientists to address issues with both the manufacture and disposal of plastic materials, without creating additional unforeseen impacts on human and environmental health.

See Also: Fast Food; Government Role in Green Health; Industrial Ecology; Personal Consumer Role in Green Health; Petrochemicals; Phaseout of Toxic Chemicals; Private Industry Role in Green Health; Solid Waste Management; Taxation of Unhealthy Products.

Further Readings

American Chemistry Council. "Plastic Beverage Bottles." http://plasticsinfo.org/s_plasticsinfo/sec_level3_collapsed.asp?CID=657&DID=2594 (Accessed May 2011).
Gerngross, Tilman U. and Steven C. Slater. "How Green Are Green Plastics?" *Scientific American*, 283 (2000).
Plastic Bottle Institute of the Society of Plastics Industry. "About Plastics." http://www.plasticsindustry.org/aboutplastics/?navItemNumber=1008 (Accessed April 2010).
Stein, R. S. "Plastics Can Be Good for the Environment." *NEACT Journal*, 21 (2002).
Thomas, V. and T. Spiro. "An Estimation of Dioxin Emissions in the United States." *Toxicological and Environmental Chemistry*, 50 (1995).

Carrie Nicole Wells
Clemson University

PRESCRIPTION DRUG ADDICTION

Addiction (termed *substance dependence* by the American Psychiatric Association) is defined as a maladaptive pattern of substance use leading to clinically significant impairment or distress, characterized by several of the following characteristics: tolerance; well-defined physiological symptoms upon withdrawal; escalating use; persistent desire or unsuccessful attempts to stop use; increasing time spent obtaining, using, and recovering from use; disregard for important social, occupational, or recreational activities; and continued use despite the knowledge of physical or psychological problems related to substance use. A rapidly growing area of concern is addiction to prescription medications. Such abuse can be described as medications taken by someone other than the patient for whom the medication was prescribed, or taken in a manner or dosage other than what was prescribed. Abuse can be a forebearer of and lead to addiction.

Prescription drugs are licensed medications, regulated by legislation and requiring written instructions from a healthcare provider to a pharmacist. Although any medication has the potential to be abused, certain types or classes of medications are abused more frequently than others. In 2006, 16.2 million Americans age 12 and older had taken a prescription pain reliever, tranquilizer, sedative, or stimulant for nonmedical purposes at least once in the year prior to being surveyed.

Opioids are pain medications (analgesics) that act within the brain, spinal column, and gastrointestinal tract to bind to specific receptors and alter the way a person experiences pain. In addition, opioids act in a specific area of the brain that alters the way a person experiences pleasure. This alteration leads to a sense of euphoria when the medication is used. When properly managed, opioids are safe, effective, and rarely cause addiction. Examples of opioid medications include codeine, oxycodone, hydrocodone, morphine, hydromorphone, meperidine, fentanyl, and propoxyphene.

Repeated exposure to opioid medications, such as long-term use for chronic pain, can lead to a physical dependence upon the medications. This is different than addiction to the medications. The body adapts to long-term use, resulting in tolerance to the medication. The person taking the medication requires a larger dose to achieve the same effect. With larger doses, the person is more likely to experience side effects such as drowsiness and constipation. The most concerning side effect of large doses of opioids is respiratory depression, wherein the drive to breath diminishes. This respiratory depression can be worsened when the opioids are combined with other medications that can cause respiratory depression, too, such as alcohol, barbiturates, or benzodiazepines. The combined respiratory depression can be life threatening. Without proper supervision by a healthcare provider, this tolerance can lead to abuse behavior and eventually addiction. When dependence develops, the healthcare provider also needs to be aware of the potential for withdrawal symptoms upon cessation of the medication. Symptoms of withdrawal can include restlessness, muscle and bone pain, insomnia, diarrhea, vomiting, cold flashes with goose bumps (called "cold turkey"), and involuntary leg movements.

Treatment for addiction to opioid medications begins with a period of detoxification from the drug. During this time, the person is weaned off of the medication in a controlled manner to minimize the physical and psychological symptoms of withdrawal. Effective detoxification processes utilize a combination of medications and behavioral therapy to treat the addiction and potential relapse. Several different medications are used in the treatment of opioid addiction. Methadone, a synthetic opioid, can eliminate withdrawal symptoms and relieve cravings. It has been used in the treatment of heroin and opioid addiction for more than 30 years. A more recent addition to the market is buprenorphine, another synthetic opioid. Naltrexone, a long-acting opioid receptor blocker, is used once a person is through the withdrawal period and can help prevent relapse. It occupies the same receptor as the opioid medications without the ability to produce pain relief or euphoria. Naltrexone requires a highly motivated individual to be effective, as it does not affect the psychological addiction and is easily subverted by the recipient missing or skipping doses. Finally, naloxone, a short-acting opioid receptor blocker, is used to treat acute overdoses. Its short yet powerful period of action prevents its use as a long-term treatment option.

Central nervous system (CNS) depressants (e.g., tranquilizers or sedatives) are medications that slow brain function. CNS depressants are divided into three groups. Barbiturates are medications used to promote sleep. Benzodiazepines are medications used in the treatment of anxiety disorders, panic attacks, convulsions, and occasionally sleep disorders. The final group is made up of newer sleep medications that act on a subset of benzodiazepine receptors and appear to have a lower risk of abuse and addiction. All of these medications work by producing drowsiness or a calming effect.

As with opioid pain medications, tolerance can develop with CNS depressants. Tolerance to the hypnotic (sleep-inducing) effect typically develops much more quickly than the anxiolytic effects. It is due to this fact that all these medications are only indicated for short-term use in the management of sleep disorders. As with the opioids, tolerance can eventually lead to escalating dosages and dependence. Dependence can develop rapidly with short-acting, high-potency benzodiazepines such as alprazolam compared to a long-acting, low-potency benzodiazepine such as diazepam.

Withdrawal symptoms from CNS depressants are mainly anxiety and insomnia. However, serious symptoms can develop with acute withdrawal, such as seizures and delirium tremens. Symptoms are usually more likely and more pronounced with short-acting, high-potency CNS depressants, therefore, it is of utmost importance to gradually taper the dose of these medications to avoid the development of acute withdrawal symptom. Prolonged symptoms of anxiety, depression, and insomnia have been known to occur, particularly with long-term use of these medications.

CNS depressants are rarely the only drug of abuse. Instead, they are usually part of polydrug abuse. Common drugs or substances abused with benzodiazepines are opioids or alcohol for synergistic or enhancing effects, or cocaine and other stimulants to temper the stimulation of these drugs. Because of the high potential of polydrug abuse and the availability of alternatives for the management of anxiety and insomnia, the use of CNS depressants is steadily declining.

Stimulants are medications used to increase alertness, attention, and energy. These medications are prescribed predominantly in the treatment of attention deficit disorder (ADD), narcolepsy, and, rarely, depression that has failed to respond to other treatments. Stimulants work by mimicking the effects of neurotransmitters within the brain. When starting them at low doses and increased gradually, a healthcare provider can achieve a therapeutic effect without producing euphoria or increasing the risk of addiction. As with CNS depressants, stimulants are best used in combination with behavioral therapy and education.

Abuse of stimulant medications is rarely done with prescription strengths of the drug and rarely involves someone who has been prescribed the medication for ADD because they are highly controlled substances requiring close monitoring by a healthcare provider. Abusers and people addicted to prescription stimulants often have histories of abuse with street amphetamines, which are much higher doses and produce more of a "high." The intense euphoria from the drugs quickly produces cravings for more and rapidly leads to addiction. Addicts learn to fake symptoms of ADD because prescription stimulants are more easily obtained than street amphetamines. Addicts usually crush the tablets or dissolve them in water to achieve a more rapid onset of effect; they rarely inhale or inject the medications.

There are currently no proven medication treatments for addiction to prescription stimulants. Detoxifications from the drug and behavioral therapy are the mainstays of treatment and have been shown effective for treating addiction. Behavioral therapy seeks to identify and correct areas of potential return to addictive behavior in family, social, and work contexts. Newer programs incorporate incentive programs to facilitate retention in treatment programs and promote abstinence from stimulants.

To lessen the burden both on individuals and society as a whole, patients and healthcare providers need to work together to prevent abuse and addiction. Healthcare providers need to screen all patients for substance abuse history and observe for potential clues to medication abuse, such as increasing dosage or frequency of use. Providers should communicate with pharmacists about any potential abuse behaviors they may have observed in their patients, and pharmacists must notify physicians of patients obtaining medications from other providers. Patients must carefully follow directions for the use of their medications and read all information provided by the pharmacist.

See Also: Anti-Depressant Drugs; Personal Consumer Role in Green Health; Pharmaceutical Industry.

Further Readings

American Academy of Family Physicians. http://www.aafp.org (Accessed January 2010).

Arana, G. W. and S. E. Hyman. *Handbook of Psychiatric Drug Therapy*, 2nd ed. Boston: Little, Brown, 1991.

Ashton, H. "Protracted Withdrawal Syndromes From Benzodiazepines." *Journal of Substance Abuse Treatment*, 8 (1991).

Attention Deficit Disorder Association. http://www.add.org (Accessed January 2010).

Gold, M. S., N. S. Miller, K. Stennie, and C. Populla-Vardi. "Epidemiology of Benzodiazepine Use and Dependence." *Psychiatric Annals*, 25 (1995).

National Family Partnership. http://www.nfp.org (Accessed January 2010).

National Institute on Drug Abuse. http://nida.nih.gov (Accessed January 2010).

National Survey on Drug Use and Health. http://www.samhsa.gov (Accessed January 2010).

Sisson, R. and N. H. Azrin. "The Community Reinforcement Approach." In *Handbook of Alcoholism Treatment Approaches: Effective Alternatives*, R. K. Hester and W. R. Miller, eds. New York: Pergamon Press, 1989.

Wilens, T. E., S. V. Faraone, J. Biederman, and S. Gunawardene. "Does Stimulant Therapy of Attention-Deficit/Hyperactivity Disorder Beget Later Substance Abuse? A Meta-Analytic Review of the Literature." *Pediatrics*, 111/1 (2003).

Rance McClain
Independent Scholar

PRIVATE INDUSTRY ROLE IN GREEN HEALTH

The concept of green health is all-encompassing, touching every aspect of our lives, from the air we breathe and the Earth we walk on, to the food and water we ingest, to the materials used in creating our built environment, to the energy that drives our homes and transportation, to the care that we give to our bodies. In the quest to embrace a healthy, sustainable way of life, both individual commitment to green living and government funding of green technologies and practices are vital, but it is private industry's role in green health that has the capacity to be the most transformative, on local, national, and global scales.

Whether spurred by government programs and initiatives or by the simple desire to minimize environmental harm and protect worker and consumer safety, businesses are embracing everything from recycling and reusing to innovative methods of raw material extraction, manufacturing, distribution, and packaging. By acknowledging and responding to the finite, fragile nature of our environment and its natural resources, private industry can lead the way toward our collective embrace of a sustainable existence. From healthcare facilities to concrete manufacturers to furniture stores to alternative energy entrepreneurs, we see impassioned examples of private industry shaping a sustainable future. Of course, whether for ideological, political, or financial reasons, the private industry sector includes countless businesses and corporations that continue to engage in processes that degrade the Earth and our environment, but there are still many inspiring examples of private industry embracing a holistic view of our world and its natural resources.

One such example can be found in the concrete industry, which is the largest user of natural resources in the world, annually requiring billions of tons of sand, stone, and water in the mining and manufacturing process as well as massive amounts of energy to excavate and transport both raw materials and the finished product. In Maryland, the Department of Transportation's State Highway Administration (SHA) is working with local construction companies to use recycled pavement for highway projects. One of those companies, P. Flanigan and Sons, has a facility that creates 100 percent recycled crushed aggregate base (GAB); this GAB is used on roadways prior to paving them and has been approved by the SHA for use in its upcoming projects. In addition to preserving natural aggregate and its surrounding environment, the benefits of using GAB also include saving fuel that would be used in mining and transport, which also reduces vehicle emissions, and conserving shrinking landfill space. Use of GAB is a vital component in the quest to foster sustainable growth while creating vibrant, livable communities. Another breakthrough related to the cement industry has come from Calera, a company that says that not only has it come up with a way to reduce the energy costs of producing cement, but it has also found a way to capture and sequester in cement the carbon dioxide emissions from coal and gas power plants. The company opened its first demonstration site in California in August 2009.

While perhaps not as obviously energy-intensive as industries such as concrete, petroleum, and aluminum, healthcare ranks as the second most energy-intensive industry in the United States, imprinting the world with a massive carbon footprint. The American healthcare sector, which includes healthcare facilities, scientific research, and the production and distribution of pharmaceutical drugs, accounts for almost 10 percent of carbon dioxide emissions in the United States. Part of the reason healthcare's carbon footprint is so large is its reliance on conventional, nonrenewable energy sources, such as oil and coal. With more than 5,000 hospitals and healthcare facilities in the United States, green health initiatives in this sector are vital in the quest to create a sustainable existence. La Maestra Community Health Center, located in San Diego, is a shining example of private industry's embrace of green health initiatives. La Maestra is expected to earn LEED (Leadership in Energy and Environmental Design) Gold certification and will be the first health clinic to garner this status. In planning and building La Maestra, designers and sponsors had several goals: providing a high level of comfort, using less energy, and introducing the concept of green building to the community in which the clinic was built. Corporate sponsors include Kaiser Foundation, Bank of America, and San Diego Gas and Electric. La Maestra is expected to serve 180,000 patients annually. Some of La Maestra's features include the following:

- High-performance insulated glass
- Rooftop solar photovoltaic array that will generate 15 percent of its energy needs
- An energy-management system
- Use of recycled and sustainable materials
- Use of materials with low levels of volatile organic compounds (VOCs)
- High-efficiency landscape system and use of drought-resistant and native plants

While La Maestra is a somewhat unique example of the ways that private industry, healthcare, and sustainability can interconnect, Practice Greenhealth is a membership and networking organization for members of the healthcare community that have made a commitment to sustainable practices. Participants include hospitals, healthcare systems, businesses, and other stakeholders who wish to bring a more ecofriendly approach to providing healthcare services. Practice Greenhealth members include both not-for-profit as well as private healthcare providers.

Ecofriendly initiatives are also evident in many other business sectors, such as the telecommunications industry. For example, AT&T's Industrial Ecology Faculty Fellowship provides funding to universities that seek to advance knowledge of how the environment is impacted by Information and Communications Technology (ICT). The program's goals include the following:

- Understanding how ICT can affect the environment
- Recognizing how businesses can apply research findings to ICT products and services to reduce environmental impact
- Nurturing university faculty and students who can contribute to solving environmental problems

Businesses as diverse as Levi Straus, Sara Lee, and Rhino Records have also embraced green health initiatives by working with Natural Logic, a U.S.-based consulting firm that provides guidance to firms who are seeking to apply ecological theory to industrial methods. Natural Logic CEO Gil Friend, who has been named one of the top 25 Clean Tech Leaders, argues that by applying the principles of nature's evolutionary systems we can create economic systems that are efficient, adaptive, and sustainable. The aforementioned firms and dozens of others have embraced these notions by working with Natural Logic to stay fiscally robust while causing the least harm to the environment. Natural Logic, by helping private industries to assess their carbon footprints, to understand the energy impact of their supply chains, and to recognize methods for diminishing harmful practices, provides much-needed guidance for companies interested in embracing green health initiatives.

Another business that has embraced green health concepts is Sunny Delight, maker of an orange-flavored beverage. In 2009, Sunny Delight announced that they had achieved the goal, three years early, of going "zero waste"; that is, all six of Sunny Delight's manufacturing plants now send absolutely no waste to landfills. The company has also cut energy use by 8 percent and water use by 9 percent; further, Sunny Delight has cut their carbon emissions by 17 percent. Other green health concepts embraced by Sunny Delight include the following:

- Going paperless in their corporate offices
- Encouraging their vendors to submit invoices electronically
- Raising the juice content of their product
- Plans to cut the calories of their juice 45 percent by 2015
- Acquisition of a company that produces acai berry drink—the acai is sustainably harvested in the Amazon

As more and more people come to see the green health mindset as the proper framework for creating communities that are dynamic, sustainable, and resilient, we are beginning to see more and more green technology breakthroughs that are poised to profoundly impact how we go about our daily lives. 2008 was an especially potent year for green technology as Barack Obama campaigned, in part, on pledges to create green-collar industries and jobs. Economic troubles in the United States since then may curb some of the early enthusiasm shown for green health initiatives, yet there are still many green entrepreneurs who are focused on creating the kinds of industries and products that stand to usher in a new, green era.

For example, Thomas Hinderling, a Swiss researcher, has been working on creating "solar islands," several miles across, that are capable of producing hundreds of megawatts of relatively cheap power. The solar islands, which will be located in oceans, will be composed of a plastic membrane fitted with solar concentrating mirrors that float above the water. The mirrors will heat liquid and turn it into steam, which will drive a turbine that generates energy. The idea has been met with some skepticism, but Hinderling is confident that his idea has the potential to radically alter our dependence on nonsustainable energy sources.

In addition to proposals of floating solar islands, an "older technology"—solar thermal—has regained popularity as firms including eSolar, BrightSource, and Ausra have unveiled plans to construct solar thermal plants in deserts. These plants, like Hinderling's proposed floating solar islands, use sunlight-reflecting mirrors to turn liquids into steam; the steam, in turn, will drive turbines, just like in Hinderling's islands and in coal-fired electricity plants. This technology first made a splash in the 1980s, though falling fossil-fuel costs made it less attractive at that time; in 2008, oil prices reached an all-time high, and Obama's green technology campaign platform gained widespread support, though oil prices dropped again in 2010, some wondered if green technology would again fall by the wayside. Alternative energy champions lament the fact that green energy initiatives are so vulnerable to fluctuating oil prices, and they argue that even with low oil prices green energy initiatives makes sense in the long term, not just environmentally, but economically.

San Jose, California–based Nanosolar has garnered a lot of interest and support for their solar panels, which are made out of cheap plastic as opposed to more expensive silicon. In the past, solar energy has been seen as prohibitively expensive, but with Nanosolar's manufacturing process and their breakthrough in creating the cheapest solar coating ever created, it seems as if they have created a power-generating technology that can compete with coal. Their plant in California is the largest cell factory in the world, and their facility in Germany is the world's largest panel-assembly factory in the world. Their PowerSheet cells reduce the cost of production from $3 a watt to 30 cents per watt—which is even cheaper than coal.

In the world of transportation, Project Better Place, led by Shai Agassi, is focused on complete replacement of gasoline-powered automobiles with fully electric ones; not hybrids, and not flex fuel-driven engines, but completely electric cars. His plan calls for the creation of an entirely new green infrastructure and has garnered interest from Australia, Israel, Denmark, China, Canada, Hawaii, and San Francisco. Agassi calls the infrastructure electric recharge grid operator (ERGO), and it would consist of a network of charging stations that would allow drivers to plug in and charge their car's batteries, anywhere, anytime; drivers would subscribe to plans, much like with cellular phones, that range from unlimited miles, to a maximum number of miles each month, to plans that allow drivers to pay as they go. Further, Agassi says there would eventually be plugs for charging in people's homes, at their offices, and at shopping malls. And, for those who do not have the time to wait for re-charging, they could pull into a battery exchange where a hydraulic lift would remove the spent battery and replace it with a fully charged one. Some question whether or not Agassi's plan is just shifting the environmental burden to the electric utility, though Agassi has pledged to buy energy made only from renewable sources, such as wind farms or solar-powered plants. Currently, Agassi has raised nearly a half a billion dollars in committed capital and in February 2010, the company's first electric vehicle demonstration center opened in Israel.

While Agassi's vision will obviously take time to come to fruition, there have been other hints that corporate and institutional buyers are ready to embrace a range of fuel-efficient options for their fleets. Hybrid-electrics, biodiesel, diesel, and electric vehicles are beginning to be embraced more and more by large companies that rely on fleets to deliver their goods. For example, Coca-Cola's delivery fleet of hybrid-electric diesel vehicles is the largest in North America, Frito-Lay added more than 1,000 fuel-efficient Sprinter delivery vehicles to its fleet, and San Francisco, Boston, and Phoenix are home to several all-hybrid taxi companies. Further, UPS has purchased nearly 2,000 alternative-fuel vehicles globally, the U.S. Postal Service exchanged 6,500 old delivery vehicles for a mix of hybrids and flex-fuel vehicles, and FedEx added almost 100 retro-fitted delivery trucks to its fleet; these trucks produce 96 percent fewer particulates and 75 percent fewer smog-causing emissions than traditional vehicles.

Other examples of private industry embracing green health include big-box retailer IKEA, which stands out as a company that is committed to sustainability. Some of the ways IKEA has embraced green health include the following:

- Eliminating the use of plastic bags
- Investing $77 million in clean technology ventures
- Recycling 84 percent of the waste produced by its stores
- Ensuring that nearly all its products—71 percent—are recyclable
- Embracing the strictest emissions policies in the world

By focusing on four broad areas, IKEA is striving to become even more sustainable. The four areas are: suppliers, products and materials, climate change, and community involvement. IKEA suppliers must do the following:

- Comply with national legislation
- Use no forced or child labor
- Not practice discrimination
- Pay at least minimum wage and offer overtime compensation
- Provide a safe and healthy work environment
- Take responsibility for waste, emissions, and chemical-handling
- Allow third-party auditing of their practices

Further, IKEA requires the following of products and materials:

- Third-party audit of its wood suppliers
- Limited use of raw materials to eventually phase out solid wood
- Use of as much renewable and recyclable materials as possible

IKEA's climate change policies include the following:

- Business travel reduction
- Public transportation to some of its stores
- Free bike trailer rentals at some of its stores
- Launched "IKEA goes renewable" in 2006, focused on radically decreasing carbon dioxide emissions
- Committed to reducing energy consumption by 25 percent (compared to 2005)
- Long-term objective of powering all IKEA stores, warehouses, offices, and factories with 100 percent renewable energy

Community involvement implemented by IKEA includes the following:

- Works with World Wildlife Federation, Save the Children, and United Nations Children's Fund (UNICEF), among others
- Works to improve cotton-cultivation practices
- Practices reforestation and strives to maintain rainforests
- Provides in-kind donations in emergency situations
- Prevents child labor and works to improve children's rights in India
- Annually provides one-year scholarships to 22 Eastern Block students to study forestry in Sweden

Another big retailer, Walmart, gained attention in 2009 by launching the Sustainability Index. The project began with a 15-item questionnaire that Walmart sent to thousands of its suppliers. The questionnaire was based in four broad categories: energy and climate; material efficiency; natural resources; and people and community. For this Walmart has also pledged to create life cycle–based standards for thousands of its products. An incredibly difficult task Walmart has enlisted the aid of academics and other experts to help create the metrics and standards needed to evaluate their products. This group of experts belongs to the Sustainability Consortium, which was seed-funded by Walmart.

As all of the above examples show, green health initiatives have been embraced by a wide variety of members of the private industry community. It can be daunting to make the decision to alter business practices in the name of ecofriendliness, and many businesses that desire to make such changes are often confused as to how to best bring their company's practices in line with notions of sustainability. GreenBiz.com, launched in 2000, provides much needed support for those companies that have decided to follow green health practices. By making available a vast array of resources for companies interested in green health, ranging from news of how other companies have embraced green health to information on best practices, GreenBiz provides vital information, learning opportunities, and other resources designed to help any company become more environmentally responsible while remaining profitable.

GreenBiz is part of the Greener World Media family, which is the world's first and only mainstream media company focused exclusively on sustainability issues. Their report, "State of Green Business 2010," offers a detailed look at the emerging green economy. Recessions have historically limited interest in green health initiatives, and the current recession is no different—interest in sustainability has waned in the past few years as crippling economic conditions have worked to diminish what felt like a wellspring of interest just a few years ago. Still, as the Green Business Report illustrates, there is much to celebrate for those interested in green health. The report's researchers and authors found that the current recession has not had as profound a negative impact on green professionals as have past recessions, and, in fact, in some ways, green health has managed to thrive in private industry. Many see this as evidence of the ways that green health initiatives, which often begin as altruistic endeavors, have now become, for a growing number of companies, a fundamental part of their business practices. Whether this can be attributed to continued altruism, or to impassioned desire to embrace sustainability, or to the desire to cut costs and improve community standing, notions of green health are steadily becoming business as usual for many corporations. While green practices may have once been seen as an economic liability, many members of the private industry community now see green health as a valid way to remain competitive.

A new spirit of collaboration has also emerged among companies interested in green health practices. For example, Nokia, IBM, Sony, Pitney-Bowes, and others launched the

Eco-Patent Commons in 2008 to foster the contribution of environmental patents to the public domain. To date, more than 100 "IP-free" (intellectual property) technological innovations have been made openly available to all participants. Similarly, Best Buy, Nike, and other companies created the GreenXchange in order to foster the sharing of green ideas regarding packaging, product design, and manufacturing. GreenXchange is partnered with the nonprofit Creative Commons, which has designed licenses that allow creators of intellectual property to share their work with others.

The private industry sector, then, has provided many positive examples of how to incorporate green health thinking into business practices. Further, organizations like Practice Greenhealth, Natural Logic, and Greener World Media provide much-needed guidance and resources to those members of the private industry community interested in sustainability. Private industry can also look to the U.S. government for support in pursuing green health initiatives. As part of the 2009 American Recovery and Reinvestment Act, the U.S. government set aside billions of dollars to be dispersed in loans and grants available to private industries that are interested in generating renewable energy, such as wind and solar. These loans are intended both to spur private sector expenditures in renewable energy and to make high-risk investment and research in renewable energy less chancy. And on the global level, The World Council for Sustainable Development (WBCSD) is an association of 200 companies committed to sustainable development. The council provides a framework for members to research sustainable development and to share knowledge and best practices. Further, The council sponsors forums and provides guidance for working with governments as well as with nongovernmental and intergovernmental organizations. Made up of members from more than 30 countries and from 20 major industrial sectors, WBCSD also has relationships with 60 national and regional business councils from around the globe. WBCSD focuses on energy and climate, development, ecosystems, and the role of business in society. The council's stated goals are the following:

- Be an advocate for sustainable development
- Participate in policy development that nurtures sustainable human progress
- Build and encourage the business case for sustainable development
- Demonstrate the business contribution to sustainable development solutions and share leading edge practices among members
- Contribute to a sustainable future for developing nations and nations in transition

And so, it is clear that notions of green health have become an integral component of planning for a sustainable, resilient future. The examples provided above illustrate the fact that many members of the private industry sector take their role as shapers of a collective, healthy future very seriously. While there is still a daunting amount of work to be done to achieve true, green health, there is much to praise when it comes to assessing the role of private industry in green health.

See Also: Alternative Energy Resources (Solar); Automobiles (Emissions); Climate Change; Electricity; Government Role in Green Health; Hospitals (Carbon Footprints).

Further Readings

GreenBiz.com. "Energy, Clean Technology (Cleantech), and Climate Change." http://www .greenbiz.com/business/browse/energy-climate (Accessed August 2010).

GreenBiz.com. "State of Green Business 2010." http://stateofgreenbusiness.com (Accessed August 2010).

Pelosi, Nancy. "American Recovery and Reinvestment Act." http://www.speaker.gov/newsroom/legislation?id=0273 (Accessed August 2010).

World Business Council for Sustainable Development. "Vision 2050." http://www.wbcsd.org/templates/TemplateWBCSD5/layout.asp?MenuID=1 (Accessed August 2010).

Tani Bellestri
Independent Scholar

Radiation Sources

Radiation is energy that is emitted from a source and is radiated or transmitted in the form of rays, waves, or particles. Radiation travels through space and has the potential to penetrate different materials.

Ionizing Versus Nonionizing Radiation

Electromagnetic radiation can be described by either its frequency f, its wavelength λ, or its photon energy E. Wavelength is the distance between successive peaks or troughs of a wave and is inversely proportional to frequency. The higher end of the electromagnetic spectrum has very short frequencies on the scale of the size of atoms, and this end of the electromagnetic spectrum has the highest energy, while longer frequencies have lower energies.

There are two distinct types of radiation, nonionizing radiation, which is the type of radiation emitted by light sources, radio transmitters, and microwaves, to name but a few sources. These come from the lower end of the electromagnetic spectrum and have insufficient energy to ionize—which can potentially damage human health.

Nonionizing radiation does not have enough energy per quantum to ionize target atoms or molecules, and as such it can only excite electrons to higher energy states. Nonionizing radiation can cause some health effects, however. Some nonionizing radiation can cause thermal heating of the skin and body, resulting in burns or thermal damage. Furthermore, exposure to high levels of light in the optical portion of the electromagnetic spectrum can cause damage to the eye or blindness.

Ionizing radiation is different in that it can produce ions, which are charged particles. It does this by detaching electrons from atoms or molecules—this process is called ionization, and it is relevant in sustainability terms, as ionizing radiation in sufficient doses has the potential to damage human health. Ionizing radiation must have a sufficiently high energy to be able to interact with the atoms of a target.

This rest of this article will concentrate on ionizing radiation, as it has greater potential to impact human health.

Ionizing Radiation Particles

Alpha Radiation

Alpha radiation consists of a helium-4 (4He) nucleus that moves at high speed and can penetrate a sheet of paper.

Beta Radiation

Beta radiation consists of electrons moving and can be halted by a thin sheet of aluminum.

Gamma Radiation

Gamma radiation consists of high-energy electrons and can be halted by dense materials such as lead or concrete by absorption.

Sources of Ionizing Radiation

Radiation sources can be divided into natural and man-made. We call the radiation that we receive through exposure to a variety of different radiation sources in everyday life *background radiation*. At sea level, the average background radiation is 26mrem (millirems). There are also man-made sources of radiation that are used for a variety of industrial, scientific, medical, and energy generation uses.

Types of Ionizing Radiation

Cosmic Radiation

Radiation that comes from the sun, stars, and outer space is known as cosmic radiation. The Earth's atmosphere acts as a filter against us receiving high doses of cosmic radiation. Exposure to cosmic radiation depends on altitude; those at higher altitudes will therefore receive higher doses of cosmic radiation. Cosmic radiation contributes about 13 percent of the background radiation level on Earth.

Airline flight crews, who spend considerable time at high altitudes, receive a greater dose of cosmic radiation than those of us who spend most of our time on the ground. A member of a cabin crew spending 1,000 hours in space can expect to receive 2 to 5 milliSievert (mSv) of additional cosmic radiation per annum in addition to an annual dose of 2 to 3 mSv for those on the ground.

Radon

Radon is the second-leading cause of lung cancer after tobacco smoke and the most prevalent natural radiation source. Some areas are particularly susceptible to the production of radon gas due to the composition of the local geology. In areas where radon gas is found, building codes will often stipulate measures to prevent the accumulation of radon in basements and cellars.

Industrial and Medical Radiation Sources

In medical imaging, the health benefits of being able to understand and diagnose illness without the need for any invasive procedures far outweighs any potential consequences associated with exposure to very low levels of radiation.

X-Rays

X-radiation is commonly used in medical and industrial imaging. Hard X-rays can penetrate solid objects and thus find applications in diagnostic radiographic imaging and crystallography. In medical applications, "soft" X-rays are often unwelcome, as they are totally absorbed by the body—therefore, a sheet of thin aluminum is used to filter them out. X-rays are generated by an x-ray tube, which is a type of vacuum tube. Inside this tube, electrons produced by a hot cathode are accelerated by way of a high voltage differential. When these electrons hit the anode, a metal target, X-rays are created.

It is worth noting that the definitions of X-rays and gamma rays have changed recently. As a result of the development of shorter wavelength x-ray sources and longer wavelength gamma ray emitters, their wavelength bands have begun to overlap and it is no longer meaningful to distinguish between them on the basis of wavelength. Therefore, we now distinguish them by origin: X-rays are electrons emitted outside the nucleus of an atom, while gamma rays are emitted by the nucleus of an atom.

Radioisotopes

Radioisotopes are commonly used in medical imaging. A tracer is injected or ingested into the body and is used to provide high-quality medical imagery of bones and soft tissues. The tracers used in medical imaging produce gamma rays.

Consumer Products

There are a number of consumer products that contain sources of radiation. Old gas mantle lamps used thorium impregnated fibers to produce the glow of the gas lamp; while not in common use now, they are sometimes used in gas-powered camping lamps for use in remote areas. Older antique watches with luminous dials also contain radium or tritium, which made their numbers and dials glow in the dark. Smoke detectors that rely on ionizing radiation will often contain small quantities of Americium, used to detect smoke.

Atmospheric Testing of Nuclear Weapons

Although testing nuclear weapons in the atmosphere is now outlawed by the majority of nations, the radiation resulting from the extensive testing of nuclear devices throughout the 1940s to 1960s remains in the environment. Some of the contamination as a result of nuclear testing is local in area, confined to the immediate surroundings of where nuclear devices were detonated; however, some of the radiation has travelled much farther afield and is termed *fallout*. The background radiation arising from the detonation of nuclear weapons peaked in 1963 and was estimated to be around 0.15 mSv per year worldwide. The Limited Test Ban Treaty on detonating nuclear weapons aboveground was signed in

1963, and as of the turn of the 21st century, the radiation dose from this source was estimated at 0.005 mSv per year.

Power Generation

In burning coal, radioactive elements that were buried underground and contained within the coal are released in the form of fly ash as the coal burns. Fly ash can contain significant quantities of uranium and thorium, as well as the products of uranium decomposition—radium, radon, and polonium.

Industrial Radiation

There are a number of industrial processes that use radioisotopes for monitoring and inspection. An example is the manufacture of thin metal sheet, where a radioisotope placed on one side of the sheet and a detector placed on the other are used to monitor the thickness of a sheet that is being rolled. By measuring the emissions from the radiation source, it is possible to measure the thickness of the sheet continuously and adjust the process accordingly. Industrial radiation sources account for a miniscule proportion of our exposure to radiation.

The Effect of Radiation on Human Health

High doses of radiation have the potential to be very harmful to human health—even fatal. The type of radiation, length of exposure, and the particular area exposed influence the severity of the damage caused by radiation. Depending on the level of exposure, effects of radiation poisoning can manifest themselves within days, months, or, for some delayed effects, years.

There is the potential for delayed affects, including an increased risk of cancer and the risk of harm to unborn children. Even relatively small amounts of radiation have the potential to raise the lifetime risk of cancer. Safe levels of radiation exposure are a cause of ongoing debate among scientists and health professionals.

History of Radiation and Human Health

The natural sources of radiation have been in existence since the Earth was formed; however, in the last century, as we have discovered valuable applications for radiation, we have added man-made sources of radiation to their number.

Radiation was discovered in the late 19th century. The first recorded notes about radiation's impact on human health were made by Nikola Tesla, who subjected his hand to X-rays in 1896; however, he attributed the burns he developed incorrectly to ozone rather than the radiation source. It was not until later, in 1927, that Herman Joseph Muller noted the effect that radiation had on genetic material, for which he would later be awarded the Nobel Prize in 1946. Throughout the late 1920s and early 1930s, a number of radium-containing products were marketed as "quack" medical cures, as illustrated in the notable case of Eben Byers. In 1945 and 1946, in two separate incidents, scientists working on nuclear weapons died in incidents involving high levels of radiation exposure.

Radiation Units

In some older publications, radiation is measured in units known as "rem"—which stands for Röntgen equivalent man. The rem unit accounts for a very large unit of radiation, so as a result, the term *millirems*, abbreviated *mrem*, is often used instead to measure doses of radiation on the scale of exposure during, for example, an X-ray. The use of this term is now discouraged, although it is still used in many texts.

Rems and millirems easily convert into the Systeme Internationale units system, where the sievert is used as the unit of radiation.

$$1 \text{ rem} = 0.01 \text{ Sv} = 10 \text{ mSv} = 10000 \text{ } \mu\text{Sv}$$

$$1 \text{ millirem} = 0.00001 \text{ Sv} = 0.01 \text{ mSv} = 10 \text{ } \mu\text{Sv}$$

Human Exposure to Radiation

According to the National Council on Radiation Protection & Measurements (NCRP), the background radiation that the average human receives comes from the following sources:

- *Natural background radiation:* 55 percent
- *Medical X-rays:* 11 percent
- *Radiation inside the body:* 11 percent
- *Cosmic radiation:* 8 percent
- *Rocks and soil:* 8 percent
- *Nuclear medicine:* 4 percent
- *Consumer products:* 3 percent
- *Other*: less than 1 percent
- *Nuclear industry:* 0.05 percent

There is still much work to be done to understand the effect of radiation on the human body. For many years, the view has been that the relationship between radiation dose received and the potential for deleterious effects to human health is linear, known as the linear no-threshold (LNT) model. This view is based on the presupposition that a single charged particle of radiation has a given probability of causing damage to genetic material, and as the subject is exposed to increasing levels of radiation, the chances of ill health resulting from damage to genetic material increases linearly. However, there are a number of authors who contest that the human body's response to radiation is nonlinear and that the dangers of low-level radiation are underestimated by the current model. Some authors even contest that low levels of radiation exposure stimulate the body's DNA damage repair system and that exposures to low levels of radiation may even be beneficial to human health. While this theory or hormesis is controversial, it is clear that there is a lack of consensus about the effects of low-level radiation on human health.

- *0.05–0.2 Sv (5–20 rem)*: No symptoms. According to the LNT model of exposure, there is some potential for increasing the risk of genetic mutation and cancer.
- *0.2–0.5 Sv (20–50 rem)*: No noticeable symptoms apart from a temporary decrease in the count of red blood cells present.
- *0.5–1 Sv (50–100 rem)*: Mild radiation sickness, headaches, and a temporary decrease in the immune system's ability to resist infection. In males, it is possible for them to become temporarily infertile.

- *1–2 Sv (100–200 rem)—Light Radiation Poisoning (LD 10/30)*: Those suffering from light radiation poisoning are subject to a 10 percent risk of fatality within 30 days of exposure. Three to six hours after exposure there is a chance of nausea, occasionally with vomiting. This can last up to a day and can begin from three to six hours after exposure. This is followed by a phase known as the latent phase, during which little change is apparent. After this there will be further symptoms, such as fatigue, illness due to compromised immune system response, and an increased risk of infection. Males may be temporarily sterile.
- *2–3 Sv (200–300 rem)—Moderate Radiation Poisoning (LD 35/30)*: Nausea is commonly unavoidable at the upper end of this range, with a high risk of vomiting. Symptoms present themselves between one and six hours after irradiation and tend to last between one and two days. Following that there is a period known as the latent phase that lasts between one and two weeks. Following this period, further symptoms such as hair loss and fatigue may occur. There will be a decrease in white blood cells (leukocytes) and the body's immune system may be compromised. There is a chance of permanent female sterility.
- *3–4 Sv (300–400 rem)—Severe Radiation Poisoning (LD 50/30)*: Symptoms present themselves as with the 2–3 Sv dose, with the addition of uncontrollable oral bleeding and bleeding under the skin and in the kidneys following the latent phase.
- *4–6 Sv (400–600 rem)—Acute Radiation Poisoning (LD 60/30)*: With this level of radiation poisoning, there is a 60 percent chance of fatality within a month. Between half an hour and two hours after exposure, symptoms will present themselves. Urgent intense medical care is required for any chance of survival. Symptoms will usually last for two days, after which there is a latent phase followed by symptoms commensurate with poisoning at the 3–4 Sv level but with a greater level of activity. This level of radiation poisoning may result in female sterility. If death occurs, it will commonly happen between two and 12 weeks following exposure; recovery can take several months to a year before the patient returns to health.
- *6–10 Sv (600–1,000 rem)—Acute Radiation Poisoning (LD 100/14)*: With this level of radiation poisoning, there is a near 100 percent chance of fatality within a fortnight. Survival is possible but dependent on a significant investment of medical care and attention. A bone marrow transplant is required, as this is largely or completely destroyed by the radiation poisoning. Internal gastric and intestinal tissue is severely damaged, and there is high potential of fatal internal bleeding. Symptoms start between 15 and 30 minutes after exposure and last for a couple of days. This is followed by a phase during which there is a high chance of death by infection or internal bleeding. It is likely that only a partial recovery is possible, and this may take several years.
- *10–50 Sv (1,000–5,000 rem)—Acute Radiation Poisoning (LD 100/7)*: There is a guarantee of fatality within a week with acute radiation poisoning. Symptoms spontaneously present themselves between five and 30 minutes after exposure. Immediately powerful fatigue and nausea set in. This is followed by a period of a couple of days of well-being, informally referred to as the "walking ghost" phase. Cells in the gastric and intestinal tissue die, causing loss of blood and fluids and diarrhea. Circulation begins to break down, leading to delirium and coma. Currently the only treatment available is palliative pain management.
- *More than 50 Sv (>5,000 rem)*: There are a few documented cases of workers at nuclear power plants receiving doses in excess of 50 Sv. Cecil Kelley, a worker at Los Alamos National Laboratory, received a dose of between 60 and 180 Sv and survived for 36 hours, while a worker at Wood River , Rhode Island, survived for 49 hours after receiving a dose of 100 Sv.

See Also: Government Role in Green Health; Low-Level Radioactive Waste; Occupational Hazards; Radon and Basements.

Further Readings

Forshier, Steve. *Essentials of Radiation, Biology and Protection.* Florence, KY: Delmar Cengage Learning, 2008.

Lombardi, Max H. *Radiation Safety in Nuclear Medicine*, 2nd ed. Boca Raton, FL: CRC Press, 2006.

National Council on Radiation Protection and Measurements. "Ionizing Radiation Exposure of the Population of the United States." Report No. 93, 1987. http://www.ncrponline.org/ Publications/93press.htm (Accessed August 2010).

Pollycove, M. "Nonlinearity of Radiation Health Effects." http://www.ncbi.nlm.nih.gov/pmc/ articles/PMC1533290 (Accessed August 2010).

Gavin D. J. Harper
Cardiff University

Radon and Basements

Radon is a chemical element (symbol Rn) with an atomic number of 86. Radon is a naturally occurring, odorless, colorless, radioactive gas produced by the radioactive decay of radium-226. Soils, rocks, and particularly granite are the major sources of radium in nature. Radon is generated from the normal radioactive decay of uranium. Uranium has been around since the Earth was formed, and one of its most common isotopes has a very long half-life (4.5 billion years), which is the amount of time required for one-half of uranium to break down. Uranium, radium, and thus radon will continue to exist indefinitely at about the same levels as they do now. A series of tiny radioactive particles are generated as radon gas decays.

Rn is one of the heaviest substances that remains as a gas in normal conditions and would be considered a health hazard. It is one of the decay products of uranium-238. It is generated by the radioactive decay of its "parent isotope" radium-226 with a half-life of 1,620 years. Radon-222 decays to its "daughter isotope" polonium-218 with a half-life of 3.05 minutes. Radon-222, which has a half-life of 3.8 days, is the most stable isotope. Therefore, radon-222 is the radon isotope of most concern to public health because of its longer half-life (3.8 days). When either the gas or these particles are breathed into the lungs, some are deposited and will continue to emit radiation. Lifelong exposures to high levels of radon can cause lung cancer.

The majority of the mean public exposure to ionizing radiations is by radon gas. It is often the single largest contributor (about 55 percent) to an individual's background radiation dose and is certainly the most variable from location to location—and is actually a function of location, varying with the composition of the underlying soil and rocks. Radon gas from natural sources can accumulate in buildings and other structures, especially in confined areas such as basements. Radon can also be found in some spring waters and hot springs.

Radon (at concentrations encountered in mines) was recognized as carcinogenic in the 1980s, in view of the lung cancer statistics for miners. Although radon may present significant risks, thousands of people annually travel to radon-contaminated mines for deliberate exposure to help with symptoms of arthritis without any serious health effects.

Following a highly publicized event in 1984, national radon safety standards were set, and radon detection and ventilation became a standard homeowner concern. Here, a radon vent tube made of PVC is included in a new home.

Source: iStockphoto.com

Detecting Radon

How is radon detected? Actually, there are two types of inexpensive indoor radon-level detectors. The short-term detectors use activated charcoal to adsorb radon from the air and are typically used for tests with duration of two to seven days. The long-term alpha-track detector consists of a piece of special plastic inside a container. It is typically used for tests of 91 days or more. The detectors are sent to special laboratories for analysis when the collection time has expired.

Stanley Watras in 1984 discovered the possible danger of radon exposure in dwellings when an employee at the Limerick nuclear power plant in Pennsylvania set off the radiation alarms on his way to work for two weeks while authorities searched for the source of the contamination. They found that the source was high levels of radon—about 100,000 Bq/m^3 (2,700 pCi/L)—in his home's basement, and it was not related to the nuclear plant. Following this highly publicized event, national radon safety standards were set, and radon detection and ventilation became a standard homeowner concern, though typical domestic exposures are two to three order of magnitude lower (100 Bq/m3, or 2.5 pCi/l). Beginning in the late 1980s, this led to activists forming campaigns to raise awareness of radiation resulting from radon.

Radon as a terrestrial source of background radiation is of particular concern because, although on average it is very rare, this intensely radioactive element can be found in high concentrations in many areas of the world. Some of these areas, including Cornwall and Aberdeenshire (U.K.), Ramsar (Iran), and Kuala Lumpur (Malaysia), have high-enough natural radiation levels. These sites would already exceed legal radiation limits—the natural topsoil and rock deliver a high dose of radon. People living in these areas are exposed to a high dose of radon and can receive up to 10 mSv per year background radiation.

Health Effects of Radon

This discovery has led to a health policy problem: what is the health impact of domestic exposures to radon for the concentrations (100 Bq/m^3) typically found in some buildings?

Breathing high concentrations of radon can cause lung cancer. Thus, radon is considered a significant contaminant that affects indoor air quality worldwide. According to the

U.S. Environmental Protection Agency (EPA), radon could be the second most frequent cause of lung cancer, after cigarette smoking; and radon-induced lung cancer is the sixth leading cause of cancer death overall, causing 21,000 lung cancer deaths per year in the United States.

Any home can have a radon problem. The average person receives more radiation from radon each year than from all other sources. Almost all risks come from breathing air containing radon and its decayed products.

Air pressure inside homes (basements) is slightly lower than in the ground, creating a negative pressure, like a vacuum, that draws in radon from several feet away into the basement through openings and pores in concrete. Warm air inside homes moves upward like inside a chimney, and this chimney effect reduces air pressure in the basement. When the ground is soaked with rain, the radon gas in the ground moves to a warm opening such as a basement. This stack effect causes radon inflow that easily migrates into the home.

Concrete cures and passes moisture to the surfaces, creating a network of capillaries (pores). Almost half of the water used in poured concrete mix is surplus and has to evaporate. The pores allow a passageway for radon gases, water vapor, and liquid water to enter the basement. Heavy radon gas accumulates in basements and on lower floors. According to the residential radon lung cancer study completed in Iowa, the first floor of a home receives 40 percent of its air from the basement level. Sealing cracks in the foundation along with the basement walls, floors, or molding helps to minimize the entry of radon gas into the home. Covering all crawl spaces with a heavy polyethylene barrier and sealing it to the foundation wall also helps. It is also useful if the sump pits and floor drains are closed.

There is an official limit on radon levels that demands an action plan. Radon Act 51, passed by the U.S. Congress, set the natural outdoor level, which averages 0.4 (pCi/L), as the target radon level for homes. Therefore, a home with 4 pCi/L radon level needs to be fixed since it is within the action limit. The EPA has made it clear that the 4 pCi/L action limit is not a safe level. Therefore, the lower the radon level in your home, the lower your family's risk of lung cancer.

Unlike radon levels in homes, occupational radon limits are governed by law and regulations. The Miners Safety and Health Act (MSHA) applies to underground miners. Their annual exposure is limited to less than 4 WLM (Working Level Months), equivalent to 33 pCi/L, during working hours. The Occupational Safety and Health Act (OSHA) limits cumulative radon exposure in the workplace to 30 pCi/L, based on 40 working hours per week. Assuming the highest radon level in modern mines, the average person receives 4 pCi/L in their own home over 12 years—the same radiation dose as if he or she worked in a uranium mine for five years.

See Also: Cancers; Environmental Protection Agency (U.S.); Indoor Air Quality; Personal Consumer Role in Green Health; Radiation Sources.

Further Readings

Blaugrund, Andrea. "Confusion Mounting Over Radon." *Gainesville Sun* (April 4, 1988).
Catelinois, O., et al. "Lung Cancer Attributable to Indoor Radon Exposure in France: Impact of the Risk Models and Uncertainty Analysis." *Environmental Health Perspectives*, 114/9 PMID 16966089; PMC: 1570096 (2006).

Harrison, Kathryn and George Hoberg. "Setting the Environmental Agenda in Canada and the United States: The Cases of Dioxin and Radon." *Canadian Journal of Political Science*, 24/1 (1991).

United Nations Scientific Committee on the Effects of Atomic Radiation. "Publications." http://www.unscear.org/inscear/en/publications.html (Accessed May 2010).

U.S. Environmental Protection Agency. "A Citizen's Guide to Radon." http://www.epa.gov/radon/pubs/citguide.html (Accessed May 2010).

U.S. Public Health Service, in collaboration with the U.S. Environmental Protection Agency. "Toxological Profile for Radon." Agency for Toxic Substances and Disease Registry, December 1990.

Hassan Bagher-Ebadian
Independent Scholar

RECREATIONAL SPACE

People are increasingly acknowledging the influence humans have on the well-being of the planet, as the growing interest in environmental preservation evidenced by the green movement attests. The converse of this, that the environment significantly influences human health and well-being, is equally true. One arena that evidences the sundry ways in which physical environmental characteristics can influence human health is recreational space. Indeed, recreational space can impact personal health in several domains. The World Health Organization's definition of health explicitly encompasses more than the mere absence of disease and subsumes physical, mental, and social well-being. Recreational space is linked with well-being in each of these areas.

Recreational spaces include gardens; athletic fields and courts; open grass or sand spaces; paved and unpaved trails for walking, biking, and horseback riding; water features; playgrounds; picnic and group entertainment facilities; and space for cultural and educational programming. When established and maintained in urban settings, the health benefits of recreational space are undeniable. Recreational space provides a place for relaxation and relief from psychological stress, contributes to the safety of urban communities, and encourages physical activity (e.g., walking or bicycling for leisure or transport rather than relying on automobiles). Thus, recreational spaces not only provide physical, social, and mental health benefits but also confer additional economic and environmental health benefits.

Psychological Health

Recreational space provides salutogenic psychological/mental health benefits. Landscape architects have long recognized the importance of recreational space for the psychological health of urban populations. Contributing to mental health through fun and inspiration is an aspect of the mission statements of several park services. The California Department of Parks and Recreation strives "to provide for the health, inspiration, and education of the people," and the National Park Service aims for its parks to supply "enjoyment, education, and inspiration of this and future generations."

The psychological benefits of open and recreational space have been a growing area of research since Yi-Fu Tuan published his theory of topophilia in 1974. Topophilia is a person's innate attraction to safe, open topography. Many safe, open environments, including outdoor and natural recreational spaces, are considered restorative environments (i.e., places that afford visitors a chance to reduce psychophysiological stress and to restore mental resources that have been drained by directed attention fatigue). Recreational space (especially natural, outdoor recreational space) has consistently been identified as a highly restorative environment. Recreational spaces provide individuals opportunities to be away from stressors and directed attention. These spaces also incorporate amenities and design elements that fascinate visitors and offer an extent or breadth of opportunities and aesthetics that are compatible with visitors' needs. Among youth, use of natural recreational space is correlated with increased cognitive development. Recreational space also has therapeutic effects for individuals with attention deficit disorder.

Recreational space is especially effective in promoting psychological restoration because recreational spaces provide venues for participation in physical activity, which is also known to reduce stress and redirect attention in a beneficial manner, adding to the mental health benefits conferred by these naturally restorative environments. Participating in physical activity and visiting a recreational space is often a social event as well. Affiliating with a group of friends and/or teammates offers additional psychological health benefits.

Social Health

Recreational spaces provide common ground for people to come together and build social ties within their community. Researchers have found that green space and vegetation in cities facilitates stronger ties between neighbors. Compared to people living in barren spaces, individuals who live closer to green spaces participate in more social activities, know more of their neighbors, and feel a stronger sense of community belonging. Proximity to recreational spaces is also associated with lower crime rates, higher feelings of safety, and a closer-knit community.

The capacity for recreational spaces to promote social health benefits is grounded in a social and environmental justice framework. Some key characteristics of successful recreational spaces include (1) the ability to be used in multiple ways by many people and groups, (2) community participation and ownership of the space, and (3) the ease of accessibility and connectivity to surrounding areas. People of different ages and backgrounds utilize recreational spaces for many different reasons and at many different times throughout the day. Recreational spaces strengthen communities by reducing social isolation of vulnerable populations, increasing informal social control over spaces, and improving the spaces' safety. Community participation in city planning has been critical across municipalities in the shaping of public spaces that facilitate social benefits and increase the amount of green space in a city. City governments and residents increasingly acknowledge social justice issues embedded within resident proximity to parks. In order for a recreational space to successfully confer these health benefits, it must first be accessible to the community. If a recreational space is inaccessible due to either financial or transportation-related barriers, the health benefits of the space cannot be reaped.

Community involvement in recreational space begins early in the planning phase. The construction and revitalization of recreational spaces provides unique opportunities for community engagement. For example, the Great Park construction in Irvine, California, involved transforming a large military base and airport into a major regional open space, and community participation proved integral in planning the park. Public participation in such processes enhances social health and builds relationships through informal, face-to-face meetings, which can establish increased understanding and trust.

The green movement has provided additional reasons for recreational spaces to serve as sites for social activities. Volunteering to clean up local parks and open spaces is a common way for students and concerned citizens to engage in service learning and for schools, companies, and residents to connect with each other and with members of the local community. Other programming in recreational spaces (e.g., summer camps and sports leagues) facilitates socialization, feelings of ownership of recreational space, and appreciation for the natural environment.

Physical Health

Another reason that recreational space is important to social health is also a primary reason that recreational spaces are beneficial to physical health—they provide places for people to play and be active. Recreational space has reemphasized the importance of play not only for children (e.g., through informal sports activities as well as formal camps and leagues) but also for adults (e.g., running groups and sports clubs often meet and exercise in public recreational spaces). Research on after-school programming, sports leagues, and physical education suggests that participation in sports and play can be a direct or indirect facilitator of crucial social skills like conflict resolution, leadership, and teamwork. Moreover, engaging in play and sport is associated with enhancements of both physical health and social cohesion. Physical activity is associated with lower rates of chronic illness (e.g., obesity, cardiovascular disease, diabetes, and cancer) and increased energy levels, increased immune function, and many other physical health benefits. Proximity and access to open recreational space has also been associated with reductions in community health disparities.

Access to environmental resources like parks, fields, and trails is a robust predictor of physical activity behavior. Individuals with convenient access to environmental resources are more likely to engage in physical activity, and this finding has been reported in child, adolescent, and adult populations from nations around the world. So substantial is this link between recreational space and health that interventionists aiming to promote physical activity and/or reduce obesity have begun targeting environmental changes, as this strategy provides a more cost-effective and impactful public health measure than targeting individual factors like exercise motivation or self-efficacy.

Additional and Overlapping Health Benefits

Recreational space is associated with positive psychological, social, and physical health outcomes including restoration from directed attention, increased positive affect/reduced negative affect, reduction in perceived stress levels, reduced resting blood pressure, increased quality of life, cognitive benefits, residential satisfaction, and an enhanced sense of place and identity. It is important to note that these benefits are not independent; they can and do co-occur—recreational spaces can provide the same individuals with benefits in all three of these domains and can also confer additional health benefits beyond those that fall neatly into one of these three categories. For example, in addition to the individual health benefits associated with recreational spaces, these spaces also promote environmental health in many ways. Appropriately designed recreational spaces can serve as storm water runoff abatement systems by including grassy swales and wetland areas for runoff drainage or as carbon sinks with the inclusion of leafy plants to increase carbon dioxide intake; they can also reduce the heat island effect of higher temperatures around heavily paved and built urbanized areas. Proper design and planting of flora provide for biodiversity in the community and bring in diverse wildlife from butterflies and bumblebees to hummingbirds and hawks. To this end, the Audubon Society has incorporated a sanctuary program whereby golf courses, one type of large recreational space, can be certified as natural habitats for numerous bird species. Recreational spaces also provide opportunities for intergenerational enjoyment of nature and a forum for sharing and valuing the natural environment.

Community gardens are among the recreational spaces capable of conferring multiple health benefits to humans and also to the environment. Community gardens are sources of healthy, local foods and have become central in the growing food justice movement focused on local food systems. Community gardens are places for neighborhood residents to gather, socialize, and create a community resource together. The garden vegetation positively impacts local air quality and provides a place for runoff. Community gardens also serve as educational venues for schools and communities, displaying sustainable food production practices that promote individual and environmental health. In addition, community gardens can also contribute to individuals' physical health—the Centers for Disease Control and Prevention (CDC) has reported that engaging in lifestyle activities, including gardening, for 30 minutes per day is sufficient for satisfying physical activity requirements associated with positive health benefits.

Conclusion

Unlike many other measures taken to improve individual health (e.g., prescription drugs, spa treatments, vacations), the health benefits associated with recreational spaces are not typically accompanied by a steep economic cost. Indeed, recreational spaces contribute to economic as well as individual health. Creating or expanding a public recreational space increases a community's resources and green spaces, which in turn increases property values and contributes to the revitalization of neighborhoods, attracting or retaining businesses, tourists, and residents. Recreational space offers pleasant scenery to shoppers, storekeepers, and residents while enhancing perceived and objectively assessed safety and providing for the health of the individual, the community, and the environment.

See Also: Mental Exercises; Physical Activity and Health; Topophilia; Urban Green.

Further Readings

Frumkin, Howard. "Healthy Community Design." *CDC Factsheet* (June 2008). http://www
.cdc.gov/healthyplaces/healthy_comm_design.htm (Accessed June, 2009).

Hartig, Terry. "Green Space, Psychological Restoration, and Health Inequality." *The Lancet*,
372 (2008).

Ho, Ching-Hua, Laura Payne, Elizabeth Orsega-Smith, and Geoffrey Godbey. "Parks,
Recreation and Public Health." *Parks & Recreation*, 38/4 (2003).

Hug, Stella-Maria, Terry Hartig, Ralf Hansmann, Klaus Seeland, and Rainer Hornung.
"Restorative Qualities of Indoor and Outdoor Exercise Settings as Predictors of Exercise
Frequency." *Health & Place*, 15/4 (2009).

Kaplan, Rachel and Steven Kaplan. *The Experience of Nature: A Psychological Perspective.*
Cambridge, UK: Cambridge University Press, 1989.

Tuan, Yi-Fu. *Topophilia: A Study of Environmental Perception, Attitudes, and Values.* Englewood Cliffs, NJ: Prentice Hall, 1974.

Daniel J. Graham
University of Minnesota

J. Aaron Hipp
Washington University in St. Louis

Victoria Lowerson
University of California, Irvine

RECYCLED WATER

Recycling, or reuse of previously used, water is becoming increasingly prevalent in areas where fresh potable water is scarce. Perhaps the most infamous type of water recycled is indirect potable reuse, sometimes referred to as "toilet-to-tap." But not all recycled water is used for drinking water. Other uses of recycled water include direct nonpotable reuse and greywater.

Domestic wastewater is made up of blackwater and greywater. Blackwater is the wastewater from toilets. Greywater is the wastewater from washing laundry, baths and showers, and sink water with food wastes removed. The majority of the disease risk from untreated wastewater comes from blackwater, due to the large amount of pathogens naturally present in feces. Greywater can be reused for noncontact uses, which do not involve bathing or washing, such as irrigation or toilet flushing. A bathroom design that uses sink greywater for flushing toilets, popular in Japan, is now used in some homes in the United States. Reusing greywater not only saves potable water, but it reduces the amount of wastewater entering sewers through reuse. Greywater often requires filtration and disinfection for reuse.

Not all water treatment is the same. The requirements for the treatment of wastewater are not as stringent as for drinking water. Most wastewater today requires tertiary treatment, which includes solids removal, biological treatment, and additional processes to remove nutrients, metals, or chemicals. Tertiary-treated wastewater meets the requirements for release into the environment and can be utilized for nonpotable uses, such as irrigation. This water is often referred to as "reclaimed" water. Within the United States, areas experiencing sustained water shortages may reuse treated wastewater.

The state of California has mandated that, where treated wastewater is available, it should be utilized for green-belt irrigation instead of potable water. San Diego, which depends mostly on imported water, adopted a water reclamation ordinance in 1989. This program has issued official rules and regulations for the reuse of water in the city that are meant to prevent human consumption of the reclaimed wastewater, prevent cross contamination between potable water and reclaimed water systems, and isolate the reclaimed water from contamination by other sources. One of the most famous elements of these rules is the requirement that reclaimed water be conveyed in purple piping, providing a means of easily distinguishing the nonpotable reclaimed water from potable water. This program allows the irrigation of green spaces within San Diego, while reducing the demand on the precious resource of potable water.

In contrast to nonpotable recycling uses of water, recycled potable water must be treated to drinking water standards. After traditional tertiary wastewater treatment, additional treatment will most likely occur through the use of reverse osmosis membranes, which reject impurities at the molecular level, followed by advanced oxidation to inactivate remaining pathogens. Recycled potable water is generally not injected directly into drinking water distribution systems, but goes through indirect potable use (IPU). In the United States, IPU has been used as recharge water to augment drinking water sources, such as groundwater and reservoirs. This water is mixed with the background surface or ground water, which then undergoes the normal drinking water treatment process. Perhaps the most famous IPU project in the world is in Singapore.

Because freshwater is scarce on the island nation of Singapore, IPU was explored as a means of reducing reliance on the importation of water from Malaysia. Singapore began treating its wastewater with reverse osmosis and ultraviolet light, and branded it "NEWater." In 2002, a review by the Singapore National Water Agency approved NEWater for blending into the raw water reservoirs, prior to drinking water treatment. Initial amounts introduced (3 million gallons per day) represented about 1 percent of the total water consumption, with plans in place to increase the amount of IPU water to 2.5 percent of the daily water consumption by 2011. NEWater is also used for direct nonpotable use in industrial settings, such as for wafer fabrication. By reusing recycled wastewater, through IPU and nonpotable uses, Singapore will be able to reduce its reliance on imported water in the future.

There are a variety of ways to recycle wastewater, each of which requires the appropriate level of treatment. Greywater can be used for noncontact purposes with little more than the removal of solids. Toilets that reuse greywater can reduce overall freshwater demand. Tertiary-treated wastewater can be used for nonpotable reuse, such as irrigation of green spaces within cities. Nonpotable reuse can allow the development of parks and golf courses without increasing the demand on limited freshwater supplies. Wastewater treated with advanced techniques may be used to augment water supplies going into drinking water plants through indirect potable use, which can provide relief to areas experiencing water shortages.

See Also: Bottled Water; Chlorination By-Products; Fertilizers; Groundwater; Reverse Osmosis; Water Scarcity; Waterborne Diseases.

Further Readings

City of San Diego Water Department, Recycled Water Program. "Rules and Regulations for Recycled Water Use and Distribution Within the City of San Diego." (September 2008). http://www.sandiego.gov/water/pdf/rulesandregs.pdf (Accessed September 2010).

Lepisto, Christine. "WaterSaver Technologies Aqus Uses Sink Greywater for Toilet." *Design & Architecture (Bathroom)* (2006). http://www.treehugger.com/files/2006/10/watersaver_tech.php (Accessed September 2010).

New Mexico State University. "Safe Household Use of Greywater." Guide M-106. College of Agriculture, Consumer and Environmental Sciences, New Mexico State University. Revised by Marsha Duttle, February 1994.

PUB, Singapore's National Water Agency. "Singapore Water Reclamation Study: Expert Panel Review and Findings." (June 2002). http://www.pub.gov.sg/newater/AboutNEWater/Documents/review.pdf (Accessed September 2010).

Michelle Edith Jarvie
Independent Scholar

REGIONAL DUST

Most dust comes from natural sources such as volcano ash, salty ocean aerosols, and desert sand. Dust is also produced by eroded roads, construction sites, volcanoes, and wildfires, and it contains pollutants from factory and auto emissions and oil, gas, and mineral mining. These man-made particulates mix with natural-source dust, creating a potent cocktail that has the ability to impact human health and fragile ecosystems. Dust is most prevalent in regions with features such as arid deserts, desiccated farmland, and barren river and sea basins, though the regional dust these terrains produce does not remain static; regional dust travels from one area to another, across continents and oceans—sometimes, you can even see unfathomably huge dust clouds from space. Scientists have been studying dust for more than 150 years, and in the past several decades, they have become increasingly interested in researching the ways that dust alters the environment and impacts human and ecosystem health.

Scientists estimate that as many as two billion metric tons of dust travel through the Earth's atmosphere every year. The Aral Sea, located in Uzbekistan and Kazakhstan, has decreased so dramatically in size that massive dust clouds, containing both pesticides and herbicides, are common to the area.

Source: Wikipedia/Staecker

The most prolific sources of dust are the deserts of Asia, Africa, and the Middle East, with the Sahara and the Sahel regions of north Africa believed to be the most significant sources of airborne sediments. Florida and other southern and eastern states exhibit high concentrations of African dust, and scientists estimate that 13 million metric tons of African dust fall on South America's North Amazon Basin each year. Mongolia's Gobi Desert and China's Taklimakan Desert create regional dust that, as it moves east through China's heavily industrialized regions, mixes with man-made pollutants, blanketing Beijing and other parts of China before crossing the Pacific Ocean into the United States. Scientists have found that there are days when nearly one-third of the air above Los Angeles and San Francisco can be traced back to Asia; further, most of Hawaii's soil originated in China and central Asian deserts. Overgrazing, deforestation, and certain agricultural practices contributed significantly to China's desertification, and a large regional dust storm here can deliver 4,000 metric tons per hour of Asian dust to the Arctic; pesticides and herbicides originating in Asia have been found in animal tissue and human breast milk among the Arctic's indigenous populations.

In the United States, abandoned agricultural fields and military training grounds as well as dry lands created by deforestation, livestock grazing, and farming create regional dust storms that can be seen via satellite imagery from space. Dry lake beds such as Owens Lake in Southern California also serve as a source of airborne dust. It is estimated that about 8 million metric tons of lake-bed dust are carried into the atmosphere each year from Owens Lake—just one of countless dry lake and sea beds that mark the Earth. Lake Chad, in North Africa, and the Aral Sea, located in Uzbekistan and Kazakhstan, have both decreased so dramatically in size that the formation of massive dust clouds above them is a common feature; the dust has been found to contain both pesticides and herbicides, and DDT has been found in local inhabitants' breast milk.

Scientists estimate that as many as 2 billion metric tons of dust travel through the Earth's atmosphere every year. The drier the atmosphere, the longer particulates can remain aloft and cruise the world—some linger in the air for more than three years. Regional dust is moved most effectively via weather systems such as the Santa Ana winds, which are hot, dry winds that blow across Southern California during the fall and winter months, and "harmattan," a dry wind from the Sahara that blows toward the west coast of Africa during the winter season. Harmattan winds carry Saharan dust southward, across Nigeria, where rising desertification adds more dust to harmattan winds. Harmattan brings with it colds, flu, and asthma attacks. Similar to the Santa Ana winds and harmattan, the "shamal" are winds that blow persistently during the summer months over Iraq and the Arabian Gulf. The shamal sends dust and sand into the air that can remain aloft for days.

Scientists use trajectory computer modeling to trace regional dust back to its place of origin, though on-site monitoring of particulate levels provides the most reliable evidence of regional dust levels and their travels. Scientists have found that microbes can survive at altitudes as high as 77 kilometers above the Earth's surface and that fungus can survive even after being subjected to ultraviolet rays, low temperatures, repeated freezing and thawing, and high vacuum levels. Pollen grains have been discovered at nearly 20 kilometers above the Earth's surface, and researchers have found that the number of airborne microorganisms dramatically increases following a dust event. Thus, it is clear that regional dust and sediment are not only hardy but can also be transported across great distances, significantly impacting the Earth's air quality, no matter where such dust originates.

The role of dust in the environment is complex, with some of its effects being benign and even useful and others being quite harmful. For instance, scientists believe that the sulfates contained in mammoth dust clouds actually work to cool the Earth by reflecting sunlight away from the planet. Dust also carries useful nutrients to the ocean that help sustain microscopic marine plants that decrease carbon dioxide in the atmosphere; dust also feeds soil around the world, helping to nurture agriculture and sustain vegetation. On the other hand, dust also acts to spread pollutants around the globe, and the solar-heat–absorbing soot in dust clouds warms the Earth. Further, when dust blocks sunlight, it reduces evaporation and rainfall, stunting crop yields. Still, most scientists agree that the overall effect of dust is to cool the Earth, though, as China and other countries work to lower or eliminate their sulfate emissions, global temperatures may climb higher and faster than predicted.

In her book *The Secret Life of Dust*, Hannah Holmes examines regional, globetrotting dust. Referring to China's dust as the "Asian Express," Holmes notes that not only does dust in China kill one million people a year while also hindering crop growth there, but that this export from China also seriously impacts pollution levels in the United States. Further, the Saharan Dust River—so vast that sailors were aware of it 150 years ago—creeps across the Caribbean and into Alabama, Mississippi, Louisiana, Texas, Illinois, and eventually the entire Eastern seaboard. Interestingly, this Saharan dust has actually had positive impacts— for example, loading Jamaica with so much bauxite that Jamaican bauxite mines are among the most productive in the world. Holmes also writes about cancer-causing dusts and dusts that lead to such diseases as miner's asthma, black lung, and silicosis, and she reports that the dust inside our houses contains poisons, pesticides, dinosaur bones, tire rubber, lead, molds, bacteria, smoke, and chemicals. Many theorize that the rise of dust, both inside and outside our homes, is the cause for the worldwide explosion of asthma; since 1970, the number of people with asthma has grown by 50 percent every decade.

Dust is classified according to size: inhalable dust, including silicon, calcium, iron, and aluminum, can be trapped in the nose, throat, and upper respiratory tract; and respirable dust, which is smaller and includes gases and volatile organic compounds, can be drawn into the lungs. Inhalable dust can cause irritation of eyes, ears, and nasal passages and can injure mucous membranes and skin, while respirable dust can cause bronchitis, asthma, and cardiovascular problems. It is estimated that particulate pollution kills as many as 9,000 Southern Californians each year, and the Environmental Protection Agency (EPA) estimates that air pollution contributes to tens of thousands of American heart-related deaths each year. The EPA is working to set air quality standards and to incorporate approved methods for monitoring pollution levels, and the European Union is also embracing policies to minimize air pollution. Globally, dust is linked to millions of deaths each year, and it exacerbates heart problems and is linked to lung cancer and emphysema. African dust has a direct effect on human health and is reported to be linked to meningitis in sub-Saharan Africa, and researchers are currently studying a possible connection between African dust events and high rates of asthma in the Caribbean.

Dust, then, is not merely dirt but is also composed of herbicides, pesticides, and numerous microorganisms including bacteria, viruses, and fungi. Regional dust, aided by weather systems, carries microbes and chemicals around the globe, seriously impacting both human health and ecosystems. Further research, rigorous monitoring of air quality, and new methods for controlling airborne sediment are necessary to diminish the negative effects of regional dust.

See Also: Asthma; Climate Change; Environmental Protection Agency (U.S.); Lung Diseases; Particulate Matter.

Further Readings

Griffin, Dale W., Christina A. Kellogg, Virginia H. Garrison, and Eugene A. Shim. "The Global Transport of Dust." *American Scientist*, 90/4 (2002).

Holmes, Hannah. *The Secret Life of Dust: From the Cosmos to the Kitchen Counter, the Big Consequences of Little Things.* Hoboken, NJ: John Wiley, 2001.

Merefield, John. "Dust to Dust." *New Scientist* (September 21, 2002).

Tani Bellestri
Independent Scholar

Reproductive System Diseases

The reproductive system is unique in its form and function compared with other organ systems in the human body. Two striking differences distinguish males from females in terms of reproductive system structure and function. The male reproductive system consists of essential (gonads) and accessory organs and glands (epididymides, vasa deferentia, ejaculatory ducts, seminal vesicles, and prostate gland) and supporting structures (scrotum, penis, and spermatic cord). The female reproductive system consists of essential (gonads) and accessory organs (uterine tubes, uterus, and vagina), external genital (vulva), and additional sex glands (mammary glands). The female reproductive system is also intimately involved with nurturing and developing a fetus, while the male's is not. Regarding function, the reproductive system exists mainly to sustain human survival from one generation to the next. Hormones produced by the sex organs are essential in the development of structural and functional differences between males and females. While differences exist between the male and female reproductive systems, there are many similarities, as well. For example, tissues for both come from the same embryologic structure, and some of the hormones that affect these two systems are identical.

The female and male reproductive systems are susceptible to environmental toxicants. Toxic chemicals can enter the system via ingestion, inhalation, and skin contact with water, food, air, and soil. The reproductive system can be environmentally exposed by nonregulated landfill areas, sewage discharge, and runoff into waterways, or occupationally exposed to toxins from industrial processes of plants or the application of pesticides during farming. Once those toxins enter into the human system, the amount and duration of these toxins become the two major factors that determine the degree of adverse effects to the reproductive system. Sexual behavior, onset of puberty, fertility, gestation time, and pregnancy outcome are among the potential factors affected by reproductive toxicity in females. The male reproductive system can be affected by toxins that alter sperm count, sperm morphology, and sexual behavior. Important effects of environmental/occupational exposure to toxicants on the reproductive system are discussed below.

Heavy Metals

The toxic effects of heavy metals on the reproductive system have become a major global health concern. Metals can be ingested, inhaled, and/or absorbed into the blood system where they are distributed to and accumulate in the reproductive organs. Broad-spectrum irreversible toxic actions at cellular and molecular levels have been observed in the reproductive systems of both experimental animals and humans. Heavy metals can directly induce pathogenic effects in the tissues of organs and can indirectly alter the functions of reproductive systems by compromising hormone actions.

Exposure to several specific metals—lead, mercury, cadmium, and manganese—have been linked to adverse reproductive outcomes in humans, wildlife, and laboratory animals. The potential toxicity of metals could cause alteration in sperm morphology, count, and motility in males, as well as causing biochemical disruptions of enzymes and hormones. Humans and wildlife are exposed to lead in water, food, soil, and air. In rodents, lead suppresses the follicle-stimulating hormone and luteinizing hormone, affects gonadotropin-receptor binding in the ovary, and alters steroid metabolism. In women, lead has been linked to an increased risk of spontaneous abortions, miscarriages, intrauterine fetal deaths, or preterm deliveries. In men, toxicity of metal is manifested in the male reproductive system by deposition of lead in the testes, epididymis, vas deferens, seminal vesicle, and seminal ejaculate. Lead has an adverse effect on sperm count and retards the activity of sperms. In addition, motility, prolonged latency of sperm melting, and increased incidence of teratospermia, both in exposed persons and experimental animals, were observed after lead exposure. One of the mechanisms associated with such effects is that lead affects calcium and potassium channels located in essential organs of the reproductive system, and that, in turn, induces abnormalities in male testes and sperm function.

Cadmium is a pollutant associated with several modern industrial processes, such as welding, soldering, mining, and ceramics. In animal studies, cadmium causes testicular growth and maturity at different stages from the norm. Also, the metal can decrease gonadal development in animals with retarded germ cell migration into the ridges, resulting in depleted populations of germ cells, defective maturation of gametes, and subfertility in male offspring. In female rodents, cadmium causes cleft palate, anencephaly, and anopthalmia. Also, it decreases production of human chorionic gonadotropin (hCG) and inhibits placental transfer of oxygen and nutrients to the fetus. In humans, a link between cadmium exposure and adverse reproductive outcomes has not been well studied. However, given the well-documented effects of cadmium on reproduction in laboratory animals, it is compelling to consider whether similar effects may occur in humans.

Humans and wildlife are exposed to three different forms of mercury: organic, elemental, and inorganic. Organic mercury is used as a fungicide in paints. Elemental mercury is used in dental amalgam fillings, batteries, and gold mining. Inorganic mercury is used in electrical equipment, fungicides, and antiseptics. All three forms of mercury are thought to adversely affect human health. In women, organic mercury has been linked to an increased risk of spontaneous abortion and birth defects such as microcephaly, cerebral palsy, and mental retardation. Elemental mercury has been linked to an increased risk of spontaneous abortion and menstrual irregularities. In men, mercury can affect testicular tissue, specifically Leydig cell membranes, and disrupt spermatogenic cells by allowing chromatin to divide improperly in the acrosome.

A number of mechanisms of metal toxicity on reproductive tissues have been suggested, including ionic and molecular mimicry, interference with cell adhesion and signaling, oxidative stress, apoptosis, and cell cycle disturbance. Although the overall effect of metal toxicity on reproductive tissues or organs is likely to be due to a synergism of several mechanisms, it is possible that one mechanism may predominate in a specific cell type.

Endocrine Disruptors_

During the last two decades, there has been widespread growing concern and public debate over the potential health effects resulting from exposure to endocrine disruptors. Endocrine disruptors are exogenous chemicals that mimic hormones, interfere with the production, release, transport, metabolism, or elimination of hormones that regulate developmental processes and support endocrine homeostasis in an organism. Chemicals with known estrogenic or androgenic properties include some pesticides, phthalates, phytoestrogens, and certain industrial waste products. Humans are exposed to endocrine disruptor chemicals (EDCs) in the workplace, home, community, or through food and water consumption.

Effects of EDCs on human hormones can range from minor to serious depending on the specific endocrine receptor and the amount of exposure. A wealth of data from animal studies has shown that in utero or prenatal exposure to xeno-oestrogens (DDT, mono-n-butylphthalate, bisphenol A, diethylstilbestrol or DES) causes hypospadia, undescended testes, cryptorchidism, reduced sperm counts, or in the worst case, intersex conditions, teratomas, and Leydig cell tumors. Also, exposure to the dioxin 2,3,7,8-tetrachlorod-ibenzo-p-dioxin (TCDD) and dioxin-like PCB is associated with an increased prevalence and severity of endometriosis and recurrent miscarriages. Substantial evidence of adverse developmental effects caused by endocrine disruptors comes from observations made in wildlife after accidental environmental disasters. For example, a well-known case of demasculinization and reproductive failure of alligators in Lake Apopka was caused by a spill of DTT, a weak oestrogenic compound, which is metabolized to a potent anti-androgen DDE, a metabolite of DDT.

Such clear effects of EDCs on reproductive organs have not yet been fully proved for humans. However, some epidemiological studies have shown an increased risk to reproductive health by some EDCs. EDC exposure has been found to be associated with increasing the risk of female reproductive dysfunctions (e.g., reduced fecundability and miscarriage). Studies have shown a positive association between the body burden of persistent chlorinated EDCs (PCB, DDT, and their metabolites, other chlorinated insecticides), and gynecological problems, especially recurrent miscarriages. Nevertheless, conflicting observations do exist in regard to the role of EDC exposure. A study on Japanese patients with a history of recurrent miscarriages has not found an association with serum levels of industrial products such as PCBs, hexachlorobenzene (HCB), or DDE. Also, the effects of phthalates on the reproductive function of women and men apparently have not been reported, although several animal studies indicated adverse affects on their reproductive function.

The major limiting factor in drawing any conclusions about female reproductive system effects and EDCs is the scarceness of actual exposure data. In most studies to date, exposure can only be inferred and not actually measured. Another problem lies in the sample sizes, which are often too small to allow detection of an effect. Although there is evidence for geographical differences and temporal trends in some aspects of human reproduction, there has been no systematic attempt to look for evidence that the mechanisms behind these changes could involve endocrine pathways. Despite these drawbacks, the biological plausibility of possible damage to human reproduction derived from exposure to EDCs seems strong when viewed against (1) the background of known influences of endogenous and exogenous hormones on many of the processes involved and (2) the evidence of adverse reproductive outcomes in laboratory animals exposed to EDCs.

Solvents

Solvents are a variety of chemicals widely used in electronics, dry cleaning, auto repair, glues, and paints. Exposure to solvents can come through inhalation or absorption by the skin and transport to essential and accessory reproductive tissue via the bloodstream. Solvents known as 2-bromopropane, tetrachloroethylene, perchloroethylene, toluene, xylene, and styrene all have been linked to adverse reproductive outcomes in laboratory animals and humans.

Tetrachloroethylene and 2-bromopropane are used in dry cleaning plants. Workers in such facilities are at risk for exposure to high levels of solvents via inhalation and skin absorption. Studies have shown that dry cleaning workers have a higher rate of menstrual disorders, infertility complications, and disruption to the reproductive and endocrine system. Specifically, women exposed to 2-bromopropane showed secondary amenorrhea, primary ovarian failure with high follicle-stimulating hormone levels, hot flashes, and undetectable estradiol levels. Exposed men displayed azoospermia, some degree of oligospermia, or reduced sperm motility. One mechanism of pathogenic effect comes via high concentration exposure to 2-bromopropane, which inhibits and/or damages the reproductive blood supply. The 2-bromopropane mechanism causes moderately atrophied testes, decreased germ cell concentration, and severely atrophied seminiferous tubules.

Toluene, a petroleum distillate and one of the most abundantly produced chemicals in the United States, is a solvent for paints, lacquers, thinners, and adhesives. Readily absorbed from the lungs, usually by inhalation, exposure reduces fetal weight, delays skeletal development in laboratory animals, and may increase the risk of spontaneous abortion in women.

Xylene delays fetal development, reduces birth weight, lowers serum progesterone and estrogen level, and prevents ovulation in rodents. It also increases the risk of spontaneous abortion and induces caudal regression in humans. Another solvent, glycol ether, can contribute to spontaneous abortion and a decrease in fertility in laboratory animals and may increase the risk of low-motile sperm count and decreased semen concentrations in humans. A possible mechanism of glycol ether is a direct toxicity on epididymis tissue or an indirect toxicity leading to hemolysis or microangiopathy.

An epidemiological study showed that occupational exposure to organic solvents can increase risk of infertility. The most common primary factor responsible for infertility was attributed to ovulatory dysfunction. However, the absence of data about dosage and duration of exposure and possible recall bias from confounders of epidemiological studies weaken the ability to make a firm conclusion on the link between exposure to solvents and adverse reproductive effects.

Pesticides

Pesticide factory workers and/or those engaged in manufacture, formulation, and application of pesticides are exposed to the chemical via different routes while handling or spraying. The general population also suffers exposure to pesticides or their metabolites to some extent. The negative effects of pesticide exposure on semen quality had been shown as early as the 1970s. A classic example of a reproductive toxicant is 1,2-dibromo-3-chloropropane (DBCP). Its spermatotoxic effect in rats was discovered in the 1960s, but its deleterious effect on human spermatogenesis was not discovered until 1977. Other pesticides, such as methoxychlor and chlordecone, also produce adverse effects on both the male and female reproductive system.

Organophosphate pesticides (i.e., parathion, malathion, and diazanon), herbicides, and fungicides have been well studied in laboratory animals. In rats, organophosphate pesticides inhibit the growth of ovarian follicles, induce premature ovulation, decrease the levels of luteinizing hormone and progesterone in the blood, and cause poor oocyte development. Herbicides have been known to cause fetotoxicity, pseudopregnancy, ovarian regression, and anovulation. Specifically, methoxychlor accelerates vaginal opening, induces abnormal estrous cyclicity, inhibits luteal function, blocks implantation, reduces fertility, and decreases litter size. Chlordecone or kepone, a synthetic insecticide and known carcinogen, has adverse effects on spermatogenesis related to decreased sperm concentrations, percentage of motile sperms, and low percentages of morphologically normal sperm.

In humans, organochlorine pesticides are found in breast milk and adipose tissue. Some investigators believe that these chemicals are linked to an increased risk of breast cancer, spontaneous abortion, and preterm delivery. However, future studies are required to clarify such links because most previous investigators have used small sample sizes, indirect measures of exposure, or ecological data.

Air Pollution

Some air pollutants in outdoor and indoor ambient air have been linked to reproductive problems. In the past 30 years, outdoor levels of some pollutants, such as particulates, sulfate oxides, and carbon monoxide, have been declining in many U.S. and eastern European cities due to emission controls on vehicles, heating systems, power generation, and industry; however, many outdoor air quality problems still exist in the developed world and are worsened by increasing use of fossil fuels and industrial chemicals. Air pollution can enter the body via the airway and pass from the lungs into the bloodstream, which conveys toxins to the male and female reproductive system. Animal studies have definitively shown that certain air pollutants affect reproductive health, including a higher number of implantation failures, lower fetal and placental weights, and changes in functional morphology of the placenta.

Epidemiological studies report that air pollution may decrease male fertility, alter sperm quality, and result in prematurity. Particulate matter and polycyclic aromatic hydrocarbons generated from vehicle emissions are associated with decreased sperm concentration, morphology, and DNA breakage. Sulfur oxides generated from burning coal, vehicle emissions, and oil refineries are linked with abnormalities in the testis reproductive function. Also, ozone was observed to link with abnormal sperm concentrations. In addition, there is a growing body of epidemiologic literature reporting an association between ambient pollutants and reproductive outcomes, particularly birth weight and gestational duration. For example, outdoor carbon monoxide and sulfur oxides significantly increased the risk of preterm births. Formaldehyde levels have been linked with increased rates of preterm births. Also, higher levels of outdoor NO_2 (nitrogen dioxide) have been associated with a significant increase in sudden infant death syndrome, as contrasted to higher CO (carbon monoxide) levels without that reproductive outcome.

Smoking

Cigarette smoking has been directly and indirectly related to reproductive health outcomes. Smoking has been a risk factor in ovarian cancer and prostate cancer. Many chemical compounds found in tobacco products have malignant/deleterious effects on the male and female reproductive pathways. For instance, cigarette smoking prevents the ovarian follicle from maturing properly by increasing gonadotropin dose and duration of controlled ovarian hyperstimulation. Oocytes affected by smoking have a decrease in maturity, fertilization rate, embryos per cycle, and embryo quality. In males, carcinogenic chemicals found in cigarettes reduce sperm production and increase sperm morphology abnormalities due to secretory dysfunction of the Leydig and Sertoli cells.

Specific compounds in tobacco products and smoke from their combustion, such as dioxins, polycyclic aromatic hydrocarbons, and metals, have been getting research attention.

Most of the toxic effects on the reproductive system are induced by dioxins and polycyclic aromatic hydrocarbons mediated by the aryl hydrocarbon receptor (AhR). The resulting compounds produce an AhR-ligand complex, which translocates into the nucleus and forms a heterodimer along with a coactivator. Subsequently, a potential condition arises to initiate or induce carcinogenesis. Metals such as zinc and cadmium are found in tobacco products. Those metals can affect the plasma membrane $Ca2^+$-ATPase activity of sperms. In a recent study, levels of cadmium had decreased $Ca2^+$-ATPase activity to the point of decreasing sperm motility. The interference of $Ca2^+$-ATPase activity may potentially activate a pathway for carcinogenesis and produce low sperm viability. The evidence at hand has demonstrated that tobacco products and smoke from their use produce deleterious effects on the reproductive system. However, the potential of cigarette smoke for activating the dioxin or metal pathway needs further evaluation to fully elucidate the mechanism on pathogenic effects.

Conclusion

Animal studies have found that certain endocrine disruptors, metals, solvents, and pesticides are linked with increased risk of adverse effects on both male and female reproductive systems. At the same time, scientific knowledge about the harmful effects of environmental and occupational toxicants on human development and reproductive functions remains limited and indirect. Particularly, the biological exposure monitoring data with effect parameters are insufficient to indicate exactly which chemicals or compounds are responsible for reproductive impairment. Also, there is still lack of concrete protocols as to how to proceed with understanding reproductive toxicology on humans. For example, there are no definitive methods for classifying the effect of specific chemical toxins on humans as done for carcinogens. Nevertheless, epidemiological studies on the association between environmental/occupational exposure and the reproductive system, including growth and development, are relatively new. Some studies showed an association between exposure to toxicants and certain reproductive outcomes. Accumulated studies provide vague evidence that suggest a weak association, while others fail to find evidence of a relationship between environmental toxicants and adverse reproductive health effects in humans. This stands in contrast to animal studies that clearly demonstrate that outcome. With this body of evidence and the potential for grave harm in mind, environmental toxicants should be considered a risk factor for adverse biological effects on the human reproductive system.

See Also: Automobiles (Emissions); Biological Control of Pests; Centers for Disease Control and Prevention (U.S.); Dental Mercury Amalgams; Dry Cleaning; Men's Health; Particulate Matter; Pest Control; Women's Health.

Further Readings

Caserta, D., L. Maranghi, A. Mantovani, R. Marci, F. Maranghi, and M. Moscarini. "Impact of Endocrine Disruptor Chemicals in Gynecology." *Human Reproduction Update*, 14 (2008).

Chowdhury, A. R. "Recent Advances in Heavy Metals Induced Effect on Male Reproductive Function—A Retrospective." *Alameen Journal of Medicine Science*, 2 (2009).

Curtis, L., W. Rea, P. Smith-Willis, E. Fenyves, and Y. Pan. "Adverse Health Effects of Outdoor Air Pollutants." *Environmental International*, 32 (2006).

Kumar, S. "Occupational Exposure Associated With Reproductive Dysfunction." *Journal of Occupational Health*, 46 (2004).

Sharara, F. I., D. B. Seifer, and J. A. Flaws. "Environmental Toxicants and Female Reproduction." *Fertility and Sterility*, 70 (1998).

Hueiwang Anna Cook Jeng
Roberto Carlos Mendez
Old Dominion University

Reverse Osmosis

Reverse osmosis is a method of liquid filtration that removes larger molecules from smaller molecules by forcing the liquid through a membrane. The process usually involves high pressure and a membrane with holes that will only allow the smaller particles to pass through. Jean Antoine Nollet first described reverse osmosis in 1748. Studies led to the production of freshwater from seawater in the 1950s, and today over 10,000 desalination plants are in operation worldwide with a combined capacity of over 35 million cubic meters per day. Reverse osmosis should not be confused with filtration or straining. Reverse osmosis uses pressure, 600 to 1,000 pounds per square inch (psi), to achieve the osmotic process and therefore is known as the "reverse" of normal osmosis.

Common uses of reverse osmosis include the following:

- Water purification for drinking and industrial use, and wastewater purification for industrial, agricultural, and landscape use
- Water purification for patients undergoing kidney dialysis that will remove waste products that their diseased kidneys cannot remove
- Concentration of food juices without damage to proteins and enzymes
- Production of whey protein isolate from cheese manufacturing and concentration of milk to reduce the cost of shipping
- Removal of unwanted compounds in the wine industry
- Prevention of water spotting during the final rinse before the air blower dryers in the car wash industry
- Removal of water before the final boiling stage in maple syrup production
- Purification of tap water to remove chemicals that can kill sensitive organisms and cause algae growth in reef aquariums
- Desalination of sea or brackish water to obtain drinking water
- Household purification of home drinking water
- Military purification of water to remove nuclear, biological, and chemical agents in order to provide safe drinking and industrial water
- Hydrogen production to prevent mineral deposits on the electrodes

All methods of reverse osmosis require an energy input. Large desalination plants such as those in the Middle East rely on oil. They may be located near power plants to reduce energy transmission costs and to provide the power plant with cooling water. Pretreatment will remove larger solids and bacteria and adjust the pH of the water to prevent mineral scaling during the actual reverse osmosis process.

Most household reverse osmosis units are designed to go under the sink. They have low back pressure and therefore use large amounts of water, recovering only 5 to 15 percent of the water that enters the system. To produce 5 gallons of treated water, a home unit may discharge 40 to 90 gallons per day to the sewer or septic system. Under test conditions, a home unit can remove up to 92 percent of nitrate-nitrogen, 60 to 99 percent of total dissolved solids, 60 to 98 percent of sulfates, and 60 to 93 percent of sodium. A large-scale municipal or industrial system will have similar levels of efficiency for removal of nitrate-nitrogen, dissolved solids, sulfates, and sodium. The big difference between the home system and the larger system is that close to 50 percent of the water that enters the system will be recovered because the large-scale systems use higher pressures for performing reverse osmosis.

Concentration of food juices, maple syrup, production of whey, and reducing the weight of milk for shipment reduce manufacturing and transportation costs and greenhouse emissions. Reverse osmosis also preserves valuable proteins and enzymes that would otherwise be destroyed by a heat concentration process. Studies have shown that reverse osmosis of orange juice will increase its vitamin C concentration up to fourfold without changing sensory attributes or taste. Reverse osmosis can remove lactose from milk for those who are lactose intolerant. Milk solids can be increased from 5 to 6 percent to 10 to 30 percent of the final product by removing excess water and, at the same time, keeping milk temperatures between 40 and 80 degrees C. This manipulation of solids can fortify cheeses, increase the cream content of ice cream, yield low-fat or fat-free yogurts, and aid in the production of skim milk. During maple syrup production, 40 to 55 percent of the water is removed. This reduces not only the total energy consumption but also the exposure of the final syrup to high temperatures. Reverse osmosis in maple syrup production reduces the sap to finished product ratio from 55:1 to 11:1.

Reverse osmosis is used in the wine industry to remove water from wine must, the freshly pressed juice containing skins, seeds, and stems. It can also be used to remove excess alcohol from finished wine. This may become more important in the future because global warming increases sugar content of ripe grapes and thus the alcohol content of the finished wine. Reverse osmosis is also used to remove 4-ethylphenol and related compounds, by-products of Brettanomyces yeast contamination. These compounds give wine an undesirable Band-Aid, barnyard, or horse stable bouquet.

The potential exists for reverse osmosis to be used in additional industrial and commercial settings, where it could reduce production, transportation, and energy costs.

See Also: Food Allergies; Groundwater; Recycled Water; Water Scarcity.

Further Readings

Crittenden, John, Rhodes Trussel, David Hand, Kerry Howe, and George Tchobanoglous. *Water Treatment Principles and Design*, 2nd ed. Hoboken, NJ: John Wiley 2005.
Wangnick/Global Water Intelligence. "Worldwide Desalting Plants Inventory." Oxford, UK: Global Water Intelligence, 2005.
"Water Review: Residential Reverse Osmosis." Lisle, IL: Water Quality Research Council, 1991.

Western Australia Water Corporation, Perth. "Desalinisation." Canberra: Australian State of the Environment Committee, Department of the Environment and Heritage, 2006.

Paul Richard Saunders
Canadian College of Naturopathic Medicine

RURAL AREAS

Rural areas are commonly characterized as large isolated regions in which there is a low ratio of inhabitants to open land. As numerous studies indicate, rural areas are difficult to define with great precision. While the transition from urban city to rural countryside is usually abrupt in low-income countries, it is much more gradual in high-income countries, making it difficult to define urban–rural boundaries. An additional challenge is that nations do not use the same statistical criteria for rural and urban populations because there is no consensus on such boundaries. For instance, in Japan, a population of fewer than 30,000 people is considered rural, whereas Albania classifies a group of more than 400 inhabitants as an urban population. The main economic activities associated with rural areas involve the production of agriculture and raw materials.

Over the past few decades, the rural countryside in high-income countries has slowly been depopulated. To mitigate this process, agricultural development specialists have suggested methods of increasing productivity within rural areas in order to limit migration of large numbers of farm workers off the land.

Source: U.S. Department of Agriculture, Agricultural Research Service/Scott Bauer

With the onset of climate change, farmers and rural communities are feeling a large portion of these environmental changes as many viable agricultural businesses are failing to yield produce.

Classification of Rural Areas

There remains limited consensus on the specific classification of what defines a rural area, although such a term often conjures up images of small towns, farms, and open spaces. Defining rural parameters can have major implications on provision of healthcare services and development of national policies, laws, and research. Researchers and government agencies use such definitions for statistical consistency and accuracy when conducting their studies. To date, a number of definitions have been identified that emphasize different variables, such as population density, geographic isolation, and/or economic output. When defining a rural area, four salient considerations include its administrative boundaries, land-use patterns, population density, and overall economic influence.

When defining a rural area, the key is to identify a rural–urban definition that best fits the needs of a specific activity. For instance, an administrative definition of rural emphasizes defining urban along municipal and other jurisdictional boundaries for determining program eligibility. On the other hand, land-use and population density data are more ideal when developing infrastructure and water and sewer services. For programs requiring the coordination of efforts within broader market areas, such as area-wide transportation planning assistance, a definition based on economic concepts may be more appropriate.

In the United States, there are three government agencies whose rural definitions and classifications are in wide use—the U.S. Census Bureau, the Office of Management and Budget, and the Economic Research Service of the U.S. Department of Agriculture. Of the three, the Census Bureau has taken the lead in developing a working definition of rural by defining what is urban, and then defining rural by exclusion. Based on this rationale, the Census Bureau defines an urbanized area as consisting of adjacent, densely settled census block groups that meet a minimum population density of at least 50,000 people. Subsequently, the Census Bureau defines all other areas as rural. Similar to this, the Organisation for Economic Co-operation and Development (OECD) defines a rural area based merely on the population density. However, a wide variety of literature has alluded to the inadequacy of such classification, suggesting that other variables, such as rural–urban commuting area (RUCA) codes, be considered for the urban–rural typology rather than a focus on population density calculations.

Depending on the population threshold and the choice of boundary, populations defined as rural may vary substantially. In 2000, 21 percent of the U.S. population was designated rural using the Census Bureau's land-use definition. Yet, raising the population size threshold for the land-use definition from 2,500 to 50,000 increases the rural population from 21 percent to 32 percent. Lowering the threshold for the economic definition from 50,000 to 10,000 decreases the rural population from 17 percent to 7 percent. Such alterations can pose large implications for the provision of goods and services and economic benefits provided to rural communities.

Rural Society and Economy

In the past, rural societies were characterized by their adherence to farming as a way of life. These cultures were not market oriented, since rural dwellers sought subsistence over surplus. Farmer families were traditionally large to include a greater percentage of males to work in the fields. Generally, however, as the children became older, there was not enough productive land for them to support their own families, forcing them to migrate to urban centers. Over the past few decades, the rural countryside in high-income countries has slowly been depopulated. As a result, an acceleration of urbanization has been occurring, often creating large slums in many urban centers. To mitigate this process, agricultural development specialists have suggested methods of increasing productivity within rural areas in order to limit migration of large numbers of farm workers off the land. Recommendations include improvements in soil technology and changes in seed stocks, irrigation, and drainage.

In the mid-20th century, the rural electrification project was implemented to raise the standard of rural living and slow the extensive migration of rural Americans to urban centers. Under the program, more than 98 percent of the farms in the United States have been equipped with electric power. The Rural Electrification Administration, which implemented the program, provided low-interest loans to farm cooperatives for the construction and operation of power plants and power lines in rural areas. Rural electrification established city conveniences, such as electric lighting and radio, to areas of low population density and enabled the automation of a number of farm operations. However, although rural electrification did contribute to bridging the gap between urban and rural life, it did not succeed in checking the movement of farm workers to cities.

In general, rural families tend to rely more heavily on earnings, less on public welfare, and more on jobs in the informal rural economy than families residing in urban centers. Consequently, earnings comprise a larger share of income in rural families and, therefore, a key solution to hinder rural poverty is to improve these earnings through added employment opportunities or earning supplements. Local economic development will occur in areas that have effective institutions and organizations as well as an active process for public participation. To promote sustainable rural economies, governments should encourage the development of regional trade associations, local industrial districts, producer cooperatives, and other forms of locally based entrepreneurship initiatives.

In terms of exports, little information exists on the extent to which exports benefit rural areas as well as which types of businesses participate in export markets, although the main economic activities associated with rural areas involve the production of agriculture and raw materials. The relative isolation of rural locations and distance to ports are barriers that make it more difficult for rural businesses to participate in world markets. Given today's increasingly globalized economy, the extent of participation in export markets by rural businesses is an important indicator of future prospects for the rural economy.

Rural Health

Residents in rural areas often report poorer health than their urban counterparts with respect to several chronic health outcomes including diabetes and heart disease. Additionally, reports have indicated that rural dwellers may have higher mortality rates and, consequently, a lower life expectancy. Some of this can be attributed to the fact that rural populations have less access to healthcare services, compared with those living in urban zones.

Typically, rural areas suffer from a chronic shortage of healthcare professionals, including physicians and other medical specialists. In the United States, more than 20 million people live in rural areas that have a shortage of physicians to meet their basic healthcare needs. To fill this gap, nurses and nurse practitioners have played a large role in providing primary care. This has become a dominant trend, particularly in low-income countries. To counter some of these dilemmas, governments have been spearheading incentive policies for physicians to engage in rural medical training through enhanced benefits and wages. Additionally, medical schools have adopted a selective admission policy to enhance a primary care choice in underserved communities.

The Impact of Global Climate Change on Rural Areas

Rural communities have been experiencing the impact of climate change. A large portion of these effects has been predominantly experienced by farmers whose agricultural businesses have failed to produce yield. In recent decades, rural communities have sought to diversify their economies so that they are less dependent on traditional farming. The range of rural products has been broadened to food production, tourism, and other service industries and small-scale manufacturing. Increasing fluctuations in weather patterns are leading to more severe floods, droughts, heat waves, and wildfires that have fostered a loss of agricultural productivity and resulted in food shortages.

Shifts in climate will bring different changes to different regions. Some areas may see greater natural resources because of increased rainfall. However, the poorest regions are least able to adjust to new conditions and are experiencing the negative consequences of failed crops. In many African countries, such warmer and drier conditions have curtailed growing seasons. Yields from rain-fed agriculture are expected to fall as much as 50 percent in some poor African countries, and fisheries production will also likely decline. Rural regions in Latin America are also expected to be affected as climate change will increase the saline content of the soil, which will reduce crop productivity. As previously productive lands become more arid, Latin America could also see greater desertification.

Rural communities tend to rely heavily on climate-sensitive resources such as agricultural land and local water supplies and engage in climate-sensitive activities such as arable farming and livestock husbandry. Climate change has reduced the availability of these local natural resources and limited the options for rural households that depend on natural resources for consumption or trade.

Some pilot projects in rural communities, such as ECOPERTH in northern Ontario, Canada, have proven successful in limiting the negative impacts of climate change. Rural communities have been engaging in community awareness campaigns and implementing environment-friendly initiatives, including using pesticide-free products, conserving rainwater, promoting local food production, and walking to work campaigns, among others. Before any projects are implemented, rural communities must be actively involved in the planning process in order to meet their community needs.

Climate Change and Rural Migration

It has been noted that the varying patterns of climate change have also resulted in altering migration patterns. For example, in Burkina Faso and many other African countries experiencing drought, residents of dry rural areas have been migrating to rural regions with greater rainfall. Migration has also been a survival strategy used by Ethiopian households in times of environmental stress. Other survival strategies include using food reserves, seeking local nonfarm employment, borrowing food, selling livestock, or selling farm and household equipment. Yet, once these livelihood options are exhausted, people often migrate to a new rural or urban area.

To address the impacts of climate change, governments will likely need to take policy action to reduce climate-related migration and support rural development of agriculture impacted by such conditions, particularly in rural regions of low-income countries. Although residents of low-income countries have contributed little to climate change, they are suffering disproportionately from the effects.

See Also: Chemical Pesticides; Climate Change; Health Disparities; Suburbs.

Further Readings

Cromartie, John and Shawn Bucholtz. "Defining the 'Rural' in Rural America." http://www .ers.usda.gov/AmberWaves/June08/Features/RuralAmerica.htm (Accessed February 2010).

"ECOPERTH: A Small Rural Community Takes Action on Climate Change." http://dsp-psd
.pwgsc.gc.ca/Collection/NH18-23-82E.pdf (Accessed February 2010).

Meze-Hausken, Elisabeth. "Migration Caused by Climate Change: How Vulnerable Are
People in Dryland Areas?" *Mitigation and Adaptation Strategies for Global Change*, 5/4
(2004).

U.S. Department of Agriculture. "Measuring Rurality: Rural–Urban Commuting Area
Codes." http://www.ers.usda.gov/briefing/rurality/ruralurbancommutingareas (Accessed
February 2010).

Kadia Petricca
University of Toronto

S

Seasonal Flu

The seasonal flu is an infectious disease caused by RNA viruses of the family Orthomyxoviridae (the influenza viruses) that affects birds and mammals. The name influenza comes from the Italian *influenza*, meaning "influence" (Latin: *influentia*). Influenza is a viral infectious disease that is considered one of the most important causes of respiratory tract infections. Symptoms are similar to those of the common cold, however, are more severe, and their beginning is generally abrupt. Influenza can affect a significant number of persons of all age groups during the occurrence of an epidemic. This disease frequently requires medical attention and hospitalization, contributing significantly to economical losses, excessive numbers of hospitalizations, and deaths. The ability of influenza A and B viruses to suffer gradual antigenic changes in their two surface antigens, the hemagglutinin (H) and the neuraminidase (N), complicates vaccination against this disease. Epidemics of influenza were responsible for an average of 36,000 deaths per year in countries such as the United States during the 1990s, affecting all age groups but mainly children less than 2 years old and adults older than 65 years.

Influenza viruses of types A and B are associated with the epidemics and outbreaks typical of influenza "seasons" (seasonal influenza). Influenza type C is thought to be associated primarily with milder common cold–like illnesses. Type A viruses are further subdivided into subtypes (e.g., H1N1 and H3N2). During an influenza season, one of the two influenza A subtypes or influenza B can be predominant, while in other years, all three viruses may be found (this is considered in the production of vaccines for protection against influenza viruses). Influenza viruses undergo rapid genetic evolution that eventually results in changes to the virus's antigenic characteristics. Influenza vaccine virus strains are selected each year to make sure that the vaccine is matched as closely as possible to the currently circulating influenza strains. Other subtypes of influenza A viruses occur in animals, and all 16 HA and 9 NA subtypes are found in birds (mainly in water fowl); interspecies transmission (1918 pandemic) and viral reassortment (1957, 1968 pandemics) may give rise to new subtypes able to infect and be easily transmissible between humans and cause the next pandemic.

Influenza is mainly transmitted by droplets disseminated by unprotected coughs and sneezes. Short-distance airborne transmission of influenza viruses may occur particularly in crowded, enclosed spaces. Hand contamination and direct inoculation of virus is another possible source of transmission.

From a geographical point of view, influenza is a global disease, occurring according to the season in specific regions. In temperate regions, influenza is a seasonal disease occurring typically in winter months: it affects the northern hemisphere from November to April and the southern hemisphere from April to September. In tropical areas, there is no clear seasonal pattern, and influenza circulation is year-round, typically with several peaks during rainy seasons.

The administration of the influenza vaccine for children (up to 6 years old) is recommended at a minimum age of 6 months for trivalent inactivated influenza vaccine (TIV) and at age 2 years for live, attenuated influenza vaccine (LAIV). Influenza vaccine can be administered annually to children aged 6 months through 18 years. For healthy, nonpregnant individuals (i.e., those who do not have underlying medical conditions that predispose them to influenza complications) aged 2 through 49 years, either LAIV or TIV may be used. Children receiving TIV should receive 0.25 mL if aged 6 through 35 months or 0.5 mL if aged 3 years or older. Two doses can be administered (separated by at least four weeks) to children aged younger than 9 years who are receiving influenza vaccine for the first time or who were vaccinated for the first time during the previous influenza season but received only one dose. The timing of the vaccination is important. This should be done before the start of the influenza season, but this depends on where the individual is geographically located. However, vaccine for visitors to the opposite hemisphere may not be obtainable before arrival at the travel destination. For travelers in the highest risk groups for severe influenza who have not been or cannot be vaccinated, the prophylactic use of antiviral drugs such as zanamivir or oseltamivir is indicated in countries where they are available. Amantadine and rimantadine may also be considered when the circulating strains are known to be susceptible. However, the latter drugs are not active against influenza B, and high frequencies of resistance in H3N2 and less often H1N1 viruses make them unreliable for prevention at the time of this writing.

In addition to the vaccine, other preventive measures can be taken. Whenever possible, avoid crowded, enclosed spaces and close contact with people suffering from acute respiratory infections. Hand washing after direct contact with ill persons or their environment may reduce the risk of illness. Ill persons should be encouraged to practice cough etiquette (maintain distance, cover coughs and sneezes with disposable tissues or clothing, wash hands).

In the event of an influenza pandemic, the world would be in a situation where potential vaccine supply will fall short by several billion doses from global needs. In response to this, the World Health Organization is preparing global action plans with specific short-, medium-, and long-term activities designed to increase influenza vaccine production and surge capacity, to identify key obstacles and driving forces, and to estimate funding needs, as well as to strengthen the engagement and collaboration of key partners and stakeholders in order to reduce the impact of epidemics and a potential pandemic.

See Also: Acquired Immune Deficiency Syndrome; Airborne Diseases; Bird Flu; Centers for Disease Control and Prevention (U.S.); Tuberculosis.

Further Readings

Centers for Disease Control and Prevention. "Immunization Schedules." http://www.cdc.gov/vaccines/recs/schedules/default.htm (Accessed April 2009).

Franco-Paredes, Carlos, Alfonso J. Rodríguez Morales, and José Ignacio Santos-Preciado. "Clinical and Epidemiological Aspects of Influenza." *Latin American Medical Student Science and Research*, 11 (2006).

Kieny, Marie Paule, Alejandro Costa, Joachim Hombach, Peter Carrasco, Yuri Pervikov, David Salisbury, Michel Greco, Ian Gust, Marc LaForce, Carlos Franco-Paredes, José Ignacio Santos, Eric D'Hondt, Guus Rimmelzwaan, Ruth Karron, and Keiji Fukuda. "A Global Pandemic Influenza Vaccine Action Plan." *Vaccine*, 24 (2006).

World Health Organization (WHO). "International Travel and Health 2010." Geneva: WHO, 2009.

Alfonso J. Rodriguez-Morales
Universidad de Los Andes, Universidad Central de Venezuela

SEVERE ACUTE RESPIRATORY SYNDROME

The spread of severe acute respiratory syndrome (SARS) in 2003 represented the first global infectious disease epidemic of the new century, affecting individuals in 29 countries. From February to May 2003, over 8,000 SARS cases were identified worldwide, with a 10 percent fatality rate. The epidemic had significant economic repercussions in many sectors, but those related to hospitality, tourism, and the service industry were directly impacted by a dramatic decrease in travel to SARS-affected areas. Due primarily to the adoption and strict adherence to traditional public health measures such as contact tracing, isolation and quarantine, travel restrictions, screening at international borders, and the implementation of stringent infection prevention measures in hospitals, the epidemic was halted by July 2003. The epidemic demonstrated how developments that uniquely define our contemporary globalized existence, such as intensified interconnections of places and people throughout the world and the highly mobile nature of our world today, are conducive to the rapid spread of infectious disease. In addition, the generally effective response to SARS highlighted another aspect of globalization, namely, the advent and role that an extensive communications and information network may play in combating a developing international infectious disease threat.

The earliest cases of SARS had been identified in Foshan, China, in November 2002, but its potential as a disease threat had not been recognized at that time. Sporadic outbreaks of a respiratory illness then referred to as "atypical pneumonia" continued sporadically throughout southern China for the next year. The global spread of SARS was initiated, however, when an elderly medical professor who had been treating patients with atypical pneumonia in the province of Guangdong unknowingly became infected and subsequently traveled to Hong Kong to attend a relative's wedding in mid-February 2003. During his stay at the Metropole Hotel, the disease spread to 11 other hotel guests, each of whom continued their respective travels to various other locations throughout the world. It is not known exactly how the disease spread to these other hotel guests, but environmental transmission through a contaminated elevator is suspected. A few days later, local disease outbreaks arose almost simultaneously in those destinations where the infected guests traveled, particularly impacting major world cities, including other parts of Hong Kong as well as Toronto, Singapore, and Hanoi. Although most of these outbreaks were limited to the hospital settings (i.e., nosocomial transmission), there were at least three major community outbreaks: one in the Amoy Gardens apartment complex in Hong Kong involving 300 people; the second in the Pasir Panjang wholesale market that resulted in 14 cases; and a cluster of 31 cases within a closely knit religious community in Toronto.

The Etiology of SARS

SARS seriously damages the alveoli of lungs, thus interfering with the process of respiratory gas exchange. In early April, the causative agent of SARS was discovered to be a novel strain of the coronavirus—the viral family in which one of the most prevalent causes of the common cold is found. The airborne virus is transmitted directly from person to person on respiratory droplets when an infected individual coughs or sneezes but can survive on some surfaces for several hours, thereby enabling environmental transmission as well. As is true of many viruses that affect human beings, SARS could be traced to an animal reservoir, namely, the palm civet cat (although some scientists also implicate the horseshoe bat). The palm civet cat is used for human consumption in parts of southern China, and it is suspected that the crossover of the virus from the civet to human beings probably occurred in the live animal markets of that region—as indicated by the fact that as many as one-third of the early cases of SARS involved food handlers.

The incubation period of the virus (i.e., the period between exposure and symptom development) is two to 10 days, with the individual being most contagious at the time he or she is most ill. These two biological characteristics had significant implications for the nature of disease diffusion. First it meant that air travel became an important factor in the spread, since the incubation period of SARS was shorter than the travel time between most global cities of the world. Consequently, seemingly healthy individuals who were in a position to travel could unknowingly carry SARS to international locations. Second, because the time at which the individual was most infectious coincided with the greatest experience of illness, he or she would more likely admit him or herself to the hospital at that time he or she posed the greatest public health threat. For this reason, disease outbreaks would more likely occur in the hospital setting rather than in the community. A third factor that played a role in the spread of SARS was the phenomenon of "superspread," where some infected individuals, for unknown reasons, were able to infect an inordinate number of other susceptible individuals. Thus, for example, one young woman returning home from the Hong Kong Metropole Hotel infected 20 others in a Singapore hospital.

The Global Response to SARS

During the early stages of the pandemic, the government of the People's Republic of China did not reveal to the international community the extent to which the mysterious illness was spreading. This shroud of secrecy could not be maintained, however, as reports of the mysterious disease were transmitted informally through cell phones and the Internet. One of these reports was captured by Canada's Global Public Health Information Network (GPHIN)—a computerized search engine that monitors websites, news wires, local online newspapers, public health e-mail services, and electronic discussion groups in six languages—and forwarded to the World Health Organization (WHO). On the basis of this lead, the WHO began more extensive investigations and, with the emergence of subsequent outbreaks outside China, on March 15, 2003, issued an unprecedented "worldwide health threat" including travel advisories to those intending to travel to SARS-affected areas. The WHO also coordinated the international response to SARS by officially calling upon its network of 120 partners in its Global Outbreak Alert and Response Network (consisting of national government agencies and scientific institutions) to focus on the virological, clinical, and public health dimensions of the SARS epidemic. Central to these efforts was the unprecedented sharing of data, resources, and expertise from around the globe through the formation of virtual networks predicated on satellite broadcasts, tele-conferencing, and webcasts. It was on the basis of this rapid real-time information sharing and analyses that a clinical case definition of SARS was developed and agreed upon by international consensus within the remarkable span of one week, while the viral agent and its genetic code were identified in the unprecedented span of several weeks (as opposed to months). Such developments highlight one of the greatest successes of the global outbreak response. Many scientists predict that a major global influenza epidemic will ensue in the future and, in this light, it was clear that SARS served as a sort of trial run that provided many lessons as to how governments, scientists, and communities may respond more effectively to future public health threats.

See Also: Airborne Diseases; Bird Flu; Nosocomial Infections; Seasonal Flu.

Further Readings

Abraham, Thomas. *Twenty-First Century Plague: The Story of SARS.* Baltimore, MD: Johns Hopkins University Press, 2007.

Ali, S. Harris and Roger Keil, eds. *Networked Disease: Emerging Infections in the Global City.* Oxford, UK: Wiley-Blackwell, 2008.

Centers for Disease Control and Prevention. "Severe Acute Respiratory Syndrome." http://www.cdc.gov/ncidod/sars (Accessed September 2010).

Drexler, Madeline. *Emerging Epidemics: The Menace of New Infections.* New York: Penguin, 2009.

Fidler, David P. *SARS, Governance and the Globalization of Disease.* New York: Palgrave Macmillan, 2004.

S. Harris Ali
York University

Sexually Transmitted Diseases

Sexually transmitted diseases (STDs) or sexually transmitted infections (STIs) are a major public health problem with multiple significant impacts on the world population, particularly among young people. STDs are among the first 10 causes of unpleasant diseases in young adult males in developing countries and the second major cause of unpleasant diseases in young adult women.

Adolescents and young adults (15–24 years old) make up only 25 percent of the sexually active population but represent almost 50 percent of all newly acquired STDs. In general, sexually transmitted diseases can be considered today as epidemics and present enormous health and economic consequences. In this setting, is important to consider that many times STDs can occur simultaneously in one individual or in one couple (e.g., infection due to *Chlamydia* and *Neisseria gonorrhoeae*).

From an epidemiological and public health perspective, an adequate screening for STDs should be done on a routine basis in every part of the world, particularly in those countries with high incidence of such conditions and known significant prevalence of risk factors among the population (e.g., high rates of prostitution). Specifically, screening should be directed at the most vulnerable groups, such as pregnant women in the antenatal programs (including at least a test for human immunodeficiency virus [HIV], hepatitis B virus, and syphilis). The burden of STDs at this level is also important given the potential impact on mothers and their newborns. Congenital STDs are also a public health problem of relevant significance. They carry significant medical, economic, societal, and emotional burden, although this has been poorly characterized. Inexplicably, the fight to eradicate many of the congenital STDs has failed to attract international attention compared to HIV. The magnitude of this disease burden globally rivals that of HIV infection in neonates but receives little attention in some countries where the seroprevalence is even higher. Despite national policies on antenatal testing and the widespread use of antenatal services, sexually transmitted disease screening is still implemented only sporadically in many countries, leaving these diseases undetected and untreated among many pregnant women. Even with high antenatal care coverage, the quality of the screening should be regularly monitored. Weak organization of services and the costs of screening are the principal obstacles facing programs. Antenatal screening and treatment programs are as cost effective as many existing public health programs, such as measles immunization. This needs to be reinforced even in areas of low prevalence to avoid the unnecessary burden of STDs in pregnant women and children.

Clinically, many sexually transmitted diseases are asymptomatic and therefore can be difficult to control. For this reason, it is important not just to make passive diagnosis at any healthcare center but also to actively search for such diseases at the community level, particularly in highly risky places (e.g., prisons and incarcerated people), in highly risky populations, and in highly vulnerable groups.

Beyond the importance of individual diagnosis of any sexually transmitted diseases, of utmost importance is the reporting of these conditions, passive and active, for epidemiological surveillance purposes. This is done to ensure that persons who are infected will be quickly diagnosed and appropriately treated in order to control the spread of infection, and also so that partners are notified, tested, and appropriately treated. Today, most STDs are not just preventable but also very easily treated with antibiotics (e.g., penicillin, azithromycin).

Underreporting and STDs

It is estimated that reported cases of sexually transmitted diseases represent only 50 to 80 percent of reportable STD infections in the United States, reflecting limited screening and low disease reporting, although this could be clearly different among other developed countries (e.g., higher in some European nations) and developing countries (e.g., lower in African countries). However, as expectable even inside some nations, screening can significantly change among different states, in part given by their own sexually transmitted disease prevalence.

High-risk sexual behavior is a strong contributive factor of this process, as it often leads to teenage pregnancies and the HIV infection/acquired immunodeficiency syndrome (AIDS). One possible explanation for this behavior is that people do not have enough information about the transmission of sexually transmitted diseases or ignore the precautions required for safe sex. Sometimes cultural and religious issues significantly limit access to the knowledge or access to the STD-prevention program because they are seen in those contexts as contraceptive measures, when, in fact, avoiding morbidity and mortality due to STDs is a primary goal and avoiding conception is secondary. Historically, the epidemiology of sexually transmitted diseases has been based on individual attributes and behavior. However, STDs constitute a good example of diseases that depend on personal contacts for dissemination, which is particularly important among young people, especially in changing countries where social networks and relationships are significantly evolving, allowing more close interactions between people in a casual way, even in countries where this kind of behavior was absent 10 or 20 years ago (e.g., some countries in Latin America or Asia). Because approximately 60 percent of new HIV infections worldwide occur in young people, all these social and psychological aspects related to the behavior of this age group should be taken into account in national STD-prevention programs. The frequency of high-risk behaviors among youths may also be influenced by the opportunity to engage in them, particularly the amount of time that youths are unsupervised by adults, which could be higher in elder youths and in those coming from families with any kind of social or family problems. Sex education and health education are crucial in order to increase the awareness and knowledge of young people in general and of the sexually active population—even more so today, when new educative technologies such as computers and the Internet can be used in this process. At this point, it should also be remembered that recently, in some countries, incidence and prevalence of sexually transmitted diseases in people older than 65 years have significantly increased.

Preventing the Spread of STDs

Through early diagnosis and treatment of those individuals who have acquired sexually transmitted diseases, we can effectively prevent the spread of more complex STDs, such as HIV/AIDS. Recently, several studies have indicated that individuals infected with STDs are 5 to 10 times more likely than uninfected individuals to acquire or transmit HIV through sexual contact. Many examples of this have been highlighted, such as syphilis, which facilitates the acquisition of HIV. The breaking of the genital tract lining or skin creates a portal of entry for HIV and, hence, HIV-infected individuals with other STDs are more likely to shed HIV in their genital secretions.

To date, the condom is the most effective method available for protection against sexually transmitted diseases. If women are well educated about the dangers of STDs and that they may be prevented with a condom, they can emphasize to the men the need for its use in sexual intercourse. Additionally, one of the greatest potential means for prevention of HIV and other STDs today lies in using a topical microbicide. Microbicides are applied to vaginal or rectal microbicide surfaces with the goal of preventing or at least significantly reducing the transmission of sexually transmitted diseases.

It is important to control STDs, and education and prevention can be the key for this process. Prevention can be achieved through education of the population, identification of symptomatic and asymptomatic people, and effective diagnosis and treatment of these patients and their partners.

Everybody in public health and science understands the need for vaccines in the context of sexually transmitted diseases in order to significantly increase the impact of preventive programs. In general, fortunately, in the public health setting, past and ongoing attempts to create vaccines against sexually transmitted pathogens have met with varying success (e.g., hepatitis B, human papillomavirus).

Sexually transmitted diseases can be caused by a wide range of different organisms, including bacteria (e.g., *Neisseria gonorrhoeae*), fungi (e.g., *Candida albicans*), viruses (e.g., hepatitis B virus), and even parasites (e.g., *Trichomonas vaginalis*). Independently of this, it should be remembered that many of them can be present at the same time in one individual or in one couple, as stated above.

It is expected that with new biomedical technologies, such as molecular biology and genomics, the physiopathology and molecular aspects of sexually transmitted diseases can be better understood. This could lead to improved diagnostic tools as well therapeutics and, particularly, vaccines for this complex of diseases that represent a significant burden of morbidity and mortality in the world.

See Also: Acquired Immune Deficiency Syndrome; Antibiotic Resistance; Antibiotics; Centers for Disease Control and Prevention (U.S.); Tuberculosis; Vaccination/Herd Immunity.

Further Readings

Da Ros, Carlos T. and Caio da Silva Schmitt. "Global Epidemiology of Sexually Transmitted Diseases." *Asian Journal of Andrology*, 10 (2008).

Dickson-González, Sonia M., Alfonso J. Rodríguez-Morales, Libia Jiménez, Rosa A. Barbella, Jorge Vals, Afra Molina, and José D. Mota-Gamboa. "Maxillo-Facial Rosai-Dorfman Disease in a Newly Diagnosed HIV-Infected Patient." *International Journal of Infectious Diseases*, 12 (2008).

Guerrero-Lillo, Lisette, Jorge Medrano-Diaz, Francisco Perez, Carmen Perez, Angela Bizjak-Gomez, Juan Silva-Urra, and Alfonso J. Rodriguez-Morales. "Sexual Behavior and Knowledge About HIV/AIDS and STIs Among Health Sciences Students From Chile." *Sexually Transmitted Infections*, 83 (2007).

Hammett, Ted M. "Sexually Transmitted Diseases and Incarceration." *Current Opinion in Infectious Diseases*, 22 (2009).

Hoover, K., M. Bohm, and K. Keppel. "Measuring Disparities in the Incidence of Sexually Transmitted Diseases." *Sexually Transmitted Diseases*, 35 (2008).

López-Zambrano, Maria A., Gustavo Briceño, and Alfonso J. Rodriguez-Morales. "Trends in the Screening of HIV and Syphilis in Pregnancy Among Women Under Antenatal Care in Central Venezuela." *International Journal of Infectious Diseases*, 12 (2008).

Navas, Rosa M., Reinaldo Parra, Maivelys Pacheco, Jimena Gomez, Iris Bermudez, and Alfonso J. Rodriguez-Morales. "Congenital Bilateral Microphthalmos After Gestational Syphilis." *Indian Journal of Pediatrics*, 73 (2006).

Ramachandran, R. and P. Shanmughavel. "Role of Microbicides in the Prevention of HIV and Sexually Transmitted Diseases—A Review." *Current HIV Research*, 7 (2009).

Santos Périssé, André Reynaldo, and José Augusto da Costa Nery. "The Relevance of Social Network Analysis on the Epidemiology and Prevention of Sexually Transmitted Diseases." *Reports of Public Health (Brazil)*, 23 (2007).

Starnbach, Michael N. and Nadia R. Roan. "Conquering Sexually Transmitted Diseases." *Nature Reviews Immunology*, 8 (2008).

Vásquez-Manzanilla, Omira, Sonia M. Dickson-Gonzalez, and Alfonso J. Rodriguez-Morales. "Congenital Syphilis and Ventricular Septal Defect." *Journal of Tropical Pediatrics*, 55 (2009).

Vásquez-Manzanilla, Omira, Sonia M. Dickson-Gonzalez, José G. Salas, Alfonso J. Rodriguez-Morales, and Melissa Arria. "Congenital Syphilis in Valera, Venezuela." *Journal of Tropical Pediatrics*, 53 (2007).

Vásquez-Manzanilla, Omira, Sonia M. Dickson-Gonzalez, José G. Salas, Luis E. Teguedor, and Alfonso J. Rodríguez-Morales. "Influence of Mother VDRL Titers on the Outcome of Newborns With Congenital Syphilis." *Tropical Biomedicine*, 25 (2008).

World Health Organization (WHO). *International Travel and Health*. Geneva, Switzerland: WHO, 2009.

Zeña-Castillo, Dina, Edward Mezones-Holguin, Glauco Valdiviezo-García, Arnaldo La-Chira-Albán, Alfonso J. Rodriguez-Morales, and Sonia Dickson-Gonzalez. "Impact of Hospital-Associated Anxiety and Depression on the CD4 Counts in Naïve HIV/AIDS Patients From Northern Peru Locations." *International Journal of Infectious Diseases*, 13 (2009).

Alfonso J. Rodriguez-Morales
Universidad de Los Andes,Universidad Central de Venezuela

SKIN DISORDERS

The skin is the largest and one of the most vulnerable organs of the body. While infrequently life threatening, skin disorders can be uncomfortable and may cause chronic disabilities. Furthermore, because the skin is so visible, skin disorders can lead to psychological stress. There are many disorders of the skin that require clinical care by a physician or other health-care professional.

Common Skin Disorders

Some of the more common skin disorders are listed below with a brief description of each for general information, bearing in mind that a diagnosis should always be made in person by a physician or healthcare professional.

Eczema

Eczema is a disease in the form of dermatitis, or inflammation of the epidermis. This encompasses a broad range of skin conditions. Eczema includes dryness, rashes, redness, skin edema or skin swelling, itching, crusting, flaking, blistering, cracking, bleeding, and/or areas or temporary skin discoloration. Eczema is often found on the flexor aspect of joints.

Psoriasis

Psoriasis is a chronic, noncontagious autoimmune disease that affects the skin and joints. It commonly causes red scaly patches to appear on the skin, and features areas of inflammation and excessive skin production. These patches are called psoriatic plaques, and frequently occur on knees and elbows but can affect any area including scalp and genitalia. Psoriasis is likely to be found on the extensor aspect of joints.

Skin Cancer

Skin cancer generally develops in the epidermis, so a tumor is usually clearly visible. This makes most cancers detectable in early stages. The most common skin cancers are basal cell cancer, squamous cell cancer, and melanoma; each of these is named after the type of skin cell from which it arises. Melanoma is the least common, but the most serious. Nonmelanoma skin cancers are the most common.

Contact Dermatitis

Contact dermatitis is a skin reaction resulting from exposure to allergens or irritants. (Phototoxic dermatitis occurs when the allergen or irritant is activated by sunlight.) Contact dermatitis in a localized rash with inflammation present in epidermis and dermis. It results in large, burning, and itchy rashes. It can take days to fade, and only if the skin no longer comes in contact with the allergen or irritant.

Sebaceous Cysts

Sebaceous cysts are slow-growing, harmless bumps under the skin. The cysts usually contain dead skin and other skin particles. Can usually be treated by a physician by puncturing and removing the contents. If the cyst reappears it may have to be surgically removed. However, if a cyst becomes infected, treatment may include administering antibiotics and then surgically removing it.

Keratosis Pilaris

Keratosis pilaris is a common skin disorder characterized by small, pointed pimples. The pimples usually appear on the upper arms, thighs, and buttocks. The condition worsens in the winter and clears up in the summer. There is no known cause, but it tends to run in families.

Dry Skin

Dry skin is a very common condition, usually characterized by irritated skin and itchiness. Frequent bathing can aggravate dry skin, as can cold air in the winter. Dry skin may become flaky or scaly. However, dry skin symptoms may resemble other skin conditions or result from other disorders such as under active thyroid or acquired immune deficiency syndrome (AIDS).

Mastocytosis

Mastocytosis is the abnormal growth of mast cells in the body. The most common form of mastocytosis is when mast cells accumulate on the skin, causing reddish brown spots or bumps. In rare case, it can affect other parts of the body, such as the stomach or intestines.

Pressure Sores

Pressure sores are injured skin and tissues. They are usually caused by sitting or lying in one position for too long, putting pressure on certain areas of the body. The pressure can reduce blood supply to the skin and tissues beneath. When blood supply gets too low, a sore may form. Pressure sores are also called bedsores, pressure ulcers, and decubitus ulcers.

Rosacea

Rosacea is a chronic condition characterized by facial erythema, redness, and sometimes pimples are included as part of the definition. Rosacea typically affects the face, but also less commonly affects the neck, chest, ears, and scalp. In some conditions, additional symptoms appear such as red-domed papules and pustules; red, gritty eyes; burning and stinging sensations; and dilation of superficial blood vessels on the face. Rosacea affects both sexes but is more common in women. Treatment, if wanted, usually involves topical medications to reduce inflammation.

Acne Vulgaris

Acne vulgaris is characterized by noninflammatory follicular papules or comedones and by inflammatory papules, pustules, and nodules in more severe forms. It affects the areas of skin with the densest population of sebaceous follicles; these areas include the face, the upper part of the chest, and the back. Acne lesions are commonly referred to as pimples, blemishes, spots, zits, or acne. Acne is common during adolescence, and for most people it diminishes and disappears or decreases when one reaches one's early 20s. However, every case is different and some individuals will continue to suffer well into their 30s, 40s, and beyond. Acne can cause scarring to the skin, and has psychological effects, such as reduced self-esteem and, in some cases, depression. Early treatment is advocated to lessen the overall impact to individuals.

Intertrigo

Intertrigo is an inflammation of the body folds, or adjacent areas of skin. It sometimes refers to a bacterial, fungal, or viral infection that develops at the site of broken skin due to inflammation. Intertrigo usually develops from the chafing of warm moist skin in the areas of armpits, inner thighs, genitalia, underside of belly, under breasts, behind ears, and the web spaces between toes and fingers. It usually appears red and raw looking. It also may itch, ooze, and be sore. It is more common among overweight individuals, individuals with diabetes, people who use medical devices like artificial limbs, and those restricted to bed rest or diaper use.

Tinea Infections

Tinea infections are more commonly known as athlete's foot, jock itch, and ringworm. It is a fungus that can grow on the skin, nails, or hair. The infection can spread throughout the body and is named differently as it spreads, such as tinea corporis (body and limbs), tinea cruris (groin), and tinea pedis (feet). In tinea corporis, the infection spreads out in a circle, leaving normal-looking skin in the middle, making it look like a ring. At the edge of the ring, the skin is lifted up by the irritation and looks red and scaly. In tinea pedis, it causes scaling, flaking, and itching of the affected skin. Blisters and cracked skin also may occur. In tinea cruris, the affected areas may appear red, tan, or brown with flaking, rippling, peeling, or cracking skin.

Shingles

Shingles is another name for a condition known as herpes zoster. It is an infection that results from reactivation of the same virus that causes chickenpox, called the varicella virus. It causes a painful itching rash. After one has had chickenpox, usually as child for most people, the virus stays in the body in certain nerve cells. The immune system keeps the virus in these cells. However, as people age, their immune systems weaken and the varicella virus may escape from the nerve cells and cause shingles. Most people who get shingles are more than 50 years old or have a weakened immune system.

Treating Skin Disorders

Many skin disorders are treated with creams, lotions, tablets, and in severe cases, intravenous antibiotics. It should be noted, however, that there are a few things that can naturally help prevent and heal the skin. Exercise is commonly acknowledged as helping to keep the body healthy and strong. It increases blood flow everywhere, which helps skin cells and all other cells in the body regenerate faster. Quitting smoking can greatly help an individual's skin. Smoking decreases oxygen in the skin, and increases free radicals—highly reactive chemical particles that can cause cell damage and increase the breakdown of collagen. Sugar, the culprit in a host of other health issues, also is bad for the skin. Sugar has been known to increase the risk of acne, and also is now associated with increased aging.

Hydration is elemental to prevention of dry skin. Proper amount of water intake, as well as oils (such as fish or flaxseed oil) improves the skin's barrier and helps it retain more moisture. Tea bags and cucumber slices, a recognizable sign of home skin care, do more than just provide relaxation. Cucumber's cooling effects helps reduce blood flow to the affected area, and it has an anti-inflammatory capability. Tea bags, such as chamomile and black tea, serve as anti-inflammatories, and the caffeine helps to shrink blood vessels and reduce redness. Sleep is another important factor, as sleep helps the body regenerate cells. In sleep-deprived people, the body has less opportunity to create new skin cells. Too little sleep also can lead to increased levels of cortisol, a hormone produced by the adrenal glands, and high levels of sugar in the blood. A diagnosis and treatment plan should always be made by a physician or healthcare professional, but simple changes such as exercising more, quitting smoking, proper hydration and sleep, as well as cutting back on consumption of sugar can help skin disorders.

See Also: Antibiotics; Cancers; Men's Health; Mental Health; Women's Health.

Further Readings

Lamberg, Lynne and Sandra Thurman. *Skin Disorders*. New York: Chelsea House Publishers, 2000.

Marks, Ronald. *Facial Skin Disorders*. Boca Raton, FL: CRC Press, 2007.

Schneiderman, Paul I. and Marc E. Grossman. *A Clinician's Guide to Dermatologic Differential Diagnosis, Volume 1: The Text.* New York: Informa Health Care, 2006.

Alessandra Guimaraes
University of Sint Eustatius School of Medicine

SMOG

The term *smog* was introduced by Harold Antoine Des Voeux in 1905. A member of the Coal Smoke Abatement Society in London, Des Voeux coined the term to describe the combination of smoke and fog that was visible in several cities throughout Great Britain. During the early 1900s, smog was applied to the emissions resulting from the burning of coal and other raw materials. During that time, coal was used primarily for generating energy, and raw materials were burned to produce chemicals such as soda ash for use in consumable products including soap, detergents, and glass. The smoke resulting from the combustion of coal and raw materials in British industrial cities mixed with fog from the North Sea.

Several cities, including New York, face pollution from both automobiles and combustion, which emit significant amounts of soot and other trace gases into the atmosphere, resulting in poor visibility.

Source: Centers for Disease Control and Prevention/Dr. Edwin P. Ewing, Jr.

These pollutant episodes were characterized by an extreme reduction in visibility and a stench of sulfurous emissions. Such pollution fog was experienced in London for more than a century, was usually gray in color, and was referred to as "gray smog" or "London-type" smog.

In contrast to "gray smog" or "London-type" smog is "Los Angeles–type" smog. In the Los Angeles basin of southern California, the smog-generating conditions are totally different from Great Britain and other European cities. The major source of pollutants in Los Angeles is from light-duty motor vehicles emitting exhaust and evaporative gases into the atmosphere. The climate is highly semi-arid, and the Los Angeles basin is bordered by mountains to the north and east. These mountains serve as barriers inhibiting air movement, reducing the dispersion of pollutants. Exhaust emissions from motor vehicles, in conjunction with abundant sunlight and poor air dispersion conditions, contributes to the area's smog development. This version of smog is generally called "Los Angeles–type" or "brown smog." The brown color of the Los Angeles smog is due to the presence of nitrogen dioxides (NO_2). Also, because the smog produced in Los Angeles is mainly due to photochemical reactions, it also is called "photochemical smog." Photochemical smog refers to smog resulting from a combination of atmospheric trace gases, each one of which is, initially, the product of reactions involving a mix of nitrogen oxides and hydrocarbons with normal atmospheric gases.

Due to rapid industrialization, several cities in the world, in particular developing countries in Asia, face pollution from both automobiles and combustion. In these countries, the resulting smog from pollution episodes can be intermediate, between the types of smog found in London and Los Angeles. Also, pollutant activities from automobiles and combustion emit significant amounts of soot and other trace gases into the atmosphere, resulting in poor visibility. Soot is emitted primarily during coal, diesel fuel, jet fuel, and biomass burning. Since black carbon absorbs all wavelengths of white light, transmitting none, smog sometimes appears black when soot particles dominate against the background sky.

Similar to smog, another common term used in pollution studies is *haze*. Although haze and smog refer to poor visibility conditions, haze generally refers to large-scale air pollution episodes. For example, the term *Asian brown cloud* (ABC) was coined by researchers of the Indian Ocean Experiment and United Nations Environment Programme to refer to a large amount of brown haze that appeared over the south Asian region and the tropical Indian Ocean, Arabian Sea, and Bay of Bengal. Although huge, the ABC event is purely a temporary phenomenon, and occurs only during January through March in that region.

The main chemical constituents of photochemical smog are a mixture of gases and aerosol particles. Smog contains more high molecular weight organics, particularly aromatic compounds, than the background air. During the early 1950s, Arie Haagen-Smith elucidated the chemistry of the photochemical smog. After thorough chemical investigation, Haagen-Smith concluded that ozone was a primary constituent of photochemical smog. Also, he found that ozone caused eye irritation, damage to materials, and respiratory problems. Researchers found that ozone may aggravate chronic lung diseases, such as emphysema and bronchitis. Also, studies suggest that children are much more susceptible to smog-related lung damage than adults. Further, studies in animals suggest that ozone may reduce the immune system's ability to fight off bacterial infections in the respiratory system. In plants, ozone causes several types of symptoms including chlorosis (yellowing of leaf tissue due to lack of chlorophyll) and necrosis (death of plant cells and tissues). Several additional symptoms were reported with ozone exposure to plants, including bronzing and reddening of leaf tissues. Photochemical smog is the most common problem in several cities of the world and strict regulatory measures are needed to control it.

See Also: Fungi and Sick Building Syndrome; Lung Diseases; Ozone; Particulate Matter.

Further Readings

Golish, Thad. *Air Quality*, 4th ed. Boca Raton, FL: Lewis Publishers, 2003.

Jacobs, Chip and William J. Kelly. *Smogtown: The Lung-Burning History of Pollution in Los Angeles*. Woodstock, NY: Overlook Press, 2008.

Jacobson, Mark Z. *Atmospheric Pollution: History, Science and Regulation*. Cambridge, UK: Cambridge University Press, 2002.

Wise, William. *Killer Smog: The World's Worst Air Pollution Disaster*. Lincoln, NE: iUniverse.com, 2001 [1968].

Krishna Prasad Vadrevu
Ohio State University

SMOKING

By the 1940s, when American scientists first detected a correlation between cigarette smoking and lung cancer (German scientists had found the same correlation in the 1920s and spearheaded the first antismoking campaign in modern history), humans already had a relationship with tobacco that reached back thousands of years. Widespread cultivation of tobacco began about 5,000 years ago; in addition to daily use (chewing, drying and inhaling, smoking), tobacco was used as an insecticide, medicinally as an analgesic or antiseptic, and for various ritualistic practices. Though tobacco has been celebrated for far longer than it has been denigrated as an unhealthy habit, condemnation of tobacco smoking has deep roots, too, reaching back to early Christians who viewed indigenous peoples' smoking as evil and satanic. Still, by the 1600s, tobacco was touted across

In addition to daily use—including chewing, drying and inhaling, and smoking—tobacco has been used as an insecticide, as an analgesic or antiseptic, and for various ritualistic practices. U.S. Department of Agriculture scientists studied the natural insecticidal compounds found in these wild tobacco plants, eventually identifying the active substances to be sugar esters.

Source: U.S. Department of Agriculture, Agricultural Research Service/Alvin Simmons

Europe for its medicinal properties as well as for the simple, base pleasure of smoking, chewing, or snuffing it.

The cigarette, which is the most common tobacco-delivery system today, has its roots in Aztec practices of rolling tobacco in maize husks; this method was embraced by Spaniards and eventually by the French, who came into contact with *cigaritos* when the French army occupied Spain early in the 19th century. That same century, in the United States, it was accidently discovered that heat-cured tobacco was milder and more flavorful than was smoke-cured tobacco; then, in 1880, a Virginian made the first cigarette-rolling machine, which could produce 200 cigarettes a minute. These two innovations—heat-cured tobacco and mass manufacture—dramatically increased tobacco smoking in the United States.

Today, there are more than one billion tobacco smokers in the world, and it is estimated that smoking kills almost five million people worldwide annually. Despite the fact that women comprise only 20 percent of the world's smokers, tobacco use among women is increasing, as tobacco companies have intensified their marketing efforts to capture this burgeoning market. Still, men are more likely to smoke than women, poor people are more likely to smoke than the wealthy, and people in developing countries are more likely to smoke than those in developed countries. In the United States, over 40 million people smoke tobacco, resulting in the deaths of almost 500,000 people per year and causing serious illness in nearly nine million others; tobacco use is considered the single most preventable cause of disease, death, and disability in the United States. People with psychiatric disorders, particularly depression, schizophrenia, and post-traumatic stress disorder, are more likely to smoke tobacco than are those without psychiatric diagnoses, and those with psychiatric diagnoses smoke almost half the cigarettes consumed in the United States.

Nicotine is the addictive element in tobacco, and in small doses it acts as a stimulant. It constricts arteries, making it harder for the heart to pump blood, and repeated exposure is linked to cardiovascular disease, heart attacks, and high blood pressure. Nicotine is also linked to stomach ulcers, acid reflux disorders, and strokes. In addition to nicotine, tobacco contains more than 19 known cancer-causing chemicals, which are most commonly referred to as *tar*, as well as more than 4,000 other chemicals. The health risks associated with smoking include the following:

- Blood clots and brain aneurysms
- Strokes
- Blood clots in the legs, which can travel to the lungs
- Coronary artery disease
- Cancers of the lungs, mouth, larynx, esophagus, pancreas, cervix, bladder, and kidneys
- High blood pressure
- Decreased blood flow to the penis, which can cause erection problems
- Emphysema, chronic bronchitis, asthma
- Low birth weight for babies, premature labor, miscarriage, and cleft lip
- Tooth and gum diseases
- Loss of sight due to macular degeneration
- Harm to sperm, which contributes to infertility

Secondhand smoke creates health risks for nonsmokers, which include the following:

- Heart attacks and heart disease
- Lung cancer
- Eye, nose, throat, and lower-respiratory-tract irritation

The effects of secondhand smoke on infants and children include the following:

- Asthma
- Virus-caused upper-respiratory infections
- Ear infections and pneumonia
- Lung damage
- Sudden infant death syndrome

In the United States, secondhand smoke is believed to be responsible for nearly 50,000 deaths per year and for causing up to 300,000 instances of lower-respiratory-tract infections in children under 18 months old. In addition to the alarming health effects of smoking for both smokers and nonsmokers, tobacco use is estimated to carry a financial burden of $200 billion a year in medical expenditures and lost productivity. A study by the Centers for Disease Control and Prevention found that there are higher levels of cancer-causing chemicals in certain U.S.-brand cigarettes than in cigarettes from Canada, the United Kingdom, or Australia. That is not to say that cigarettes from other countries are "healthy," only that they contain less cancer-causing chemicals than most U.S.-brand cigarettes. Despite the overwhelming evidence of its negative consequences, the addictive nature of nicotine and the physical effects of it and the other active ingredients in cigarettes, such as heightened heart rate and alertness as well as the release of dopamine, make quitting smoking an incredibly difficult task for smokers.

While it is true that cigar smokers have lower rates of cancer and other diseases associated with tobacco smoking than do cigarette smokers, cigar smokers still have higher rates of heart disease, lung disease, and lung cancer than do nonsmokers. Further, smoking cigars exposes the lips, mouth, tongue, throat, and larynx to cancer-causing chemicals, just as cigarette smoking does; oral and esophageal cancer risks are similar among cigar smokers and cigarette smokers. Cigars are incredibly high in nicotine—a single cigar can contain as much nicotine as a pack of cigarettes, and even if the cigar smoker does not inhale, nicotine is absorbed through the mouth's lining. Finally, cigars contain more tar and more toxins than do cigarettes, and cigar smoke contains a higher level of cancer-causing substances than does cigarette smoke.

Smoking tobacco through non-water-based pipes, though less risky than smoking tobacco cigarettes, is still associated with deaths from various types of cancers and other pulmonary diseases, and pipe smoking is considered just as harmful, if not more so, than smoking cigars. The hookah is a water pipe made to smoke fruit-flavored tobaccos. Hookahs are an integral part of Arabic culture, and they most likely originated in ancient Persia and India, though today, hookah cafes can be found around the globe. People typically smoke from hookahs in groups, passing the hose from person to person. Though many people argue that hookah smoking is less harmful than other forms of tobacco smoking (due to the alleged filtration supplied by the water), smoking from a hookah still delivers nicotine, and, because of the depth of inhalation and length of the smoking sessions, hookah smokers likely absorb higher concentrations of the toxins found in cigarette smoke; a one-hour session typically involves inhaling up to 200 times the volume of smoke that would be inhaled from a single cigarette. Thus, hookah smokers are at risk for the same diseases associated with other forms of tobacco smoking, including various kinds of cancers and reduced lung functioning as well as fertility problems, clogged arteries, and heart disease.

The past several years have seen the rise of electronic cigarettes, or "e-cigarettes," which are battery-powered vaporizers that deliver inhaled doses of nicotine. Marketed as an alternative to smoking tobacco, the e-cigarette does not rely on combustion of tobacco and produces no actual smoke; one draws on it in the same way as drawing on a cigarette yet inhales an atomized liquid—which can come in a variety of flavors, including menthol, strawberry, coconut, and coffee—and expels a nearly odorless vapor. While some e-cigarettes do not deliver nicotine, most do, and for this and related reasons, the Food and Drug Administration and the World Health Organization, among other entities, have been reluctant to embrace the e-cigarette as a healthy alternative to smoking and have challenged claims that they function as a legitimate smoking cessation aid.

In 2004, global cigarette production reached 5.5 trillion—about 868 cigarettes for every single human being on the planet. Smoking tobacco also has profound environmental consequences and, beyond the effect on air quality, producing tobacco has the following effects:

- Takes up space for food crops
- Contributes to deforestation
- Requires heavy doses of pesticides, which affect local water supplies and fragile ecosystems
- Taxes the soil, leaching out potassium and rendering the land nearly useless for food crops

- Consumes massive amounts of paper
- Creates millions of tons of solid and chemical waste

While the movement to curb tobacco smoking is embraced by many individuals as well as various state, national, and independent organizations, the fact remains that there are more than one billion tobacco smokers worldwide, half of whom will likely die from this insidious habit.

See Also: Asthma; Cancers; Cardiovascular Diseases; Centers for Disease Control and Prevention (U.S.); Men's Health; Oral Diseases; Women's Health.

Further Readings

Centers for Disease Control and Prevention. "Tobacco Use: Targeting the Nation's Leading Killer: At a Glance 2010." http://www.cdc.gov/chronicdisease/resources/publications/AAG/osh.htm (Accessed August 2010).
MedlinePlus. "Smoking." http://www.nlm.nih.gov/medlineplus/smoking.html (Accessed August 2010).
Random History. "The Long Tobacco Road: A History of Smoking From Ritual to Cigarette." http://www.randomhistory.com/2009/01/31_tobacco.html (Accessed August 2010).

Tani Bellestri
Independent Scholar

Solid Waste Management

As humans have organized society toward greater productivity and prosperity, solid wastes have increased due to ever-higher levels of consumption of goods and services. Solid waste is typically an unwanted by-product from many of our daily activities, easily generated and often difficult to eliminate. Waste is best defined as material from a process that is discarded or unusable. It may occur as a liquid waste, gaseous waste, biodegradable waste, or solid waste. Some wastes are considered a pollutant, disrupting natural ecosystems, while others like biodegradable waste are more in balance with the system or process from which they are created. For this article's purpose, solid waste is commonly referred to as household trash or garbage, discarded or abandoned, including the often-used interchangeable term *municipal solid waste* (MSW). Several strategies exist to manage MSW, including landfilling, recycling/reusing, incineration, and source reduction.

Prior to the modern industrial age, human populations were much smaller and, hence, their solid waste outputs were a fraction of what they are today. Solid waste in ancient times consisted primarily of ash from fires, animal carcasses, and discarded tools. It was not until people gathered in ever-larger groups that solid waste became a concern. The first recorded landfill site was created around 3000 B.C.E. in the Cretan capital of Knossos. Waste was placed in large holes in the ground and covered with dirt. In response to the black plague of the mid-14th century, Britain employed "rakers" to collect refuse on the street, place it in carts, and remove it from the population centers. These two events mark the most widely adopted strategy of solid waste management to this day—collection of waste for removal and disposal in a landfill.

Modern solid waste management in the United States consists mainly of disposing of waste directly on land, incinerating it, or placing it in a sanitary landfill. The sanitary landfill process is currently the most widely used method to manage MSW. The waste industry employs over 360,000 with annual revenues of approximately $43 billion.

Modern solid waste management in the United States consists mainly of disposing of waste directly on land, incinerating it, or placing it in a sanitary landfill. The landfill option is currently the most widely used.

Source: iStockphoto.com

Life Cycle

Americans generated approximately 250 million tons of garbage in 2008 or 4.5 pounds person/day. Fifty-four percent of this waste was discarded in landfills, 33 percent was recycled or recovered, and 13 percent was incinerated. Paper, plastics, and food scraps comprise the majority of MSW. Modern MSW management begins with the collection of discarded household materials ranging from food scraps to appliances. Most local governments in the United States have responsibility for providing solid waste management services. It is collected at the source by municipalities or contracted to private collection services to be transported to a central location termed a transfer station. Once waste is placed outside the home or an establishment for collection, local government obtains ownership. After transport to the transfer station, it is unloaded from collection vehicles and held until it can be reloaded onto larger trucks, trains, or barges for shipment to its final destination of landfills or other treatment or disposal facilities.

Landfills

The latest figures show the United States has over 2,000 active municipal landfills (over 1,800 in the continental United States, around 300 in Alaska, and 10 in Hawaii as of 2008). The purpose of the landfill rather than a "dump" is to isolate waste from the surrounding environment. In addition to being the repository of waste, solid waste sanitary landfills are designed to protect the environment from contaminants in the waste stream. Modern municipal solid waste landfills have several restrictions and design parameters to prevent such damage. These include siting the landfill away from environmentally or geologically sensitive areas, collection and removal of leachate (liquid that drains from the landfill), and monitoring groundwater. The landfill is designed with a combination of clay and plastic liners to keep the waste and its by-products from contaminating adjacent soil and subsurface water. Further, the landfill employs the practice of compacting and covering the waste with several inches of soil each day to control dispersal, odor, and vermin. A network of collection systems below and above ground collects leachate, storm water, and methane gas, which is either vented or flared off. As a section of the landfill reaches capacity, it is permanently capped with a layer of plastic, which in turn is covered with several feet of compacted soil. The Earth is then planted with vegetation to discourage erosion and runoff.

Because of these practices to isolate the landfills' contents, there is little aerobic activity, causing much of the waste to remain intact, staving off decomposition for several decades. Modern landfills may remain in operation for more than 40 years.

Reuse

Recycling is the transformation of would-be waste into materials of value; 33 percent of the MSW stream is recovered in the form of recycling or composting. There are over 8,600 curbside recycling collection programs in the United States. The most visible example of these programs is the recovery and reuse of aluminum, plastic, and glass beverage containers. About 43 percent of steel, glass, aluminum, paper, and plastic containers and packaging were recovered and recycled in 2007. Products with the highest recovery rate include lead acid batteries, newspapers, and corrugated boxes.

Recycling otherwise waste material requires collecting, sorting, cleaning, and, eventually manufacturing of a new product. Collection of recyclables mainly occurs through curbside programs or at drop-off centers. The collected materials are sorted and sold to various buyers, who in turn manufacture the raw material into a new product. Recycled material is the constituent for many new or similar end-use products such as beverage containers, paper products, and even new fabrics made from certain plastics.

Composting is another form of reuse of waste. It is the purposeful biodegradation of organic matter by microorganisms creating a humus-like soil end product. The primary use of composted material is applying it as a valuable soil amendment for agriculture. Yard trimmings and food scraps account for approximately 26 percent of MSW composition. Many municipalities have success in composting yard trimmings, agricultural wastes, and sewage sludges. Over 60 percent of yard wastes were recovered and composted in 2007. There are nearly 3,800 composting facilities, both public and private. Additionally, there has been a surge in households composting yard trimmings and food wastes, diverting these from traditional MSW management strategies.

Incineration

Landfilling, as a disposal method for MSW, is becoming increasingly prohibitive as available appropriate land resources are in decline. Incineration of MSW as an alternative is prevalent in societies where available land is scarce. Japan incinerates nearly 75 percent of its MSW, while the practice has remained steady at 13 percent in the United States for the past several decades. The primary benefit of incineration is that it reduces waste volume, resulting in a 75 percent reduction of waste to ash for disposal in a landfill. Additionally, combustors can convert water into steam to fuel heating systems or to generate electricity. A typical waste-to-energy plant generates about 550 kWh of electricity from a ton of waste. As of 2008, there were 87 municipal waste-to-energy (WTE) facilities in the United States, processing approximately 95,000 tons per day. In contrast to mass burning of MSW, many WTE facilities are equipped to recover recyclables introduced into the combustion chamber.

Incineration of MSW produces a variety of pollutants. Combustion of MSW emits CO_2, carcinogenic persistent organic pollutants (POPs) like dioxins and furans, heavy metals, and particulates. An array of pollution control technologies attempt to reduce the gases emitted into the air. These include filters to remove particulate matter, scrubbers to neutralize NO_x and SO_x, activated carbon to absorb heavy metals, and high temperature to trigger thermal breakdown of POPs. Burning waste at extremely high temperatures has the added benefit of destroying disease-causing agents and some undesirable chemical compounds.

Source Reduction

A final pillar of municipal solid waste management relies on reducing the amount of waste before it enters the MSW stream. Source reduction or waste prevention entails designing, manufacturing, purchasing, or using materials in ways that reduce the amount of trash created. Households may reduce their waste source by reducing food waste, reducing consumption of disposable goods, and purchasing reused products. Buying in bulk is also an effective strategy to reduce waste at the household level. Manufacturers of goods in many instances will find it economically beneficial to apply source reduction. Minimizing the packaging for products not only saves the purchase of additional raw material, but cuts down on transportation costs due to the reduction in size and weight of products.

Integrated Solid Waste Management

The waste management strategies described herein represent the several alternative methods of managing solid wastes in our society. An integrated solid waste management (ISWM) program provides waste management planners a comprehensive disposal, recycling, composting, and waste prevention plan to effectively deal with MSW. ISWM considers how to effectively prevent, recycle, and manage solid waste in order to protect human health and the environment. It takes into consideration local variables, then selects and combines the most appropriate waste management strategies. Development of an ISWM program begins with the identification of what waste streams can be reduced first, then recycled, and ultimately disposed in a landfill.

As solid waste management practices evolve, disposal ultimately becomes the least preferred strategy.

See Also: Air Filters/Scrubbers; Environmental Protection Agency (U.S.); Particulate Matter; Plastics in Daily Use.

Further Readings

Cheremisinoff, Nicholas P. *Handbook of Solid Waste Management and Waste Minimization Technologies*. Oxford, UK: Butterworth-Heinemann Elsevier Science, 2003.
Kreith, Frank and George Tchobanoglous. *Handbook of Solid Waste Management*, 2nd ed. New York: McGraw-Hill Professional, 2002.
U.S. Environmental Protection Agency. "Municipal Solid Waste Generation, Recycling, and Disposal in the United States: Facts and Figures for 2008." http://www.epa.gov/epawaste/nonhaz/municipal/pubs/msw2008rpt.pdf (Accessed March 2010).

David M. Filiberto
Independent Scholar

STOMACH ULCERS AND *HELICOBACTER PYLORI*

Helicobacter pylori is a common gram-negative microaerophilic (it requires oxygen) bacteria of the stomach and duodenum that can cause mild to significant inflammation and is strongly linked to several gastric tract cancers. The 2005 Nobel Prize for Physiology and Medicine was awarded to Drs. Robin Warren and Barry Marshall for their work that showed *H. pylori* could cause stomach ulcers, stomach cancer, and duodenal ulcers.

H. pylori occurs in one-half to two-thirds of the human population, primarily in the upper gastrointestinal tract. It is more prevalent in the developing world and is thought to be decreasing in the developed world. The method of transmission is unknown, but most people probably become infected in childhood, perhaps through food or water. It can form a nonculturable coccoid form that may play a significant role in its survival and transmissibility. There are many strains, but only three have been completely genome sequenced.

Approximately 50 to 70 percent of the strains most commonly seen in the developed and developing world have the ability to secrete a peptide that creates inflammation and thus pose a greater risk for peptic ulcers and gastric cancers. The cancers most associated with *H. pylori* are MALT (mucosa-associated lymphoid tissue) lymphoma, pancreatic cancer (the second-deadliest cancer in the world and the fifth most common cancer in the United States), stomach cancer, and esophageal cancer. The presence of *H. pylori* increases the risk of pancreatic cancer twofold and MALT sixfold. The two strains most commonly associated with these gastric cancers are CagA and BabA. Testing positive for the *H. pylori* CagA positive strain increases the risk of cancer sixfold, compared to testing positive for the CagA negative strain.

H. pylori colonizes the stomach. To survive the highly acidic stomach environment (pH of 3–4), it burrows through the stomach's mucous layer into its epithelial cell layer. The bacteria can sense different pH levels of the stomach and swim to the most neutral pH (pH = 7) it can find—the epithelial cell layer. *H. pylori* possesses four to six flagella, or tails, that help it move deeper into the epithelial tissue layer. The flagella also help it to move throughout the gastric tract more freely in search of a safe environment to inhabit. The bacteria produce urease to help it neutralize the local stomach's acid environment. It also produces ammonia that is toxic to the epithelial cells and several other compounds that damage the epithelial cells. These secreted compounds create an environment favorable for the growth and colonization of *H. pylori*. It is thought that some strains can produce free radicals and increase the rate of host cell mutations. Another theory is that the local inflammation caused by the compounds it secretes produces locally elevated levels of TNF-alpha and interleukin-6. The pattern of inflammation causes chronic gastritis that allows the acid and pepsin in the stomach environment to overwhelm the mechanisms that protect the stomach and duodenal mucosa. These various hypotheses could explain why the greater risk for inflammation leads to a greater risk for ulcer formation and gastric tract cancers. The overall risk for stomach cancer is estimated at about 1 percent for any person who has *H. pylori*.

Diagnosis

Diagnosis of *H. pylori* requires a careful interview and medical history of the patient's symptoms. Physical examination is then performed on the abdomen, cervical lymph nodes, oral cavity, pulse and blood pressure, and other areas relevant to other patient symptoms. If signs and symptoms suggest an infection, then a blood antibody test, a carbon urea breath test, an endoscopy with biopsy and rapid urease test, a histological examination of the biopsy, and a microbial culture of the sample can be ordered. However, none of these tests are 100 percent accurate. Therefore, gastroenterologists often do more than one test on a given patient.

Typical patient symptoms include abdominal pain; acid reflux, heartburn, or dyspepsia; iron and/or B-12 deficiency anemia, bad breath, chest pain, constipation or diarrhea, gastritis, nausea and vomiting, bloating, feeling full after a small amount of food, lack of appetite, and dark or tar-colored stools. Less typically, patients may experience anxiety, depression, fatigue, headaches, gastric premenstrual syndrome, sinusitis, hives, sleep disturbances, and weight changes. If an ulcer develops and begins to bleed, the stools will become black or tar-colored, fatigue will develop, and the red blood cell count will be decreased. Patients with these symptoms should seek medical care immediately.

Treatment

H. pylori does not respond to a single drug, and some strains appear to be more resistant. The usual treatment is two antibiotics and a proton pump inhibitor (PPI)—triple therapy. The PPI decreases stomach acid production in order to allow the tissues damaged by the infection to heal. Up to 90 percent of individuals who take triple therapy are healed, but up to 20 percent must do the treatment protocol for a second and longer period of time in order to eradicate the *H. pylori*.

The triple therapy medications cause side effects in up to 50 percent of treated patients. Metronidazole and clarithromycin can cause a metallic taste. Consumption of alcohol with metronidzole can cause skin flushing, headache, nausea, vomiting, sweating, and tachycardia. Bismuth is used in some protocols and can cause black stools and constipation. Diarrhea and stomach cramps are common with most of the triple therapy protocols.

See Also: Antibiotic Resistance; Cancers; Degenerative Diseases; Gastroenteritis.

Further Readings

Graham, K. S. and D. Y. Graham. *Contemporary Diagnosis and Management of H. pylori-Associated Gastrointestinal Disease.* Newtown, PA: Handbooks in Health Care, 2002.

Hompes, D. *Overcoming H. pylori.* E-published, 2008. http://h-pylori-symptoms.com/shop (Accessed May 2010).

Hunt, R. H. and G. N. J. Tytgat. *Helicobacter pylori: Basic Mechanisms to Clinical Cure.* London: Axcan Pharma Medical, 1998.

Paul Richard Saunders
Canadian College of Naturopathic Medicine

Streptococcus Infections

Streptococcus is a genus of spherical gram-positive bacteria. In these bacteria, cellular division occurs along a single axis, thus they grow in chains or pairs. Certain *Streptococcus* species are responsible for many diseases, including the common streptococcal pharyngitis ("strep throat"), meningitis, bacterial pneumonia, endocarditis, erysipelas, and necrotizing fasciitis. However, many streptococcal species are nonpathogenic. Streptococci are also part of the normal commensal flora of the mouth, skin, intestine, and upper respiratory tract of humans.

Individual species of *Streptococcus* are classified based on their hemolytic properties. Alpha hemolysis is caused by a reduction of iron in hemoglobin, giving it a greenish color on blood agar. Beta-only hemolysis is complete rupture of red blood cells, giving distinct, wide, clear areas around bacterial colonies on blood agar.

Group A streptococcal bacteria and infections are a form of beta hemolytic *Streptococcus* bacteria responsible for most cases of streptococcal illness. Group A causes strep throat, scarlet fever, impetigo, toxic shock syndrome, cellulitis, and necrotizing fasciitis. It also causes pneumonia, tonsillitis, septic arthritis, osteomyelitis, meningitis, and sinusitis. Group A not only causes acute infections but also is responsible for nonsuppurative post-infectious sequelae such as rheumatic fever and glomerulonephritis.

The physical findings include erythema, edema, and swelling of the pharynx. Tonsils are enlarged, and grayish-white exudates may be present. Upon palpation, enlarged lymph nodes at the mandibular angle are found. Patients may develop chills and fever. Patients may all present with small papule that turn into pustules and develop a thick crust. To date, the number one treatment of choice is penicillin.

These bacteria are spread through direct contact with mucus from the nose or throat of persons who are infected or through contact with infected wounds or sores on the skin. Ill persons, such as those who have strep throat or skin infections, are most likely to spread the infection. Persons who carry the bacteria but have no symptoms are much less contagious. After a person has been treated with an antibiotic for 24 hours or longer, the infected person generally loses the ability to spread the bacteria. However, it is important to complete the entire course of antibiotics as prescribed.

To reduce the spread of group A streptococcal infections, hands should be washed thoroughly, especially after coughing and sneezing and before preparing or eating food. A patient with a sore throat should consult a doctor so that a test can be performed to determine if it is strep throat. If the test is positive, the person should stay away from work, school, or daycare for 24 hours after beginning antibiotics. All wounds should be kept clean and monitored for possible signs of infection, such as redness, swelling, drainage, and pain at the wound site. A person with an infected wound site and a fever should seek medical attention immediately.

Group B streptococcal bacteria is a form of beta hemolytic *Streptococcus* bacteria best known as the cause of postpartum infection and the most common cause of neonatal sepsis. Group B streptococci are found commonly in the gastrointestinal tract and have been found to colonize the urethra in both men and women without causing infection. Group B can also colonize the upper respiratory tract. Colonization by group B *Streptococcus* is also observed in wound and soft tissue cultures, in the absence of obvious infection.

Primary group B streptococcal infections without an obvious source are a common presentation in adults, while some suggest that the clinical presentation may be that of classic sepsis with shock and may carry a high mortality rate. Sustained infections may indicate endocarditis or an infected catheter. Group B streptococci can cause acute destructive endocarditis, which may require emergency valve replacement.

Urinary tract infections are a common manifestation of group B streptococcal infections and are observed in both pregnant and nonpregnant adults. Other presentations of group

B streptococcal infection include pneumonia, skin and soft-tissue infections, septic arthritis, osteomyelitis, meningitis, peritonitis, and endo-ophthalmitis. Group B *Streptococcus* remains sensitive to penicillin and ampicillin.

Asymptomatic carriage in gastrointestinal and genital tracts is common. Intrapartum transmission via ascending spread from the vagina occurs. The mode of transmission of disease in nonpregnant adults is unknown. The degree of seriousness related with group B streptococcal infections increases with age. The average age of cases in nonpregnant adults is about 60 years old. Most adult group B streptococcal infections occur in adults who already have serious medical conditions. These serious medical conditions include diabetes mellitus, liver disease, history of stroke, history of cancer, or bedsores. The rates of serious group B strep diseases are more common among residents of nursing facilities and among bedridden patients who are hospitalized. Group B streptococcal infections of nonpregnant adults may be acquired after hospital procedures like surgery or recent trauma.

Group B strep infections can cause meningitis rarely in adults—an infection of the fluid and lining surrounding the brain. The most common problems in adults are bloodstream infections, pneumonia, skin and soft tissue infections, and bone and joint infections, and are some of the ways group B strep infection can present itself.

Unlike group A streptococcal infections, group B streptococcal infections are tested with a sample of sterile body fluids, such as blood or spinal fluid. When the bacteria are grown from cultures of those fluids, group B *Streptococcus* can be diagnosed. Group B strep bacteria are usually treated with penicillin or other common antibiotics, the same as group A streptococcal infections. Sometimes, depending on the severity of the damage, soft tissue and bone infections may need surgery. Treatment for both group A and group B streptococcal infections are personalized, based on medical history, and no two patients will have identical courses of treatment.

Standard infection control measures, particularly for patients who are hospitalized or in nursing homes, help reduce the risk of bacterial infections, including those caused by group B strep. For pregnant women, it is very important to get tested prior to labor. Between weeks 35 and 37, doctors will test for the group B streptococcal bacteria, which is present in the vagina and rectum in a large number of females. If the pregnant woman tests positive for the bacteria, doctors will administer penicillin (or other antibiotic if allergic to penicillin) intravenously. This allows the baby to be born without contracting group B streptococcal infection.

Conclusion

Streptococcal infections are caused by *Streptococcus* bacteria and can result in a variety of conditions. Strep bacteria are commonly found on the skin of the human body and in the throat, intestines, and other organs without causing an infection. Disease may result when the bacteria invade parts of the body where they usually are not present or as a result of immunosuppression. Strep bacteria may be transmitted through the air and from person to person. Group B streptococcal disease may also be transmitted from a woman to her baby during birth. It is one of the most common causes of infection in newborns and may lead to bacteremia or meningitis. Most strep infections are treated with antibiotics. When patients follow the recommendations of their healthcare professionals, symptoms typically pass quickly. Some strep bacteria have developed antibiotic resistance, but infections may be prevented with a vaccine. Hand washing is also a helpful and important method of preventing strep infections.

See Also: Airborne Diseases; Antibiotic Resistance; Antibiotics; Children's Health; Women's Health.

Further Readings

McIntyre, James A. and Marie-Louise Newell. *Congenital and Perinatal Infections: Prevention, Diagnosis and Treatment.* Cambridge, UK: Cambridge University Press, 2000.
Orefici, Graziella. *New Perspectives on Streptococci and Streptococcal Infections.* Stuttgart, Germany: Gustav Fischer Verlag, 1992.
Stevens, Dennis L. and Edward L. Kaplan. *Streptococcal Infections: Clinical Aspects, Microbiology, and Molecular Pathogenesis.* New York: Oxford University Press, 2000.

Alessandra Guimaraes
University of Sint Eustatius School of Medicine

SUBURBS

A suburb is a community in an outlying section of a city or, more commonly, a nearby, politically separate municipality with social and economic ties to the central city. In the 20th century, particularly in western Europe, Australia, Japan, China, and the United States, population growth in urban areas has spilled increasingly outside the city limits and concentrated there, resulting in large metropolitan areas where the populations of the suburbs taken together exceed that of the central city. The term *suburb* usually refers to a residential area adjacent to an urban center such as a large city, or separate residential communities within commuting distance of a city. In Western countries, suburbs grew as a result of improved road and rail transport that enabled residents to commute to work.

While some suburbs have a degree of political autonomy, most suburbs are politically, socially, and economically linked to the city. Most Western countries' suburbs have lower population density than inner-city neighborhoods. Usually suburbs are dependent on central cities economically and culturally; usually, but not always, they are independent of those cities politically. American literature on suburbs describes suburbanization

More than 1.5 billion people live in the cities of the global South. Residents of the slums and shantytowns that ring these massive cities are often exposed to health and environmental risks. Here, poor families living at the edge of the wetlands around Asuncion Bay, in Asuncion, Paraguay.

Source: National Biological Information Infrastructure/ Andrea Grosse and John P. Mosesso

as a process involving changes in gender roles and relationships, in patterns of childrearing, and in relationships between men and women that accompanied the urbanization of the United States and the larger processes of which it was a part. Margaret Marsh refers to the processes of industrialization and immigration and the development of the modern corporate economy as paramount to the creation of American suburbs.

Developing countries' suburbs followed divergent patterns of development, in part affected by colonial and postcolonial developments as well as general processes of globalization and industrialization. Urban transformation is rapidly taking place in the less developed countries of Asia, Oceania, Africa, and Latin America and the Caribbean. Developing countries' population is expected to increase from its current level of 1.6 billion (at the time of this writing) to 3.5 billion by 2015. Both cities and suburbs in developing countries are centers of production, employment, and innovation, and suburbs may be less distinguishable from the main urban areas than in developed countries. Urban social scientists note that rapid urbanization has had many negative consequences on developing countries' suburbs: a sharp increase in the incidence of urban poverty, extreme industrial pollution due to the concentration of most manufacturing activities in major metropolitan areas, inadequate access to housing and basic urban services, and the degradation of the urban environment. According to Josef Guglar, a specialist on urban transformations of the developing countries, at this time nearly two-thirds of the world's urban population—more than 1.5 billion people—live in the cities of the global south. Within little more than a generation, their number will triple. While it is difficult to generalize about the typologies and characteristics of the developing world's suburbs, often referred to as "slums" or "shanty towns," it is generally observed that their inhabitants are often exposed to more health and environmental risks than those living in Western developed suburbs.

The environmental and health impacts of living in the suburbs are just as diverse as the types, characteristics, and locations of suburbs. While health and environmental risks may be shared between city and suburban dwellers, sometimes there are cases of great health and environmental disparities between the residents. Living in the suburbs may present the residents with an array of health and environmental risks and benefits distinct from those city dwellers are exposed to due to particular sociodemographic characteristics of the residents, lifestyles, infrastructure, and other conditions. Depending on the type of suburb inhabited, residents' lifestyles and environment may range from "healthy and wealthy" (considering "sheltered communities" of the affluent Western residents) to "poor and risky" (considering high levels of pollution and crime in the slums or shanty towns adjacent to Western cities or in developing countries).

Since the wealthier suburbs in Western countries usually include a relative abundance of adjacent land area, suburban landscapes may be greener than inner-city ones. More affluent suburbs may include parks and recreation areas that would allow some flora and fauna species to peacefully coexist with human residents. On the other hand, commuting to the suburbs by means of private transport, such as personal cars, may actually imply that the air pollution and disruption of landscape by roads would be detrimental to the existence of plant and animal species present in that area. However, suburban lifestyles affecting residential health and environment may differ greatly between, for example, Western and developing cities' suburbs.

Less affluent suburbs or rapidly urbanizing areas in developing countries that may be more polluted may have fewer green areas and higher density of residential buildings and poorly organized infrastructure. Such poor suburban areas may be characterized by social problems, including overpopulation, unemployment, and crime on the one hand and environmental threats such as hazardous waste and inadequate quality of drinking water on the other hand. In Western countries, some suburbs exhibit characteristics of the "ghettos" where ethnic minorities or the urban poor suffer from inferior living conditions. Particularly, developing world suburbs may be affected by a number of health and environmental risks. Respiratory problems and chronic diseases such as asthma may be more widely spread in these areas due to air pollution. Poor sanitation in combination with many other social and environmental characteristics of poor environment and lifestyle are responsible for higher rates of cardiovascular diseases and cancer instances. Many poor suburbs, especially in the developing countries, lack any green areas and exhibit a high degree of environmental degradation.

See Also: Asthma; Cancers; Cardiovascular Diseases; Cities; Health Disparities.

Further Readings

Donaldson, Scott. *The Suburban Myth*. New York: Columbia University Press, 1969.
Guglar, Josef, ed. *The Urban Transformation of the Developing World*. New York: Oxford University Press, 1996.
Jenks, Mike and Rod Burgess. *Compact Cities: Sustainable Urban Forms for Developing Countries*. London: Spon Press, 2000.
Marsh, Margaret. *Suburban Lives*. New Brunswick, NJ: Rutgers University Press, 1990.
McKee, David L. *Urban Environments and Emerging Economies*. Westport, CT: Praeger Publishers, 1994.
Tortajada, Cecilia. "Challenges and Realities of Water Management in Megacities: The Case of Mexico City Metropolitan Area." *Journal of International Affairs*, 61 (2008).

Helen Kopnina
University of Amsterdam

Supplements

The term *dietary supplements* is often used in the fields of conventional, natural, and integrative medicine to describe a category of over-the-counter vitamins, minerals, herbs, and homeopathic remedies. It is estimated that some 45 to 68 percent of American adults use supplements in some form, often with the idea that such use will prevent or cure disease, provide a competitive edge in sports, or complement a healthy lifestyle. Increasingly, even conventional physicians have begun suggesting supplements to their patients, particularly those products that have undergone clinical studies with positive results, such as simple multivitamin and mineral formulas and omega-3 fatty acids. The very nature of the word *supplement* indicates that it does not replace adequate nutrition from whole foods. Several studies indicate that conservative approaches to holistic health—including the use of dietary supplements—may have a positive impact on overall wellness, but they are not a replacement for nutrition from food or advice from a licensed medical professional. Though vitamins, minerals, botanicals, and homeopathic remedies have been safely used for centuries, the dietary supplement industry in the United States has met with criticism from concerned parties who claim that the safety and efficacy of the products remains undemonstrated, and several critics have called for stricter regulation. This has opened a widely publicized debate among various groups, and it remains to be seen what the outcome will be.

Persuasive evidence exists that most people do not get required amounts of necessary vitamins and minerals from their diets alone; the standard American diet, for example, lacks magnesium (a mineral found in fresh, leafy greens), which is necessary for efficient digestive, muscular, and cardiovascular function, and magnesium supplementation frequently aids consumers with constipation, muscle pain, and rapid heartbeats. Evidence also suggests that vitamin C can be effective for building a strong immune system, and recent studies indicate that probiotics, or the healthy bacteria that make up much of our digestive systems, may have wide-ranging health benefits. The U.S. Department of Agriculture (USDA) has established guidelines for the minimum amounts of each vitamin and mineral required for human health, and a comprehensive guide can be found on its website. Supplements have been used for centuries; in 1899, the first edition of *Merck Manual of Diagnosis and Therapy*, now in its 17th edition, was published and included information about prescribing natural remedies to patients. Looking back now, the reasons for eliminating remedies like borax and arsenic, potentially fatal toxins that the *Merck* manual frequently suggested, become clear.

Supplements and Health

Indeed, the rapid evolution of medicine throughout the 20th and 21st centuries has demonstrated, particularly in Europe, the need for integrating nutrition, diet, and lifestyle with supplements and, if necessary, pharmaceuticals. Although the American supplement industry has been widely criticized for making and promoting what critics call "sugar pills," evidence exists in countless clinical trials that supplementation is often necessary to promote good health. For example, recent studies, demonstrated by a search of online medical journals, link the consumption of omega-3 fatty acids (found in fish and fish oil supplements) to a reduced instance of cardiovascular disease, dementia, and inflammation in otherwise healthy adults. What becomes clear in these clinical trials is that quality is more important than quantity, and that patients, practitioners, and consumers should educate themselves about the pros and cons of any remedy, natural or chemical. There is evidence that most dietary supplements are harmless or even beneficial, but taking supplements, particularly in conjunction with prescription drugs, can be dangerous.

The supplements industry is currently regulated by the U.S. Food and Drug Administration (FDA), which places the same restrictions on supplements as it does on food. According to the FDA, any product taken by mouth that contains an ingredient meant to supplement the diet, including vitamins, minerals, herbs, amino acids, enzymes, organ tissues, glandulars, and metabolites, falls under the category of "dietary supplement." In 1994, Congress passed the Dietary Supplement Health and Education Act (DSHEA), which requires supplement manufacturers to conduct rigorous safety testing before marketing products to consumers, prohibits the use of health claims on labels and in advertisements, and requires a specific type of label that clearly marks the product a "dietary supplement."

This situation has created a debate; some groups call for stricter regulation of supplements, while others believe that any regulation is too much. Following a series of adverse effects from the amino acid l-tryptophan and the herbs ephedrine and kava in the mid-1990s, several groups have admonished the FDA for negligence, calling for supplements to be regulated like drugs, which typically take anywhere from eight to 10 years to be tested, approved, and marketed. In return, supplement manufacturers have argued that many supplements are, in fact, food; a good example is brewer's yeast, which contains the B family of vitamins, in addition to minerals like chromium and magnesium. If supplements are to be regulated in the same manner as pharmaceuticals, these manufacturers claim, a consumer wishing to purchase brewer's yeast for cooking would have to obtain a prescription.

Some advocates of supplement safety in the United States call for a system similar to the European Union's, which has strict guidelines for dietary supplements, calling on manufacturers to submit a safety profile for each blend, to clearly label safe upper doses, and to label possible risks and side effects, much like drugs in the United States. Though vitamins and minerals are considered food in the EU, herbal medicines are strictly regulated. The regulatory differences can be summed up this way: in the United States, an herbal product must prove itself dangerous before the FDA orders its removal from shelves; in the European Union, it must prove itself safe before it can be sold. Doctors in Europe frequently prescribe supplements as alternatives to drugs, but they do so knowing the potential side effects and interactions.

How Much Is Enough?

Most recently, the Codex Alimentarius (or "food code") Commission, established by the World Health Organization (WHO) and World Trade Organization (WTO) in 1962 to protect the safety of globally traded food supplies, has been working to establish guidelines for the safe production and consumption of supplements. This has met with harsh criticism from proponents of natural medicine, who claim that Codex violates the individual's right to choose how to treat or prevent health conditions; Codex, however, maintains that it has no intention of making supplements illegal or available only by prescription. According to the FDA and Codex, the role of Codex is to police supplement labeling and to make illegal the practice of misleading consumers into believing false information about supplements.

The scrutiny that the American supplement industry has faced has been met with vocal opposition from industry leaders, who believe that too much regulation is just as harmful as too little. Beginning in 2007, the FDA required supplement manufacturers to report adverse effects (Adverse Event Reporting Program, or AER). According to the FDA's website, in 2008, it received 1,080 adverse event reports, of which 672 were deemed serious. The pharmaceutical industry received 526,000 AERs in 2008, of which more than 300,000 were considered serious. Between 1969 and 2002, 75 FDA-approved drugs were removed from the market due to safety concerns. Makers of quality supplements point out that these numbers indicate the overall safety of vitamins, minerals, botanicals, and homeopathic remedies, particularly in comparison with synthetic drugs.

Several consumer watchdog groups exist that conduct independent testing on various formulas. At the time of this writing, many false claims on supplement labels have been uncovered. According to these tests, many supplements appear to contain less—or more— vitamins and minerals than their labels suggest, prompting consumers to switch to more reputable brands. Such information makes it clear that, when it comes to dietary supplements, all brands are not equal.

Though unregulated, the U.S. supplement industry is a booming business. While entire categories of supplements remain untested for clinical efficacy, the industry continues to experience growth and grossed some $25 billion in 2007. Most doctors now recommend taking a daily multivitamin and mineral complex, and many are beginning to recognize the therapeutic benefit of some botanical compounds. Ultimately, it is up to the consumer to decide what supplements to use, and the advice of a licensed healthcare practitioner is advisable.

See Also: Cardiovascular Diseases; Men's Health; Personal Consumer Role in Green Health; Pharmaceutical Industry; Women's Health.

Further Readings

Codex Alimentarius. "Frequently Asked Questions." http://www.codexalimentarius.net/web/index_en.jsp (Accessed July 2009).

Hecht, Esther. "Seller Beware: What Do Tough European Supplement Laws Say About Safety in the U.S.?" *Better Nutrition* (March 2003). http://findarticles.com/p/articles/mi_m0FKA/is_3_65/ai_97823105 (Accessed June 2009).

Murray, Michael and Joseph Pizzorno. *Encyclopedia of Natural Medicine*, 2nd ed. New York: Three Rivers Press, 1998.

U.S. Department of Agriculture. "Dietary Guidance for Individuals." http://fnic.nal.usda.gov/nal_display/index.php?info_center=4&tax_level=3&tax_subject=256&topic_id=1342&level3_id=5140 (Accessed June 2009).

U.S. Food and Drug Administration. "Tips for the Savvy Supplement User: Making Informed Decisions and Evaluating Information." http://www.fda.gov/Food/DietarySupplements/ConsumerInformation/ucm110567.htm (Accessed July 2009).

Kate Birdsall
Michigan State University

Supplying Water

Water is a vital resource used for industrial, agricultural, residential, and municipal needs. However, only 2.5 percent of water on the Earth is fresh, and merely a third of that is accessible due to amounts stored in polar ice caps and glaciers. Water is considered a renewable resource as a result of the water cycle and is available in most regions with the proper technology. Rural areas of the world often draw their water by hand from any source available including wells, rivers, lakes, and collected rainwater. In the United States and other developed countries, however, more sophisticated methods are used to obtain water. The water supplying these countries is often treated and distributed to those industries, farms, and communities in need. The infrastructure necessary to supply water is expensive to build and maintain and a constant expense to both governmental entities and consumers. Infrastructure used to supply water, as well as government regulations, can contribute to a sustainable water supply.

Groundwater

The water supplying communities and cities in the United States comes from a variety of sources, including groundwater wells, dams, reservoirs, and surface water intakes.

Groundwater, which is found naturally in subterranean geological caverns and formations, is the most abundant form of available freshwater. Rainwater, or occasionally glacial water, that is not held by the soil trickles down until it encounters impervious rock or clay. There the water creates pockets of groundwater as it accumulates. This downward action is called percolation, and the large pockets of subterranean water are called water tables due to their lateral growth pattern. Groundwater must be pumped to the surface for use. Although it is possible to recharge groundwater deposits through the percolation process, water is being removed too quickly in many places, decreasing the amount of groundwater available. Overdrawing groundwater results in "falling" water tables and, ultimately, the complete exhaustion of groundwater in certain locations.

Communities and scientists have attempted to address groundwater sustainability. Groundwater sustainability revolves around maintained use of groundwater to conserve its use for an indefinite amount of time. Some methods used to foster sustainable groundwater supplies are regulations or taxes that limit pumping, improving the efficiency of water use, and the recycling of wastewater for purposes other than consumption, such as cooling systems and irrigation. Some municipalities have taken advantage of the natural filtration process that groundwater undergoes. As water percolates through the soil, microorganisms, plant fibers, roots, and plants filter or absorb contaminants. An example of a city utilizing this method is New York City. Water used by New York City is filtered by the Catskill Mountains located northwest of the city. In recognition of the role they play in water filtration, the Catskills are now protected under the U.S. Environmental Protection Agency (EPA).

Surface Water

Surface water is where most supplies of potable water originate. Although surface water is not as abundant as groundwater, it is easier—and less expensive—to access. Water from dams, reservoirs, and other surface sources is withdrawn with the use of intake pipes. Surface water supplies are dwindling quickly as increasingly more water is needed to meet growing human demands. Usable surface water is also becoming harder to locate due to pollution. Surface water is easily contaminated with runoff and waste containing disease-causing pathogens or chemical adulteration. However, new technology has been able to create surface water through more and larger dams and reservoirs. Not only do these dams create a water source, but they also provide a source of hydroelectric energy, which is created by harnessing the power of water. Dams have also created problems by displacing millions of people and drowning local ecosystems.

Water Treatment

As water from both the Earth's surface and underground is collected, it is processed through water treatment plants. Drinking water must meet standards placed according to country and any site-specific needs. Regulations placed by the EPA set standards for public water systems, while the U.S. Food and Drug Administration (FDA) regulates bottled water. Regulations include testing for pH level, nitrogen, iron, and bacteria levels. Not all countries have water treatment systems. As a result of lack of water treatment facilities, millions of deaths are caused by waterborne illnesses due to contamination in developing countries. The United Nations Millennium Development Goals, the result of an international collaborative effort, has as one of its objectives decreasing by half the number of people without access to sanitary drinking water.

Water treatment plants are expensive to build and maintain. At these plants, solids are first removed from water. Solids can include a variety of wastes, such as debris from surface water such as sticks, garbage, and plant matter. Alum is mixed into the water, which attracts debris particles. The water then moves to a large tank where the solid particles are allowed to separate and settle to the bottom and clean water is allowed to flow out. Further filtration follows. Chlorine is added for disinfection, and the water then passes through an additional sand filter. Finally, lime is added to increase the water's pH, and fluoride is sometimes included to improve dental health. From the treatment plants, water is transported to communities. Water can then be stored in tanks to provide for fluctuating water usage throughout the day and or be distributed directly to the community.

Regulation and Risks

The water supply for communities and businesses must be managed and regulated for sustainability. Some suggest placing constraints on water consumption to achieve the goal of a sustainable water supply. As groundwater is overpumped along coastal regions, saltwater intrusion takes place. Saltwater intrusion occurs as freshwater is pumped out of the water tables, sucking in saltwater to replace the removed water. Saltwater is not usable by humans or for agriculture and industry. Saltwater intrusion can render a water table useless.

Although water is considered a renewable resource, it is not inexhaustible. Overuse of rivers and lakes has created dry riverbeds, wetlands, and lakebeds. The fragile habitats of waterfowl, fish, and other aquatic life are destroyed. As our water supplies run out, we, too, are left in drought and with unusable fields and lands.

See Also: Bottled Water; Chemical Pesticides; Government Role in Green Health; Tap Water/Fluoride; Waterborne Diseases.

Further Readings

Alley, W. M. "Tracking U.S. Groundwater: Reserves for the Future?" *Environment*, 48/3 (2006).

Morrison, J. "How Much Is Clean Water Worth?" *National Wildlife*, 43/2 (2005).

Wright, R. T. *Environmental Science*. Upper Saddle River, NJ: Pearson Prentice Hall, 2008.

Stephen T. Schroth
Claire C. Turner
Knox College

Swimming Pools

Swimming pools, artificially constructed repositories of water intended for swimming or water-based recreation, are popular across the United States. Many municipalities or other local governments operate public swimming pools. Many private homes also have

Public heath pathogens such as viruses, bacteria, protozoa, and fungi can be present in swimming pools and cause diseases to swimmers. Common diseases caused by contaminated water include diarrhea; otitis externa, commonly known as swimmer's ear; skin rashes; and respiratory infections.

Source: Centers for Disease Control and Prevention

swimming pools, especially in warmer climates. The water in swimming pools may play host to a variety of health hazards, such as bacteria, viruses, algae, and insect larvae. To control these hazards, those responsible for maintaining swimming pools often use chemical disinfectants such as chlorine, bromine, or mineral sanitizers, as well as additional filters. Concerns about the health risks of chemicals used in pool sanitation have led to the development of chemical-free alternatives to chlorine and other substances. Many of the alternative methods of sanitizing swimming pools also have negative environmental repercussions. As a result, no consensus exists as to sustainable methods to sanitize swimming pools.

Water Contaminants

Swimming pools have been popular since ancient times. Ancient Greeks and Romans built swimming pools for athletic training, for recreation, and for military exercises. By the mid-19th century, indoor swimming pools became popular in Great Britain, and after the advent of the modern Olympics in 1896, this trend spread to the United States. Interest in competitive swimming grew after World War I, and many high schools, colleges, and universities added swimming pools for use by competitive teams. Residential swimming pools became popular after World War II, and the postwar economic boom also saw the construction of many municipal swimming pools for public use. Swimming pools can be built outdoors or indoors, above ground or in ground. Many swimming pools are heated to provide added comfort and to extend their periods of use.

It is imperative that the water in swimming pools be safe. To that end, organizations such as the U.S. Centers for Disease Control and Prevention (CDC) and the World Health Organization (WHO) have published guidelines for maintaining the safety of swimming pools and standards for minimizing microbial and chemical hazards. Swimming pool contaminants endanger water safety and are introduced from both environmental sources and swimmers. Outdoor swimming pools are particularly affected by environmental contaminants such as rain containing microscopic algae spores, droppings from birds, windblown debris, and the like. Both outdoor and indoor swimming pools are subject to contaminants introduced by swimmers. Swimmer-introduced contaminants can include perspiration, cosmetics, suntan lotion, urine, saliva, and feces. The interaction between swimming pool disinfectants and water contaminants can additionally produce a mixture of chloramines and other hazards. Public heath pathogens such as viruses, bacteria, protozoa, and fungi can be present in swimming pools and cause diseases to swimmers. Common diseases caused by contaminated water include diarrhea; otitis externa, commonly known as swimmer's ear; skin rashes; and respiratory infections.

Conventional Sanitation Methods

When swimming pool water contains low levels of bacteria and viruses, the spread of diseases and pathogens among users is greatly reduced. Well-maintained, properly functioning pool filtration and recirculation systems are the first barrier to contaminants large enough to be filtered. Removing filterable contaminants rapidly reduces their impact on the disinfection system and limits the formation of chloramines, restricts the formation of disinfection by-products, and optimizes sanitation effectiveness. Pumps and mechanical filters are useful in removing bacteria, algae, and insect larvae from swimming pool water. Chemical disinfectants, such as chlorine and bromine, are the most popular means of sanitizing swimming pools. Chlorine is the more popular of the two and is both convenient and economical. Both chlorine and bromine are members of the halogen family and have the ability to destroy and deactivate a wide array of bacteria and viruses. Swimming pools treated with chlorine-releasing compounds are often supplemented with cyanuric acid, which is a granular stabilizing agent that extends the active chlorine residual half-life by four to six times.

Despite its effectiveness, many swimmers do not like chlorine because it can irritate the skin and cause other reactions. Chlorine has also been implicated in the destruction of Earth's ozone layer, as chlorine-containing molecules such as chlorofluorocarbons have been found in the upper atmosphere. Additionally, chlorine can react with urea in urine and other wastes from humans to produce chloramines, which can lead to respiratory problems, including asthma. This reaction occurs when an insufficient amount of chlorine is used to disinfect a contaminated pool. These problems have caused some responsible for swimming pool sanitation to seek alternatives to chlorine and other chemical treatments.

Alternative Sanitation Methods

Those who elect not to use chlorine or bromine to sanitize swimming pools often select chemical-free, electronic oxidation sanitation methods. Electronic oxidation water sanitation, however, relies upon metals that can be toxic to aquatic life even in very small quantities. Electronic oxidation of the water molecule is used to generate the natural oxidizers atomic oxygen, hydrogen peroxide, hydroxyl, and molecular oxygen. These all have a higher oxidation reduction potential than chlorine. Electronic oxidation's benefits include speed, as it can generate more oxidizers in one minute than chlorine or other popular chemical methods can generate in an hour. When electronic oxidation is combined with low levels of copper ionization, it provides a very effective method of swimming pool sanitization that is chlorine free. Such a method, however, is not environmentally friendly because the metals produced are far more lasting than chlorine and are toxic to a wide range of aquatic life in very small amounts.

See Also: California's Green Chemistry Initiative; Children's Health; Chlorination By-Products; Cost-Benefit Analysis for Alternative Products.

Further Readings

Doheny, Kathleen. "Swimming Pool Chemicals May Carry Cancer Risk." *WebMD* (September 13, 2010). http://www.webmd.com/cancer/news/20100913/study-swimming-pool-chemicals-may-pose-risks (Accessed September 2010).

Health and Safety Executive (HSE). "Managing Health and Safety in Swimming Pools." London: HSE. http://www.hse.gov.uk/entertainment/swimpools.htm (Accessed September 2010).

Stringer, Ruth and Paul Johnston. *Chlorine and the Environment: An Overview of the Chlorine Industry*. Amsterdam, Netherlands: Springer, 2009.

Stephen T. Schroth
Brianna R. Collishaw
Knox College

T

TAP WATER/FLUORIDE

Drinking water should be free from disease germs and hazardous contaminants. In order to provide this, water is filtered, is cleaned, and has chemicals added to it. In the case of tap water, the water is first purified and then rendered fit for consumption by introducing minute quantities of chlorine and passing ammonia through it. This destroys harmful bacteria that may have slipped through the filters, such as the organisms of typhoid and paratyphoid fevers, cholera, dysentery, jaundice, and parasitical worms that breed in water. All these chemicals are filtered out before tap water reaches millions of homes, and experts are paid to taste it and ensure that it is palatable.

Good drinking water should contain minerals necessary for our bodies. Fluoride, for example, is one such mineral, which in small proportions is excellent, especially for the teeth. The same mineral in excess is said to cause harm to the body. However, fluoride is not added to water to destroy bacteria, and not all of it is eliminated by filtering or by other means. When fluoride is added to tap water, it is meant to remain in the water and enter the bloodstream of those who drink it. Even when less than 1 part per million (ppm) is ingested, the cumulative effect tends to be harmful over a period of time.

Fluoride is assumed to be safe because some water supplies contain fluoride ions derived from the calcium fluoride that is naturally present in it. People consuming this water for years have apparently shown no adverse health effects. Water experts and water supply authorities often argue that when sodium fluoride is added artificially, the fluoride ions remain precisely the same as those that appear naturally. When fluorine occurs in water naturally, it is nearly always found in hard water, so that for every part of fluorine present, there are between 200 and 300 parts of calcium, which limit, control, and partly nullify the highly active fluoride. However, when fluoride is added to soft water, there may be a presence of only 10 parts of calcium to one of fluoride, a proportion insufficient to nullify the poisonous effects of the fluoride ions over long periods.

Where water contains fluorine naturally, the fluorine is present in the form of calcium fluoride. Calcium fluoride is a very inert and insoluble substance, and because of this, there is considerable difficulty in adding it to a drinking water supply. In case of tap water or drinking water supplies, it is impracticable to add calcium fluoride to water because it is virtually insoluble; and so, water is dosed with sodium fluoride. Sodium fluoride is both soluble and active, and it is this that is added to tap water supplies. Water experts assert that sodium fluoride in water has the ability to ionize and the resultant reaction would be the same as if calcium fluoride were present, and theoretically, this is so. But in practice, it is not so and laboratory experiments have shown that the ionization of calcium fluoride and that of sodium fluoride in the presence of other calcium salts differ widely. Those who wish sodium chloride to be added to water maintain that when added in the proportion of 1 ppm, the chemical is completely ionized and, therefore, loses its normal biological properties and reacts on living cells precisely as inertly as calcium fluoride.

Fluoridation of public water supplies is considered to have been one of the 10 most important public health measures in the last century and is recommended by international health organizations. According to the World Health Organization (WHO), consumption of water containing fluoride in excess of 1.5 milligrams/liter can cause dental fluorosis, and if the fluoride quantity is too much in excess, the body faces the problem of skeletal fluorosis. Dental fluorosis is the first visible sign of excess fluoride consumption and is shown as white or brown mottled teeth. Pitting of teeth may also occur. Though health experts assure that fluorides preserve the enamel on the teeth of growing children, the same mineral has little effect for children in the age group 14 to 16 years. Thus the question remains as to why subject the entire population to forcible fluoridation, for it merely means that yet another destructive element is being taken into the system, which may cause or help to accelerate some degenerative diseases like cancer, diabetes, or hardening of the arteries.

However, everyone is not affected in the same way by fluoridation. Special attention should be paid to the accumulation of fluoride during pregnancy, in the mother and the fetus; and in mothers who suffer from affections of the heart, liver, kidneys, and bones or have a tendency to abort. Sufferers from allergies, diabetes, rheumatoid arthritis, all neoplasms (especially cancer), digestive disorders, glandular diseases, and diseases of the heart, liver, kidneys, bone and teeth, blood and blood vessels, and nervous system will react more violently to fluorides than people who are healthy.

Tap water coming from surface water sources is generally free from excess fluoride. However, if a bored well is the source of water or if water is bought from private tankers, it is a safe practice to check a sample of the water for excess fluoride content. If excess fluoride is detected, it must be removed before the water is used for drinking or cooking. Boiling the water does not remove fluoride.

Since most rural habitations and many urban habitations depend on deep groundwater, some method of removal is necessary to check the presence of fluoride. Various methods of defluoridation of water are available. These include household-level defluoridation using activated alumina to community-level defluoridation plants. While these work well under careful supervision and ownership, many have failed to survive the test of time simply due to apathy or difficulties in maintenance.

Rooftop rainwater harvesting in sump tanks can provide enough drinking and cooking water for a whole year that is free from chemical contaminants such as fluoride, arsenic, and nitrates. Taking steps to recharge groundwater through rainwater harvesting is also wise. These are long-run measures to reduce fluoride. Shrinking groundwater levels have been reported as one cause for the sudden increase in fluoride in groundwater. Sustainable use of groundwater resources, keeping the aquifers clean and charged, and using the dynamic groundwater are measures that will go a long way in ensuring water for all in a climate-changing world. Water wisdom lies in realizing the true ecological value of water and taking steps to protect this precious resource.

See Also: Arsenic Pollution; Bottled Water; Centers for Disease Control and Prevention (U.S.); Environmental Protection Agency (U.S.); Groundwater; Oral Diseases.

Further Readings

Centers for Disease Control and Prevention. "Achievements in Public Health, 1900–1999: Fluoridation of Drinking Water to Prevent Dental Caries." *Morbidity and Mortality Weekly Report*, 48/41 (1999).

Centers for Disease Control and Prevention. "Ten Great Public Health Achievements—United States, 1900–1999." *Morbidity and Mortality Weekly Report*, 48/12 (1999).

McDonagh, M. S., P. F. Whiting, P. M. Wilson, A. J. Sutton, I. Chestnutt, J. Cooper, K. Misso, M. Bradley, E. Treasure, and J. Kleijnen. "Systematic Review of Water Fluoridation." *British Medical Journal*, 321/7265 (2000).

World Health Organization. "World Oral Health Report" (2003). http://www.who.int/oral_health (Accessed March 2010).

Mohua Guha
International Institute for Population Sciences

TAXATION OF UNHEALTHY PRODUCTS

The United States has a long history of taxing unhealthy products, colloquially referred to as "sin taxes." The most common products with excise taxes are alcohol and cigarettes. Some states and localities also tax (or propose to tax) other unhealthy products such as soda, candy, and fast food, which leads to much controversy. Proponents argue that they generate revenue to offset the additional demand on public resources caused by the negative effects of consumption; opponents argue that these taxes are regressive and place an undue burden on the poor.

The first excise tax was proposed on alcohol by Alexander Hamilton in 1791. Although defeated by the Whiskey Rebellion, it demonstrates that alcohol has long been viewed as a luxury and as a product needing to be controlled. In addition to federal taxation on spirits, wine, and beer, states and some local governments also levy taxes on these products. Health advocates argue that these taxes serve as behavior deterrents and encourage lower consumption and healthier lifestyles. However, the data do not support this, causing others to argue that this revenue should be earmarked for responsible drinking campaigns and/or treatment programs. Only a few states are currently doing this. Other programs funded by alcohol taxes are education, corrections, and tourism promotions. The dollars that are not earmarked go to the general fund and indirectly help pay for public health programs.

Taxation of cigarettes and other tobacco products does serve as a deterrent as well as raise revenue specifically for healthcare programs. The federal tax was raised to $1.01 per pack on April 1, 2009, and helps fund the State Children's Health Insurance Program (SCHIP). The average state cigarette tax is $1.34 per pack. Counties and cities also impose local sales tax and cigarette-specific taxes. These cigarette-specific taxes range from a few cents to more than $3. These taxes tend to be lower in states where tobacco production is a major industry and job producer.

The United States has taxed tobacco since the Civil War, when it was classified as a luxury. Today, some economists argue that the cumulative taxes are unfair to the poor because cigarette smoking is more common among lower-income people. This shift has largely occurred during the past few decades. While it is associated with increased awareness of the detrimental health effects of smoking, it is also linked to laws prohibiting indoor smoking as well as the tax hikes. These collectively have created a stigma against smoking to which the middle and upper classes and the educated have responded. Antismoking advocates estimate that the federal tax increase alone could reduce smoking-induced deaths by 900,000 lives. In addition to these taxes saving lives, proponents argue that smokers use more healthcare resources and therefore should pay into the system through this method. It is widely known that smoking is directly linked to multiple diseases including hypertension, cancer, emphysema, and birth defects. It is also widely accepted that secondhand smoke (cigarette smoke breathed by nonsmokers) has negative health consequences.

Taxation to support public health programs helps to offset the cost of healthcare resources necessary to treat disease and also to fund campaigns to discourage smoking. Additionally, these taxes may serve as behavioral deterrents for teenagers. Smoking has long been considered "cool," but taxes are helping make the product cost prohibitive. Studies are inconclusive, but some teenagers who do smoke will likely not become long-term habitual smokers as a result of the economic impact associated with cost. Antismoking advocates estimate that higher cost, due to the tax burden, could stop two million teens from becoming habitual smokers.

Although smoking rates have declined, obesity is a growing health problem at a time that healthcare is becoming more and more expensive. Legislators are looking at new taxes to raise revenue as well as to encourage changes in consumption habits. Proposals to tax sodas (and other sugary drinks) and fast food have been introduced as a method to reduce obesity. There has also been discussion of taxing high-fructose corn syrup, which is the sweetener in many unhealthy products. This action would be extremely controversial as the U.S. Department of Agriculture subsidizes corn production. Candy is already taxed in some states (e.g., Colorado, Illinois). These excise taxes are not substantial enough to alter behavioral choices but add money to state coffers.

With regard to green health, the taxation of unhealthy products can be linked to reduced environmental impact. For example, the decreased smoking rate helps improve air quality, and a reduction in demand on healthcare resources will result in less material waste. The healthcare industry is a major source of garbage generation due to the need to maintain sterility of products and of single-use items to protect patient and worker. Because of the environmental impact, packaging is now being evaluated for taxation. Some European countries have already levied packaging taxes. A similar measure in the United States could generate substantial revenue. It is unknown if these taxes would be associated strictly with unhealthy products or with particular packaging types. Regardless, such taxes would encourage innovative technology and reduction of unnecessary packaging material, which would have enormous positive consequences in the healthcare industry as well as reduce the environmental burden associated with unhealthy products.

See Also: Children's Health; Fast Food; Indoor Air Quality; Obesity; Smoking; Wine and Other Alcohols.

Further Readings

Chaloupka, Frank J. "How Effective Are Taxes in Reducing Tobacco Consumption?" http://tigger.uic.edu/~fjc/Presentations/Papers/taxes_consump_rev.pdf (Accessed February 2010).
O'Donoghue, Ted and Matthew Rabin. "Optimal Sin Taxes." http://www.econ.berkeley.edu/~rabin/shortsin.pdf (Accessed February 2010).

Waisanen, Bert. "Taxing Behavior: Personal Vices Have Been Around a Long Time, but the Price for Those Habits and Other Controversial Behaviors Is About to Go Up." *State Legislatures*, 30/6 (June 2004).

Denese M. Neu
HHS Planning & Consulting, Inc.

TOPOPHILIA

As part of his work on human geography, Yi-Fu Tuan developed a theory of topophilia. Topophilia, literally "love of place," is defined as humans' innate affection for place and environment. It is, above all, an individual's perception and instinctive reaction to environmental settings, typically associated with more natural environments. In the context of green health, topophilia highlights the affective bond and salutogenic psychological response to natural environment settings.

As a theory, topophilia aims to address how humans respond to their environment, the relationship between this response and health, and how urbanization affects the topophilic response. Human response to the environment is personal but can be contextualized for a deeper theoretical understanding. This contextualization can occur across system levels. At the individual level, the response to environment is highly based on perception, and perception is the culmination of the senses. There is a distinct visual acuity within environments. There are aspects that fascinate and draw attention; there are facets that distract and appear incoherent. In outdoor environments, changes in topography and built or natural features may perk up visual perception. With indoor environments, response is often limited to the contents within the room but may also include the view field out of a window.

The tactile sense can influence environmental perception through the arduous task of walking up a steep, rocky path; the act of digging and planting a garden; or the memories of sand between the toes. The auditory sense provides awareness of things seen and unseen: the crashing waves, birds chirping, traffic sounds, and so forth. Smell is an imperative sense, as it has strong memory attachment. The smell of a flower or distinct environment can take a person back to a previous experience and time. In perceiving an environment with these four senses, a response and memory are formed. The memory can be invoked with the reexperiencing of any of these senses. This can be an unconscious response to an environment. Associated feelings resurface, and affective bonds are realized: topophilia.

Human perception can also be described at the group level. There are age, gender, and cultural differences in the response to environments. Value is added to specific attributes of environments. Various tribes of American Indians responded to and valued different aspects of their environments. American Indians in the southwestern United States highly valued the scarce water sources and the ethereal peaks and sandstone formations of their desert habitat. Those tribes across the prairies and Great Plains valued the open expanse and great horizons. Eastern American Indians greatly valued the many streams and rivers flowing from the mountains to the sea. This environment provided a mode of transportation and an abundant food source.

In terms of gender and cultural norms, the daily environments and mental maps of men and women can vary greatly. A study of rural women in Spain found the home and surrounding environment to be a place of both topophilia and topophobia. Some women described their homes and setting with affection, citing recreation, relaxation, escape, and creation. Others described feelings of imprisonment and seclusion, more aligned with fear or aversion to place, or topophobia.

Perception and value of place also vary across age. The foundations of topophilic environments are developed across childhood and often returned to and thought of fondly as one ages. Perception changes a great deal as one ages. For example, the forest is perceived as less dark and scary as one ages, and the mountain is not perceived as so high.

Topophilia and the Environment

Topophilia and the environment are intrinsically linked. People, regardless of group and individual differences, have psychological and physical responses to different environments. Psychological response is bound primarily in visual perception but also includes the olfactory and auditory senses. Visually, topophilia is based on aesthetic appreciation and valuation. Environmental psychology—and the theories of attention restoration and psychophysiological stress recovery—underscores the importance of aesthetically appealing environments. Studies reveal a preference for familiar, natural environments, as opposed to urban and built environments. Parks, mountains, the riverside, and the seashore are often listed as favorite and preferred places, regardless of age group. Familiar and favorite places become environments of attachment and affection. These are environments with consistent appeal to which people often return. Topophilic environments also provide a physical and psychological sense of being away; one is removed from stressors and can relax and enjoy a favorite and preferred environment.

The human physical association and connection with the natural environment is perhaps most salient but is becoming less frequent. The natural environment has become more associated with recreation than vocation. Even recreation within natural environments is declining and has brought about terms such as *nature-deficit disorder*. Federal and state agencies are encouraging more children to get outside and play. Outdoor play and experiences provide leisure-time exercise and also allow young minds to explore their natural surroundings and make sense of their environments. The experiences in natural environments can lead to greater ecological and environmental appreciation, increasing knowledge and value and, in turn, promoting pro-environment and green behaviors. The experience of natural environments—especially favorite natural environments—has been linked to a variety of salutogenic health outcomes.

Topophilia and Health

The theory of topophilia is inherently connected to green health. The realization of a psychological and physical connection with the natural environment often leads to greater pro-environment behaviors and green choices. These behaviors may include recycling, installing solar panels, choosing to bike to work, and so forth. Natural and open spaces in urban environments promote greater walking and bicycling for leisure and for transportation, which in turn decrease harmful vehicle emissions, increase physical activity, and build neighborhood social capital. Each of these behavior changes can reduce an individual's and a community's carbon footprint, helping to mitigate global climate change.

There are beneficial psychological and physical health outcomes associated with affection for place. Fascinating, coherent environments that provide a sense of scope and extent have been shown to restore directed attention, reduce blood pressure, and increase positive affect. Across human–environment studies, the natural environment has been shown to facilitate such health benefits significantly more than urban and built environments. One need not literally be in the topophilic environment to experience benefits. Studies have shown that simple views of nature out a window provide salutogenic health outcomes. Imitations of topophilic environments can also provide psychological health benefits. Zen gardens, with rocks representing mountains and raked sand representing the sea, provide a tranquil environment and experience. Likewise, those proficient with bonsai trees spend time emulating topophilic environments and trees on a miniature scale. Virtual and photo-represented topophilic environments have also been associated with health benefits.

Most human–environment studies are based on the foundations of topophilia and a person's innate affection for certain environments. The psychological benefits of stress relief and restoration from directed attention fatigue are emphasized by the World Health Organization's (WHO's) 2002 finding that the world's number one burden of disease is neuropsychiatric disorders including depression, post-traumatic stress syndrome, and alcohol and substance abuse. Psychological health benefits of topophilia have been elucidated in a study associating topophilic attributes of environments with the WHO's Quality of Life instrument. In the study, synesthetic tendencies—or those environmental attributes associated with the senses—accorded to hypothetical ecosystem components were found to be most associated with a higher rating of quality of life. Other attributes highly associated with quality of life were ecodiversity, especially bodies of water and mountains/hills, and environmental familiarity, which included spaciousness and identifiability.

There are activities that provide both psychological and physical health benefits in topophilic environments. Gardening allows for a physical relationship with nature. Calories are burned through the process of tilling the soil, planting the garden, and harvesting the produce. Gardening is also cited for its psychological benefits, allowing for the redirection of attention and reduction of stress. (The new field of ecopsychology uses gardening and other experiences with nature as a psychological healing tool.) Urban community gardens provide community-level benefits by building sense of place and pride, which in turn can establish a topophilic bond between community members and their urban environment. Whether personal or community based, gardens provide a visually aesthetic environment as well as a source of healthy food items. Fishing can provide similar experiences and health benefits. More physical recreational activities such as hiking, mountain biking, and kayaking also provide the dual health benefits of psychological and physical experience within topophilic natural environments.

Topophilia and Environmental Change

In the second half of the 20th century there was an exodus from the city to suburbia. This mass move was associated with topophilia. City dwellers sought nature but desired to hold on to the convenience of the city. The overriding affection for nature drew citizens to the border between the country and the city. In suburbia, residents could tend their gardens, take walks in parks and tree-lined neighborhoods, but still remain close to the commerce of the city center. Over time, adverse health and environmental effects have been associated with the move to suburbia. Time in cars greatly expanded as people began commuting from the suburbs to the city; this led to a more sedentary lifestyle and increased carbon dioxide pollution. Though there may be topophilic elements associated with the home environment, the act of commuting via automobile contributes greatly to global climate change and the associated environmental degradation. Such a lifestyle change (even for the benefits of home-based topophilia) is unsustainable. Unsustainable actions are the greatest threat to favorite and preferred places—those places with the greatest topophilic bond.

Unsustainable practices in environments previously associated or thought of as topophilic can be seen in tourist environments, especially around the world's beaches. Hawaii has long been seen as a topophilic environment with its wide beaches, crashing waves, and luscious green coastal mountains. However, the attraction to such an environment came with huge increases in tourism, which included the building of many oceanfront properties, hotels, and commercial space. Now, certain beaches are no longer seen as attractive and natural but instead are hotels with a sandy front. What was once a topophilic environment has lost the perceived positive effect.

This, of course, does not have to be the circumstance with every aesthetically appealing environment that people seek to visit and experience. Whether in the mountains, by the sea, or in between, sustainable measures can be applied and adapted. Ecotourism is an example. Sustainable building and management of topophilic environments allows for the human experience of the environment without costing the environment its attractive charm.

Topophilia and Green Health in the 21st Century

With a majority of the world's population now living in cities, the relationship between topophilia and health is as pertinent as ever. Those living in urbanized areas are in need of topophilic environments to support their psychological and physical health. Public open space provides a suitable topophilic environment for many. Cities throughout the world are encouraging green design and also topophilic design. This includes green roofs that reduce the heat-island effect and reduce storm water runoff. The roofs also provide a green and natural respite for workers within the buildings. Similarly, living walls—vertical walls literally covered with vegetation—are also gaining popularity.

There is a great human search for the natural environment within today's urban matrix. There is motivation to incorporate nature and natural settings in many aspects of cities, but to be successful—in terms of topophilia and green health—designs and efforts must be sustainable. As residents begin to physically and psychologically perceive the environments inherently preferred, affective bonds and salutogenic health outcomes will naturally follow.

See Also: Mental Health; Recreational Space; Suburbs; Urban Green.

Further Readings

Gonzalez, Beatriz Munoz. "Topophilia and Topophobia." *Space and Culture*, 8/2 (2005).
Kauko, Tom. "Sign Value, Topophilia, and the Locational Components in Property Prices." *Environment and Planning*, 36 (2004).
Ogunseitan, Oladele A. "Topophilia and the Quality of Life." *Environmental Health Perspectives*, 113/2 (2005).
Ruan, Xing and Paul Hogben, eds. *Topophilia and Topophobia*. New York: Routledge, 2007.
Tuan, Yi-Fu. *Topophilia: A Study of Environmental Perception, Attitudes, and Values*. Upper Saddle River, NJ: Prentice Hall, 1974.

J. Aaron Hipp
Washington University in St. Louis

TUBERCULOSIS

The persistence of tuberculosis (TB) as a public health problem is as much a social phenomenon as it is a biological one. As far back as the time of Hippocrates in classical antiquity, TB was commonly referred to as "consumption" because of the characteristic deterioration of the body associated with the disease. Evidence extracted from the mummies of ancient civilizations has revealed that the TB bacillus has plagued human beings for tens of thousands of years. Still in the early 21st century, it is estimated that there are 8 to 12 million new cases of TB worldwide each year. The disease exhibits one of the highest infection rates of all known diseases. The causative agent of TB was discovered by Robert Koch in 1882 to be the rod-shaped bacterium *Mycobacterium tuberculosis*. The bacteria may infect the lungs and spread to others through sneezing, coughing, spitting, or talking. Notably, only a smaller number of the bacilli need to be inhaled for a

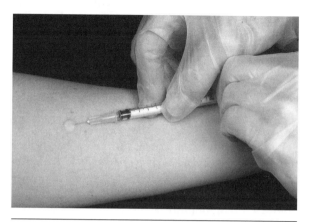

A technician conducts a tuberculin skin test. Recent tuberculosis (TB) outbreaks in cities such as New York and London in the 1980s and 1990s have shown that TB is not simply a disease affecting the developing world.

Source: Centers for Disease Control and Prevention/ Gabrielle Benenson

person to become infected. Although pulmonary TB is the most well-known variant, TB may also affect the brain, bones, skin and lymph nodes, genitals, and other tissues, which can be eaten away by the bacteria. In addition to weight loss, other symptoms include breathlessness, coughing, fever, malaise, anemia, the disruption of metabolic functions, and psychological disturbances. Pulmonary TB may be identified through sputum (i.e., phlegm from deep inside the chest) microscopy, chest X-rays, or a skin test.

The Etiology of Tuberculosis

The development of TB follows various stages based upon the progression of various defense mechanisms initiated by the human immune response system. Initially, toxic chemicals released by the TB bacteria induce the human body to produce characteristic swellings or lesions at sites where the bacteria become clustered in the infected tissue. During the later stages, the immune system attempts to prevent the spread of TB to other parts of the body by enveloping the lesions in a thick fibrous coating of collagen, while bacteria that do escape the encapsulation are destroyed by immune response agents known as macrophages. The encapsulation eventually forms a hard grain called a granuloma that appears as a small dot on a chest X-ray. The resultant walled-off bacterial clusters can lie dormant in this latent phase for many years. At this stage, the individual is otherwise healthy and noninfectious and is classified as a nonactive case. In roughly 5 to 10 percent of the latent cases, the disease may become activated if the individual's immune system becomes weakened or damaged later in time. During this active stage, the bacteria may overcome the body's defense mechanisms and proliferate in the lungs, where they can be spread more easily to others through respiratory secretions, or be discharged into the circulatory and lymph system, where they can spread to other parts of the body. In terms of developing and implementing effective public health responses to TB, it is important to distinguish between cases that are the result of recent transmission (i.e., secondary cases) and those resulting from reactivation of old infections, especially since the original site of infection may be very distant from where the latent case currently resides, as well as very distant in the past.

A TB vaccine was developed in 1921 (the Bacille Calmette-Guerin vaccine) and has been used in children's mass vaccination programs sponsored by the World Health Organization (WHO) throughout the world. The main treatment strategy involves five primary drugs. Due to the development of multidrug-resistant antibiotic strains (MDRTB), a secondary group of drugs has been developed, although these tend to be more toxic, have severe side effects such as nausea and abdominal pain, are less effective, and cost up to 100 times more than the first-line drugs (essentially making them unaffordable to those in the developing world). Furthermore, MDRTB treatment involves a daily multidrug regime that must be followed strictly for at least two years, compared to the standardized daily dose of the same first-line drug given over a six- to nine-month duration.

The number of cases of drug-resistant strains has risen dramatically in recent times; in 1984, only 10 percent of the cases worldwide were drug resistant, but by 1995, 52 percent of the cases were resistant to at least one first-line drug. Nonadherence to the prescribed drug regime is the main contributing factor to this trend, as haphazard or inconsistent use induces antibiotic resistance. To counter this tendency, in the early 1990s, the WHO adopted the program of directly observed therapy (DOTS) as a global strategy. The key feature of DOTS is that the patient must take the anti-TB drug in front of the healthcare worker at the prescribed time so as to ensure complete compliance. Although DOTS has been found to be effective in substantially decreasing rates of drug-resistant TB, this program, along with an exclusive emphasis on chemotherapy more generally, has tended to shift attention away from the structural factors that are inextricably involved with TB as a multifaceted public health problem.

The Social Dimension of Tuberculosis

Thomas McKeown found that a dramatic decline in TB morbidity and mortality rates between 1840 and 1940 occurred before the invention and introduction of antibiotics, and could not thus be attributed to medical intervention alone. He attributed this decline to improved social conditions based on better nutrition, improvements in sanitation, less crowded living conditions, and a higher standard of living more generally. Indeed, the natural history of TB reveals that the disease can become more widespread because of its interaction with a multiplicity of social, economic, and political factors such as poverty, inequalities in wealth leading to social deprivation and marginalization, crowding, poor sanitation, malnutrition, homelessness, migration, and inadequate access to medical care. As such, TB has often been referred to as a "social disease" or "poverty's penalty" and it follows from this recognition that effective TB control cannot rely exclusively on medical interventions; rather, the social context must be considered.

The strong relationship between lived social conditions and TB is understandable given that this disease is particularly influenced by an individual's immune system capacity. For example, malnutrition, the stress of war, or the debilitating effects of human immunodeficiency virus (HIV) can all have deleterious effects on the ability of the immune system to keep TB in its inactive form. Moreover, those who are socially marginalized may face problems in absorbing anti-TB drugs, as these populations are more likely to be burdened with gastrointestinal abnormalities (e.g., HIV-positive people, alcoholics, drug users, and residents of overcrowded housing), as supported by the fact that the ulcer bacterium *Helicobacter pylori* is found most frequently in residents of overcrowded housing. Further complicating the relationship is that those who are socially marginalized, particularly recent immigrants, the homeless, alcoholics, the incarcerated, and the mentally ill, are the very same people who experience the greatest difficulties in accessing the mainstream healthcare system.

Recent TB outbreaks in cities such as New York and London in the 1980s and 1990s have shown that TB is not simply a disease of the developing world. The reasons for these outbreaks inextricably involve larger structural changes. For example, significant reductions in public health programs resulted in the persistence of low levels of TB in New York City, as programs for tracking TB cases and ensuring that active cases finished their long course of prescriptions were dismantled. This, in turn, facilitated drug resistance, which further complicated struggles to contain TB. Other structural changes that contributed to the TB outbreaks involved the inequitable distribution of resources and wealth within the city. This is seen, for example, in reductions in public housing investment and the concomitant increase in support for private homeowners and the private rental sector in London. With these changes, an acute shortage of affordable housing arose and the poor were literally priced out of the housing market and rendered homeless or forced into overcrowded, temporary, or substandard accommodations. The resultant influx of understandably stressed and desperate people into overcrowded and poorly ventilated housing clearly contributed to the TB outbreaks. Thus, in the case of New York City, it is not surprising to learn that about 80 percent all MDRTB cases could be traced back to prisons and homeless shelters.

Finally, it should be kept in mind that in light of the significant role that social, political, and economic factors play in the etiology of TB, the public health response to this disease must not focus exclusively on biological factors (such as the compounding factor of HIV, or mutations that lead to drug resistance) or on individualizing factors such as noncompliance with the prescribed drug regime. Both of these perspectives divert attention away from the underlying causes of TB, namely, the inequalities that structure society. It is these inequalities that must first be addressed if a truly effective TB prevention and control plan is to be developed.

See Also: Acquired Immune Deficiency Syndrome; Airborne Diseases; Antibiotic Resistance; Health Disparities; Immune System Diseases; Lung Diseases.

Further Readings

Centers for Disease Control and Prevention. "Tuberculosis (TB)." http://www.cdc.gov/tb/faqs/default.htm (Accessed September 2010).

Gandy, Matthew and Alimuddin Zumla. *The Return of the White Plague: Global Poverty and the "New" Tuberculosis.* New York: Verso, 2003.

Roberts, Charlotte and Jane Buikstra. *The Bioarchaeology of Tuberculosis: A Global View on a Reemerging Disease.* Gainesville: University Press of Florida, 2008.

Rothman, Sheila M. *Living in the Shadow of Death: Tuberculosis and the Social Experience of Illness in American History.* Baltimore, MD: Johns Hopkins University Press, 1995.

S. Harris Ali
York University

ULTRAVIOLET RADIATION

Ultraviolet (UV) radiation is defined as that portion of the electromagnetic spectrum between X-rays and visible light, with wavelengths that range between 40 and 400 nanometers (nm). This wavelength interval falls in the energy range of 3 to 30 electron volts (eV). The UV spectrum is divided into the following five categories:

- Vacuum UV with wavelength of 40 to 190 nm
- Far UV with wavelength of 190 to 220 nm
- UVC with wavelength of 220 to 290 nm
- UVB with wavelength of 290 to 320 nm
- UVA with wavelength of 320 to 400 nm

The sun is our primary natural source of UV radiation. Artificial sources include germicidal lamps, mercury vapor lamps, tanning booths, black lights, curing lamps, halogen lights, high-intensity discharge lamps, fluorescent and incandescent sources, and some types of lasers (excimer lasers, nitrogen lasers, and third harmonic Nd:YAG lasers). Different sources of UV radiation might have different effects and hazards depending on their wavelength range.

The biological effects of UV radiation involve the following:

- UVC, far UV, and vacuum UV are almost never observed in nature because they are absorbed completely in the atmosphere. Therefore, any changes in ozone layer will cause a big change in UV shielding of the atmosphere.
- Germicidal lamps are designed to emit UVC radiation because of its ability to kill bacteria. In humans, UVC is absorbed in the outer dead layers of the epidermis. Accidental overexposure to UVC can cause corneal burns, commonly termed welders' flash, and snow blindness, a severe sunburn to the face. While UVC injury usually clears up in a day or two, it can be extremely painful.
- UVB is typically the most destructive form of UV radiation because it has enough energy to cause photochemical damage to cellular DNA, yet not enough to be completely absorbed by the atmosphere.
- UV plays an important role in the synthesis of vitamin D. Therefore, humans need UVB for metabolism. However, harmful effects can cause erythema (sunburn), cataracts, and

development of skin cancer. Individuals working outdoors are at the greatest risk of UVB effects. Most solar UVB is blocked by ozone in the atmosphere, and there is concern that reductions in atmospheric ozone could increase the prevalence of skin cancer, since the shielding effect is reduced as result of atmospheric ozone reduction.

UVA is the most commonly encountered type of the UV light in nature. UVA exposure has an initial pigment-darkening effect or tanning followed by erythema if the exposure is more excessive than normal exposure. As the photon energy increases, the chance of its absorption decreases. Therefore, since the energy of UVA falls in 320 to 400 nm, atmospheric ozone can absorb very little of this part of the UV spectrum.

The photochemical effects of UV radiation can be exacerbated by chemical agents. There are some chemical agents that make the human body more sensitive to UV radiation. Any accidental UV overexposure can injure the victims due to the fact the UV is invisible and does not produce an immediate reaction after absorption. Labeling on UV sources usually consists of a caution or warning label on the product or the bulb packaging cover or a warning sign on the entryway. Actually, the intensity of the UV source and length of exposure influence the level of the accidental damage. Hazard communication training is especially important to help prevent accidental exposures in the workplace.

Exposure guidelines for UV radiation have been established by the American Conference of Governmental Industrial Hygienists (ACGIH) and by the International Commission on Non-Ionizing Radiation Protection (ICNRP). Although handheld meters to measure UV radiation are commercially available, expert advice would be more helpful and is recommended to ensure selecting the correct detector for the UV wavelengths emitted by the natural and artificial sources.

In summary, UV radiation has numerous useful applications but increased awareness and control of UV hazards are needed to prevent accidental overexposures.

The level of UV radiation that reaches the Earth's surface can vary, depending on a variety of factors. Each of the following factors can increase the risk of UV radiation overexposure and its consequent health effects. UV absorption level depends on the following factors:

- Although the ozone layer absorbs most of the sun's UV rays, the amount of absorption is a function of the time of the year and also other natural phenomena. The UV absorption level is decreased as the ozone layer is thinned due to the release of ozone-depleting substances that are being widely used in most industries.
- The time of the day will affect UV absorption. When the sun reaches its highest in the sky around noon, its rays have the least distance to travel through the atmosphere. Therefore the UVB levels are at their highest. In the early morning and late afternoon, since the sun's rays have to pass through the atmosphere at an angle, their intensity is greatly reduced. The sun's angle also varies with the seasons, causing the intensity of UV rays to change. Therefore, the UV intensity increases to its highest during the summer months.
- Latitude is another factor for the level of UV exposure. The sun's rays are strongest at the equator, where the sun is most directly overhead and UV rays have to travel the least distance through the atmosphere; the shorter travel distance brings down the chance of absorption and increases the UV exposure on Earth. Also, ozone is naturally thinner in the tropics compared to the mid and high latitudes, so there is less ozone to absorb the UV radiation as it passes through the atmosphere.
- Weather conditions directly affect the level of UV exposure, and clouds increase or reduce the UV levels, but not completely. The thickness of the cloud acts as the shielding layer.

Some shiny surfaces with high reflection properties, such as snow, glass, steel, sand, grass, or water can reflect much of the UV radiation that reaches them. Because of this reflection, UV intensity can be deceptively high, even in shaded areas.

See Also: Cancers; Children's Health; Light Bulbs; Men's Health; Radiation Sources; Skin Disorders; Women's Health.

Further Readings

Altmeyer, Peter, Klaus Hoffmann, Markus Stücker, and O. Braun-Falco. *Skin Cancer and UV Radiation.* New York: Springer, 1997.

Centers for Disease Control and Prevention. "NIOSH Workplace Safety and Health Topics: UV Radiation." http://www.cdc.gov/niosh/topics/uvradiation (Accessed September 2010).

Harrison, R. M. and R. E. Hester. *Causes and Environmental Implications of Increased UV-B Radiation (Issues in Environmental Science and Technology).* London: Royal Society of Chemistry, 2000.

Hassan Bagher-Ebadian
Independent Scholar

UNITED NATIONS ENVIRONMENT PROGRAMME

While environmental problems were originally viewed as local problems, they are now recognized as interconnected, global problems; states and countries realize that they are unable to single-handedly address serious environmental problems. The United Nations Environment Programme (UNEP) developed out of the recognition that a central coordinating mechanism was needed in order to encourage collective action and create a common vision. UNEP acts as the chief organization at the international level to address and ultimately solve inherently global environmental programs; it is the core institution for the global environment within the United Nations (UN) system. UNEP's aim is to offer political and conceptual leadership and encourage shared partnership in assisting peoples and nations to increase their health and overall quality of life without compromising the health of future generations. By continuously monitoring environmental matters regionally and worldwide, UNEP can bring emerging issues to the attention of governments and elicit an international call to action when deemed appropriate. It can develop methods to avoid or reduce global environmental health risks, establish international joint norms, and work to avoid or settle environment-related disputes between and among member states.

UNEP was established in 1972, emerging out of the June 1972 Stockholm Conference on the Human Environment. Beginning in the late 1960s, the UN system had begun to develop a holistic mindset of the environment, but this first Conference on the Human Environment was a pivotal event in bringing the environment to the forefront of the international agenda. At the 1972 Conference on the Human Environment, a collection of 131 countries agreed on a set of foundational principles and signed a declaration announcing that the UN would be the key arm to protect and improve the environment. A few months later, the UN General Assembly, via Resolution 2997 (XXVII), approved the establishment of UNEP. It was designed to offer sufficient resources and authority to coordinate planned and ongoing programs and activities in order to effectively tackle global environmental matters. By the close of 1972, the UN General Assembly had unanimously voted Maurice Strong to head UNEP. He was the ideal candidate, having served as the secretary general for the 1972 Stockholm Conference on the Human Environment.

Organization

In terms of structure, UNEP is composed of a governing council with 58 UN member states and a secretariat with 890 staff members. Each seat on the governing council is allocated based on geographic region and serves a three-year term. This group acts as a leadership body, promoting cooperation and collaboration among UN member states on environmental issues. In addition to this diplomatic role, the governing council is the main developer of UN policy guidelines related to the environment. The secretariat, on the other hand, is responsible for overseeing the implementation of UNEP policies through programs in addition to the UNEP budget. Of the 890 staff members in the secretariat, approximately 500 are international staff, and the remaining staff are hired locally. The budget consists of over $100 million, which is almost wholly earned from the member states. Seven different divisions exist within UNEP: communication and public information; early warning and assessment; environmental policy implementation; technology industry and economics; environmental law and conventions; global environment facility coordination; and regional cooperation.

The UNEP headquarters is located in Nairobi, Kenya. Achim Steiner has served as the executive director of UNEP since 2006. Prior to Steiner, Klaus Topfer had led UNEP for nine years, serving two executive terms. Another notable leader of UNEP is Dr. Mostafa Kamal Tolba, who served for 17 years, from 1975 to 1992.

The charge of UNEP is challenging; environmental problems are often hard to conceptualize, spread over many nations, and exist in an undefined timeline. Sovereignty concerns and economic considerations also make international programs difficult to launch. Global policy efforts can range from mere survival to actual promotion of a healthy quality of life within a sustainable community. Furthermore, nations accustomed to defending and protecting their sovereignty are often slow to respond to an international call for global collaboration. Yet, a global recognition does exist that environmental factors (e.g., viruses, climate change) do not operate according to the artificial lines drawn by sovereign national jurisdictions.

UNEP is involved in a wide range of environment-related issues to protect the health of terrestrial and marine ecosystems as well as the atmosphere. Six priorities guide UNEP's environmental efforts in the 21st century: climate change; disasters and conflicts; ecosystem management; environmental governance; harmful substances; and resource efficiency. The climate change priority focuses on increasing the capability of individuals, communities, and nations to work toward a low-carbon existence as well as to gain knowledge of the Earth's changing state and climate science in general. Through the disasters and conflicts priority, UNEP aims to reduce environmental threats to human health caused by disasters and conflicts. As part of the disaster and conflicts priority, UNEP offers four main services to member states: post-crisis environmental assessments, post-crisis environmental recovery, environmental cooperation for peace making, and risk reduction for disasters and conflicts. The priority of ecosystem management involves managing ecosystems in a way that meets future ecological and human health needs. In terms of the environmental governance priority, UNEP advocates for informed decision making based on internationally agreed upon objectives. As part of the harmful substances priority, UNEP oversees the sound management of chemicals as well as education about them. For instance, UNEP has been actively involved in monitoring the international trade of chemicals that could harm human health by promoting international procedures and rules on the export and import of these banned pesticides, toxic wastes, and drugs, particularly to third world countries. Last, the resource efficiency priority of the 21st century works to foster more sustainable production and consumption of natural resources in both developed and developing countries. In addition to the efforts in these six priority areas, UNEP also concentrates on critical thematic areas such as health and environment, biodiversity, poverty and environment, and freshwater. For example, the health and environment thematic area seeks to highlight the systemic interconnectedness between climate change and health.

Role in Projects and Policy

UNEP plays a significant role in funding many environment-related development projects. An example of one such project is UNEP's Marshlands Project in the Middle East that works to protect the largest wetland ecosystem in the Middle East through environmentally sound planning. The project, titled Support for Environmental Management of the Iraqi Marshland, began in 2004 after UNEP had alerted the global community of the destruction of approximately 90 percent of the marshlands. UNEP funds the project by supporting strategy formulation, monitoring the marshland conditions, and providing water and sanitation options for the health of the people in the area.

UNEP has also played a chief role in creating and developing international environment conventions. For example, one of the most lauded accomplishments of UNEP is the 1987 Montreal Protocol, which is an international agreement to protect the ozone layer. Not only did the Montreal Protocol help solve a critical environmental problem, but it also may have established a foundational ground rule for all global policy moving forward: in order to elicit a global response, an equitable distribution of the opportunity costs is imperative when attempting to regulate shared resources.

Finally, UNEP publishes numerous newsletters, reports, and atlases. For example, UNEP published the Fourth Global Environment Outlook (GEO-4) assessment in late 2007, which is an extensive report on human well-being, the environment, and development. It is aimed at an audience of concerned world citizens and policy makers. Such reports often warn the audience of health hazards. For instance, a 2008 UNEP report cites a study indicating that the shallow pits left behind by gem miners in Sri Lanka are breeding grounds for mosquitoes and, ultimately, malaria. UNEP tries to reduce such a health hazard through activities such as clearing potential mosquito breeding sites, providing mesh screens for homes, and organizing the delivery of mosquito-eating fish and trees and plants that repel mosquitoes.

While UNEP juggles countless roles and conducts innumerable activities, it is generally recognized as being particularly successful in a few notable realms: monitoring and scientific assessment, beginning policy processes for international environmental agreements, and acting as an international ally in many countries where environmental activism is marginalized, providing a forum where environmental activists can discuss issues with their counterparts.

Critiques

Despite the strengths of UNEP, many critics exist as well. Some analysts argue that UNEP only responds when impending catastrophes present themselves or when severe degradation occurs, as opposed to approaching environmental initiatives through a more proactive stance. Part of this problem is because of a dearth of financial resources; a number of critics assert that the voluntary contributions from the UN member states fail to reach what is needed to make a difference in environmental world health for current and future generations. These critics state that financial support is one of the most significant challenges currently facing global environment protection.

In 2007, dissatisfaction with UNEP became so rampant that 46 countries supported a document read aloud by French president Jacques Chirac called a "Paris Call for Action." The document urged that UNEP be dismantled and be replaced by a more influential organization to be called the UN Environment Organization (UNEO), which could be modeled after the World Health Organization (WHO). The 46 countries that signed the declaration consisted of primarily European Union countries, but did not include the countries with some of the highest greenhouse gas emissions (e.g., United States and China).

Some critics also believe that UNEP has fallen short in serving as the leader of all international environmental organizations, contending that international environmental leadership is distributed across too many international organizations, including other UN bodies, Bretton Woods institutions, the World Trade Organization, and independent secretariats and governing bodies of the many environmental organizations. Examples of UN bodies include the UN Educational, Scientific and Cultural Organization (UNESCO), the UN Commission for Sustainable Development, and the UN Development Programme (UNDP). An example of a Bretton Woods institution involved in international environmental efforts is the World Bank, as it offers significant financial support to global environmental programs and projects. Individuals who argue for centralized power within UNEP maintain that this distribution of leadership has engendered too many treaties and projects that squander scarce resources and lead to unnecessary competition and duplication, competing with UNEP programs rather than complementing them.

Still others believe that UNEP was never intended to be a central operating agency. This camp believes that sharing a policy and leadership role with many other organizations ultimately accomplishes more for the environment, contending that international leadership capability has only been strengthened with additional organizations being involved. They believe that distributed leadership allows policy responses to be orchestrated more effectively and at multiple levels: regional, national, and international.

Despite the criticism that UNEP has received, the organization strives to work toward the highly challenging goal of enhancing world health by protecting the global environment. International policy and initiatives are always daunting, but they are much needed in order to preserve the health of current and future generations.

See Also: Government Role in Green Health; Health Disparities; International Policies; World Health Organization's Environmental Burden of Disease.

Further Readings

Ivanova, Maria. "Understanding UNEP: Myths and Realities in Global Environmental Governance." Ph.D. dissertation, Yale University, 2006.

Unatawale, Mukund G. "Global Environmental Degradation and International Organizations." *International Political Science Review*, 11 (1990).

United Nations Environment Programme. "About UNEP: The Organization." http://www
.unep.org/Documents.Multilingual/Default.asp?DocumentID=43 (Accessed July 2009).

Annie W. Bezbatchenko
New York University

URBAN GREEN

"Urban green" refers to an approach to ecological predicaments that arise most frequently in a metropolitan setting and to green spaces within cities. As increasing proportions of the population elect to live in metropolitan areas, urban green has become an important approach to reducing society's ecological footprint and increasing the quality of life of city residents. Urban green approaches seek to consider economic, social, and environmental concerns when formulating solutions that allow populations to live in an environmentally friendly, sustainable manner. In considering urban green policies and projects, concepts such as energy efficiency, light, air quality, noise levels, and citizens' health and well-being are considered.

Parks and other urban green spaces, such as Tiergarten in Berlin, help moderate local climates by slowing wind and water and providing homes and businesses shade that assists in the conservation of energy. Green spaces also improve nearby property values.

Source: iStockphoto.com

Access to green space has always been a consideration in urban planning, but during the 1990s, highlighting the natural environment in land use and other decisions grew increasingly popular. A growing understanding that natural ecosystems provide important functions to cities helped promote urban green as a way to support long-term sustainability. An urban green approach is predicated upon analyzing the function of the natural environment and then using planning policy and regulations to safeguard critical natural areas. Urban green seeks a sustainable and efficient use of natural resources. Urban green is also highly contextual in practice. Major metropolitan areas such as London, New York, and Los Angeles, for example, have different needs and concerns than smaller cities, although all may seek more environmentally sustainable ways of dealing with problems related to water usage. In the United States, the Environmental Protection Agency (EPA) has partnered with cities to improve water quality through wide-ranging control of storm water runoff. Seeking to reduce stress on traditional drainage infrastructure, the EPA has sponsored improved storm water management, focusing on such innovations as permeable alley paving in city alleys in places such as Chicago and Los Angeles.

Urban green also applies to placement and management of parkland and other shared spaces within cities. The benefits of green spaces are considerable, augmenting and improving the human environment in many ways. Green spaces filter air, water, and sunlight as well as provide habitat for wildlife and recreational areas for humans. Parks and other green spaces, such as Central Park in New York City and Tiergarten in Berlin, help moderate local climates, slowing as they do wind and water and providing homes and businesses shade that assists in the conservation of energy. In addition to health benefits, green space improves values of property adjoining it.

Private initiatives also are part of the urban green movement. Many landlords, hospitals, universities, and other entities have sought Leadership in Energy and Environmental Design (LEED) certification for buildings they construct or remodel. LEED certification was begun in 1998, since which time it has involved over 14,000 projects worldwide. LEED certification seeks to define and standardize green practices, promote environmentally friendly design processes, stimulate consumer awareness of sustainable building practices, and transform the building market through recognition of ecological leadership. Although LEED-certified buildings operate more efficiently than those merely built to code, they are more expensive to construct. Higher costs of construction are mitigated over time, however, as maintenance and operational costs are reduced. Municipalities sometimes encourage LEED-certified projects through waivers of property taxes, density bonuses, reduction or elimination of building fees, or other assistance. As architects and builders have become more familiar with LEED-certified building practices, the initial costs to attain this status have decreased.

The urban green movement has also embraced the practice of cultivating, processing, and distributing food within or near metropolitan areas. This phenomenon, known as urban farming, increases food safety and food security by increasing the amount of fare available to city residents and enabling fresher fruits, vegetables, and other foodstuffs to be offered for sale. City farming has a variety of ancillary benefits that buttress the urban green movement, including economic and social, providing employment opportunities for many looking for work. Urban farming also permits greater control over pesticides and encourages healthier eating habits. Globally, a variety of cities, including Beijing, China; Chicago, Illinois; Havana, Cuba; Los Angeles, California; and New York City have implemented programs that encourage urban farming. Municipal encouragement of urban farming includes setting aside land for cultivation, promoting farmers markets, and providing training and tools of production to those interested.

Urban green is a response to environmental concerns that is highly popular because of its propensity to include governmental, business, and individual initiatives in its process. Concerned with finding ways to make communities more environmentally sustainable and healthier, the urban green movement encompasses both small and large actions, thus making it accessible to those cities interested in becoming greener at various levels of sustainable implementation. Urban green also allows multiple avenues for entry into more environmentally friendly actions, including regulation, incentives, and direct support. Urban green projects are as diverse as the cities they serve.

See Also: Cities; Education and Green Health; Government Role in Green Health; Health Disparities; Organic Produce; Physical Activity and Health; Recreational Space.

Further Readings

Baycan-Levent, T., R. Vreeker, and P. Nijkamp. "A Multi-Criteria Evaluation of Green Spaces in European Cities." *European Urban & Regional Studies*, 16/2 (2009).

Benedict, M. A. and E. T. McMahon. *Green Infrastructure: Linking Landscapes and Communities.* Washington, DC: Island Press, 2008.

Irvine, K. N., P. Devine-Wright, S. R. Payne, R. A. Fuller, B. Painter, and K. J. Gaston. "Green Space, Soundscape and Urban Sustainability: An Interdisciplinary, Empirical Study." *Local Environment,* 14/2 (2009).

Stephen T. Schroth
Jason A. Helfer
Luke L. Karner
Knox College

Vaccination/Herd Immunity

Vaccination is the medical procedure in which protection is intended through the induction of immunological responses to specific agents. Vaccination is the administration of antigenic material to produce immunity to a disease. In some cases, antigens of different agents are placed together in the same vaccine, thus protecting against various diseases at the same time (e.g., viral trivalent or MMR—mumps, measles, and rubella). Some vaccines can prevent disease (e.g., measles or yellow fever vaccines), and others can mitigate the effects of infection by a pathogen (e.g., BCG vaccine or Bacillus Calmette-Guerin for tuberculosis).

Vaccination with effective antigenic materials has revolutionized the interventions in public health since Edward Jenner's discovery in 1796. Jenner tested the possibility of using a cowpox vaccine as an immunization for smallpox in humans for the first time; he is considered the father of immunology and vaccinology. Vaccination (Latin: *vacca*, cow) is so named because the first vaccine was derived from a virus affecting cows (cowpox virus), which provides a degree of immunity to smallpox in humans.

Despite today's safe and effective vaccines, which are widely available for many diseases—considered vaccine-preventable—many of them continue to be a public health problem because of a wide range of cultural and religious issues in many parts of the world. This is particularly true in Africa and Asia. Diseases such as poliomyelitis were eradicated from the Americas in 1991—the last indigenous case occurred in a 3-year-old boy, Luis Fermin Tenorio, from Junin in northern Peru. Polio is still a significant public health problem, however, given the virus's transmission and its morbidity and mortality in children. In 2008, 1,651 confirmed cases of polio were reported: 798 in Nigeria, 559 in India, and 117 in Pakistan, among other endemic countries.

In this context, it is very important to introduce the concept of herd immunity (or community immunity), which describes an immunological protection that occurs when the vaccination of a fraction of the population provides protection to unprotected individuals. Herd immunity theory proposes that in diseases passed from person to person, it is more difficult to maintain a chain of infection when large numbers of a population are immune. The higher the proportion of individuals who are immune, the lower the likelihood that a susceptible person will come into contact with an infected individual.

From the public health perspective, vaccinations and herd immunity are of utmost importance in the prevention of many diseases, as well as in the preservation of health, particularly in children but also in adults.

Annually, in the United States, the Advisory Committee on Immunization Practices (http://www.cdc.gov/vaccines/recs/acip), the American Academy of Pediatrics (http://www.aap.org), the American Academy of Family Physicians (http://www.aafp.org), and the Centers for Disease Control and Prevention (http://www.cdc.gov) develop the Recommended Immunization Schedules. In 2009 the following vaccines for persons aged 0 through 6 years were included in those recommendations: hepatitis B virus vaccine (HepB), to be applied at minimum age of birth; rotavirus vaccine (RV), to be applied at minimum age of 6 weeks; diphtheria and tetanus toxoids and acellular pertussis vaccine (DTaP), to be applied at minimum age of 6 weeks; Haemophilus influenzae type b conjugate vaccine (Hib), to be applied at minimum age of 6 weeks; pneumococcal vaccine, to be applied at minimum age of 6 weeks for pneumococcal conjugate vaccine (PCV) and 2 years for pneumococcal polysaccharide vaccine (PPSV); influenza vaccine, to be applied at minimum age of 6 months for trivalent inactivated influenza vaccine (TIV) and 2 years for live, attenuated influenza vaccine (LAIV); MMR, to be applied at minimum age of 12 months; varicella vaccine, to be applied at minimum age of 12 months; hepatitis A vaccine (HepA), to be applied at minimum age of 12 months; and meningococcal vaccine, to be applied at minimum age of 2 years for meningococcal conjugate vaccine (MCV) and for meningococcal polysaccharide vaccine (MPSV).

Additionally, recommended vaccines as of 2009 for persons aged 7 to 18 years are tetanus and diphtheria toxoids and acellular pertussis vaccine (Tdap), to be applied at minimum age of 10 years for Boostrix and 11 years for Adacel; human papillomavirus vaccine (HPV), to be applied at age 10 to 19 years; meningococcal conjugate vaccine (MCV), to be applied at age 11 or 12 years, or at age 13 to 18 years if not previously vaccinated; influenza vaccine, to be administered annually to children age 6 months through 18 years; pneumococcal polysaccharide vaccine (PPSV), hepatitis A vaccine (HepA), hepatitis B vaccine (HepB), and inactivated poliovirus vaccine (IPV)—for children who received an all-IPV or all-oral poliovirus (OPV) series, a fourth dose is not necessary if the third dose was administered at age 4 years or older; and MMR and varicella vaccines.

Sometimes vaccines can be associated with clinically significant adverse events. These events should be reported to the Vaccine Adverse Event Reporting System (VAERS). Guidance about how to obtain and complete a VAERS form is available at http://www.vaers.hhs.gov.

See Also: Acquired Immune Deficiency Syndrome; Airborne Diseases; Antibiotics; Centers for Disease Control and Prevention (U.S.); Seasonal Flu; Tuberculosis.

Further Readings

Berard, Marion and David F. Tough. "Qualitative Differences Between Naïve and Memory T Cells." *Immunology*, 106 (June 2002).

Centers for Disease Control and Prevention. "Immunization Schedules." http://www.cdc.gov/vaccines/recs/schedules/default.htm (Accessed April 2009).

Fine, Paul E. "Herd Immunity: History, Theory, Practice." *Epidemiologic Reviews*, 15 (2003).

Hammarsten, James F., William Tattersall, and James E. Hammarsten. "Who Discovered Smallpox Vaccination? Edward Jenner or Benjamin Jesty?" *Transactions of the American Clinical and Climatological Association*, 90 (January 1979).

John, T. Jacob, and Reuben Samuel. "Herd Immunity and Herd Effect: New Insights and Definitions." *European Journal of Epidemiology,* 16 (July 2000).

Rifakis, Pedro Miguel, Jesus Alberto Benitez, Jose De-la-Paz-Pineda, and Alfonso J. Rodriguez-Morales. "Epizootics of Yellow Fever in Venezuela (2004–2005): An Emerging Zoonotic Disease." *Annals of the New York Academy of Sciences,* 1081 (2006).

Torres, Jaime. "Hepatitis B and Hepatitis Delta Virus Infection in South America." *Gut,* 38 (1996).

World Health Organization (WHO). "International Travel and Health." Geneva: WHO, 2009.

Alfonso J. Rodriguez-Morales
Universidad de Los Andes, Universidad Central de Venezuela

WATERBORNE DISEASES

Waterborne diseases is the grouping term for those human diseases that are caused by organisms that can be transmitted by contaminated food and water (waterborne transmission). Such diseases include amebiasis, cholera, cryptosporidiosis, enteroviroses, giardiasis, hepatitis A and E, leptospirosislisteriosis, salmonellosis, and typhoid fever. Another potential source of waterborne infection is contaminated recreational water (e.g., important for schistosomiasis transmission). In some special cases, other transmission ways can occur (e.g., by skin in leptospirosis, Chagas disease). Waterborne diseases represent an important environmental threat that is currently a relevant aspect for the global public health, which should be considered in the new perspectives of a more ecologically green world in which such risks should be significantly reduced or even eliminated. The safety of food,

Guinea worm disease is a parasitic infection that can be spread through contaminated drinking water. This pond in Ogi, Nigeria, which served as a source of water for this woman in 2004, had been treated to prevent breeding by water fleas, which harbor Guinea worm larvae.

Source: Centers for Disease Control and Prevention

drink, and drinking water depends mainly on the standards of hygiene applied locally in their growth, preparation, and handling. There is a high risk of contracting such organisms or microbial agents in countries with low standards of hygiene and sanitation as well as poor infrastructure for controlling the safety of food, drink, and drinking water.

Water quality is regulated by a complex system of federal and state legal provisions and agencies, which has been poorly studied. Some authors have surveyed state and territorial agencies responsible for water quality about their laws, regulations, policies, and practices related to water quality and surveillance of cryptosporidiosis related to drinking water. The development and current status of federal drinking water regulations and identification of conflicts or gaps in legal authority between federal agencies and state and territorial agencies have been assessed in some studies. However, it is important to consider the court-imposed limitations on federal authority with regard to regulation of water quality. Recommendations are made for government actions that would increase the efficiency of efforts to ensure water quality, protect watersheds, strengthen waterborne disease surveillance, and protect the health of vulnerable populations.

The etiological organisms related to waterborne diseases can be bacteria (including enteric and aquatic bacteria), fungi, viruses (enteric), and parasites (enteric protozoa).

Of these, bacteria are considered among the most important. Fortunately, waterborne outbreaks of bacterial origin (particularly typhoid fever) in developing countries have declined dramatically in the last decades. Therefore, some early bacterial agents such as *Shigella sonnei* remain prevalent, and new pathogens of fecal origin such as zoonotic *Campylobacter jejuni* and *Escherichia coli* O157:H7 may contaminate pristine waters through wildlife or domestic animal feces. The common feature of these bacteria is the low inoculum (a few hundred cells) that may trigger disease. The emergence in early 1992 of serotype O139 of *Vibrio cholerae* with epidemic potential in southeast Asia suggests that serotypes other than *V. cholerae* O1 also could initiate an epidemic. Cholera significantly affected some countries in the Americas in the same decade (e.g., Peru, Ecuador, and Colombia).

Some new pathogens include environmental bacteria that are capable of surviving and proliferating in water distribution systems. Other than specific hosts at risk, the general population is refractory to infection with ingested *Pseudomonas aeruginosa*. The significance of *Aeromonas spp.* in drinking water to the occurrence of acute gastroenteritis remains a debatable point and has to be evaluated in further epidemiological studies; however, reports have indicated its importance and its disease burden in human beings, particularly in some vulnerable risk groups.

Legionella and *Mycobacterium avium* complex (MAC) are environmental pathogens that have found an ecologic niche in drinking and hot water supplies. Numerous studies have reported Legionnaires' disease caused by *L. pneumophila* occurring in residential and hospital water supplies. *M. avium* complex frequently causes disseminated infections in AIDS patients, and drinking water has been suggested as a source of infection; in some cases, the relationship has been proved.

Another bacterium of growing international interest for its clinical spectrum is *Helicobacter pylori*. More and more reports show that the pathogen DNA can be amplified from feces samples of infected patients, which strongly suggests fecal-to-oral transmission. Therefore, it is possible that *H. pylori* infection is waterborne, but these assumptions need to be substantiated.

Giardiasis has become the most common cause of human waterborne disease in the United States over the last 30 years. However, as a result of the massive outbreak of waterborne cryptosporidiosis in Milwaukee, Wisconsin, affecting an estimated 403,000 persons, there is increasing interest in the epidemiology and prevention of new infectious disease caused by *Cryptosporidium spp.* as well as monitoring water quality. The transmission of *Cryptosporidium* and *Giardia* through treated water supplies that meet water-quality standards demonstrates that water-treatment technologies have become inadequate and that a negative coliform no longer guarantees that water is free from all pathogens, especially from protozoan agents. Substantial concern persists that low levels of pathogen occurrence may be responsible for the endemic transmission of enteric disease.

In addition to *Giardia* and *Cryptosporidium*, some species of genera *Cyclospora*, *Isospora*, and of the family *Microsporidia* are emerging as opportunistic pathogens and may have waterborne routes of transmission.

More than 15 different groups of viruses encompassing more than 140 distinct types can be found in the human gut. Some cause illness unrelated to the gut epithelium, such as hepatitis A virus and hepatitis E virus. Numerous large outbreaks have been documented in the United States between 1950 and 1970, and the incidence rate has strongly declined in developing countries since the 1970s. Hepatitis E is mostly confined to tropical and subtropical areas, but recent reports indicate that it can occur at a low level in Europe. A relatively small group of viruses have been incriminated as causes of acute gastroenteritis in humans, and fewer have proven to be true etiologic agents, including rotavirus, calicivirus, astrovirus, and some enteric adenovirus. These enteric viruses have infrequently been identified as the etiologic agents of waterborne disease outbreaks because of inadequate diagnostic technology, but many outbreaks of unknown etiology currently reported are likely due to viral agents. Actually, Norwalk virus and Norwalk-like viruses are recognized as the major causes of waterborne illnesses worldwide.

The global burden of infectious waterborne disease is considerable. Reported numbers highly underestimate the real incidence of waterborne diseases. The most striking concern is that enteric viruses such as caliciviruses and some protozoan agents, such as *Cryptosporidium*, are the best candidates to reach the highest levels of endemic transmission because they are ubiquitous in water intended for drinking. They are highly resistant to relevant environmental factors, including chemical disinfecting procedures. Other concluding concerns are the enhanced risks for the classic group of debilitated subjects (very young, old, pregnant, and immuno-compromised individuals) and the basic requirement to take specific measures aimed at reducing the risk of waterborne infectious diseases in this growing, weaker population.

In places where waterborne transmission is significant, people should take precautions with all food and drink to minimize any risk of contracting a food-borne or waterborne infection and the consequently clinical complications related to such diseases.

It is to be expected and possible that in the near future, new organisms, from any type, as well other old organisms, can produce new waterborne diseases that will deserve proper epidemiological, microbiological, and clinical studies for proper management.

Common Waterborne Diseases

Cholera

Caused by *Vibrio cholerae* bacteria, serogroups O1 and O139, infection occurs through ingestion of food or water contaminated directly or indirectly by feces or vomitus of infected persons. Cholera affects only humans; there is no insect vector or animal reservoir host.

Giardiasis

Caused by the protozoan parasite *Giardia intestinalis*, also known as *G. lamblia* and *G. duodenalis*, infection usually occurs through ingestion of the organisms' cysts in water (including both unfiltered drinking water and recreational waters) or food contaminated by the feces of infected humans or animals.

Hepatitis A

Caused by the hepatitis A virus, it is acquired directly from infected persons by the fecal–oral route or by close contact, or by consumption of contaminated food or drinking water. There is no insect vector or animal reservoir (although some nonhuman primates are sometimes infected).

Hepatitis E

Caused by the hepatitis E virus, this is a waterborne disease usually acquired from contaminated drinking water. Direct fecal–oral transmission from person to person also is possible. There is no insect vector. Hepatitis E virus has domestic animal reservoir hosts, such as pigs.

Leptospirosis

Caused by various spirochetes of the genus *Leptospira*, infection occurs through contact between the skin (particularly skin abrasions) or mucous membranes and water, wet soil, or vegetation contaminated by the urine of infected animals, notably rats. Occasionally, infection may result from direct contact with urine or tissues of infected animals or from consumption of food contaminated by the urine of infected rats.

Listeriosis

Caused by the bacterium *Listeria monocytogenes*, listeriosis affects a variety of animals. Food-borne infection in humans occurs through the consumption of contaminated foods, particularly unpasteurized milk, soft cheeses, vegetables, and prepared meat products such as pâté. Listeriosis multiplies readily in refrigerated foods that have been contaminated, unlike most food-borne pathogens. Transmission also can occur from mother to fetus or from mother to child during birth.

Schistosomiasis

Caused by several species of parasitic blood flukes (trematodes), of which the most important are *Schistosoma mansoni*, *S. japonicum*, *S. mekongi*, and *S. haematobium*, infection occurs in freshwater containing larval forms (cercariae) of schistosomes, which develop in snails. The free-swimming larvae penetrate the skin of individuals swimming or wading in water. Snails become infected as a result of excretion of eggs in human urine or feces.

American Trypanosomiasis or Chagas Disease

Caused by the protozoan parasite *Trypanosoma cruzi*, infection is usually transmitted by bloodsucking triatomine bugs ("kissing bugs"). Oral transmission by ingestion of unprocessed, freshly squeezed sugar cane in areas where the vector is present also has been reported. During feeding, infected bugs excrete trypanosomes, which can then contaminate the conjunctiva, mucous membranes, abrasions, and skin wounds, including the bite wound. Transmission also occurs by blood transfusion when blood has been obtained from an infected donor. Congenital infection is possible, due to parasites crossing the placenta during pregnancy. *T. cruzi* infects many species of wild and domestic animals as well as humans.

Typhoid Fever

A disease caused by *Salmonella typhi* (*Salmonella enterica* serovar Typhi), the typhoid bacillus infects only humans. Similar paratyphoid and enteric fevers are caused by other species of *Salmonella*, which infect domestic animals as well as humans. Infection is transmitted by consumption of contaminated food or water. Occasionally, direct fecal–oral transmission may occur. Shellfish taken from sewage-polluted beds are an important source of infection. Infection occurs through eating fruits and vegetables fertilized by night soil and eaten raw, and milk and milk products that have been contaminated by those in contact with them. Flies may transfer infection to foods, resulting in contamination that may be sufficient to cause human infection. Pollution of water sources may produce epidemics of typhoid fever when large numbers of people use the same source of drinking water.

See Also: Acquired Immune Deficiency Syndrome; Antibiotics; Bottled Water; Centers for Disease Control and Prevention (U.S.); Dehydration; Gastroenteritis; Groundwater; Recycled Water; Supplying Water; Swimming Pools; Tap Water/Fluoride; Vaccination/ Herd Immunity.

Further Readings

Dickson-Gonzalez, Sonia M., Marleny Lunar de Uribe, and Alfonso J. Rodriguez-Morales. "Polymorphonuclear Neutrophil Infiltration Intensity as Consequence of *Entamoeba histolytica* Density in Amebic Colitis." *Surgical Infections*, 10 (2009).

Fernandini-Paredes, Gino G., Edward Mezones-Holguin, Rolando Vargas-Gonzales, Eugenio Pozo-Briceño, and Alfonso J. Rodriguez-Morales. "In Patients With Type 2 Diabetes Mellitus, Are the Glycosylated Hemoglobin Levels Higher for Those With *Helicobacter pylori* Infection Than Those Without Infection?" *Clinical Infectious Diseases*, 47 (2008).

Gostin, L. O., Z. Lazzarini, V. S. Neslund, and M. T. Osterholm. "Water Quality Laws and Waterborne Diseases: Cryptosporidium and Other Emerging Pathogens." *American Journal of Public Health*, 90 (2000).

Rodríguez, Cruz N., Rosa Campos, Bileida Pastran, Ivette Jimenez, Ada Garcia, Pilar Meijomil, and Alfonso J. Rodriguez-Morales. "Sepsis Due to ES L Producing *Aeromonas hydrophila* in a Pediatric Patient With Diarrhea and Pneumonia." *Clinical Infectious Diseases*, 41 (2005).

Rodriguez-Morales, Alfonso J. "Chagas Disease: An Emerging Food-Borne Entity?" *Journal of Infection in Developing Countries*, 2 (2008).

Rodriguez-Morales, Alfonso J. "Ecoepidemiology and Satellite Epidemiology: New Tools in the Management of Problems in Public Health." *Peruvian Journal of Experimental Medicine and Public Health*, 22 (2005).

Rodriguez-Morales, Alfonso J., Rosa A. Barbella, Cynthia Case, Melissa Arria, Marisela Ravelo, Henry Perez, Oscar Urdaneta, Gloria Gervasio, Nestor Rubio, Andrea Maldonado, Ymora Aguilera, Anna Viloria, Juan J. Blanco, Magdary Colina, Elizabeth Hernández, Elianet Araujo, Gilberto Cabaniel, Jesus Benitez, and Pedro Rifakis. "Intestinal Parasitic Infections Among Pregnant Women in Venezuela." *Infectious Diseases in Obstetrics and Gynecology*, 14 (2006).

Rodriguez-Morales, Alfonso J., Lisette Echeverria, Carmen N. Mora, Susmira Guevara, Dorania Plaza, Cruz N. Rodriguez, Andrea G. Rodríguez, Ada Garcia, Bileida Pastran, Ivette Jiménez, Pilar Meijomil, Ildefonso Tellez, and Carlos Franco-Paredes. "Antimicrobial Susceptibility of Bacterial Strains Isolated From Recreational Swimming Pools in Two Provinces of North-Central Venezuela." *Journal of the Medical and Surgical Society of the Emergency Hospital Perez de Leon*, 39 (2008).

World Health Organization (WHO). "International Travel and Health 2010." Geneva: WHO, 2009.

Alfonso J. Rodriguez-Morales
Universidad de Los Andes, Universidad Central de Venezuela

WATER SCARCITY

Between 2000 and 2030, Asia's urban population is projected to grow from 1.4 billion to 2.6 billion; such rapid growth can dramatically increase per capita use of freshwater. Here, a muddy slum in Agra, India, with the Taj Mahal in the far distance.

Source: iStockphoto.com

The era of globalization is going to confront, among many other things, the challenge of the growing population, deterioration of environmental quality, and scarcities of certain resources that are indispensable necessities for the sustenance of life on Earth. Of all the planet's renewable resources, the availability of freshwater is likely to pose the greatest challenge because of an increased demand with population rise and economic activities and shrinking supplies due to overexploitation and pollution. The availability of this key natural resource in acceptable quality, in adequate quantity, in the appropriate places, and at the time needed is very important. On a planet whose surface is more than two-thirds covered by water, the illusion of abundance has clouded the reality that renewable freshwater is an increasingly scarce commodity because only a very small proportion of this is effectively available for human use. The water available on Earth is in finite quantity that has not changed over millennia. This has to be juxtaposed with increasing demands from a growing population. The population of the world, currently around six billion, is expected to exceed eight billion by 2050. Aside from sheer numbers, the processes of urbanization and development are also expected to vastly increase the demand for freshwater. This situation of a finite supply and a growing demand leads to the projections of water scarcity, which could be severe in some parts of the world.

Worldwide, three-fourths of all current population growth has been experienced in urban areas, and particularly cities are gaining an estimated 55 million people per year. The world's population is quickly becoming urbanized as people migrate to cities for better opportunities. In 1950, less than 30 percent of the world's population lived in cities. This number grew to 47 percent in 2000 (2.8 billion people), and this is expected to grow to 60 percent by 2025. Developed nations have a higher percentage of urban residents than do less developed countries. However, urbanization is occurring rapidly in many less developed countries, and it is expected that most urban growth will occur in less developed countries during the next decades. Urbanization and rapid growth in urban population can dramatically increase per capita use of freshwater. Fast population growth with accelerated urbanization, combined with scarce water supplies, means that governments all over the world often cannot supply enough water to meet demand. According to a World Bank study, of the 27 Asian cities with populations over 1 million, Chennai and Delhi are ranked as the worst-performing metropolitan cities in terms of hours of water availability per day, while Mumbai is ranked as second-worst performer and Calcutta, fourth worst.

Globally, all future population growth will take place in cities, especially in Asia, Africa, and Latin America. In Asia and Africa, this growth will signal a shift from a rural to an urban base, changing a millennia-long trend. Between 2000 and 2030, the urban population in Africa and Asia is projected to double. Asia's urban population will grow from 1.4 billion to 2.6 billion. Africa's urban population will surge to more than twice its size, from 294 million to 742 million. Latin America and the Caribbean Islands will see urban population rise from 394 million to 609 million. By 2030, 79 percent of the world's urban dwellers will live in the developing world's towns and cities. Africa and Asia will account for almost seven out of every 10 urban inhabitants globally, and poor people will make up a large part of future urban growth. How to provide clean, potable drinking water to the cities, especially to the urban poor, is the challenge that civic bodies face all over the world.

Water Scarcity Today

Water scarcity already affects every continent, and four out of every 10 people in the world are hit by the crisis. The situation is getting worse due to population growth, urbanization, and the increase in domestic and industrial water use. Many countries do not have sufficient water to meet demand, resulting in aquifer depletion due to overextraction. Moreover, the scarcity of water is accompanied by deterioration in the quality of available water due to pollution and environmental degradation. Water scarcity forces people to rely on unsafe sources of drinking water. In the developing world, more than 25 percent of the urban population does not have access to safe drinking water. About 50 percent of the urban population suffers from poor sanitation and waterborne diseases. Poor water quality can increase the risk of diseases including cholera, typhoid fever, tuberculosis, salmonellosis, other gastrointestinal viruses, and dysentery. It is estimated that these illnesses are responsible for 60 percent of deaths in urban areas. Not only are there 1.1 billion people without sufficient and safe drinking water, but the United Nations also acknowledges that 2.6 billion people are without adequate water for sanitation (e.g., waste disposal).

Untreated sewage and lack of sanitation lead to breeding of microbes, bacteria, and fungus. This in turn leads to the eutrophication of surface water bodies.

The United Nations recommends that people need a minimum of 50 liters of water a day for drinking, washing, cooking, and sanitation. In 1990, over a billion people did not have even that. Providing universal access to that basic minimum worldwide by 2015 would take less than 1 percent of the amount of water we use today, but still we are a long way from achieving that. By 2025, nearly 2 billion people will be living in countries or regions with absolute water shortages, where water resources per person will fall below the recommended level of 500 cubic meters per year. The problem is that with the growth of human population, there is a growing concern for adequacy of freshwater, which has further been aggravated by an uneven regional distribution of water over space and time. A third of the world's population lives in water-stressed countries now. By 2025, this is expected to rise to two-thirds.

Water scarcity is aggravated by six principal factors:

- Reluctance to treat water as an economic as well as a public good, resulting in inefficient water use practices by households, industries, and agriculture
- Excessive reliance in many places on inefficient institutions for water and wastewater services
- Fragmented management of water between sectors and institutions, with little regard for conflicts between social, economic, and environmental objectives
- Inadequate recognition of the health and environmental concerns associated with current practices
- Environmental degradation of water sources, in particular, reduced water quality and quantity due to pollution from urban or land-based activities
- Inadequate use of alternative water sources; alternative water sources other than groundwater and surface water rarely explored

In recent years, there has been a growing perception of a looming water scarcity. Water has suddenly become a favored subject of debate all over the world. The United Nations Development Programme (UNDP), the World Bank, and the Asian Development Bank are seriously concerned about the projected global water scarcity. Currently, there is a fashionable thesis that future wars will be fought over water, not oil. That is a debatable proposition, but the prognosis of acute water scarcity in the not-too-distant future cannot easily be disputed. A series of institutions and networks have sprung up to deal with this and related matters: World Water Commission, World Water Council, Global Water Partnership, and so on. Several exercises have been going on to build up national, regional, and global "Water Visions" for 2025.

Given the fact that various national governments do not have the money or the interest to replace/repair installations and infrastructure for water supply, there is great temptation to privatize supply and distribution of water. Multinational corporations are assiduously trying to strike deals with governments to gain access to the lucrative water market. There are several cases of privatization of river water, overdrawing of water for private companies, and privatization of urban water supply, which have triggered stiff resistance from citizens' groups and social movements. There is an increasing awareness that water is a public good and cannot be privatized. It is also felt that water is a fundamental right and that the state has the responsibility to supply a minimum amount of water per citizen.

The emerging scarcity of water has also raised a host of issues related to sustainability of the present form of economic development, sustained water supply, equity and social justice, and water financing, pricing, governance, and management. Sustainable use of any resource is defined as the present generation's utilization of a natural reserve in a manner that will not compromise the ability of future generations to use the same resource and the span of sustainability can be prolonged for a reasonably long period with judicious operation and maintenance of the system. However, few human activities are sustainable in the long run. For water, which is a renewable resource, it is possible to develop sustainable usage policies provided the quality of water is not compromised. We therefore must adopt a new approach to water resources management in the new millennium so as to overcome these failures, reduce poverty, and conserve the environment—all within the framework of sustainable development.

Measures to Overcome Water Scarcity

Most of the traditional water-harvesting systems in urban areas have been neglected and fallen into disuse, worsening the urban water scenario. One of the solutions to the urban water crisis is rainwater harvesting—capturing the runoff. In essence, harvesting water means harvesting the rain. In the urban context, wastewater recycling is another answer. Recycled water can be used for toilets, gardening, air conditioning, industrial and household use, and groundwater aquifer recharging. Along with adequate sewage treatment plants to recycle water, it is also necessary to have pipes laid to carry the recycled water to consumers. Apart from water recycling, it should be mandatory for all buildings in urban areas to have rainwater harvesting systems. Finally, concrete city pavements are the worst offenders as far as recharging of underground aquifers is concerned. Authorities must avoid this practice, and if need be, dismantle existing pavements to allow for groundwater recharging. Citizens need to take this up as an urgent public cause. Some of the key themes to develop a coherent water strategy will clearly have to move toward, and revolve around, the following issues:

- Protection of forests, soil, and water resources
- Promotion and coordination of traditional and environmentally friendly technologies in agriculture and water conservation
- Sustainable use of this renewable natural resource
- Water conservation measures at the domestic level
- Ensure recharging of groundwater to meet increasing dependence on groundwater
- Harvesting of rainwater
- Improvement of irrigation technology to avoid overuse and loss in water conveyance
- Better international cooperation in research and technology to evolve permanent mechanisms for monitoring climate change impacts
- Develop better institutional capacity for water resources management
- Special task force and special funds for extreme climate conditions—emphasis on urban hydrology
- Conservation and management strategies to cope with any extremes in water balance
- Foster an awareness of water as a scarce resource and its conservation as an important principle—through nongovernmental organizations (NGOs) and CSOs
- New management approaches—empowering people for equitable sharing of water, creating a political will and good governance, and developing and sharing knowledge and technology to improve water resources management

See Also: Bottled Water; Dehydration; Gastroenteritis; Groundwater; Recycled Water; Supplying Water; Waterborne Diseases.

Further Readings

Guha, M. and Kamla Gupta. "Water Resources in India: Critical Issues Related to Availability and Sustainable Use." *IASSI Quarterly*, 25/3 (2007).
Paramasivan, G. and J. Sacratees. "Water Scarcity—Issues and Challenges." *Kurukshetra*, 57/5 (2009).
Srinivas, Hari. "An Integrated Urban Water Strategy." http://www.gdrc.org/uem/water/urban-water.html (Accessed March 2010).

Mohua Guha
International Institute for Population Sciences

WINE AND OTHER ALCOHOLS

The environmental movement has impacted most contemporary industries, including the production of wine and other alcohols. Conventional use of pesticides, herbicides, and fungicides by wine- and alcohol-producing entities has the potential to contaminate beers and wines so produced. As the process of winemaking has become increasingly influenced by scientific measurements, residues from such chemicals have been found in conventionally made wines as well as in the vines and soil in which they were grown. These chemicals can increase the risk of cancer and have lasting detrimental impacts on the environment. Also, worker safety can be negatively impacted by traditional practices, thus making reduction of hazardous practices an issue of consideration for breweries and wineries seeking to produce green alcohols. The term *green wine* refers to various degrees of certification and production of environmentally oriented wines. This should not be confused with Vinho Verde, a Portuguese wine that is named for its young age rather than its color.

Some wineries now use plastic stoppers and metal caps to seal their bottles, instead of cork. However, cork has some advantages in that it is biodegradable and cork harvesting does not involve killing the trees but rather skinning their bark, as shown in this photograph from Portugal's Alentejo region.

Source: iStockphoto.com

Wineries and other alcohol producers are increasingly marketing their products to consumers who fear that their consumption of wine and other alcohols comes at the expense of both the natural environment and workers' health.

Most wine is sold in bottles, which appeal to both environmental advocates and green activists. One consequence of green wine's appeal involves wineries' decision to utilize plastic stoppers and metal caps to seal their bottles rather than using the bark harvested from cork trees to stop the containers. This decision stems from the financial benefits offered by metals and plastics, which are cheaper to produce, as opposed to cork, which must be imported from regions such as southwestern Europe, including Spain and Portugal. While this growing trend has had devastating financial ramifications for these regions, they still maintain the upper hand in consumers' perceptions. Aficionados maintain that the most reputable wines and champagnes exclusively use authentic cork stoppers. However, this perception is slowly fading as reputable wineries have adopted synthetic bottle closures. Also, this new trend helps reduce the possibility of cork taint, which is the occasional contamination of unopened wines and champagnes by the natural elements found in cork and oak barrels. From an environmental perspective, however, although plastic and metal bottle stoppers are less attractive and undoubtedly less expensive, these closures are not biodegradable like their natural counterparts. Also, it should be noted that the cork harvesting process does not involve the killing of trees but rather the skinning of their bark.

Producers of wine and other alcohols wish to make environmental claims in their product advertisement. With this desire, however, comes the complicated and expensive nebula of regulating such claims so that credible certification may exist. Hence, a common distinction in labeling found in the field of green wines exists between advertising a wine as "organic" versus merely being "made with organic grapes." The organic label represents compliance with the vast number of sometimes seemingly incidental regulations that nomenclature dictates, such as prohibiting the use of metal fences in favor of wood. The main distinction between organic wine and wine made with organic grapes entails the decision in the latter case to use additives and preservatives in the winemaking stage after harvest in order to effect changes in taste, alcohol content, and shelf life of the wine. Though wines bearing either label must generally adhere to sustainable practices in the stages of viticulture, or grape growing, they differ in the stages of vinification, or winemaking, largely by a winery's decision to use preservatives such as sulfur dioxide and potassium sorbate to enhance shelf life and prevent malolactic fermentation.

One aspect of producing environmentally sustainable wine and other alcohol that is less clear from labeling practices involves the decision to use machinery when harvesting grapes and separating stems and skin from the grapes, or doing this by hand. Traditionally, the more reputable wines and champagnes have done this work with hired manual labor. Many wineries in the United States, however, now utilize manual labor for only the tasks of grape harvesting and de-stemming, but using machines for grape skinning. Wineries that utilize machines as part of the harvesting process heighten the risk of including moldy or otherwise unsuitable grapes in the mash. Also, machines used in harvesting heighten the risk of accidentally killing or displacing small animals that often build nests in vineyards.

Biodynamic wine also is increasing in popularity. Biodynamic wine refers to wines produced using a holistic approach to agriculture that views the entire farm as a living entity, balancing the interrelationship of soil, plants, and animals. Biodynamic practices share with organic farming a focus on using manure and composts on soil and excluding the use of artificial chemicals. Biodynamic agriculture differs from other approaches insofar that it favors nine different preparations alleged to aid fertilization. Biodynamic agriculture has two field preparations for the stimulation of humus formation: one using cow horns filled with manure that are buried in the ground, the other employing quartz-filled cow horns that are left in a field during the growing season and then removed. Compost preparations involve adding herbs commonly used for medicinal remedies to a compost pile, including yarrow and chamomile blossoms, stinging nettle, oak bark, dandelion and valerian flowers, or horsetail. Some vintners who have adopted biodynamic approaches claim to have greatly improved the health of their fields, increasing biodiversity, crop nutrition, and soil fertility, and decreasing the amount of disease, insect, and weed problems. Biodynamic wine also has been judged superior to conventionally made wine in several taste tests. Biodynamic wine's proponents emphasize its benefits for vineyard health and wine taste, although some critics maintain that organic farming practices would yield the same results.

Wine and other alcohol producers who are adherents of organic farming practices seek to balance farm profitability, environmental stewardship, and the quality of life of those involved in farming. For crops ultimately used to make beer or other alcohols, organic farming practices follow accepted standard practices. Monoculture, or the planting of a single crop, tends to be discouraged in organic farming. Viticulture, however, demands vineyards that take years, or even generations, to develop. Green winemakers seek to avoid the use of synthetic fertilizers, pesticides, herbicides, and other substances that can damage the environment. They also eschew manures, which replenish the soil but cause harmful runoff that vitiates nearby rivers and coastal waters. Organic wines are not necessarily sulfite free, for sulfite use stabilizes the vintage.

Wine and other alcohols have the potential to become more green. Organic and biodynamic methods have proven popular with consumers. Greater regulation is needed, however, to ensure standardization of claims used by producers and terms associated with green wine and other alcohols.

See Also: Cost-Benefit Analysis for Alternative Products; Fertilizers; Organic Produce; Private Industry Role in Green Health.

Further Readings

Berner, A., I. Hildermann, A. Fließbach, L. Pfiffner, U. Niggli, and P. Mäder. "Crop Yield and Soil Fertility Response to Reduced Tillage Under Organic Management." *Soil & Tillage Research*, 101/1/2 (2008).

Cozzolino, D., M. Holdstock, R. G. Dambergs, W. U. Cynkar, and P. A. Smith. "Mid Infrared Spectroscopy and Multivariate Analysis: A Tool to Discriminate Between Organic and Non-Organic Wines Grown in Australia." *Food Chemistry*, 116/3 (2009).

Joly, Nicolas. *What Is Biodynamic Wine: The Quality, the Taste, the Terroir.* East Sussex, UK: Clairview Books, 2007.

Stephen T. Schroth
Jason A. Helfer
Daniel O. Gonshorek
Knox College

WOMEN'S HEALTH

Recent research has indicated that women, compared to men, are more likely to seek out preventative healthcare; to subscribe to annual physicals; and to ask more questions regarding their diagnoses, prognoses, and treatment. While no definitive explanation for this behavior has yet been offered, some theorize that women's role as caregivers of others may compel them to take their own health more seriously. Another factor could be the fact that women have traditionally been penalized by the healthcare system simply for being female; perhaps this marginalized status has compelled women to be more proactive and assertive in their relationships with the healthcare system. Much has been written about the way that women's lives are viewed through a "masculine lens" of experience and assumptions. The tendency to view women through this masculine lens occurs both on a general sociocultural level and in more specific realms, such as in the healthcare industry. This model sees the male body and its experiences as normative, while the female body and female experiences are seen as the "other." Though many, including healthcare professionals, have worked to move beyond this gendered model, its assumptions continue to pervade our culture, making it difficult at times to even recognize when we are subscribing to these sex-based notions of normative behaviors. Women patients frequently report feeling mistreated, ignored, and not taken seriously by doctors and other healthcare professionals, which could provide a rationale for their tendency to be more assertive in their experiences with the healthcare industry. With a long history of being seen as incapable of making informed decisions regarding their own bodies, of being seen as suffering from conditions such as "hysteria," and of being excluded from clinical trials because of fears that their menstrual cycles and potential to become pregnant could skew findings, women today still face many barriers in their attempts to challenge their marginalized status in the healthcare industry.

In a broader response to this marginalized status, legislators included as part of the 2010 Health Care and Education Reconciliation Act measures to forbid sexual discrimination in health insurance—a move that many advocates for women's health celebrated by declaring: "Being a woman is no longer a pre-existing condition." Prior to the passage of this legislation, the health insurance industry openly discriminated against women, with most states allowing companies selling individual health plans to charge women more than men for the same coverage—even for policies that lacked maternity care. Even more disturbing was the practice of insurance companies to charge nonsmoking women more than men who smoked. And it is not just individual plans that engaged in these practices; insurance companies used the same pricing schemes in their group policies, though sex discrimination laws bar businesses from passing along these higher costs to their workers. This increased economic burden has been particularly difficult for small and midsize businesses with numerous female employees, leaving many such businesses either unable to offer healthcare coverage or only able to offer plans with high deductibles. Industries hit particularly hard by this include childcare, home healthcare, and many nonprofit organizations.

Other discriminatory practices in the sale of individual health insurance policies included charging extra for maternity care or not offering it at all and citing conditions such as previous Caesarean sections (C-sections) or having been the victim of domestic violence as reasons to deny health coverage to women. Not only does the new healthcare legislation mandate that maternity coverage be included in health insurance policies, it outlaws viewing previous C-sections and experiences with abuse as preexisting conditions. While certain parts of the healthcare reform bill will not be initiated until 2014, the new laws barring sexual discrimination in health insurance take effect immediately. Unfortunately, the legislation barring sex discrimination in insurance premiums only applies to individual and small-group plans; larger group plans—those serving 100 employees or more—will continue to be permitted to discriminate against women. This suggests, again, that many companies will either not be able to afford to offer insurance coverage or will only be able to offer plans with prohibitively high deductibles.

Continuing Discriminatory Practices

And so, while U.S. women certainly have some reasons to celebrate the passage of health-care reform (except for those politically opposed to healthcare reform), many are dismayed, as are many men, by the fact that insurers providing group policies will be allowed to continue their discriminatory practices. Further, many women's rights advocates are profoundly disturbed by the restrictions placed by the legislation on abortion. President Barack Obama signed an executive order mandating that no federal monies be used to fund abortions, citing the 1976 Hyde Amendment as the guiding framework for how abortion would be dealt with in relation to the healthcare bill. For example, under the new healthcare legislation, "high-risk insurance pools" will be created to offer insurance to those denied coverage by private carriers; these insurance pools are temporary measures, intended to be replaced in 2014 by "health insurance exchanges." Until then, any woman who obtains insurance coverage through the high-risk insurance pools will not have access to abortion coverage, except in cases of rape, incest, or threat to the mother's life, even if she elects to pay for such coverage out of her own pocket. In effect, then, the federal government is barring the use of private funds to obtain abortion coverage, which goes beyond the restrictions of the 1976 Hyde Amendment, which bans only the use of federal funds for abortions. Women's rights advocates see this as outrageous. Further, once the health insurance exchanges are created, which are expected to insure the bulk of U.S. citizens, insurance plan enrollees desiring abortion coverage are required to write two monthly checks—one for an abortion policy and one for all other healthcare. In addition, businesses that purchase their employees' insurance through the exchanges will be required to write two separate checks for each employee who desires abortion coverage. Many see this scheme as unnecessarily complicated and burdensome and as evidence that a woman's right to an abortion, if not being outright eliminated, has been profoundly limited.

While abortion is certainly a polarizing issue in the United States, women face other problems in their relationship with the healthcare industry that are, perhaps, more universal. For example, while men and women suffer from many of the same health problems, these issues often affect them differently, and, as previously stated, due to the "masculine lens" with which many health issues are viewed, women do not always get the best care for their conditions. One example of this relates to heart disease, which is the leading cause of death in the United States for both women and men. Coronary heart disease (CHD) is the most common type of heart disease, and it affects both men and women. However, coronary microvascular disease (MVD) primarily affects women, and it often goes undiagnosed, most likely because standard tests for CHD fail to identify MVD. Most physicians continue to rely on a framework that guides them to use the same diagnostic tools for evaluating both women and men, even as evidence continues to accumulate showing that not only do women not always present the same symptoms as do men, but that women suffer from conditions that men do not often experience. This is punctuated by the fact that while death rates for heart disease in men have rapidly declined in the past 30 years, those for women have not. More than 315,000 women die annually in the United States from heart disease.

Strokes

Women are nearly three times as likely as are men to suffer a stroke, which is the third leading cause of death for women in the United States; each year, about 55,000 more women than men experience a stroke, and more women than men die each year from strokes. Further, despite all the attention that breast cancer receives, each year strokes kill twice as many women in the United States as does breast cancer. About 425,000 women have a stroke every year in the United States, and about 82,000 die from it. A mix of life-style choices and medical conditions contribute to both stroke and heart disease, including the following:

- High cholesterol
- High blood pressure
- Diabetes
- Cigarette smoking
- Overweight and obesity
- Poor diet
- Physical inactivity
- Alcohol use

Women face particular challenges in dealing with high cholesterol. For years now, the approach for dealing with high cholesterol has been to focus on lowering high overall cholesterol levels, lowering LDL (low-density cholesterol, i.e., "bad cholesterol"), and rais-ing HDL (high-density cholesterol, i.e., "good cholesterol"). This approach has proven to be more beneficial for men than for women. Recent research shows that in women, low HDL levels are the strongest predictor of heart disease; that is, the ratio of HDL to total cholesterol is the best predictor of heart disease in women—when total cholesterol is about four times as great as HDL, the risk for heart disease is significant. So, for women with low overall cholesterol rates, yet who also have low HDL rates, there is still a high risk of heart disease. Yet the male-centered model of focusing on total cholesterol levels as the first step toward determining treatment dangerously ignores those women who have low over-all cholesterol levels combined with low HDL levels.

Embracing physical activity, a necessary component of any healthy lifestyle, is also more difficult for women than for men. Studies have shown that, for a variety of reasons, women find it difficult to incorporate daily physical exercise into their lives. Reasons given for not exercising regularly varied some with race and ethnicity as well as with region; some of the reasons given include the following:

- Guilt and shame associated with taking time to care for themselves as opposed to caring for their families
- Lack of social support
- Lack of a safe environment in which to exercise

Though men do not face these same barriers, failure to exercise regularly is a problem for both sexes; with only 16 percent of all men and women over age 15 exercising or par-ticipating in some kind of sports activity daily, there is clearly a need for both men and women to initiate a daily exercise program into their lives. Figuring out methods to over-come the unique barriers that women face in incorporating exercise into their lives is an important step toward improving the health of individual women and of society as a whole.

Cancer

Cancer is the second leading cause of death for women in the United States, claiming the lives of about 270,000 women each year. Skin cancer is the most prevalent form of cancer among both men and women. Though more women than men are diagnosed with skin cancer, more men die from it. Still, startling trends have been recognized that show that melanoma, the deadliest type of skin cancer, is actually increasing in women aged 15 to 39. About 3,000 women die each year from melanoma. Aside from skin cancer, the three most common types of cancer in women are lung, breast, and colon cancers. Cancer among women in the United States, in brief, generally involves the following forms:

- Lung cancer claims the lives of more women than any other type of cancer; about 71,000 women die of lung cancer in the United States each year. It is estimated that 80 percent of lung cancer deaths in women are related to cigarette smoking.
- Breast cancer is the most common type of cancer in women, aside from nonmelanoma skin cancers, and it is the second-leading cause of cancer deaths in women in the United States, claiming the lives of about 40,000 women each year.
- Colon cancer kills about 27,000 women each year. Considered one of the most preventable and treatable forms of cancer, many women, believing that it is a "man's disease," avoid scheduling colonoscopies, leaving them at great risk of developing and dying from this type of cancer.
- Cervical cancer is one of the most common types of cancers that affect a woman's reproductive system. It is related to certain strains of the human papillomavirus (HPV), and it claims more than 4,000 U.S. women's lives yearly.

The remaining cancer death statistics comprise other genital system cancers, lymphoma, urinary system cancer, leukemia, brain and other nervous system cancers, and myeloma cancers, as well as other and unspecified primary sites, according to the American Cancer Society.

In addition to heart disease, stroke, and cancer, the leading causes of death for women in the United States include chronic lower-respiratory diseases, Alzheimer's disease, unintentional injuries, diabetes, influenza and pneumonia, kidney disease, and septicemia. Women, like men, die from many diseases and events over which they have no control, though, also like men, their bad habits and lifestyle choices contribute to an alarming number of deaths each year—some estimate as many as half of all deaths in women over age 40 were preventable and could have been avoided by embracing healthy habits such as exercising and eating a healthy diet and shunning bad habits such as smoking. Many of women's preventable deaths are also related to misinterpreting signs of heart disease and not controlling high blood pressure.

Violence and Abuse

Another way that women's health is compromised in the United States is through violence and abuse. About 1,200 women are murdered each year by intimate partners; a third of all women's murders are committed by someone the woman knows or with whom she was in a relationship. The numbers for rape are even more alarming—as many as 600 women are raped every day in the United States, for a total of more than 200,000 sexual assaults each year. While certainly not the sole reason for mental health problems in women, these patterns of abuse profoundly impact the mental well-being of women. For example, posttraumatic stress disorder (PTSD) is an anxiety disorder that can manifest after being a victim of or witnessing a terrifying event or series of events. It is most often associated with rape, sexual abuse, and combat service, though natural disasters, transportation accidents, and violent nonsexual attacks can all trigger PTSD. About 10 percent of all women in the United States suffer from PTSD, most likely due to having experienced sexual assault as an adult or sexual abuse as a child. While men are more likely to have experienced accidents, nonsexual assaults, or combat or to have witnessed death or injury, women are much more likely to experience sexual abuse and assault, both as adults and as children—it is estimated that one in three women has or will experience at least one sexual assault in her lifetime. Researchers theorize that such trauma is likely to cause more emotional suffering than the kinds of trauma to which men are most often exposed.

Depression in women is a major problem in the United States, where about 12 million women experience clinical depression each year, and about one in eight women can expect to develop clinical depression in her lifetime. The rates of depression in women are about twice those in men, and these numbers have spurred furious debate among various factions—some that argue that women are overpathologized, some that insist that Americans, in general, and women, specifically, are overmedicated. In addition to PTSD and depression, women suffer from many other mental disorders, just as do men, including schizophrenia, anxiety disorders, obsessive-compulsive disorder, and various phobias. Eating disorders are also disproportionately experienced by women and girls in the United States, with estimates of about 10 million women and girls who suffer from one of several eating disorders, compared to about one million men and boys.

Reproductive Health

Women's reproductive health, including menstruation, pregnancy, and menopause, is an area that has taken great strides since the days when menstruation was viewed as evidence of everything from a woman's weakness to her inherent evilness, pregnancy was used as an excuse to limit women's freedom and opportunities, and menopause was seen as a depressing sign of the end of fertility. Though women still struggle with the challenges presented by these physical manifestations of femaleness, both on a personal and a cultural level, new research and changing societal norms have helped women to embrace these specifically female experiences. Still, there is much room for improvement, and alarming statistics regarding an increase in women dying in childbirth have many people demanding improvements in prenatal care.

Lesbians and bisexual women in the United States face unique challenges within the healthcare system. Many healthcare providers, even those who are not homophobic, have not had sufficient training to understand the specific health experiences of lesbians or bisexual women. Lesbians and bisexual women cite the following issues as significant to their healthcare experiences:

- Fear of negative reactions if they disclose their sexual orientation
- Healthcare providers' lack of understanding of disease risks and issues that are important to lesbians and bisexual women
- Lack of health insurance because of no domestic partner benefits
- Low perceived risk of getting sexually transmitted diseases and some types of cancer

Because of these issues, lesbians and bisexual women often avoid routine health exams and often postpone seeking medical care when health problems arise. Facing even more challenges in seeking healthcare are transgendered and transsexual people who are viewed with even more apprehension and suspicion than are lesbians and bisexual women. While significant strides have been made in gaining tolerance and rights for gay, lesbian, bisexual, and transgender people in general, they are still subject to significant incidents of verbal and physical abuse as well as more subtle forms of discrimination.

Women's health is a broad, complex subject and the information and statistics cited here vary with race, ethnicity, class, sexuality, and physical disability. Still, it is obvious that while technological advances and cultural changes have helped to improve the health of women in general, there is much room for improving not just the health of women, but the quality of the healthcare that they receive.

See Also: Cancers; Cardiovascular Diseases; Healthcare Delivery; Health Insurance Industry; Men's Health; Mental Health; Physical Activity and Health; Reproductive System Diseases.

Further Readings

BNET: CBS Interactive Business Network. "Women's Healthcare Disparities and Discrimination." http://findarticles.com/p/articles/mi_m0HSP/is_1_4/ai_66678569/?tag=content;col1 (Accessed August 2010).
Centers for Disease Control and Prevention. "Women's Health." http://www.cdc.gov/women (Accessed August 2010).
The Huffington Post. "Health Care Reform Produces Win for Abortion Foes." http://www.huffingtonpost.com/2010/07/19/health-care-reform-produc_n_650928.html (Accessed August 2010).
Jemal, Ahmedin, Rebecca Siegel, Jiaquan Xu, and Elizabeth Ward. "Cancer Statistics, 2010." *CA: A Cancer Journal for Clinicians* (July 7, 2010). http://caonline.amcancersoc.org/cgi/content/full/caac.20073v1/TBL1 (Accessed April 2011).
"Overhaul Will Lower the Costs of Being a Woman." *New York Times* (March 29, 2010). http://www.nytimes.com/2010/03/30/health/30women.html?_r=3 (Accessed August 2010).

Tani Bellestri
Independent Scholar

WORLD HEALTH ORGANIZATION'S ENVIRONMENTAL BURDEN OF DISEASE

Assessments of population health status on global and regional scales are important for setting research agendas, prioritizing interventions, and allocating resources for healthcare programs. The World Health Organization (WHO), in collaboration with the World Bank, developed a procedure for quantitatively estimating the burden of diseases attributable to environmental risk factors (environmental burden of disease, or EBD). Conceptually, the procedure identifies how much the modifiable environment contributes to the disease burden and estimates the extent of public health improvement that can be achieved through environmental risk exposure reduction. For EBD assessments, WHO relies on an extensive review of epidemiological studies and expert surveys. The standardized methodology and internal consistency make it possible to estimate the health gaps by comparing EBD assessments across different population groups.

The World Health Organization has identified several preventable environmental risk factors, including outdoor and indoor air pollution; lead; water; sanitation and hygiene;

climate change; solar ultraviolet radiation; recreational and drinking water quality; community noise; poverty; and occupational factors including carcinogens, dust, airborne particles, injuries, noise, and ergonomic stressors. The WHO provides spreadsheets generated for different risk factors and methodological support for all who wish to estimate a specific environmental disease burden for regional and local populations with available datasets.

Several composite measures have been developed to estimate the disease burden attributable to particular health-risk factors. Among the prominent quantitative measures, the Disease-Adjusted Life Years (DALYs) is perhaps best known. The DALY approach measures health gaps in a population by combining years of life lost due to a certain disease and years of life lived with a disability from the same disease. Other measures used included Quality-Adjusted Life Years (QALYs), which reflects individual preferences for time spent in different states of health; Disability-Adjusted Life Expectancy (DALE), which estimates the years of life expected to be lived without any disease; and Healthy Life Years (HeaLYs), which measures the number of life years lost prematurely combined with the amount of healthy life years lost due to morbidity.

The numerous WHO studies have done much to raise the profile of DALYs as the predominant measure for expressing disease burden at the global level. This is due in part to DALYs' broad comparability, reflection of years lived with disability, likelihood to be used in calculating the cost-effectiveness of interventions, and success record in using this measure for estimating burden of various diseases for age- and sex-specific population groups in different regions. As a composite indicator of population health status, environmental DALYs (e-DALYs) measure in units of time the gaps between the state of health in a population where ideally everyone lives into old age with the absence of disease and the state of health of the same population being exposed to a specific environmental risk factor. This gap is referred to as an attributable risk; therefore, e-DALY is estimated for a specific attributable risk factor. In general, DALY reflects the years lived with a disability (YLD) and the years of life lost to premature death (YLL) due to a disease i and is calculated as their sum:

$$\text{DALY}i = \text{YLL}i + \text{YLD}i$$

The longest life span is regarded as 82.5 years for women and 80 years for men based on a life table for the Japanese population.

Detailed assessment of disease burden using this approach has been produced for at least six environmental and five occupational risk factors. Analysis of WHO estimates of disease- and age-specific DALYs per 1,000 population globally shows that diarrhea is the biggest problem. With a burden of 94 percent attributed to the environment and clustered in sub-Saharan Africa, it kills approximately 1.5 million people annually. Children under 14 suffer the most from environmentally related diarrhea. Exposure to contaminated water has been identified as the main environmental risk factor for diarrhea. Therefore, targeted consistent modification of the environment through water sanitation along with improvements in hygiene behavior would make a significant positive impact on the population.

Based on regional and population exposure, national health statistics, and environmental specificity, WHO quantifies the extent to which an overall health problem can be attributed to environmental risk factors for its 192 members creating country profiles of EBD. It also prepares profiles of deaths, disabilities, and proportional burden of diseases attributable to environmental risk factors across 14 regions and mortality stratums and over 30 disease groups and 26 specific environmental risk factors. The global profiles of EBD illustrate that 15 African countries carry the heaviest burden of disease due to modifiable environment, with the overall proportion of disease burden attributable to environment, ranging from 37 percent in Angola to 30 percent in Liberia. Global estimates of EBD provide indications of potential health gain worldwide by eliminating exposure to environmental risks. For example, over 40 percent of global disease level can be prevented by improving water, sanitation, and hygiene; 11 percent by eliminating exposure to lead. In addition, 65 percent of cataracts could be avoided by eliminating excessive UV exposure.

EBD is a quantitative tool for the rational evaluation of health policy priorities that address specific environmental risk factors, diseases caused by those factors, and populations affected. However, several assumptions are embedded in the model due to a paucity of information. Therefore, it does not present complete knowledge about a specific disease burden. The WHO EBD estimation model strives to provide the best information based on the most recently available data and generally accepted contemporary assumptions for identifying populations at risk; for prioritizing, evaluating, and projecting health policies and interventions; and for planning preventive activities and prioritizing research.

See Also: Health Disparities; Lead Sources and Health; Waterborne Diseases.

Further Readings

Murray, C. J. L., J. A. Salomon, C. D. Mathers, and A. D. Lopez, eds. "Summary Measures of Population Health: Concepts, Ethics, Measurement and Applications." Geneva, Switzerland: World Health Organization, 2002.

Prüss-Üstün, A., et al. *Introduction and Methods: Assessing the Environmental Burden of Disease at National and Local Levels.* World Health Organization Environmental Burden of Disease Series, No. 1. Geneva, Switzerland: World Health Organization, 2003.

Prüss-Üstün, A. and C. Corvalán. "Preventing Disease Through Healthy Environments: Towards an Estimate of the Environmental Burden of Disease." Geneva, Switzerland: World Health Organization, 2006.

World Health Organization (WHO). "Environmental Burden of Disease—Country Profiles." http://www.who.int/quantifying_ehimpacts/countryprofiles/en/index.html (Accessed June 2009).

World Health Organization (WHO). "The Global Burden of Disease: 2004 Update." Geneva, Switzerland: WHO, 2008.

Natalie Milovantseva
University of California, Irvine

Green Health Glossary

A

Accident Site: The location of an unexpected occurrence, failure, or loss, either at a plant or along a transportation route, resulting in a release of hazardous materials.

Acute Exposure: A single exposure to a toxic substance that may result in severe biological harm or death. Acute exposures are usually characterized as lasting no longer than a day, as compared to longer, continuing exposure over a period of time.

Affected Public: The human population adversely impacted following exposure to a toxic pollutant in food, water, air, or soil.

Air Pollution: The presence of contaminants or pollutant substances in the air that interfere with human health or welfare, or produce other harmful environmental effects.

Asbestos: A mineral fiber that can pollute air or water and cause cancer or asbestosis when inhaled. The U.S. Environmental Protection Agency (EPA) has banned or severely restricted its use in manufacturing and construction.

B

Bacteria: Microscopic living organisms that can aid in pollution control by metabolizing organic matter in sewage, oil spills, or other pollutants. However, bacteria in soil, water, or air can also cause human, animal, and plant health problems.

Blackwater: Water that contains animal, human, or food waste.

Body Burden: The amount of a chemical stored in the body at a given time, especially a potential toxin in the body as the result of exposure.

Breathing Zone: Area of air in which an organism inhales.

Building-Related Illness: Diagnosable illness whose cause and symptoms can be directly attributed to a specific pollutant source within a building.

C

Chlorinator: A device that adds chlorine, in gas or liquid form, to water or sewage to kill infectious bacteria.

Chronic Exposure: Multiple exposures occurring over an extended period of time or over a significant fraction of an animal's or human's lifetime (usually seven years to a lifetime).

509

Clean Coal Technology: Any technology not in widespread use prior to the Clean Air Act Amendments of 1990. This act will achieve significant reductions in pollutants associated with the burning of coal.

Combined Sewer Overflows: Discharge of a mixture of storm water and domestic waste when the flow capacity of a sewer system is exceeded during rainstorms.

Commercial Waste: All solid waste emanating from business establishments such as stores, markets, office buildings, restaurants, shopping centers, and theaters.

Construction Ban: If, under the Clean Air Act, EPA disapproves an area's planning requirements for correcting nonattainment, EPA can ban the construction or modification of any major stationary source of the pollutant for which the area is in nonattainment.

Conventional Pollutants: Statutorily listed pollutants understood well by scientists. These may be in the form of organic waste, sediment, acid, bacteria, viruses, nutrients, oil and grease, or heat.

Cradle-to-Grave System: A procedure in which hazardous materials are identified and followed as they are produced, treated, transported, and disposed of by a series of permanent, linkable, descriptive documents (e.g., manifests).

Cumulative Exposure: The sum of exposures of an organism to a pollutant over a period of time.

D

Decontamination: Removal of harmful substances such as noxious chemicals, harmful bacteria, or other organisms, or radioactive material from exposed individuals, rooms, and furnishings in buildings or the exterior environment.

DES: A synthetic estrogen, diethylstilbestrol is used as a growth stimulant in food animals. Residues in meat are thought to be carcinogenic.

Diazinon: An insecticide. In 1986, the EPA banned its use on open areas such as sod farms and golf courses because it posed a danger to migratory birds. The ban did not apply to agricultural, home lawn, or commercial establishment uses.

Dioxin: Any of a family of compounds known chemically as dibenzo-p-dioxins. Concern about them arises from their potential toxicity as contaminants in commercial products. Tests on laboratory animals indicate that they are among the more toxic anthropogenic (manmade) compounds.

Disposal Facilities: Repositories for solid waste, including landfills and combustors intended for permanent containment or destruction of waste materials. Excludes transfer stations and composting facilities.

E

Ecological/Environmental Sustainability: Maintenance of ecosystem components and functions for future generations.

Effluent: Wastewater—treated or untreated—that flows out of a treatment plant, sewer, or industrial outfall. Generally refers to wastes discharged into surface waters.

Emergency (Chemical): A situation created by an accidental release or spill of hazardous chemicals that poses a threat to the safety of workers, residents, the environment, or property.

Emergency Removal Action: Steps taken to remove contaminated materials that pose imminent threats to local residents (e.g., removal of leaking drums or the excavation of explosive waste).

Emergency Suspension: Suspension of a pesticide product registration due to an imminent hazard. The action immediately halts distribution, sale, and sometimes actual use of the pesticide involved.

Endrin: A pesticide toxic to freshwater and marine aquatic life that also produces adverse health effects in domestic water supplies.

Environmental Exposure: Human exposure to pollutants originating from facility emissions. Threshold levels are not necessarily surpassed, but low-level chronic pollutant exposure is one of the most common forms of environmental exposure.

Environmental Tobacco Smoke: Mixture of smoke from the burning end of a cigarette, pipe, or cigar and smoke exhaled by the smoker.

Epidemiology: Study of the distribution of disease, or other health-related states and events in human populations, as related to age, sex, occupation, ethnicity, and economic status in order to identify and alleviate health problems and promote better health.

Exposure: The amount of radiation or pollutant present in a given environment that represents a potential health threat to living organisms.

Exposure Assessment: Identifying the pathways by which toxicants may reach individuals, estimating how much of a chemical an individual is likely to be exposed to, and estimating the number likely to be exposed.

F

Flare: A control device that burns hazardous materials to prevent their release into the environment; may operate continuously or intermittently, usually on top of a stack.

Fluoridation: The addition of a chemical to increase the concentration of fluoride ions in drinking water to reduce the incidence of tooth decay.

G

Generally Recognized As Safe (GRAS): Designation by the FDA that a chemical or substance (including certain pesticides) added to food is considered safe by experts, and so is exempted from the usual FFDCA food additive tolerance requirements.

Giardia lamblia: Protozoan in the feces of humans and animals that can cause severe gastrointestinal ailments. It is a common contaminant of surface waters.

H

Hammer Mill: A high-speed machine that uses hammers and cutters to crush, grind, chip, or shred solid waste.

Hazard: Potential for radiation, a chemical, or other pollutant to cause human illness or injury.

Hazard Assessment: Evaluating the effects of a stressor or determining a margin of safety for an organism by comparing the concentration that causes toxic effects with an estimate of exposure to the organism.

Hazard Identification: Determining if a chemical or a microbe can cause adverse health effects in humans and what those effects might be.

Hazardous Air Pollutants: Air pollutants that are not covered by ambient air quality standards but that, as defined in the Clean Air Act, may present a threat of adverse human health effects or adverse environmental effects. Such pollutants include asbestos, beryllium, mercury, benzene, coke oven emissions, radionuclides, and vinyl chloride.

Hazardous Substance: Any material that poses a threat to human health and/or the environment. Typical hazardous substances are toxic, corrosive, ignitable, explosive, or chemically reactive.

Hazardous Waste: By-products of society that can pose a substantial or potential hazard to human health or the environment when improperly managed. Possesses at least one of four characteristics (ignitability, corrosivity, reactivity, or toxicity), or appears on special EPA lists.

Health Assessment: An evaluation of available data on existing or potential risks to human health posed by a Superfund site. The Agency for Toxic Substances and Disease Registry (ATSDR) of the U.S. Department of Health and Human Services (DHHS) is required to perform such an assessment at every site on the National Priorities List.

Household Hazardous Waste: Hazardous products used and disposed of by residential as opposed to industrial consumers. Includes paints, stains, varnishes, solvents, pesticides, and other materials or products containing volatile chemicals that can catch fire, react, or explode, or that are corrosive or toxic.

Human Exposure Evaluation: Describing the nature and size of the population exposed to a substance and the magnitude and duration of their exposure.

Human Health Risk: The likelihood that a given exposure or series of exposures may have damaged or will damage the health of individuals.

Hypersensitivity Diseases: Diseases characterized by allergic responses to pollutants; diseases most clearly associated with indoor air quality are asthma, rhinitis, and pneumonic hypersensitivity.

I

Incineration: A treatment technology involving destruction of waste by controlled burning at high temperatures; for example, burning sludge to remove the water and reduce the remaining residues to a safe, nonburnable ash that can be disposed of safely on land, in some waters, or in underground locations.

Incompatible Waste: A waste unsuitable for mixing with another waste or material because it may react to form a hazard.

Infectious Waste: Hazardous waste capable of causing infections in humans, including contaminated animal waste, human blood and blood products, isolation waste, pathological waste, and discarded sharps (needles, scalpels, or broken medical instruments).

Institutional Waste: Waste generated at institutions such as schools, libraries, hospitals, prisons, and so forth.

Irradiated Food: Food subject to brief radioactivity—usually gamma rays—to kill insects, bacteria, and mold, and to permit storage without refrigeration.

Irradiation: Exposure to radiation of wavelengths shorter than those of visible light (gamma, X-ray, or ultraviolet) for medical purposes, to sterilize milk or other foodstuffs, or to induce polymerization of monomers or vulcanization of rubber.

Irritant: A substance that can cause irritation of the skin, eyes, or respiratory system. Effects may be acute from a single high-level exposure or chronic from repeated low-level exposures to such compounds as chlorine, nitrogen dioxide, and nitric acid.

L

LD 50/Lethal Dose: The dose of a toxicant or microbe that will kill 50 percent of the test organisms within a designated period. The lower the LD 50, the more toxic the compound.

Lifetime Exposure: Total amount of exposure to a substance that a human would receive in a lifetime (usually assumed to be 70 years).

Lowest Acceptable Daily Dose: The largest quantity of a chemical that will not cause a toxic effect, as determined by animal studies.

M

Maximally (or Most) Exposed Individual: The person with the highest exposure in a given population.

Medical Waste: Any solid waste generated in the diagnosis, treatment, or immunization of human beings or animals, in research pertaining thereto, or in the production or testing of biologicals.

Methoxychlor: Pesticide that causes adverse health effects in domestic water supplies and is toxic to freshwater and marine aquatic life.

Mining Waste: Residues resulting from the extraction of raw materials from the Earth.

Montreal Protocol: Treaty, signed in 1987; governs stratospheric ozone protection and research, and the production and use of ozone-depleting substances. It provides for the end of production of ozone-depleting substances such as chlorofluorocarbons (CFCs). Under the protocol, various research groups continue to assess the ozone layer. The Multilateral Fund provides resources to developing nations to promote the transition to ozone-safe technologies.

N

National Ambient Air Quality Standards (NAAQS): Standards established by the EPA that apply for outdoor air throughout the country.

National Emissions Standards for Hazardous Air Pollutants (NESHAPs): Emissions standards set by the EPA for an air pollutant not covered by NAAQS that may cause an increase in fatalities or in serious, irreversible, or incapacitating illness. Primary standards are designed to protect human health; secondary standards protect public welfare (e.g., building facades, visibility, crops, and domestic animals).

Nonhazardous Industrial Waste: Industrial process waste in wastewater not considered municipal solid waste or hazardous waste under RARA.

Nonpotable: Water that is unsafe or unpalatable to drink because it contains pollutants, contaminants, minerals, or infective agents.

No-Observed-Effect-Level (NOEL): Exposure level at which there are no statistically or biological significant differences in the frequency or severity of any effect in the exposed or control populations.

Nuclear Winter: Prediction by some scientists that smoke and debris rising from massive fires of a nuclear war could block sunlight for weeks or months, cooling the Earth's surface and producing climate changes that could, for example, negatively affect world agricultural and weather patterns.

O

Oil Spill: An accidental or intentional discharge of oil that reaches bodies of water. It can be controlled by chemical dispersion, combustion, mechanical containment, and/or adsorption. Spills from tanks and pipelines can also occur away from water bodies, contaminating the soil, getting into sewer systems, and threatening underground water sources.

Oral Toxicity: Ability of a pesticide to cause injury when ingested.

Ozone Depletion: Destruction of the stratospheric ozone layer that shields the Earth from ultraviolet radiation harmful to life. This destruction of ozone is caused by the breakdown of certain chlorine- and/or bromine-containing compounds (chlorofluorocarbons or halons) that break down when they reach the stratosphere and then catalytically destroy ozone molecules.

Ozone Hole: A thinning break in the stratospheric ozone layer. A depleted area is designated an ozone hole when the detected amount of depletion exceeds 50 percent. Seasonal ozone holes have been observed over both the Antarctic and Arctic regions, part of Canada, and the extreme northeastern United States.

Ozone Layer: The protective layer in the atmosphere, about 15 miles above the ground, that absorbs some of the sun's ultraviolet rays, thereby reducing the amount of potentially harmful radiation that reaches the Earth's surface.

P

Pandemic: A widespread epidemic throughout an area, a nation, or the world.

Photochemical Smog: Air pollution caused by chemical reactions of various pollutants emitted from different sources.

Plume: A visible or measurable discharge of a contaminant from a given point of origin. It can be visible or thermal in water, or visible in the air as, for example, a plume of smoke.

Pollutant: Generally, any substance introduced into the environment that adversely affects the usefulness of a resource or the health of humans, animals, or ecosystems.

Pollution Prevention: Identifying areas, processes, and activities that create excessive waste products or pollutants in order to reduce or prevent them through altering or eliminating a process. Such activities, consistent with the Pollution Prevention Act of 1990, are conducted across all EPA programs and can involve cooperative efforts with such agencies as the Departments of Agriculture and Energy.

Potable Water: Water that is safe for drinking and cooking.

Primary Drinking Water Regulation: Applies to public water systems and specifies a contaminant level, which, in the judgment of the EPA administrator, will not adversely affect human health.

R

Radiation Standards: Regulations that set maximum exposure limits for protection of the public from radioactive materials.

Recommended Maximum Contaminant Level (RMCL): The maximum level of a contaminant in drinking water at which no known or anticipated adverse effect on human health would occur and that includes an adequate margin of safety. Recommended levels are nonenforceable health goals.

Release: Any spilling, leaking, pumping, pouring, emitting, emptying, discharging, injecting, escaping, leaching, dumping, or disposing into the environment of a hazardous or toxic chemical or extremely hazardous substance.

Removal Action: Short-term immediate actions taken to address releases of hazardous substances that require expedited response.

Residential Waste: Waste generated in single- and multi-family homes, including newspapers, clothing, disposable tableware, food packaging, cans, bottles, food scraps, and yard trimmings other than those that are diverted to backyard composting.

Residual: Amount of a pollutant remaining in the environment after a natural or technological process has taken place; for example, the sludge remaining after initial wastewater treatment or particulates remaining in air after it passes through a scrubbing or other process.

Risk: A measure of the probability that damage to life, health, property, and/or the environment will occur as a result of a given hazard.

Risk Assessment: Qualitative and quantitative evaluation of the risk posed to human health and/or the environment by the actual or potential presence and/or use of specific pollutants.

S

Safe Water: Water that does not contain harmful bacteria, toxic materials, or chemicals and is considered safe for drinking even if it may have taste, odor, color, and certain mineral problems.

Sanitary Sewers: Underground pipes that carry off only domestic or industrial waste, not storm water.

Septic System: An on-site system designed to treat and dispose of domestic sewage. A typical septic system consists of a tank that receives waste from a residence or business and a system of tile lines or a pit for disposal of the liquid effluent (sludge) that remains after decomposition of the solids by bacteria in the tank. Septic systems must be pumped out periodically.

Site Assessment Program: A means of evaluating hazardous waste sites through preliminary assessments and site inspections to develop a hazard ranking system score.

Sterilization: The removal or destruction of all microorganisms, including pathogenic and other bacteria, vegetative forms, and spores.

Stressors: Physical, chemical, or biological entities that can induce adverse effects on ecosystems or human health.

T

Teratogen: A substance capable of causing birth defects.

Thermal Treatment: Use of elevated temperatures to treat hazardous wastes.

Toxic Cloud: Airborne plume of gases, vapors, fumes, or aerosols containing toxic materials.

Toxic Waste: A waste that can produce injury if inhaled, swallowed, or absorbed through the skin.

Treatment: Any method, technique, or process designed to remove solids and/or pollutants from solid waste, waste streams, effluents, and air emissions.

Treatment Plant: A structure built to treat wastewater before discharging it into the environment.

W

Waste: Unwanted materials left over from a manufacturing process.

Waste Treatment Plant: A facility containing a series of tanks, screens, filters, and other processes by which pollutants are removed from water.

Source: U.S. Environmental Protection Agency (http://www.epa.gov/OCEPAterms)

Green Health Resource Guide

Books

Ali, S. Harris and Roger Keil, eds. *Networked Disease: Emerging Infections in the Global City*. Oxford, UK: Wiley-Blackwell, 2008.

Allenby, Braden R. and Deanna J. Richards. *The Greening of Industrial Ecosystems*. Washington, DC: National Academy Press, 1994.

American Psychiatric Association. *Diagnostic and Statistical Manual of Mental Disorders: DSM-IV*. Washington, DC: American Psychiatric Association, 1994.

Anastas, P. T. and J. C. Warner. *Green Chemistry: Theory and Practice*. New York: Oxford University Press, 2000.

Baer, Hans and Merrill Singer. *Global Warming and the Political Ecology of Health: Emerging Crises and Systemic Solutions*. Walnut Creek, CA: Left Coast Press, 2009.

Bell, Iris. *Clinical Ecology—A New Medical Approach to Environmental Illness*. Bolinas, CA: Common Knowledge Press, 1982.

Caldicott, Helen. *Nuclear Power Is Not the Answer*. New York: New Press, 2006.

Chamberlain, N. R. *The Big Picture: Medical Microbiology*. New York: McGraw-Hill, 2008.

Chawla, O. P. *Advances in Biogas Technology*. New Delhi: Indian Council of Agricultural Research, 1986.

Chen, Joseph S., Philip Sloan, and Willy Legrand. *Sustainability in the Hospital Industry*. Oxford, UK: Butterworth-Heinemann, 2009.

Clapp, J. *Toxic Exports: The Transfer of Hazardous Wastes From Rich to Poor Countries*. Ithaca, NY: Cornell University Press, 2001.

Clark, Duncan. *The Rough Guide to Ethical Living*. London: Penguin, 2006.

Cravens, Gwyneth. *Power to Save the World: The Truth About Nuclear Energy*. New York: Alfred A. Knopf, 2007.

Crittenden, John, et al. *Water Treatment Principles and Design*, 2nd ed. Hoboken, NJ: John Wiley, 2005.

Dequeker, Jan. *Medical Management of Rheumatic Musculoskeletal and Connective Tissue Diseases*. New York: Informa Health Care, 1997.

Donaldson, Scott. *The Suburban Myth*. New York: Columbia University Press, 1969.

Elliott, David, ed. *Nuclear or Not? Does Nuclear Power Have a Place in a Sustainable Energy Future?* New York: Palgrave Macmillan, 2007.

Escohotado, Antonio. *A Brief History of Drugs: From the Stone Age to the Stoned Age.* Rochester, VT: Park Street Press, 1999.

Feldman, Mark, Lawrence S. Friedman, and Marvin Sleisenger. *Sleisenger and Fordtran's Gastrointestinal and Liver Disease.* Oxford, UK: Elsevier, 2003.

Fitzpatrick, Kevin and Mark LaGory. *Unhealthy Places: The Ecology of Risk in the Urban Landscape.* New York: Routledge, 2000.

Geary, Amanda. *The Mind Guide to Food and Mood.* London: Mind, 2004.

Godish, Thad. *Air Quality,* 4th ed. Boca Raton, FL: Lewis Publishers, 2003.

Graedel, T. E. and Braden R. Allenby. *Industrial Ecology and Sustainable Engineering.* Upper Saddle River, NJ: Prentice Hall, 2009.

Green, J. D. and J. K. Hartwell, eds. *The Greening of Industry: A Risk Management Approach.* Cambridge, MA: Harvard University Press, 1997.

Grossman, G. *Chasing Molecules: Poisonous Products, Human Health, and the Promise of Green Chemistry.* Washington, DC: Shearwater, 2009.

Guglar, Josef, ed. *The Urban Transformation of the Developing World.* New York: Oxford University Press, 1996.

Halweil, Brian. *Home Grown: The Case for Local Food in a Global Market.* Washington, DC: Worldwatch Institute, 2002.

Hillman, Mayer and Tina Fawcett. *How We Can Save the Planet.* London: Penguin, 2004.

Hinrichs, C. C. and T. A. Lyson. *Remaking the North American Food System: Strategies for Sustainability.* Lincoln: University of Nebraska Press, 2007.

Holm, Eggert and Heinrich Kasper. *Metabolism and Nutrition in Liver Disease.* Lancaster, PA: Springer, 1985.

Jacobson, Mark Z. *Atmospheric Pollution: History, Science and Regulation.* Cambridge, UK: Cambridge University Press, 2002.

Jamuna, Carroll, ed. *The Pharmaceutical Industry.* Farmington Hills, MI: Greenhaven, 2008.

Jenks, Mike and Rod Burgess. *Compact Cities: Sustainable Urban Forms for Developing Countries.* London: Spon Press, 2000.

Kaplan, Rachel and Steven Kaplan. *The Experience of Nature: A Psychological Perspective.* Cambridge, UK: Cambridge University Press, 1989.

Karwowski, Waldemar. *International Encyclopedia of Ergonomics and Human Factors.* Boca Raton, FL: CRC Press, 2006.

Kellert, Stephen and Edward Wilson. *The Biophilia Hypothesis.* Washington, DC: Island Press, 1995.

Kincheloe, J. L. *The Sign of the Burger: McDonald's and the Culture of Power.* Philadelphia, PA: Temple University Press, 2002.

Kingsolver, Barbara, Steven L. Hopp, and Camille Kingsolver. *Animal, Vegetable, Miracle: A Year of Food Life.* New York: HarperCollins, 2007.

Lamberg, Lynne and Sandra Thurman. *Skin Disorders.* New York: Chelsea House Publishers, 2000.

Madigan, Carleen. *The Backyard Homestead.* North Adams, MA: Sorey Publishing, 2009.

Marks, Ronald. *Facial Skin Disorders.* Boca Raton, FL: CRC Press, 2007.

Masters, G. M. *Introduction to Environmental Engineering and Science,* 2nd ed. Englewood Cliffs, NJ: Prentice Hall, 1998.

McIntyre, James A. and Marie-Louise Newell. *Congenital and Perinatal Infections: Prevention, Diagnosis and Treatment.* Cambridge, UK: Cambridge University Press, 2000.

Mount, Jeffrey. *California Rivers and Streams*. Berkeley: University of California Press, 1995.

Nestle, M. *What to Eat*. New York: North Point Press, 2007.

Norman, Donald. *The Design of Everyday Things*. Cambridge: MIT Press, 1998.

Orefici, Graziella. *New Perspectives on Streptococci and Streptococcal Infections*. Stuttgart, Germany: Gustav Fischer Verlag, 1992.

Palmer, Melissa. *Dr. Melissa Palmer's Guide to Hepatitis & Liver Disease*. New York: Avery, 2004.

Planck, N. *Real Food: What to Eat and Why*. New York: Bloomsbury Group, 2007.

Pollan, Michael. *The Omnivore's Dilemma*. New York: Penguin, 2006.

Pringle, Peter. *Food, Inc: Mendel to Monsanto—The Promises and Perils of the Biotech Harvest*. New York: Simon & Schuster, 2003.

Rae, J. *The American Automobile Industry*. Boston: Twayne Publishers, 1984.

Rapp, Doris J. *Our Toxic World—A Wake Up Call*. Buffalo, NY: Environmental Medicine Research Foundation, 2004.

Register, Richard. *Ecocities: Building Cities in Balance With Nature*. Gabriola Island, BC, Canada: New Society Publishers. 2006.

Reid, T. R. *The Healing of America: A Global Quest for Better, Cheaper, and Fairer Health Care*. New York: Penguin Press, 2009.

Reilly, Thomas. *Musculoskeletal Disorders in Health-Related Occupations*. Amsterdam, Netherlands: IOS Press, 2002.

Schneiderman, Paul I. and Marc E. Grossman. *A Clinician's Guide to Dermatologic Differential Diagnosis, Volume 1: The Text*. New York: Informa Health Care, 2006.

Stanley, N. F. and M. P. Alpers. *Man-Made Lakes and Human Health*. London: Academic Press, 1975.

Stevens, Dennis L. and Edward L. Kaplan. *Streptococcal Infections: Clinical Aspects, Microbiology, and Molecular Pathogenesis*. New York: Oxford University Press, 2000.

Taylor, Brian C., et al., eds. *Nuclear Legacies: Communication, Controversy, and the U.S. Nuclear Weapons Complex*. Lanham, MD: Lexington Books, 2007.

Tuan, Yi-Fu. *Topophilia: A Study of Environmental Perception, Attitudes, and Values*. Englewood Cliffs, NJ: Prentice Hall, 1974.

Schulthess, Von Gustav Konrad, and Christoph L. Zollikofer. *Musculoskeletal Diseases: Diagnostic Imaging and Interventional Techniques*. New York: Springer, 2005.

Worman, Howard J. *The Liver Disorders Sourcebook*. New York: McGraw-Hill, 1999.

Journals

Al Amen Journal of Medicine Science
American Journal of Medical Sciences
American Journal of Public Health
American Journal of Tropical Medicine & Hygiene
Asian Journal of Andrology

European Journal of Epidemiology

Indian Journal of Pediatrics
International Journal of Environment and Health
International Journal of Hygiene and Environmental Health

International Journal of Infectious Diseases
International Journal of Marine & Coastal Law

Journal of Agromedicine
Journal of Biosciences
Journal of Environmental Biology
Journal of Environmental Science & Health
Journal of Health Politics, Policy and Law
Journal of Histochemistry and Cytochemistry
Journal of Hospital Infection
Journal of Industrial Microbiology & Biotechnology
Journal of Popular Culture
Journal of the American Medical Association
Journal of Travel Medicine

Latin American Journal of Microbiology

New England Journal of Medicine

Peruvian Journal of Experimental Medicine and Public Health
The Plant Journal

Websites

Agency for Healthcare Research and Quality
www.ahrq.gov

American Cancer Society
www.cancer.org

Drug Enforcement Agency
www.dea.gov

Food and Drug Administration
www.fda.gov

Global Initiative for Asthma
www.ginasthma.org

Green Guide for Health Care
www.gghc.org

Green Highways Partnership
www.greenhighways.org

Health Reform
www.healthreform.gov

National Center for Complementary and Alternative Medicine
www.nccam.nih.gov

National Center on Minority Health and Health Disparities
www.nimhd.nih.gov

National Institute on Aging
  www.nia.nih.gov

National Osteoporosis Foundation
  www.nof.org

National Survey on Drug Use and Health
  https://nsduhweb.rti.org

Nuclear Regulatory Commission
  www.nrc.gov

Practice Greenhealth
  www.practicegreenhealth.org

U.S. Centers for Disease Control and Prevention
  www.cdc.gov

U.S. Department of Agriculture
  www.usda.gov

World Health Organization
  www.who.int

Green Health Appendix

Centers for Disease Control and Prevention: Environmental Health

http://www.cdc.gov/Environmental

This website, part of the Centers for Disease Control and Prevention of the U.S. Department of Health and Human Services, provides access to many types of information relevant to environmental health. Information may be accessed by topic (asbestos, hazardous waste sites, cruise ship health) or by type (data and statistics, publications, tools), and there is also a search interface for the website. The scope of information provided under each topic ranges from guidelines and explanations written for the general public to scientific data useful to researchers and materials for public health professionals. Many include information in several formats, including downloadable PDFs, podcasts, and video. For instance, under the topic of carbon monoxide poisoning, this website includes general information about the problem, prevention guidelines, and advice for the general public (e.g., how to avoid carbon monoxide poisoning when using a home generator or pressure washer), public service announcements, downloadable educational materials and flyers, clinical guidance to aid in recognizing cases of carbon monoxide poisoning, and links to reports and research about carbon monoxide research. The website also includes links to information about related topics (such as air pollution and respiratory health) and to information from other government agencies, including the National Center for Environmental Health.

The Green Health Center: Exploring Bioethics Upstream

http://www.unmc.edu/green

The official website for several projects at the University of Nebraska Medical Center focusing on ethical issues related to the environmental aspects of healthcare. Although research and publication is ongoing, the website focuses on two projects that have been concluded. The first is The Green Health Center, which began in 1998 by an interdisciplinary working group and opened the question of what would constitute environmentally sound healthcare practices and what ethical principles would inform examination of this question. The second project, Exploring Bioethics Upstream, was conducted between 2000 and 2004 and reviewed environmental concerns related to an academic medical center and teaching hospital, looking at the need to balance quality of care, environmental impact, and monetary cost. The website includes much information about these projects, including the main topics discussed, ethical issues, conclusions, and publications resulting from these investigations.

Health EU: Environmental Health

http://ec.europa.eu/health-eu/my_environment/environmental_health/index_en.htm

This website, part of the Public Health Portal of the European Union (EU), summarizes information about environmental health and EU activities in the field. The website is organized in seven sections: Public Health (including electromagnetic fields and health), Environment, Unexpected Events (including flooding and extreme heat), Research, Health Indicators, Health Information, and the European Environment and Health Information System. The website includes information about each project written at a layman's level as well as press releases; country-specific sections; links to relevant international organizations; a calendar of upcoming events; and links to other EU sites, including those addressing EU legislation and EU publications. Many key documents may be downloaded from the site, and a search interface allows projects to be sorted by title, organization, or year, or be searched for by keyword.

Occupational Safety and Health Administration: Ergonomics

http://www.osha.gov/SLTC/ergonomics/index.html

This website, by the Occupational Safety and Health Administration (OSHA) of the U.S. Department of Labor, includes information about ergonomics ("the science of fitting workplace conditions and job demands to the capabilities of the working population") and risk factors that may place a person in danger of developing musculoskeletal disorders. The website includes specific guidelines (downloadable in PDF format) for several industries, including poultry processing, retail grocery stores, nursing homes, meatpacking plants, and shipyards, as well as a standard protocol to guide development of standards in other industries and for specific tasks. Another section provides tools and information to help identify processes that might contribute to workplace injuries, including analyzing the job itself and reviewing injury and illness records. The website includes numerous tools intended to help find solutions to ergonomic problems in the workplace, including publications, web-based training tools, and success stories of how other companies have dealt with ergonomic programs.

World Health Organization: Quantifying Environmental Health Impacts

http://www.who.int/quantifying_ehimpacts/en

This website provides many tool and statistics regarding the environmental burden of disease and ways to measure it. A downloadable document explains the methodology the World Health Organization uses to determine the global burden of environmental disease and the factors that are included in it (such as indoor smoke, outdoor air pollution, climate change, and occupational airborne particulates). The website provides country profiles of the environmental burden of disease (downloadable as PDFs), including deaths per year and DALY (disability adjusted life years lost) per 1,000 people per year, and a separate section of the global environmental burden of disease from specific factors. Another section of the website provides practical guidance toward assessing the burden of disease at local or national levels for a number of risk factors, from lead exposure to poverty to occupational noise, including the reference paper "Comparative Quantification of Health Risks" and selected papers on the method. A final section addresses the cost effectiveness of environmental health interventions, including a downloadable background document discussing current economic evaluation methods and how they could be adapted to evaluate environmental health interventions.

World Health Organization: Regional Office for Europe: Health Impact Assessment

http://www.euro.who.int/en/what-we-do/health-topics/environmental-health/health-impact-assessment

This website, created by the Regional Office for Europe of the World Health Organization, defines Health Impact Assessment (HIA) and provides tools and advice about HIA for member states. The website also aims to increase awareness of health issues related to the environment and includes news about HIA, links to official reports and policy statements related to HIA, information about HIA activities in Europe, and studies conducted in individual regions (e.g., an impact assessment of waste treatment on human health in the Campania region of Italy). Many fact sheets, pamphlets, press releases, and reports are available for download from the website, and an interface allows searching by topic, keyword, title, or type. The website also provides access to several databases that contain statistical health information that the user may search and analyze, including the European Health for All database, the Mortality Indicator database, and the Tobacco Control database. The website also provides links to the Environment and Health Information System (ENHIS), which provides evidence-based information to support public health and environmental policies in the WHO European region.

Sarah Boslaugh
Washington University in St. Louis

Index

Article titles and their page numbers are in **bold**.